Island Challenges

John Connell • Yonique Campbell

Island Challenges

Caribbean and Pacific Futures

palgrave
macmillan

John Connell
School of Geosciences
University of Sydney
Sydney, NSW, Australia

Yonique Campbell
Department of Government
University of the West Indies
Mona, Jamaica

ISBN 978-981-95-8176-4 ISBN 978-981-95-8177-1 (eBook)
https://doi.org/10.1007/978-981-95-8177-1

This Palgrave Macmillan imprint is published by the registered company Springer Nature Singapore Pte Ltd.
The registered company address is: 152 Beach Road, #21-01/04 Gateway East, Singapore 189721, Singapore

Preface

This book seeks to examine the challenges facing the Small Island Developing States (SIDS) of the Caribbean and the Pacific, by providing a comprehensive coverage of recent trends in these states, specifically looking at livelihoods and state stability in what are becoming new times. Already, the 2020s have witnessed several significant global phenomena that have presented challenges to all countries, but especially those with relatively fragile economies. These included, initially, the COVID-19 pandemic constrained development, particularly tourism, yet offered opportunities to island states to rethink and reconsider development strategies. Consideration of an anticipated but vague 'new normal' contributed to a greater emphasis on self-reliance, and a more sustainable future, linked to a global emphasis on achieving Sustainable Development Goals (SDGs).

However, the pandemic proved only a temporary influence. Climate change, certainly not temporary, was more evident. Fuelled by climate change and El Niño, 2024 became a year of global extreme temperatures, and their visible and dramatic impacts in fires, droughts and deaths. Threats to the livelihoods of island households, especially those literally and metaphorically living on the edge, emphasised how key strands in development were tightly linked to overseas decisions and events, and occasional shocks emanating from these. Entire national economies were affected by external forces, whether security, cyclones, warfare, crime, resort chains and commodity supply chains, or migration policies. Development became more political and transactional, as a new geopolitical (dis)order emerged. Macroeconomic uncertainties were compounded by the uncertainties of the climate crisis while enthusiasm for a blue economy required new approaches to regional cooperation and global statecraft.

Consequent challenges for the island states included ongoing health concerns, disappointments over global action on climate change, threats to national and regional stability from secessionism, alongside the emerging geo-strategic significance of both regions, as China exported its Belt and Road Initiative. Few shocks have seemed of much benefit to the island states. The spectacular and disastrous volcanic eruption in Tongan seas in 2022, Hurricane Oscar in the Bahamas and Vanuatu's earthquake late in 2024, and Hurricane Melissa in Jamaica in 2025 were intermittent reminders that SIDS faced environmental hazards, some more frequent following global warming and heating. Occasionally, it seemed like a constant chronicling of disasters and their aftermaths. The book therefore seeks to offer an overview of livelihoods and security by exploring six broad and interrelated themes: state capacity to provide basic services (such as health, education, electrification); migration, remittances and aid; climate change, biodiversity and environmental hazards; agriculture and fisheries; tourism and economic diversification; state solidity and stability (with some reference to gender, corruption and equity). These themes are very loosely set in the context of the SDGs, where security has more than mere military connotations.

The book has been written in the dark shadows of the Russian invasion of Ukraine and genocide in Gaza, unpleasant reminders that little is certain in the world of geopolitics, where smaller countries face threats, and where a seemingly unpredictable shift in the European and Middle Eastern security architecture may mark an ongoing shift in a new world order, with small states uncertain bystanders. It was concluded as President Trump was initiating his unpredictable and disturbingly anti-development second regime, deportees were arriving in the Caribbean and Pacific islands, the earth was passing the 1.5 °C temperature increase mark (an overshoot beyond which scientists have predicted a tipping point into catastrophic change), and sea levels rose faster than expected, as Trump pulled out of the landmark 2015 Paris Agreement on climate change. An introspective United States, and declining aid flows, seemed of no benefit to SIDS. This may account for a shroud of pessimism—war certainly deflates the human spirit, even at a distance—but we have written in the hope that this might make some small contribution to a better world, however hard it can be to map out practicalities for positive change. However, in the final moments, the 'advisory opinion' of the International Court of Justice that nations could be held accountable for their greenhouse gas emissions offered at least a glimmer of hope.

We acknowledge and thank passing relatives, notably Church Campbell, and also Palgrave Macmillan, for their patience and tolerance, Linda Susilowati for somehow sorting out the Bibliography, Mike Siegel for the maps, Fiona Li for heroic last-minute editing assistance, Neil Coe and Phil McManus for constructive comments, and Steve Pollard for amendments to an early version and for reminding us that SIDS have private sectors. We particularly thank Courtney Lindsay for invaluable contributions to the final draft.

Sydney, Australia
Mona, Jamaica
November 2025

John Connell
Yonique Campbell

Maps

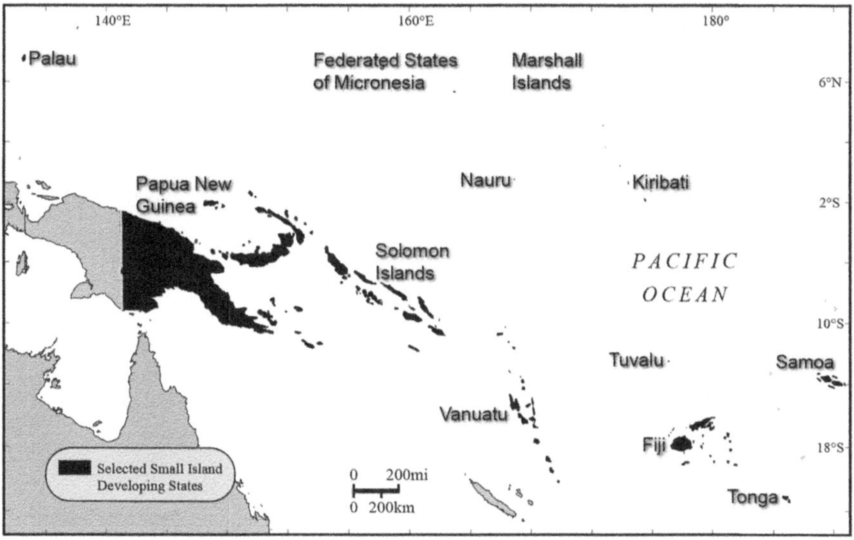

Fig. 1 The Pacific

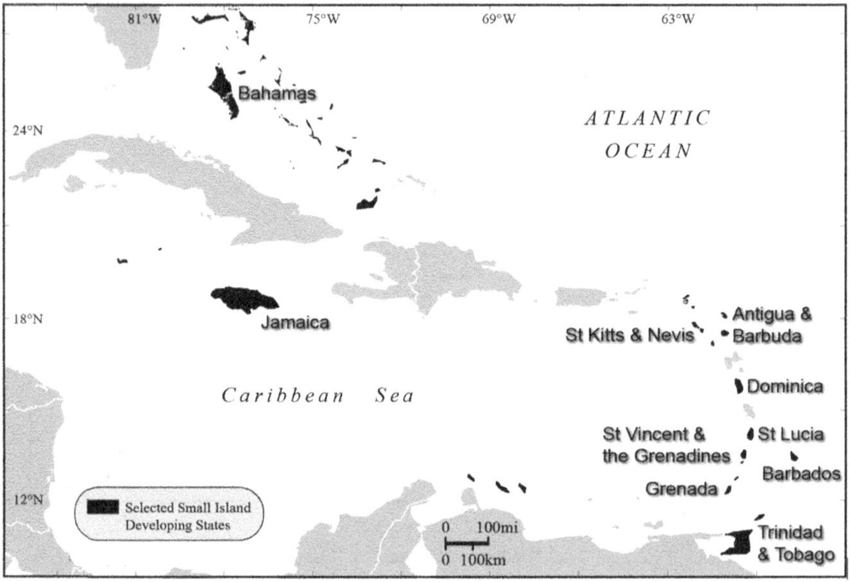

Fig. 2 The Caribbean

Competing Interests The authors have no competing interests to declare that are relevant to the content of this manuscript.

Contents

Abbreviations

ABAS	Antigua and Barbuda Agenda for SIDS
ABC	Australian Broadcasting Corporation
AI	Artificial Intelligence
AOSIS	Alliance of Small Island States
APTC	Australia-Pacific Technical College/Australia-Pacific Training Coalition
ASEAN	Association of South East Asian Nations
ATFU	Australia-Tuvalu Falepili Union Treaty
BRI	Belt and Road Initiative
BRICs	Brazil, Russia, India, China, and South Africa
CARICOM	Caribbean Community
CARIFTA	Caribbean Free Trade Association
CBI	Citizenship by Investment
CCZ	Clarion-Clipperton Zone
CDEMA	Caribbean Disaster Emergency Management Agency
CEDAW	Convention on the Elimination of all Forms of Discrimination Against Women
CHOGM	Commonwealth Heads of Government Meeting
COFA	Compact of Free Association
COP	Conference to the Parties to the United Nations Framework Convention on Climate Change
COSIS	Commission of Small Island States on Climate Change and International Law

CROP	Council of Regional Organisations of the Pacific
DSM	Deep-Sea Mining
DWFN	Distant Water Fishing Nation
ECE	Early Childhood Education
ECLAC	Economic Commission for Latin America and the Caribbean
EEZ	Exclusive Economic Zone
EIA	Environmental Impact Assessment
ENSO	El Niño-Southern Oscillation
ESCAP	Economic and Social Commission for Asia and the Pacific
EU	European Union
FAD	Fisheries Aggregation Device
FAO	Food and Agriculture Organisation
FBC	Fiji Broadcasting Corporation
FFA	Forum Fisheries Agency
FIFA	International Federation of Association Football
FIFO	Fly In Fly Out
FSM	Federated States of Micronesia
GCF	Green Climate Fund
ICC	International Criminal Court
ICJ	International Court of Justice
IISD	International Institute for Sustainable Development
IMF	International Monetary Fund
IMO	International Maritime Organisation
IO	International Organisation
IPCC	Intergovernmental Panel on Climate Change
ISA	International Seabed Authority
IUU	Illegal, Unreported and Unregulated Fishing
LIAT	Leeward Islands Air Transport
LIDS	Large Island Developing States
MDGs	Millennium Development Goals
MIRAB	Migration, Remittances, Aid and Bureaucracy
MPA	Marine Protected Area
NDCs	Nationally Determined Contributions
NFIP	Nuclear Free and Independent Pacific
NGO	Non-Government Organisation
OECD	Organisation for Economic Cooperation and Development
OECS	Organisation of Eastern Caribbean States
PAHO	Pan American Health Organization

PALM	Pacific Australia Labour Mobility Scheme
PBP	Partners in the Blue Pacific
PEV	Pacific Engagement Visa
PICs	Pacific Island Countries
PIFS	Pacific Islands Forum Secretariat
PINA	Pacific Islands News Association
PNA	Parties to the Nauru Agreement
PNG	Papua New Guinea
RAMSI	Regional Assistance Mission to Solomon Islands
RMI	Republic of the Marshall Islands
RNZ	Radio New Zealand
RSE	Recognised Seasonal Employer
SAWP	Seasonal Agricultural Workers Program (Canada)
SCF	Save the Children
SDGs	Sustainable Development Goals
SIDS	Small Island Developing States
SLR	Sea-Level Rise
SOE	State-Owned Enterprise
SPC	Secretariat of the Pacific Community
SPREP	Pacific Regional Environmental Program
SWP	Seasonal Workers Programme
TEK	Traditional environmental knowledge
TINA	There Is No Alternative
TVET	Technical Vocational Education and Training
UAE	United Arab Emirates
UHC	Universal Health Coverage
UNCLOS	United Nations Convention on the Law of the Sea
UNCTAD	United Nations (Conference on) Trade and Development
UNEP	United Nations Environment Programme
UNFCCC	United Nations Framework Convention on Climate Change
UN-OHRLLS	Office of the High Representative for the Least Developed Countries, Landlocked Developing Countries and Small Island Developing States
USP	University of the South Pacific
UWI	University of the West Indies
VDS	Vessel Day Scheme
WASH	Water, Sanitation and Hygiene
WTO	World Trade Organisation

List of Tables

1

Islands Afloat

> What words might we use for the atmosphere we all breathe, about the way humanity is changing that atmosphere's composition? What words apply when people worry about the ocean's future, about its ecosystem? … Should we draw words for discussing the earth from science, emotions, statistics or religion? Or must unspoiled areas be reduced to charts and graphs about the economic and social value of nature? (Andri Snær Magnason, *On Time and Water*, 2020: 61–62)

Small island states have rarely acquired global significance, beyond that embodied in the threat that climate change poses for the islands and their people, almost all inhabiting coastal fringes. Island states have consequently taken their place in global forums, alerting the world to the dangers of unrestrained climate change and demanding policies and practices that prevent and mitigate climate change. Within the Alliance of Small Island States (AOSIS), they have acquired strength and demonstrated agency and authority, working together and punching above their weight (Corbett, Yi-Chong, & Weller, 2018), asserting themselves as 'mice that roar' or, perhaps, in a blue economy, 'dolphins that leap'. Here then we examine the recent past and consider the future of the two largest groups of islands, those of the Caribbean and the Pacific, to consider differences and similarities as they face a range of challenges, some but not all of which are a consequence of small size and many that are not of their own making.

J. Connell, Y. Campbell, *Island Challenges*,
https://doi.org/10.1007/978-981-95-8177-1_1

Both the singularity and similarities of islands and island states are evident in their physical landscapes, from dry atolls to rainy high islands, and in a variety of human-caused changes (such as rising sea levels, plastic pollution, ocean acidification, loss of biodiversity and the diffusion of pandemics), some much influenced by political decisions at a variety of scales in increasingly neoliberal times. Hazards and politics are inherently uneven. While the Bahamas out islands may superficially look much like the atolls of Tuvalu, their histories, cultures and economies are, however, vastly different. The Bahamas was one of the first island groups to have COVID-19; Tuvalu was the last. One of us can claim to have run round an entire republic (Nauru) before breakfast, while to get from one end of Kiribati or the Federated States of Micronesia (FSM) to the other, one must travel outside the country. So, islands and island states are quite different, yet they have invariably been grouped together as 'small states' with intrinsic disadvantages, a seemingly unavoidable juxtaposition. In recent times they have been regarded as Small Island Developing States (SIDS) and this too is where we start from, despite its limitations as a classificatory scheme, but its greater value as a political grouping. The SIDS definition is inherently neo-colonial, suggesting a context where island states are peripheral and unsophisticated; Pacific island states prefer the term 'Large Ocean States', reflecting their authority over vast ocean regions (Bordner, Ferguson, & Ortolano, 2020). Nonetheless, we start from 'smallness' out of convention and simplicity, but we seek to develop a new focus on the island states that constitute the SIDS, while comparing and contrasting the Caribbean and Pacific island states, a challenging task that has rarely been undertaken but one that points to important similarities and differences and thus the need for different strategies and directions in different places. In so doing, we provide a new perspective on two significant global regions.

No universally accepted definition of what makes a country small exists, or of what makes it viable in any sense. Relevant measures can include population, size of the land or territory (including maritime zones), and national income. Nor, indeed, is there even a useful definition of island. It is therefore unsurprising that for one reason or another the states that formally constitute the SIDS are sometimes neither small in population nor land area, and some are not islands at all. Ultimately, there is no adequate definition of a SIDS (MacFeely, Peltola, Barnat, et al., 2021). That may be no bad thing where rigid terminology inhibits thinking about sustainability (Kelman, 2021). First recognised as a distinct group at the United Nations Conference on Environment and Development in June 1992, officially there are 39 SIDS (and 18 dependent territories), mostly in the Caribbean and Pacific. We focus on the independent states in the two regions and then largely follow a previous strategy, where states with populations of over 1 million or that were not islands were excluded (Connell, 2013). We have excluded Haiti, Dominican Republic and Cuba, as large island developing states (LIDS) and ignored continental Belize, Guyana, and Suriname, despite their qualities of 'islandness', but we include Jamaica and Trinidad and Tobago and make occasional passing reference to the islands of Papua New Guinea (PNG) (Table 1.1), notably to Bougainville, which aspires to secede from PNG and become an independent state. Such larger states are nonetheless of some significance for various aspects of regionalism. Exclusion does not mean that the present analysis is of no relevance to politically dependent territories, such as the Cook Islands and Montserrat (both of which are amongst the 18 SIDS associated non-UN states), or to 'non-island' states such as Guyana, Belize and Timor Leste (all officially SIDS), or much larger island states. Coverage of the SIDS will be inevitably uneven, partly because of the lack of good data, even to the extent of their being ignored in some global data sets.

The Pacific islands SIDS (Figure 1) are spread across 30 million km^2 of the tropical Pacific Ocean, a third of the world's surface, divided among 14 independent countries and, except for Papua New Guinea (PNG), which accounts for half the land area and three quarters of the population, have limited natural resources in the conventional sense, have small populations, are distant from global markets, are somewhat isolated from global trade, and have experienced a distanced colonialism, all circumstances that have slowed economic growth, hampered sustainable development, and delayed or discouraged effective independence. By contrast, the Caribbean SIDS (Figure 2) occupy a much smaller area but share many similar characteristics with those in the Pacific, notably limited resources, lack of economic diversification and susceptibility to the effects of climate change.

1.1 Facts and Occasional Figures

Statistics and summaries of basic data for the SIDS, some included here (Table 1.1), tend to focus on conventional indicators partly because it is difficult to collect (and compare and contrast) complex data, for example, on sustainability, environmental change or domestic violence. How might criminality or the quality of governance be measured? Most relevant to the Sustainable Development Goals (SDGs), how might sustainability be defined and measured? We have chosen not to attempt such an over-ambitious task, going with the flow of others' usages and loosely adhering to the Brundtland definition of 1987 where sustainability constitutes 'development that meets the needs of the present without compromising the ability of future generations to meet their own needs'. Over time attempts to develop more wide-ranging statistics on such vital themes as resilience and vulnerability, so giving rise to the Human Development Index (HDI) and the Multidimensional Vulnerability Index (MVI), have become more sophisticated and relevant, for example, in considering gender issues, but they have come no closer to thinking about development in ways that suggest new approaches or imply new strategies. Over time 'hard' measures have given way to softer qualitative measures. A limited range of measures, loosely related to ideas about development, are tabulated here to provide a basic background to the book and the countries involved enabling crude differentiation between the SIDS. For simple consistency we have based the table on the CIA's *The World Factbook*, but this is far from definitive and accurate (and provides no information on its own sources, or methodology, or what definitions are used). In two cases, the FSM and the Marshall Islands, we have adjusted what were manifestly incorrect population data (by using Pacific Community data) but have resisted the temptation to make changes elsewhere. 'Unemployment' data are of limited relevance in SIDS, such as Solomon Islands, where subsistence economies are significant (Feeny & McDonald, 2016). Remittance data are particularly doubtful. Urban populations are crude and are unhelpful in SIDS with significant urbanisation and commuting from beyond formal borders. At best these data offer rough comparisons, and all too often we have been reduced to adjectives such as 'some' and 'many' and adverbs such as 'increasingly' that have little explanatory value.

Table 1.1 The small island developing states

SIDS	Population 2024	Population growth	Population density	Urban pop %	Unemployment %	Youth unemployment %	US$GDP, p.c.	HDI	MVI	Life expectancy	Remittances % GDP
Antigua and Barbuda	102,634	1.1	213	24	11	—.	29,000	0.826	61.7	78	1.2
Bahamas	410,862	1.1	40	84	9	24	33,100	0.820	59.9	77	0.4
Barbados	304,139	0.2	647	31	8	—	19,200	0.809	57.9	79	1.3
Dominica	74,661	0.0	88	72	—	—	17,400	0.740	55.4	79	6
Grenada	114,621	0.3	345	37	24	—	16,900	0.793	61.7	76	5
Jamaica	2,823,713	0.1	262	57	4	14	10,300	0.706	49.6	76	18
St. Kitts and Nevis	55,133	0.6	180	31		—	30,400	0.838	55.0	78	3
St. Lucia	168,038	0.3	295	19	11	28	23,400	0.725	65.5	79	2.5
St. Vincent and theGrenadines	108,647	–0.2	258	54	19	42	18,300	0.772	61.8	77	8
Trinidad and Tobago	1,408,966	0.1	294	53	4	10	31,700	0.814	50.3	77	0.8
FSM	71,000	–2.5	113	23	—	19	3900	0.634	64.0	75	5
Fiji	951,611	0.4	51	59	4	15	13,700	0.729	51.7	75	9
Kiribati	116, 545	1.0	166	58	—	23	3200	0.628	59.8	69	4
Marshall Islands	54,366	–0.2	202	79	—	26	6800	0.731	58.9	75	13
Nauru	9892	0.4	597	100	—	—	12,500	0.696	68.7	69	0.6
Palau	21,864	0.4	38	82	2	6	15,800	0.797	57.0	75	0.7
PNG	10,046,233	2.3	23	14	3	4	4200	0.568	47.1	70	0.0
Samoa	208,853	0.7	78	18	4.6	20	6300	0.702	62.5	76	26
Solomon Islands	726,799	1.7	29	26	2	3	2500	0.562	55.9	77	5.4
Tonga	104,889	–0.3	145	23	2	7	6900	0.739	57.0	78	50
Tuvalu	11,733	0.8	322	66		21	5800	0.653	63.3	69	4.2
Vanuatu	318,007	1.6	27	26	5	11	3100	0.614	54.4	76	13

Source: based on *CIA, The World Factbook, 2024.* Many of the numbers in the table, especially on unemployment and economic issues, are unreliable and/or dated but included for crude comparative purposes. HDI=Human Development Index, MVI= Multidimensional Vulnerability Index.

While the Human Development Index (HDI) and the Multidimensional Vulnerability Index (MVI) are probably more useful as composite measures of comparative development ranking, their three-digit final scores suggest spurious accuracy. The HDI aggregates country scores through a vast assortment of indicators, including life expectancy, literacy rate, access to electricity, GDP per capita, exports and imports, homicide rate, multidimensional poverty index, income inequality, internet availability and many more, which in the end is not much more than the sum of the parts. No SIDS have HDI scores below 0.55, in the 'low human development' category that covers some parts of sub-Saharan Africa. Likewise, the MVI includes multiple indicators from exports, through low-elevated coastal zones to the proportion of parliamentary seats held by women. How such variables are weighted is fraught with assumptions and preferences. Data are collected in different ways in different countries and accuracy varies enormously. In an annex to its report on the development of the MVI, the United Nations lists more than 60 indicators that were considered for inclusion but excluded for various reasons, principally 'too many missing values' and 'significant [quality of] data issues' while others overlapped (United Nations, 2024). In several SIDS, statistics offices are poorly funded and statistical data rarely valued beyond the office; hence, even 'basic' population data can be inaccurate, especially where there is uncertainty over de facto and de jure classifications. It is implausible that the homicide rate is measured (or data collected or recognised) in the same way across countries. Equity fares little better. Some crucial data such as climate change and resilience are inherently difficult to measure despite improvements in means of assessment (Béné, Wood, Newsham, & Davies, 2012; Grant, Roy, Zablah, et al., 2023; Tandrayen-Ragoobur, Banerjee, Fauzel, & Matadeen, 2024).

Beyond the basic data (Table 1.1), many other possibilities for classification and ranking relate to the SDGs. Five SIDS, all in the Caribbean (Antigua and Barbuda, The Bahamas, Barbados, St. Kitts and Nevis and Trinidad and Tobago), are regarded as developed enough that they are not listed as ongoing recipients of overseas development assistance. Nauru and Palau will be included in that group in 2026. By contrast, in an otherwise lengthy list of aid recipients, just three—Kiribati, Tuvalu and Solomon Islands—are listed as Least Developed Countries. Some attach significance to particular factors; Bill Gates has argued that where education and literacy are weak, health and climate vulnerability weaken and climate resilience becomes 'nigh on impossible' (Sasingian, 2025). Other distinctions exist. It is possible to compile a Fragile States Index while Vanuatu has been listed as both the world's happiest country and the most hazard-prone country. All data are problematic and subjective in some way.

More subtle data are also lacking. While some SIDS have sponsored good indigenous research and been generous with research permits, others have not: it is therefore often difficult to make even crude comparisons between countries, for example in such key areas as fishing trends, practices and outcomes. At national scale, there is little on the extent and significance of many regional differences, though these can be considerable and vital (Cumberbatch, Drakes, Mackey, Nagdee, et al., 2020). Indeed, for the largest state we consider, its Booker Prize-winning novelist, Marlon James, has said: 'It's the biggest small country in the world – there are 15 different Jamaicas in one island' (quoted in Tate, 2025). In archipelagic states regional variations can be enormous. While small islands are often held out as 'laboratories' for assessment of climate change, often for the benefit of distant others (e.g. Pathirana, 2025), the databases in many SIDS are poor. Nonetheless, they do indicate real differences in income and population growth, for example that demonstrate the diversity of the SIDS, a theme that recurs in the following chapters. For multiple reasons we provide no other tables in the book. Although the SIDS take up some of the lower rungs of formal charts and ladders, but almost always above most sub-Saharan African states, they cover a spectrum of contexts.

1.2 Becoming Blue?

In both regions, prominence has been attached to an emergent focus on a blue economy, a rhetorical, strategic and economic direction, hinting at a rather different future, most forcibly expressed in Barbados at COP26 (the 26th annual UN Climate Change conference) when the Prime Minister of Barbados, Mia Mottley, called for reparations and a 'Caribbean

Marshall Plan' to deal with the legacies of colonialism and slavery (White, 2021). In Mottley's words:

> There was no bank account left with us at the point of independence, there was no development compact and yet there is a legitimate expectation by our people that independent governments would right the wrongs of the past and would do so quickly by giving people opportunity in this part of the world. No-one is asking for anything other than fairness at this stage ... Economic transformation and growth depend on an international compact for the Caribbean. COVID-19 has already led, in many instances, to a doubling of expenditure in health and social care support, a quadrupling in other cases, because all of a sudden you go from tourism sectors that are earning to tourism sectors with zero revenue and therefore lack the capacity to employ anyone.

Mottley pointed to contemporary economic and developmental challenges, the vulnerability of previously colonised Caribbean SIDS and the need for a shift in the 'international compact' to facilitate sustainable development. Enele Sopoaga, a former Prime Minister of the smallest Pacific state, Tuvalu, has made similar statements and been even more critical of post-colonial trends, as in his critique of Australia and New Zealand: 'Go ahead and drill and open up new coal mining or get new gas stations, but don't forget that whatever you are going to do, it's going to increase greenhouse gas emissions, which are going to sink the islands of Tuvalu and kill the people' (RNZ, 9 May 2024). Neither leader is anywhere near alone in seeking more relevant and equitable forms of development. Both, as so many others, have emphasised that sustainable development in the SIDS cannot be achieved without effective global legislation over climate change, equity and, increasingly, reparation for past damage. In the Pacific, the Boe Declaration of 2018, endorsed by Pacific SIDS leaders, identified climate change as the single greatest threat to the region, rather more so than any Chinese incursions (a view not necessarily shared by metropolitan states that fringe the Pacific, few having endorsed effective climate change strategies).

During the preparatory processes in 2011 for the Rio+20 UN Conference on Sustainable Development, the idea of development centred on a 'blue economy' was put forward. Coastal and island nations argued that maintaining ocean health was basic to sustainability, since the future resource base of SIDS was marine (Telesford, 2021). That built on established sustainability paradigms and on attempts to unify the ocean in a comprehensive framework, as in the 1994 United Nations Convention on the Law of the Sea (UNCLOS), and represented a more planned approach to ocean economies, linked to SDG14: Life Below Water (Table 1.2). As the then Secretary-General of the Pacific Islands Forum (PIF), Dame Meg Taylor, stated in 2019: 'while calling for a strategy for "the Blue Pacific Continent" we are faced with questions on how we can secure our rights to the Pacific Ocean and its resources and what innovative opportunities exist to leverage the value of our Blue Pacific as a pathway to our development' (2019: xiii). It might equally have been said of the Caribbean.

While the nature of the 'blue economy' is often deliberately vague, and scarcely innovative, central to it was the notion that oceans offer a future for SIDS, part of history and destiny, that integrates economic development and ecological sustainability. The extent to which new strategies might be required and implemented to achieve sustainability, is debatable (Telesford, 2021). For Mark Spalding, 'the blue economy was not about old versus new in time scale; it is about unsustainable versus sustainable' and raised such questions as 'are there traditional ocean activities that generate economic returns but that are benign or regenerative for our coasts and ocean? Or is it all about extraction, pollution and other activities that diminish the ocean's health?' (2016: 5). The objectives of the blue economy, simply centred on increasing the economic benefits to SIDS from sustainable use of marine resources, including fisheries, aquaculture and tourism, were to be achieved by 2030. The blue economy did suggest a cultural turn; as two Caribbean Cabinet Ministers claimed: 'The ocean is our life blood. In the Caribbean, it shapes our cultures, feeds our people, supports our economies, and connects our nations. As political leaders from small island developing states, we know the ocean's bounty and its vulnerability better than most' (James & Clarke, 2025). The blue economy also acquired strategic connotations, as in mid-2022 when metropolitan states, including Australia and the United States, established a new but ill-fated grouping, Partners in the Blue Pacific (PBP), to coordinate assistance to the region in the face of competition

Table 1.2 The sustainable development goals

1. No Poverty: Complete eradication of poverty in all of its forms everywhere
2. Zero Hunger: End hunger, promote agriculture and provide food for every person
3. Good Health and Wellbeing: Ensure health and provide care for everyone regardless of their age
4. Quality Education: Ensure quality education that is equitable and inclusive
5. Gender Equality: Eradicate gender-based discrimination and empower all women
6. Clean Water and Sanitation: Ensure and develop appropriate water supply and waste disposal systems
7. Affordable and Clean Energy: Provide access to affordable, reliable and modern energy
8. Decent Work and Economic Growth: Provide constant and inclusive enhancement of economic activities
9. Industry, Innovation and Infrastructure: Develop resilient infrastructure and enhance creativity
10. Reduced Inequalities: Further decrease and eradicate the inequalities that exist
11. Sustainable Cities and Communities: Develop cities and communities that are inclusive, safe, resilient and sustainable
12. Responsible Consumption and Production: Promote patterns of consumption and production that are sustainable
13. Climate Action: Make urgent efforts to combat climate change and its effects
14. Life Below Water: Protect and manage the oceans, seas and marine resources
15. Life on Land: Conserve and promote the sustainable use of land-based ecosystems
16. Peace, Justice and Strong Institutions: Support the achievement of peaceful and constructive solid societies
17. Partnerships for the Goals: Revamp and strengthen global partnerships for sustainable development

from China. New political complexities had come to characterise both regions. Historical forces of colonial and neo-colonial development could be obstacles to achieving a more equitable economy (Clark & Cisneros-Montemayor, 2024). Such considerations emphasised the need for development rather than growth and the crucial role of environmental management and sustainability.

1.3 Diversity and Difference?

Beyond the overworked implicit dependency of 'small' island states, a more cultural perspective has pointed to the fracturing and disempowerment of both regions and their people, the antecedents of which have a long colonial history, sometimes ideologically perpetuated in the language of aid and development. Countries and borders are colonial impositions in both regions. The SIDS exhibit considerable diversity of size, geography, economic history, location and economic development. Economies vary considerably from those where a substantial proportion of the population remain engaged in semi-subsistence agriculture (notably in the Melanesian states of PNG, Solomon Islands and Vanuatu), to those where agriculture of any kind has all but disappeared, vanquished by a tourism monoculture (The Bahamas, Antigua and Barbuda). Most economies are more mixed, with a valuable fishing industry in the Pacific, but increasingly oriented to service industries, with a significant urban bureaucratic component that has contributed to an urban bias in financing development. Few have manufacturing industries beyond limited import-substitution industries in the food and drink sector. In many SIDS the informal sector, centred on marketing, plays a valuable (if not always valued) role in absorbing unemployment.

Political systems are broadly democratic although political parties are linked to leaders rather than policy. Policy convergence in many SIDS has meant that there are rarely distinct differences between political parties and successive administrations, with most policies loosely aligned with a neoliberal, market-driven approach that has yet to absorb sustainable principles. Unstable coalitions of parties are common in Melanesian SIDS, whereas in most Caribbean SIDS, a strong two-party political system dominates. In both regions the fragility of leaders—almost always men—and parties created problems of long-term policy formation.

Long before a turn to blue, various social, political and geographical circumstances had contributed to diverse development paths and strategies. Small states have inevitably been caught up in the fluctuating fortunes and objectives of larger states. Though Tonga would claim otherwise, all were colonies at some point in the past. Some, mainly in the Caribbean, were keenly contested by colonial powers while in the Pacific a few were simply unwanted, like Kiribati and Tuvalu, the 'Cinderellas of the Empire' (MacDonald, 2022). In post-colonial times aid from former colonial powers and elsewhere has been invaluable, but this has fluctuated and usually diversified in accordance with the changing geopolitical significance of these strategically important global regions. Most recently that has meant the growing weight of China in both regions, prompting responses from other global powers. China's maritime Belt and Road Initiative (BRI) and its investment in both regions have provoked the United States (and other metropolitan states) to respond, criticise and be criticised for establishing bilateral agreements.

The larger SIDS—notably Solomon Islands, Vanuatu, Fiji and Jamaica—are mountainous (high island) countries. By contrast, the three atoll states, Kiribati, Marshall Islands and Tuvalu, are composed of many coral atolls, none more than 2 metres above sea level. Indeed, it has been suggested, rather tongue in cheek, that tourists should visit them before climate change wipes them out (Farbotko, 2010). Most SIDS consist of many islands in the Pacific, scattered over very large areas. Kiribati, with a population of just 120,000 on 24 populated atolls, crosses both the equator and the date line, and extends over 3.5 million sq kms. Outer islands in archipelagos everywhere are disadvantaged. Almost all SIDS have extensive coastlines, with the atoll states almost exclusively coastlines, offering no high land for retreat, and subject to various environmental hazards.

In this century, the effects of climate change and global warming have accentuated the 'normal' impact of hazards. Four changes are simultaneously occurring—a rise in air and sea temperatures, fluctuations in rainfall, higher sea levels and more intense cyclonic activity—a combination that has caused problems in both regions. In the early 2020s, the COVID-19 pandemic further complicated and constrained development, both as health care systems were sorely tested—and some mortality rates soared—and because SIDS experienced particularly significant financial stress, as dominant tourist economies collapsed. Caribbean islands, more reliant on tourism, were severely affected. Debt distress among SIDS, present even before the pandemic, created liquidity and solvency problems challenging fragile and unequal gains (Goujon &

Wagner, 2023). Fighting a global pandemic affected the fight against other diseases. Lost ground occurred in the fight against polio, malaria, tuberculosis, HIV and non-communicable diseases (NCDs), alongside the loss of education for many children, and rising violence against women and girls. High exposure to climate hazards further compounded vulnerability, with poor population groups, including women and people with disabilities, most at risk (Ahmed & Baroy, 2022; Hill, 2025). Managing the impact of the pandemic emerged briefly as a key focus for SIDS, demonstrating how small states with limited diversification could be disrupted by external shocks entirely beyond their control but also how 'normal service' was resumed as quickly as possible.

1.4 The SIDS as an Entity: Institutions and Agreements

The SIDS were first recognised as a distinct group at the Rio Earth Summit Conference in 1992. Two years later, a Barbados Programme of Action was formulated to support their movement towards sustainable development. The group thus emerged from an increasing global focus on environmental issues spurred by climate change. SIDS are particularly vulnerable to the marine effects of climate change, including sea-level rise, ocean warming and acidification, and greater cyclone intensity. Distinguishing the SIDS as a group of countries with significant and comparable needs is bound up with their environmental status in a context where they had limited financial and human capital to mitigate environmental and economic constraints and were thus reliant on international aid. The United Nations Conference on Trade and Development (UNCTAD), whose aegis the SIDS initially came under, stated on its website that 'the UN never established criteria to determine an official list of SIDS' while UNCTAD itself maintained a shorter, unofficial list of SIDS. Countries thus opted in. The aggregate population of the usually listed SIDS is about 65 million, slightly less than 1% of the world's population. The population of our more select group of 21 countries (excluding PNG) is little more than 7 million (Table 1.1), but collectively they have acquired considerable global significance.

In this century the SIDS have become established as a distinct group within the United Nations system under the grandly titled Office of the High Representative for the Least Developed Countries, Landlocked Developing Countries and Small Island Developing States (UN-OHRLLS), which works with the 46 Least Developed Countries, 32 Landlocked Developing Countries and 39 SIDS. UN-OHRLLS has sought to play a leading role in raising awareness of the 'special case' of SIDS at regional and global levels. This grouping pointed to the obvious implication that SIDS were disadvantaged and less developed. The 1994 Barbados Programme of Action on the Sustainable Development of Small Island Developing States, which was collectively and unanimously endorsed by SIDS, emphasised a more holistic perspective on development, involving resilience, trade, sustainable production and consumption, health and culture. These principles were regularly reaffirmed, with additional foci such as the need for progress towards achieving sustainable energy. Two decades later in 2014, the SIDS Accelerated Modalities of Action (the Samoa Pathway, formulated at an international conference in Apia, the capital of Samoa), for the decade 2014–2024, stressed broadly similar goals. It extended the environmental focus on climate change, sea-level rise and fragile environments, as threats to survival and viability, and promoted the integrated and sustainable management and conservation of natural resources and ecosystems that supported social and human development. Another decade on, the 'Antigua and Barbuda Agenda for SIDS: A Renewed Declaration for Resilient Prosperity' (ABAS) reaffirmed the objectives of the Samoa Pathway by recognising 'persistent development challenges' that centred on health systems, gender equality, energy, water and food security. It had rather less direct alignment with the SDGs and rather more with disaster risk reduction and the acquisition of climate finance. Inevitably translating such grand themes, abstractions and priorities into concerted development and improved livelihoods was conducted at a quite different scale, but the United Nations system provided a focus and framework for the needs of small states.

Emerging from the Barbados Programme of Action and the Samoa Pathway were the Paris Agreement and the Sendai Framework for Disaster Risk Reduction. Adopted in 2015 at the UN Climate Change Conference (COP21) in Paris, the Paris Agreement is a legally binding international treaty on climate change that aims to 'pursue efforts to limit the temperature increase to 1.5°C above pre-industrial levels'. Critical for the sustainable development, and perhaps even survival, of some SIDS, it faced numerous challenges, not the least being the failure of developed countries to make the necessary financial investments to mitigate and ensure alternative sources of energy (culminating in the US withdrawal from the Agreement). Developed countries include such large greenhouse-gas-emitting countries as China, the United States and Australia, each of which plays complex roles in the Caribbean and Pacific. The Sendai Framework, also adopted in 2015 and a further step towards securing sustainable goals for the SIDS, involved a set of global targets (centred on global risk governance), oriented at substantially reducing global disaster mortality by 2030. Progress with implementation failed to keep up with growing risks (Cabral-Ramírez, Niño-Barrero, & DiBella, 2025). Much like the objectives of the Samoa Pathway and its ABAS sequel, strategies constituted fine words but at some distance from the hazardous environments experienced in many SIDS and without resolving key development challenges.

Most (but not all) SIDS are also members of the Alliance of Small Island States (AOSIS), which performs lobbying and negotiating functions for the SIDS within the United Nations system. AOSIS, established in 1990, preceded the SIDS and made key inputs into the seminal Rio Earth Summit and the subsequent Barbados Plan of Action. AOSIS is a coalition of 44 mainly small island states (overlapping considerably with the SIDS), 37 of which are independent, that share similar development challenges and concerns about the environment, and especially their vulnerability to global climate change. It functions primarily as a lobby and negotiating voice for SIDS within the United Nations, working primarily through national diplomatic missions in New York, and through consultation and consensus, with no formal charter, secretariat or budget. Such bureaucratic structures have been invaluable to the SIDS in making their case for international development policies and practices that directly or indirectly contribute to sustainable national development (sometimes supported by other international and regional organisations, like the World Bank and the Asian Development Bank (ADB), but whose roles extend beyond the SIDS and, indeed, usually focus on the needs of much larger states).

1.5 The Sustainable Development Goals

In recent years the future of the SIDS has been frequently linked to their achievement or otherwise of the SDGs. Formulated in 2015, the SDGs were the complex sequels to the disappointing Millennium Development Goals (MDGs): eight goals adopted by the United Nations in 2000 to be achieved by 2015 that aimed to address global poverty, hunger, disease, illiteracy, environmental degradation and gender inequality. The more detailed SDGs followed with a target date of 2030. With 17 SDGs consisting of 169 targets and 231 indicators, across a wide range of concepts and sectors (Table 1.2), an immediate problem for SIDS was having, or collecting, data and indicators that might assess progress towards these diverse targets.

While the SDGs were a further welcome departure from economic growth towards more equitable wellbeing, that shift was criticised for not adequately challenging global power structures, being unrealistic in its timelines especially regarding ending poverty by 2030, and not providing distinctive finance to achieve the goals. The SDGs were also seen as dependent on old power structures and economic models and without a grounding in history (Esquivel, 2016; Gupta & Vegelin, 2016). Nonetheless, they pointed to valuable directions, for all countries, especially in social and environmental matters, more focused than hitherto, which might remedy past failures and enable more balanced development.

The initial and ongoing problem of defining (and then measuring) particular elements of sustainability before ever achieving them proved 'an unprecedented statistical challenge' (MacFeely, 2020). That could be entirely superficial where the goals 'were hamstrung by their commitment to the unequal power structure of the neoliberal capitalist development model' that ignored the vicissitudes of history and avoided notions of structural change while still aiming to reduce inequalities (Brissett,

2017: 18). Some such as SDG14, Life Below Water, implied marine priorities for SIDS. Others, such as ending hunger and achieving environmental sustainability, would necessitate trade-offs. The goals provided inadequate guidance on how countries might achieve and measure their progress, resulting in various suggestions on how that might be done and what should have priority (where some were in conflict). One obvious suggestion was for SIDS to focus on key sustainability indicators (social, economic and environmental), since the UN and the Intergovernmental Panel on Climate Change (set up by the World Meteorological Organisation and the UN in 1988) both recognised their vulnerability to environmental hazards (Van Beynen, Fenda, Akiwumi, & Van Beynen, 2018). Problems were attached to measuring many variables that moved country objectives away from economic issues, as even conceptualising vulnerability and resilience was difficult, and countries were being asked to measure things that had never been done before.

Beyond problems of measurement and quantification, various indices, like vulnerability, are ideological and contextual. Indeed, Charlotte Weatherill has argued that 'much of the reason for the lack of action on climate change is because vulnerability is discursively constructed as a racialized and feminized characteristic' (2025: 35). Moreover, data are usually based on such developmental markers as GDP and population, based on the logic that a country's wealth translates into a 'capacity' to mitigate and adapt to climate change. Yet GDP fails to include remittances, foreign aid and fishing licence revenues, all significant in most SIDS so that Gross National Disposable Income (GNDI) would be more useful (Howes & Chowdhury, 2025) while GDP also fails to measure the environmental impacts of economic activity, or measure work and labour outside market transactions, while prioritising questions of income and expenditure over those of wellbeing. Significantly, global GDP has more than doubled in this century whereas progress on mitigating social deprivation and achieving socio-ecological balance has slowed (Fanning & Raworth, 2025). As Charlotte Weatherill (2023: 2024) forcefully pointed out, capitalist logic valued economic growth over the harm and loss of life that climate effects cause, protecting the global economy as it is currently structured, and portraying development as the solution to climate change, as opposed to the driver of the extraction and consumption that was causing it. Humanity was losing natural protection faster than it could adapt. Vulnerability is both socially constructed, through power and politics, and inherently relational.

Nonetheless, the SIDS face a range of disadvantages that create evident vulnerabilities that impede sustainable development, however that is defined, usually formally seen as structural factors, including physical size, remoteness, limited resource base, market size, and exposure to climate risks and natural disasters. These require different approaches and financing mechanisms to move towards sustainable and equitable development. The strong negative correlation between high structural vulnerabilities and poor SDG outcomes, including extreme poverty, low life expectancy and poor subjective wellbeing, later resulted in the UN developing the MVI to measure 'structural vulnerability' and lack of resilience at the national level. The objective was to secure appropriate development finance according to country needs.

However flawed the indices, there was enough consistency to recognise that SIDS frequently appeared towards the 'difficult' and problematic end of such lists and indicators. Few such classificatory surveys sought to consider why that had occurred, and thus where and how remedies might be achieved. Questions were again raised over the need for more social variables, such as divorce, domestic violence or incarceration rates, that raised different development issues. In any case data are not always reliable, nor do they even exist, in many states (Connell, 2013: 9–14). Generally, small states and SIDS were expected to be poorly ranked and so they were. Although rare exceptions exist, such as Singapore and Malta, smallness and 'islandness' otherwise was seemingly inevitably a problem, with 'small' becoming a pejorative term implying inadequacy and unimportance. Not surprisingly, at the 2012 Pacific Forum conference it was suggested that the Pacific entities were 'large ocean' rather than 'small island' states.

Climate change and responses to it at every scale have increasingly become central to the challenges faced by SIDS and related to the SDGs. In the Pacific, the activist islander group, the Pacific Climate Warriors, for example, uses the slogan, 'Not drowning, but fighting' (later 'To Hell With Drowning') to oppose the common narrative of 'sinking island states', and enhance agency and resilience, through a strong political message that islands were not sinking but were being sunk (Chap. 2). Resolving that conjuncture required much more than development finance. Like all measures of development, conceptualisation of SIDS and SDGs had historical roots which usually had deeply colonial overtones. Indeed, it is not surprising that 18 politically dependent territories (the word 'colony' being firmly rejected) are part of the SIDS structure. For all their limitations, it is however probably the SDG themes and their relationships to a blue economy that provide the most useful way of thinking about the Pacific and Caribbean SIDS worlds.

1.6 Directions

Fundamentally, the Pacific and Caribbean SIDS have different histories and geographies but increasingly share issues and relationships. They are part of a fast-changing economic and geopolitical landscape, with unpredictable great power politics, the collapse of multilateralism and worsening economic protectionism, reducing options for SIDS influence in global forums. The climate crisis is not the only threat to livelihoods, as the COVID-19 pandemic demonstrated. As perhaps the only external shock to have affected all SIDS, and spared none, it also emphasised the need for both vigilance and resilience, in the face of inevitable uncertainties.

In the following chapters we first examine climate change and environmental hazards, and their challenges to livelihoods, ecologies and diplomacy. We trace and reflect on development problems, some longstanding, the necessity to minimise these through global action, and the responses of islanders to mitigate threats (Chap. 2). Then we look at the economies of the small states and the relationships between commodity production and tourism, and the role of an emergent blue economy, especially fisheries, whether rhetorical or real. These chapters cover the shift away from local production (and thus challenges to food security) towards tourism, a transition briefly devastated by COVID-19, and what that might have meant for 'building back better' and the wide-ranging challenges to achieving a rather different economic structure: any sense of a 'new normal' (Chaps. 3 and 4).

We then turn more directly to political and management issues in examining the challenges of service delivery and the ability of small states to meet the needs of their populations and, here particularly, the extent to which these respond to achieving the SDGs (in such areas as education, health, infrastructure, energy and IT). Other trends, such as increased corruption and crime, threaten national sovereignty and may contribute to state fractures (Chaps. 5 and 6). We then consider how regionalism may contribute to—or challenge—national development (Chap. 7). That logically extends into broader geopolitical structures where island states are inexorably involved in regional and global shifts, some beyond their direct control, that may shape different destinies. We focus particularly on the rise of China and, to a lesser extent, the fall of the United States, in the two regions, and their implications for aid (Chap. 8). The final substantive chapter examines emerging population shifts and the growing role and complexity of migration, in shaping economies through remittances and by a brain drain that challenges the management of SIDS, and which might in future become a flow much influenced by the climate crisis. Demographic destinies have shifted outwards and the role of the diaspora and remittances have become invaluable (Chap. 9). Finally, we return to the wider context of change and development and the task of meeting new challenges in an increasingly uncertain global political, economic and environmental context.

Bibliography

Ahmed, M., & Baroy, N. (2022). The spiralling debt crisis and what to do about it. *ANU DevPolicyBlog*, 22 April.

Béné, C., Wood, R., Newsham, A., & Davies, M. (2012). *Resilience: New utopia or new tyranny? Reflection about the potentials and limits of the concept of resilience in relation to vulnerability reduction programmes*. IDS Working Papers, 2012(405), 1–61.

Bordner, A., Ferguson, C., & Ortolano, L. (2020). Colonial dynamics limit climate adaptation in Oceania: Perspectives from the Marshall Islands. *Global Environmental Change, 61*, 10205.

Brissett, N. (2017). Sustainable development goals (SDGs) and the Caribbean: Unrealizable promises? *Progress in Development Studies, 18*(1), 18035.
Cabral-Ramírez, M., Niño-Barrero, Y., & DiBella, J. (2025). Lessons from the implementation of the Sendai framework for disaster risk reduction from Latin America and the Caribbean. *International Journal of Disaster Risk Science, 16*(1), 72–83.
Clark, T., & Cisneros-Montemayor, P. (2024). Colonialism and the blue economy: Confronting historical legacies to enable equitable ocean development. *Ecology and Society, 29*(3), 4.
Connell, J. (2013). *Islands at risk: Environments, economies and contemporary change*. Edward Elgar.
Corbett, J., Yi-Chong, X., & Weller, P. (2018). Climate change and the active participation of small states in international organisations. *The Round Table, 107*(1), 103–105.
Cumberbatch, J., Drakes, C., Mackey, T., Nagdee, M., et al. (2020). Social vulnerability index- Barbados – A case study. *Coastal Management, 48*(5), 505–526.
Esquivel, V. (2016). Power and the sustainable development goals: A feminist analysis. *Gender and Development, 24*(1), 9–23.
Fanning, A., & Raworth, K. (2025). Doughnut of social and planetary boundaries monitors a world out of balance. *Nature, 646*, 47–56.
Farbotko, C. (2010). 'The global warming clock is ticking so see these places while you can': Voyeuristic Tourism and Model Environmental Citizens on Tuvalu's Disappearing Islands. *Singapore Journal of Tropical Geography, 31*, 224–238.
Feeny, S., & McDonald, L. (2016). Vulnerability to multidimensional poverty: Findings from households in Melanesia. *Journal of Development Studies, 52*(3), 447–464.
Goujon, M., & Wagner, L. (2023). Fragility of small island developing states. In D. Carment & Y. Samy (Eds.), *Handbook of fragile states* (pp. 299–315). Edward Elgar.
Grant, R., Roy, S., Zablah, C., et al. (2023). Climate adaptation and resilience indices for the Caribbean region: An assessment of four leading indices. *Climate and Development*, 1–11.
Gupta, J., & Vegelin, C. (2016). Sustainable development goals and inclusive development. *International environmental agreements: Politics, Law and Economics, 16*, 433–448.

Hill, C. (2025). Realists, re-imaginers and reformers: Aid in the post-America multiverse. *ANU DevPolicyBlog*, 11 June.

Howes, S., & Chowdhury, R. (2025). GNDI: Anew approach to measuring economic performance in the Pacific. *ANU DevPolicyBlog*, 28 November.

James, K., & Clarke, J. (2025, June 11). Securing our blue future: Caribbean leadership for global marine protection. *IISD.*

Kelman, I. (2021). Words without meaning? Examining sustainable development terminology through small states and territories. *Island Studies Journal, 4*(2), 231–244.

MacDonald, S. (2022). *The New Cold War, China, and the Caribbean*. Palgrave Macmillan.

MacFeely, S. (2020). Measuring the sustainable development goal indicators: An unprecedented statistical challenge. *Journal of Official Statistics, 36*(2), 361–378.

MacFeely, S., Peltola, A., Barnat, N., et al. (2021). Constructing a criteria-based classification for small island developing states: An investigation. *Journal of Marine and Island Cultures, 10*(1), 155–195.

Pathirana, A. (2025). Small islands: Living laboratories revealing climate and sustainable development challenges. *Frontiers in Climate, 6*, 1445378.

Sasingian, S. (2025, November 10). Pacific Islands: Without literacy, climate change is nigh on impossible. *Informer.*

Spalding, M. (2016). The new blue economy: The future of sustainability. *Journal of Ocean and Coastal Economics, 2*(2), 8.

Tandrayen-Ragoobur, V., Banerjee, S., Fauzel, S., & Matadeen, J. (2024). Climate change, equity and sustainable development in small island developing states. In S. Seifi & D. Crowther (Eds.), *Equity and sustainability. Approaches to global sustainability, markets and governance* (pp. 67–90). Springer Nature.

Tate, G. (2025, February 22). It's the biggest small country in the world. *Financial Times*, 15.

Taylor, M. (2019). Foreword. In T. Aqorau (Ed.), *Fishing for success: Lessons in Pacific regionalism* (pp. xiii–xiv). Department of Pacific Affairs, ANU.

Telesford, J. (2021). Climate change and the case of Grenada's blue growth plan: Using the SDGs to propose a policy planning framework for SIDS's sustainable development. In S. Moncada, L. Briguglio, H. Bambrick, et al. (Eds.), *Small Island developing states: Vulnerability and resilience under climate change* (pp. 307–327). Springer.

United Nations. (2024). *High level panel on the development of a multidimensional vulnerability index*. UN.

Van Beynen, P., Fenda, A., Akiwumi, A., & Van Beynen, K. (2018). A sustainability index for small island developing states. *International Journal of Sustainable Development and World Ecology, 25*(2), 99–116.

Weatherill, C. (2023). Sinking paradise? Climate change vulnerability and Pacific island extinction narratives. *Geoforum, 145*, 103566.

Weatherill, C. (2025). Colonial fantasies of invulnerability to climate change. *International Feminist Journal of Politics, 27*(1), 34–55.

White, B. (2021, October 29). Barbados: The small but mighty force for climate action. *Open Democracy*.

2

Climate Change: Environments Under Pressure

Climate change and disasters (once seen as entirely 'natural') are critical and interrelated issues affecting all states, but especially small island states and their surrounding seas. In addition to human costs, in mortality and morbidity, disasters damage development prospects, worsen countries' fiscal positions and may necessitate extensive reconstruction. Almost inevitably, as climate change puts increasing pressure on SIDS, discussions of development have focused on hazards as worsening influences, as opposed to established and enduring economic issues. We unashamedly take a similar perspective but recognise the significance of other external shocks faced by island states. The interaction of climate change and disasters affects all SIDS to varying degrees, with rising temperatures predicted to increase the frequency and intensity of hazards. All islands are affected, given their coastal population concentrations, socio-economic activities and infrastructure: a coastal squeeze accentuated by land clearances. Off such coasts, an additional 'marine squeeze' has led to biodiversity loss, mainly declining fish populations, some following large-scale coastal and marine infrastructure projects, such as deepening and widening of shipping channels, that may destroy habitats, or simply clearing mangroves (Hein, Klepp, & Bercht, 2024). Such changes preclude the ability to

J. Connell, Y. Campbell, *Island Challenges*,
https://doi.org/10.1007/978-981-95-8177-1_2

access, use and benefit from coastal and marine space, affecting the livelihoods of coastal communities, raising tensions and increasing existing socio-economic inequalities. At the global level, between 68 and 135 million people could be pushed into poverty by 2030 because of climate change (Hallegatte, Bangalore, Bonzanigo, et al., 2016). A number of these are in the two regions.

Without an understanding of how colonial and neo-colonial policies endure, the drivers of climate change vulnerability cannot be understood. Island vulnerability to climate change is not just a product of sea-level rise but of a colonial history of occasionally forced migration and displacement, extraction and occupation that sometimes did great harm and undermined local adaptation practices and knowledge systems (Gahman & Thongs, 2020). Climate change and discussions of vulnerability reflect global inequalities, linked to wealth creation as a mitigation strategy for the rich but a cycle of poverty for the poor, with little regard for differential distribution within states, or between different groups of people (Weatherill, 2023). Indeed, minorities are often most at risk from hazards, whether riverbank settlers or minority ethnic groups (McAdoo, Moore, & Baumwoll, 2009; Smith & Rhiney, 2015). This has serious implications for SIDS, much affected by colonial-plantation systems, but with uneven impacts and consequences linked to ecologies and geographies.

Both regions are threatened by climate change and environmental hazards. The Pacific is often regarded as the world region most at risk from climate change, with the Caribbean not far behind, while islands are disproportionately affected by climate change, and with an adaptive capacity generally lower than mainlands (Zittis, Zoumides, Zemah-Shamir, et al., 2025). However, we start with an analysis of established threats and environmental challenges, some of which have mutated over the years, but which have long been present as challenges to tropical islands. These are natural hazards. We then turn to climate change, which, in so many respects, has influenced and exacerbated existing environmental problems.

2.1 Certain Uncertainties: Volcanoes, Earthquakes, Cyclones and ...

Both island regions have experienced multiple hazards in the past (Nunn, 2009; Taylor, Clarke, Centella, et al., 2018). The Pacific especially includes some particularly vulnerable islands and countries; Vanuatu has been classified as the most hazard-prone country in the world (Edmonds & Noy, 2018; McDonnell, 2021). Some islands have literally been blown off the map, notably Kuwae (Vanuatu), which exploded in the fifteenth century (Ballard, McDonnell, & Calandra, 2020), and others have been depopulated in the face of constant environmental threats. The eruption of Hunga Tonga-Hunga Ha'apai, 20 kms off the coast of Tonga in January 2022, was the largest eruption of the twenty-first century, generating a tsunami that reached as far as the coasts of Japan and America. Its submarine location spared lives although five died in Tonga. It was Tonga's further tragedy that vital external relief brought COVID-19 to the otherwise quarantined country for the first time. Such cataclysmic events and the eruption on Mt. Pelée in Martinique in 1902 and that in Montserrat in 1995 (where a small city and half an island were obliterated) are rare, and dramatic threats from volcanic eruptions, earthquakes and tsunamis are unusual.

Earthquakes and landslides are intermittent problems in high islands. Where deforestation occurs on steep slopes, as in Solomon Islands, FSM and St. Lucia, landslides are relatively frequent and triggered not only by heavy rainfall but also by the construction of unplanned housing (Connell, 2013a). Heavy rainfall events are likely to become more frequent with climate change, and landslides too may become more common. In the inland village of Nawairuku (Fiji), landslides were apparently unknown until the unusually severe Cyclone Winston in 2016, accentuated by intensive forest clearing on nearby slopes, resulted in the loss of entire harvests of several crops (Currenti, Pearce, Salabogi, Vuli, et al., 2019). The Caribbean has a long history of earthquakes, formally recorded as stretching as far back as 1692, when the 'pirate city' of Port Royal (Kingston, Jamaica) was destroyed. The original capital of Nevis, Jamestown, was destroyed by an earthquake and tsunami in 1690.

Earthquakes are frequent in both regions but infrequently associated with tsunamis. Volcanic activity is common in the 'ring of fire' that forms an arc through the Melanesian states, with the PNG islands and Vanuatu having some unusually active volcanoes. (By a strange quirk of fate, as this paragraph was being typed, Vanuatu was hit by a 6.6 magnitude earthquake and, by a lesser coincidence, by a serious 7.3 earthquake as the Conclusion was being edited). Fatalities and physical damage from earthquakes have increased in urban areas where building standards are inadequately maintained.

Tsunamis often follow earthquakes and most SIDS have had some experience of them. The Caribbean has had no significant tsunami for over 50 years, whereas the Solomon Islands and Samoa have experienced lethal tsunamis in this century, with several hundred deaths. In the 2007 Solomon Islands tsunami, when more than a hundred died, former migrants from Kiribati were disproportionate victims, partly because they lacked the indigenous knowledge and experience of established Melanesian residents, that raised questions, there as elsewhere, over knowledge and agency and the unpredictable entanglements of waves with colonisation, capitalism, human-animal communication, spirit beings, ancestral territory and technoscientific expertise that shaped the disaster's outcomes (Connell, 2013a; Lauer, 2023). Damage and fatalities have tended to increase as more people now live on coasts, protective mangroves removed and houses made of more dangerous materials. Seemingly illogically, low coral islands have avoided substantial damage from tsunamis. Both the Caribbean and Pacific islands are also vulnerable to storm surges, which may be becoming more frequent, but their periodicity and impact are poorly understood.

Population shifts have accompanied volcanic activity (and tsunamis) in the past half century; the third city of PNG, Rabaul (New Britain), was abandoned in 1994 after being buried in the eruption of nearby Matupit. The islands of Manam (PNG) and Ambae (Vanuatu) were effectively depopulated (the latter only temporarily) in 2005 and 2018, respectively, after particularly intense volcanic activity. After the 2022 Tongan eruption, people from the devastated island of Mango were resettled on the larger island of 'Eua, emphasising ongoing depopulation of smaller remote islands in favour of larger central islands, a function of economic, social and environmental issues (Fonua, 2023). Similar problems occurred in some Caribbean islands. In 2020, La Soufrière volcano (St. Vincent) violently erupted covering the entire island with ash, affecting agriculture, food availability and cash-crop production, necessitating the temporary evacuation of 13,000 people, most of whom were indigenous Kalinago Caribs. Tsunamis have unsurprisingly resulted in survivors relocating uphill and inland, as in Samoa, and many small-scale local moves have followed coastal erosion and flooding, usually as the result of cyclone damage, as in Fiji and Vanuatu. They suggest precedents for future moves, as more people are displaced by the effects of slow-onset climate change. Few eruption events are easily predicted or monitored.

Both regions are vulnerable to cyclones (hurricanes) and droughts, often associated with the El Niño Southern Oscillation (ENSO), and generate most environmental challenges. Almost all SIDS experience the impacts of intensified rainfall and cyclones, and periodic floods and droughts, with biodiversity and coastal zones susceptible to environmental degradation. Critically, such hazard events can affect entire countries, necessitating relief from beyond. The intensity of disasters and resilience to them vary across SIDS, and their economic effects depend on a range of variables, from intensity, location and timing, to income levels, country size, disaster type and severity (Cavallo & Noy, 2009; Raddatz, 2009). SIDS are more affected by hazards than larger, more developed economies that can better absorb the shocks and are more likely to have less affected areas that can provide support. Lack of economic diversity further heightens vulnerability to external shocks. Loss and damage attributable to the impacts of climate change are indistinguishable from those of longstanding hazards.

Damage from rapid-onset climate hazards, such as floods and cyclones, is especially severe in Pacific SIDS. Between 2015 and 2020, Pacific SIDS suffered the highest relative economic losses, with an average loss in a hazard event of nearly 9% of GDP. That average masked the extreme impact of specific disasters; when Cyclone Harold hit Tonga in 2020, economic losses amounted to nearly 28% of Tonga's GDP (ESCAP, 2024); Cyclones Pam in Vanuatu and Winston in Fiji generated similar losses and Hurricane Melissa in 2025 devastated the much larger Jamaica, with losses equivalent to around 28–32% of the country's GDP. In 2017,

Hurricanes Irma and Maria caused widespread destruction across the Caribbean with an estimated US$130 billion in losses (Moulton & Machado, 2019). In Dominica, Maria alone inflicted an unprecedented 226% GDP loss to the island. Questions of how to prepare for and recover from such physically, psychologically and financially damaging events have been at the forefront of policy discussions, but action for long-term resilience in the face of more severe cyclones has rarely proved effective. Hurricane Maria also damaged or destroyed hundreds of fishing vessels and gear, at a cost of US$2.9 million, which meant that 40% of vessels and up to half of all fishers were left unable to fish, losing income and food. In Jamaica, Hurricane Ivan in 2012 resulted in loss and damage of approximately US$833 million or 8% of GDP but GDP (boosted by reconstruction activities) is only a crude financial indicator (Mohan & Strobl, 2025). Education, health and housing bore 48% of this damage, as the hurricane set back welfare provisions. In 2024, Beryl peaked as a Category 5 hurricane affecting several Caribbean SIDS. Carriacou and Petite Martinique (Grenada), Union Island, Canouan and Mayreau (St. Vincent and the Grenadines) saw significant displacement, with economic damage in Grenada and St. Vincent and the Grenadines estimated, respectively, at 16.5% and 22% of GDP (World Bank, 2024): costly impacts that are difficult to rectify.

Climate change threatens livelihoods, economies and health, with major coastal impacts, from increased storm damage and water salinity to reduction in biodiversity, and harm to fisheries, tourism and infrastructure. In the Pacific, the atoll states are most at risk, alongside Vanuatu, whose resilience has frequently been tested in this century. In the Caribbean, more people are affected by hurricanes in the Bahamas, but the smaller eastern Caribbean SIDS have experienced serious hurricanes affecting the entire country. Climate change threatens agricultural and fisheries production and thus livelihoods and incomes, with the relatively poor and women most affected (Ngcamu, 2023). Water shortages and thus poorer quality local food have been tangible consequences in the atoll states of Kiribati and Tuvalu (Alofa & Costello, 2025), affecting water quality and public health consequences, including 'water contamination, inoperative sanitation systems, and acute food insecurity in the aftermath' (Rise, Oura, & Drewry, 2022; Shultz, Sands, Kossin, & Galea,

2020). Several Caribbean and Pacific SIDS are intermittently water stressed.

Cyclones have been a major cause of shoreline erosion, damage to coastal infrastructure and saline intrusion. That has resulted in migration away from the coast in Fiji and beach erosion close to hotels in St. Kitts and Nevis, reducing their tourist potential. The scale of the damage and social dislocation does not simply reflect the hazard outcomes, as continued structural dependency and entanglement in post-colonial relationships complicate recovery and coordination of aid, emphasising persistent colonial power dynamics (Perry, 2024; Perry & Sealey-Huggins, 2023; Sheller, 2025). Championing resilience in the face of hazards is limited but necessary whereas stressing the biopolitical regime and context offers a broader historical perspective that might eventually clarify a path towards restoration, reparations and climate change justice (Moulton, 2024; Moulton & Machado, 2019; Perry & Sealey-Huggins, 2023). Indeed, resilience implies coping with harm rather than preventing it (Weatherill, 2025). The probability of slower recovery from cyclone damage has increased as monoculture became more common, cassava replaced more wind-resistant crops, 'famine crops' were rarely planted, coastal populations increased and coastal vegetation decreased, coastal infrastructure (such as power stations, hospitals and wharfs) was constructed, population growth extended from leeward sites to windward sites, or into more low-lying areas (notably on atolls), and informal housing was constructed (Spennemann, 1996).

During the 2017 Atlantic hurricane season, three major hurricanes (Harvey, Irma and Maria) resulted in the internal and cross-border displacement of approximately 3 million Caribbean people in a single month. The entire population of Barbuda, about 1600 people, was evacuated 40 kms away to Antigua after Hurricane Irma. Subsequently that provided a bleak example of 'disaster capitalism' as it gave the Antigua government an opportunity to alter land tenure and develop an international tourism resort and golf course, to generate income from what the Prime Minister, Gaston Browne, had maligned as a 'giant welfare island'. Island priorities for roads, housing and sanitation facilities still languished 7 years later (George, 2024). In the wake of Hurricane Dorian in the Bahamas in 2019, thousands were evacuated to Nassau and neighbouring locations, where many remained in temporary shelters for months, while others moved abroad. Further changes in migration and displacement patterns are likely in both regions as extreme weather events intensify and slow-onset environmental changes place communities under increasing economic and social stress.

In already relatively dry areas, hurricanes have become accompanied by more intense droughts, with climate change causing long-term changes to the water cycle, an intensification that could make reliable supplies of fresh water more difficult to access. The global water cycle is accelerating twice as fast as global climate models have predicted, and further global warming is likely to increase rainfall intensity and hence flooding (Howden, 2022) while droughts are more frequent in both regions (Rhiney, 2015). Paradoxically, three quarters of SIDS face the risk of water shortages leading to water insecurity, a condition only exacerbated with climate change. Groundwater depletion, for example, increases the risk of saltwater intrusion, which in turn reduces agricultural productivity. Thus, deterioration of coconut quality in the Marshall Islands (and elsewhere) due to saline soil has resulted in their being less valuable as either a source of food or the basis of exports, once a key source of income for the country. This combination of effects is important yet immeasurable were there to be rigorous loss and damage assessments. Such complex issues defy straightforward analysis although climate change impacts on the substantial Fijian sugar industry have been measured, where loss and damage followed droughts despite various adaptation measures. Damage included reduced yields and income, loss of crops, resulting in the deterioration of mental, physical and emotional health; loss of hope; and uncertainty. Cascading effects included increased food insecurity risks, negative impacts on children's education, resulting in seeking of livelihood options other than sugar cane farming and encouraging younger generations to pursue other career pathways (Nand, Clissold, McNamara, & Yee, 2024; Prakash & Nakamura, 2024). While that had implications for policy interventions, costing the impacts to enable reparations and thus sustain the sugar industry in changing climate circumstance, was unusually difficult. And that had much deeper implications; over time issues of loss and damage reflected the consequences of historical and colonial responsibility rather than the immediate impacts of climate change while the concept itself remained vague and contested in academic and political circles (Handmer, Monson, & Schinko, 2025; Mohan, 2025b).

Water shortages have occurred in several SIDS (Chap. 6). Eastern Caribbean SIDS battled two tropical storms (Gonzalo in 2020 and Kirk in 2018) and a hurricane (Isaac in 2018) in 3 years, each accompanied by water shortages. In this century, both Tuvalu and Nauru have had to import water when droughts have been prolonged. Longer, more severe droughts in both regions affect rivers and reservoirs, but especially groundwater, posing distinct challenges for atolls that lack freshwater supplies other than rainfall. Caribbean drought periods in 2009–2010 and 2014–2016, the second being the most severe and widespread drought in Caribbean history, had impacts ranging from reduced crop yields, higher food prices, increased wildfires, higher incidences of plant diseases and pests to hotel cancellations (Trotman, Mahon, Meerbeeck, & Riley, 2021). Further consequences followed for the health sector stemming from poor water storage practices and the lack of clean water, including the transmission of vector-borne diseases and gastroenteritis as in Barbados (Lowe et al., 2018; Trotman et al., 2021).

Except for earthquakes, tsunamis and volcanic activity, all these broad environmental changes are affected by climate change and none in beneficial directions. Few are easily predicted, and multiple costs and contingencies are intrinsic. We now turn to the complexities of climate change and its impacts.

2.2 Climate Change

Regular scientific assessments of climate change by the IPCC are widely regarded and cited as definitive statements on climate change. The 2022 IPCC working group's report, the sixth since 1988 (Intergovernmental Panel on Climate Change [IPCC], 2022), focused on the physical science of climate change, arguing that the climate crisis was 'unequivocally' caused by human actions, pointing out that some changes were 'unprecedented', with several having become irreversible, including sea-level rise, the melting of Arctic and Antarctic ice, and the warming and acidification of the oceans. The IPCC then regarded the necessity of stabilising the climate to an increase of 1.5 °C as being possible. By 2025 that optimism had gone. That level of heating was still likely to result in increasing heatwaves, more intense storms, and more serious droughts and floods, but it pointed to the narrowing window of opportunity and how adaptation was crucial to guard against the impacts. The IPCC forecasted that sea levels would be 1.01 m higher (relative to the 1995–2014 level) by 2100 if global emissions continued unabated and stressed that sea-level rise was getting faster. It found that sea-level rose 1.3 mm per year during 1901–1971, 1.9 mm per year during 1971–2006, 3.7 mm per year during 2006–2018, and for 2013–2022 reached 4.6 mm, so that global mean sea-level was rising faster than at any time in the last 3000 years. That disguised regional variations in the rate of rise (and its impacts), and in both the northern Caribbean and western Pacific, the rate was greater than the global averages.

For SIDS, the most dramatic outcome of climate change is the greater severity of cyclones, and consequential flooding, already causing loss of life and serious economic problems in many SIDS. The IPCC's 2022 report emphasised that the difference between 1.5 and 2 degrees of global warming was substantial: every small rise in global temperatures presages much greater impacts in the climate system. While only a few countries are at risk of experiencing cyclones in categories 3–5 with winds from 180 to 250 km/h, many Pacific SIDS are likely to face more and stronger cyclones, even at just 1.5 degrees of warming, and have done so since the 2010s.

In 2021, the United Nations declared a 'Code Red for Humanity' because of climate change, with four regional climate-related changes occurring simultaneously: a rise in air and sea-surface temperatures (projected to increase by up to 2 °C by 2050), changes in rainfall (with expectations of less frequent but more intense rainfall events, with potential impacts on flooding) and higher future sea levels. Each of these varies regionally, but climate change projections suggest that sea levels may rise by as much as 1.5 m by the end of the century. Cyclones will be more severe and over a wider area, evident in the impact of Tropical Cyclone Harold that devastated parts of Solomon Islands, Vanuatu, Tonga and Fiji in 2020. Climate change has led to a 5% increase in rainfall in storms and an 8% increase in rainfall during cyclone-strength storms (Reed, Wehner, & Zarzycki, 2022). Flooding became more evident in Hurricanes

Irma and Maria in 2017, and 3 years later, Dorian in the Bahamas caused severe damage to lives and livelihoods. Vanuatu was hit by two cyclones in a single week in 2023, causing widespread destruction to homes, schools and food gardens (resulting in the government declaring a 6-month national state of emergency), when it was still reconstructing after Cyclone Pam in 2015 (the second most severe cyclone recorded in the Pacific). Such conjunctures strongly suggested that the frequency of cyclones was increasing in both regions.

The extent of damage, especially after the worst cyclones where recovery was difficult, raised moral questions about the often unequal impact and subsequent reconstruction. In the Bahamas it emphasised social divisions between Bahamians and migrant Haitians. Dorian, like most other hurricanes, represented a double environmental injustice as socio-economically disadvantaged and marginalised populations sustained disproportionate damage, with more hazardous storms exacerbating this inequity, and emphasising disaster capitalism as both a response and reconstruction that excluded marginal groups (Benett, 2023; Griffen & Robinson, 2023; Llana, 2019; Rhiney, 2020; Shultz et al., 2020; Smith, 2025; Vinke, 2025). Even where hurricanes had a 'levelling effect' with both rich and poor impacted, as in Dominica, more affluent people had more financial assets and insurance coverage to respond faster, while poorer communities (especially the indigenous Kalinago Carib population) had weaker infrastructure, less access to social capital and fewer services to support recovery, while their own local knowledge, and especially that of women, was often ignored, ironically when needs were greatest and as distant agencies were involved in reconstruction and redevelopment (Hofman, Stancioff, Richards, et al., 2021; Turner, McConney, & Monnereau, 2020). Disaster response and governance do not usually elide with environmental justice.

2.3 Heat and Health

In the 12 months prior to mid-2024, the planet's average surface temperature exceeded 1.5 °C above pre-industrial levels, the milestone of global warming that all nations pledged to avoid, reaching the highest average temperatures on record. Such temperature increases are being felt in both urban and rural areas, even by children, as in Samoa (see below), making life more difficult, especially for outdoor activities, whether work or recreation. Urban areas are hotter than non-urban areas and urban heat islands more evident. Moreover, the air can hold about 7% more water vapour for every 1 °C increase; hence, heavy rain and storms have become more frequent and more intense. Such events that provoke flooding also reduce water availability and quality.

Heat waves will be longer, more extreme and more frequent. Oceans are heating up, with every year since 2019 beating previous record years. Droughts will be more frequent, already evident in the Caribbean. Jamaica, St. Vincent and St. Kitts have had to ration water, and Trinidad and Barbados have experienced several water bans. Demand for water has outstripped supply because of changing rainfall, hurricane and drought patterns (with over 14% of the Caribbean population without water 3 months after Hurricane Maria in 2017, so reducing the ability of aquifers or groundwater to recharge, a serious issue in Barbados). In Tuvalu, severe drought driven by a La Niña event in 2011 led to a state of emergency, with groundwater contamination and water rationing, and a significant reduction in the supply of local food since watering food gardens was impossible (McCubbin, Smit, & Pearce, 2015). It caused stress, affected water security for domestic tasks (cooking and bathing), for drinking (since well water was too brackish) and for agricultural development (with coconuts failing to ripen, hence copra sales were absent, and crops [breadfruit and taro] failed). Health was affected by people drinking polluted lens water and being unable to wash themselves and their clothes as frequently as usual, with water and desalination plants shipped in from as far as New Zealand.

Data based on climate projections in the Caribbean have shown increases in temperatures, in sea levels, in extreme events and in the frequency and strength of storms and hurricanes. Global climate models suggest for the central and southern Caribbean basin, a gradual drying with up to 2% less rainfall by the mid-2020s and 6% less by the 2050s with the Lesser Antilles and southern Caribbean being the most affected. By contrast, in the Pacific island countries (PICs), the anticipated trend is for a warmer and wetter climate over the next two decades. Climate change affects every island differently, but, unequivocally, trends will prove most damaging in atolls where livelihoods are difficult, and no land is much more than a couple of metres above sea level.

In urban heat islands, the impact of climate change on health is and will be greatest, with wide-ranging impacts and implications (Gordon-Strachan, Harewood, et al., 2025). Physical and mental health in SIDS is particularly vulnerable to climate change, directly through hazards and indirectly through threats to livelihoods. Poorer mental health, apparent in several SIDS (Cloos, Belloiseau, McPherson, et al., 2023; Gibson, Barnett, Haslam, & Kaplan, 2020), is partly a function of climate anxiety. Health services are poorly equipped to cope with new pressures. Higher temperatures extend the risk of malaria and dengue spreading southwards in the Pacific. Urban heat islands boost respiratory diseases. New vector-borne diseases, including zika and chikungunya, have become more frequent. Cyclones and floods, especially in densely populated coastal areas, result in injuries, damage to health facilities and more vulnerable ecosystems and livelihoods, particularly affecting marine productivity, as ocean temperatures increase. Droughts put occasional pressure on water and food security, worsening epidemics of existing non-communicable diseases (NCDs) (Chap. 5). Those living on atolls and in overcrowded urban settlements are the most vulnerable (Connell, 2023). The health impacts of climate change constitute a slowly increasing threat to human security, but policies and practices centred on workforce development may minimise such threats and risks.

Impacts of urban heat islands are intensified with the loss of natural vegetation and its replacement by constructed materials: concrete hardening. Buildings, roads, parking areas and other infrastructure absorb the sun's heat during the day, increasing overall urban temperatures. Urbanisation also, incidentally, changes the flow and storage of water in urban catchments: an 'urban stream syndrome' largely due to human-made impervious and artificial surfaces that reduce absorption of rainwater into the soil. As a result, these hard man-made surfaces contribute to dry and hot urban soils, transform most of the rain into run-off and, after storms, contribute to flash flooding. As urban growth leads to fewer plants and more buildings and artificial surfaces, this reduces the cooling effects from plants. Shading by plants, and particularly trees, has a major influence by cooling air, soil and urban materials. Green spaces and trees, often rare in SIDS cities, notably where firewood collection takes a toll, reduce such problems and enable exercise and recreation.

2.4 Oceans and Reefs

Of the interconnected environmental threats facing the planet, the climate and biodiversity crises are most severe. Neither will be solved by ignoring the oceans, too often 'invisible' to the growing humanity at their edges. Oceans cover over 71% of the earth and are a controlling factor in the rate of climate change. They both contribute to and act as the major buffer for climate change, storing 91% of the excess heat from global warming. Accelerating increases in ocean heat and sea surface temperatures, alongside increased global acidification, have contributed to a range of other impacts: ocean heat is supercharging tropical cyclones, intensifying the hydrological cycle leading to greater and more intense precipitation events, stronger wind speeds, an increase in the size of storms and possibly a reduction in their tracking speed. The number of disasters, their intensity and financial cost and the stress to people, governments and insurers has consequently grown. Warming has also accelerated ocean circulation and the melting of polar ice (Laffoley, Baxter, Amon, Claudet, et al., 2022). Hotter oceans make weather on land more extreme, bleach coral and damage the marine life that underpins ocean ecosystems. Warming can lead to less nutrients in surface waters, which in turn leads to fewer fish. Marine heatwaves have intensified in frequency, duration and severity over recent decades, not least in the Pacific, and resulted in the movement of oceanic species like tuna to cooler waters (Weinberg, Bhardwaj, Watkins, & Kuleshov, 2024). That puts unusual pressure on fishers to adjust fishing techniques and locations, especially in atoll contexts (Moesinger, 2019b). Rapid warming of oceans thus sets off various problematic changes.

Bleaching caused by climate change is forecast to reduce global coral cover and threaten the biodiversity of coral reefs. Following a global mid-twentieth-century stasis, warming resumed on Caribbean and Pacific reefs in the early 1980s while marine heatwave events increased in both frequency and duration across the Caribbean, causing a damaging shift in the composition and functioning of coral reef ecosystems (Bove, Mudge, & Bruno, 2022; Hughes, Barnes, Bellwood, et al., 2017). Warmer oceans are also linked to weather extremes such as more intense cyclones and heavier rainfall. More frequent and more intense marine heatwaves threaten ecosystems and human livelihoods. Mass fish deaths have occurred in places, on a global scale described as 'a saline tragedy of the commons' (Heffernan, 2024), where the ocean's role as both carbon sink and source of a sixth of the world's protein is at risk. The high seas, consisting of the two thirds of the oceans beyond the control of any country, are increasingly threatened. However, as global population and demand for food grow together, oceans must provide increasing amounts of seafood.

Beyond physical changes, the use and abuse of oceans and seas has contributed to other often deleterious impacts, most obviously overfishing and pollution, part of a broader set of environmental hazards and harms, such as habitat loss and reduced ecosystem services that, separately and combined, have contributed to environmental injustice (Bennett, Alava, Ferguson, Blythe, et al., 2023). Minimising injustice is critical to the livelihoods, health, culture, rights and wellbeing of global coastal populations. To protect the oceans, marine livelihoods and ways of life, reduced carbon emissions and conservation are crucial.

2.5 Sea-Level Rise and 'Disappearing' Islands

In contrast to the rapid-onset effects of cyclones, slow-onset sea-level rise poses a steady, imperceptible but inexorable, existential threat to several SIDS and affects all of them. While sea levels are rising, especially in the western Pacific, this is directly invisible, since the rate of change is slight. They hint at what is to come, which, in due course, may have repercussions for the disappearance of islands and island states. In the Caribbean, relative to the 1995–2014 period, sea-level rise associated with climate change is predicted to increase annually by 0.1 mm in the short term (2021–2040) and 0.3 mm in the medium term (2041–2060). In the northern Caribbean, sea-level rise could be 25% higher than the global average due to other physical factors affecting land elevation so that the Bahamas will be most affected. Before 2050, Pacific Island nations such as Tuvalu, Kiribati and Fiji are expected to experience at least 15 cms of sea-level rise, regardless of whether greenhouse gas emissions change in the coming years.

Climate change poses risks, not merely to livelihoods, economies and health but to the future survival of some Pacific islands. The atoll states are most vulnerable to rising sea levels and storms. Parts of Tuvalu that experience less than five high-tide flood days a year could average 25 flood days by the 2050s while in some Kiribati islands that could mean 65 flood days by the 2050s (United Nations, 2024). A sea-level rise of 50 cm would lead to a loss of 80% of the land of Majuro atoll, the Marshall Islands' urban centre, but the habitability of this and other atoll islands would be threatened well before significant land was lost. How that habitability might be decided upon is indeterminate, based on predictions of change, past adaptation, local knowledge, cosmologies and place attachments, and not simply the material elements of human security such as housing, food and water (Scorlazzi et al., 2018; Farbotko & Campbell, 2022; Steibl, Kench, Young, Wegmann, et al., 2024). National policies may also require difficult decisions to 'sacrifice' some islands and support others.

The long sinuous coasts of atolls defy the feasibility, and cost, of seawalls (even where they are proved useful). However, few if any entire atolls are likely to disappear in this century. Remarkably, many coral islands on atolls are actually increasing in size, as storms deposit coral rubble on land, but erosion—of coasts, trees and burial grounds—attracts the most attention (Kench, Ford, & Owen, 2018; Kench, Sengupta, Ford, & Owen, 2024; Sengupta, Ford, & Kench, 2021; Steibl et al., 2024). Erosion is often compensated by deposition elsewhere, but deposition is of coral material, of no use for agriculture (but a possible barrier to waves). Atolls may become less amenable to residence (especially if other hazards intensify and islanders' expectations increase) but are unlikely to disappear, so posing no immediate problems for sovereignty. Somewhat similarly coral reefs are not decreasing in size but may well be diminishing in diversity and productivity.

Within the atoll states, and elsewhere, some small islands have disappeared—but invariably tiny uninhabited islands that are part of atolls—including islands in both the Funafuti (Tuvalu) and Tarawa (Kiribati) lagoons, and off Choiseul (Solomon Islands) and Bougainville (PNG) (Connell, 2015). So far most shoreline changes and coastal inundations have resulted from extreme events and inappropriate development (including seawalls) rather than sea-level rise alone. Thus in Solomon Islands, sea-level rise combined with high wave energy has resulted in the erosion and loss of islands and population relocation to higher islands (Albert, Bronen, Tooler, et al., 2018; Albert, Leon, Grinham, et al., 2016). In the Caribbean, only the Bahamas has a highest point lower than 100 m (just 63 m), but the country's overall topography, as well as its resources for establishing protective measures, means that disappearance would be improbable (Piguet, 2022). The Pacific is most at risk.

Atolls are particularly at risk from the impacts of climate change on local lives and livelihoods rather than through imminent disappearance. However, what has increasingly become a normalisation of loss, with atoll states widely seen to face a finite future, has presented a dilemma for these states—crudely seen as either forced migration (marked in the efforts of legal scholars to consider the nature of sovereignty over drowned homelands) or strategic local planning (but where the assumption of loss and migration complements, discourages or even prevents remedial action). Above all, the assumption of loss—and often imminent loss—implicit in the global focus on the Carteret Islands (Bougainville, PNG) is embedded in more popular media (Barnett, 2017; Connell, 2018; Weatherill, 2023). Despite accelerated global warming, the future is nowhere entirely bleak. It is erroneous and certainly unethical to write off any atoll in this century, where multiple forms of adaptation and resilience are being explored. Other narratives extend beyond any simple dichotomy, where science appears uncertain and religion offers apparent certainty (see below).

2.6 Biodiversity

Accompanying climate change, but only partly related to it, have been global losses of biodiversity. The biodiversity crisis has none of the drama of climate change with its increasingly severe and frequent extreme weather events and the displacement of people. But the immense and growing social and economic costs of biodiversity decline have become ever more evident. Protecting biodiversity is crucial to managing land and water resources sustainably and aligns with tackling climate change. If nature continues to degrade, or be degraded, economic output and food security eventually suffer. This relationship is captured in debates over assigning monetary and other values to ecosystems and ecosystem services, issues that complicate assessments of loss and damage where these spill over into losses of culture and traditional knowledge (Chandra, McNamara, Clissold, et al., 2023; Handmer et al., 2025; Thomas & Benjamin, 2020).

Throughout the SIDS, biodiversity has diminished as land has been cleared and invasive species have taken a toll. Although people may depend on wild flora and fauna to obtain food, medicine and energy, over-exploitation, invasive species, climate change, pollution and deforestation have pushed many species towards extinction, reducing the vitality and value of island ecosystems and services (Connell, 2013b). Between 1970–2020, monitored global wildlife populations all declined, with the heaviest declines amongst freshwater populations (85%), followed by terrestrial (69%) and marine populations (56%), mainly due to habitat loss and degradation, driven primarily by human food systems (with the consumption habits of wealthy nations, for food, timber and oil palm products, among the key drivers of biodiversity loss). Such declines act as early warnings of extinction risk with ecosystems more vulnerable to tipping points: thresholds when change becomes irreversible (WWF, 2024). Poorer and smaller SIDS may support areas rich in biodiversity, but with fewer means to conserve them. Reduced biodiversity and degraded landscapes increase the risk of wildfires and erosion, reducing the possibility of sustainability. Poor physical planning, poverty and natural disasters make conservation difficult for SIDS.

> Every day, our island nations are fighting unseen battles on the frontline against a vast range of invasive animal and plant species, one of the leading threats to biodiversity and island ecosystems, with the potential to have catastrophic effects on agriculture, food security and human health. (Albert

Nicholas, Deputy Prime Minister and Minister for Environment of the Cook Islands, August 2024)

Protecting biodiversity and securing healthy terrestrial and marine ecosystems are challenging.

Invasive species have occasionally been particularly damaging, whether plants, birds, animals, ants or even fish, notably lionfish in the Caribbean, that have reduced marine diversity and fisheries productivity. The presence of invasive alien species of multiple kinds has become more rapid over time to the extent that 'at no time in history has the rate of biological invasion or the diversity and volume of these invaders been so high and the consequences so great' and islands lose biodiversity at particularly high rates (Reaser, Meyerson, Cronk, et al., 2007: 98; Russell & Kueffer, 2019; Schrader, Weigelt, Cai, et al., 2024; Seebens, Meyerson, Richardson, et al., 2025). Since 2010, the spread of sargassum seaweed in the Caribbean has led to beach erosion, habitat losses, and economic and aesthetic damage to tourism and fisheries, so threatening livelihoods and incomes, being detrimental to coastal and marine ecosystems, coastal communities and their fisheries, tourism, public health and employment. Various efforts have been made throughout the region to mitigate its impacts, including the installation of floating boom barriers, manual and mechanical removal, and a search for alternative uses that would turn sargassum into an opportunity. Although this has led to the creation of innovative products and services made from sargassum, it has been a particularly costly and damaging invasion.

More prosaically, soil degradation and lost fertility can be crucial. Multiple causes of land degradation in SIDS include deforestation, land clearance for agriculture, inappropriate agricultural practices, overgrazing, population pressure, land tenure disputes and climate change. In Tonga commercial crops such as squash have resulted in increased use of fertilisers and pesticides, with one outcome a 'catastrophic eutrophication of the lagoon' where there is groundwater seepage, further emphasised by leaking septic tanks. At various sites on the main island, pesticides have also contaminated drinking water sources. In a sense, along with its squash, 'Tongatapu is exporting both its soil fertility and water quality to Japan' (van der Velde, Green, Vanclooster, & Clothier, 2007: 464). On sloping land similar practices may lead to soil erosion, nutrient loss and the increased agricultural use of marginal land, which further exacerbates erosion and sedimentation. Land degradation subsequently reduces crop productivity (Wairiu, 2017). A further consequence is river channels becoming filled with sediment, with channel capacity reduced and flooding more probable. Retaining indigenous vegetation cover is crucial to retaining biodiversity, yet vegetation cover is diminished by commodity production to the extent that resultant trade is responsible for 30% of global species threats (Lenzen, Moran, Kanemoto, et al., 2012).

COVID-19 slowed global progress on biodiversity protection (see below), yet, ironically, as biodiversity is reduced by cutting down forests and building more infrastructure, this increases the risk of pandemics. While some species are becoming extinct, those that tend to survive and thrive, such as rats and bats, are more likely to host potentially dangerous pathogens that can transfer to humans. Outbreaks of epidemic diseases that cross from animals to humans have increased in recent decades, the probable outcome of increased contact between humans, wildlife and livestock, as people move into undeveloped areas (Tollefson, 2020). Controlling and reducing deforestation and curbing the wildlife trade would reduce this possibility, but climate change is likely to push this process further (Gilbert, 2022). As climate change occurs, extinctions are more likely and will accelerate rapidly if global temperatures exceed a 1.5 °C increase. Highest-emission scenarios would threaten approximately one third of global species, with those endemic to island ecosystems being some of the most affected (Urban, 2024).

At a national scale SIDS have not always been at the forefront of supporting diversity and encouraging conservation, with government policy (or its absence) at odds with local and international environmental groups. Paradoxes persist. In 2024 the Antigua and Barbuda government's position in favour of ending a global whaling embargo was opposed by local NGOs, suggesting that more hunting would be both detrimental to the viability of struggling whale populations and negatively impact Antigua's tourism sector that promoted the country as an 'eco-conscious' destination, where whale watching was an attraction (*The Daily Observer*, 19 September 2024). The success of regional and national efforts will be influenced by the effectiveness of global governance institutions, but discussions have stalled over financing as countries struggled to find common ground, including how to protect at least 30% of the world's land and seas.

Conservation strategies at different scales have sought to slow or end losses by protecting land and marine areas. Regional programmes exist in both the Caribbean and the Pacific, many reliant on donor funding especially in the eastern Caribbean where USAID was previously involved in biodiversity conservation. A 30x30 objective became part of an overall target established by the High Ambition Coalition for Nature and People, supported by two thirds of the SIDS, that sought to protect 30% of both land and ocean resources. The Coalition is an informal group of countries within the UN Framework Convention on Climate Change (UNFCCC), founded by the Marshall Islands in 2014, and aiming at ensuring that the Paris Agreement, adopted a year later, was as ambitious as possible. The Coalition succeeded in securing some of the Paris Agreement's provisions although 'high ambition' was absent in many countries.

Since countries' national marine jurisdictions only cover a minority of the world's oceans, a United Nations High Seas Treaty (also known as the Biodiversity Beyond National Jurisdiction Treaty or the BBNJ Agreement), initiated in 2023, was established to enable a coordinated international approach to establishing substantial marine protected areas (MPAs) in the high seas, beyond national Exclusive Economic Zones (EEZs). This Agreement gradually gained enough signatories to enter into force as a Treaty in January 2026, and is potentially valuable for SIDS where any blue economy and biodiversity are inter-related, and exploitative fishing (bottom trawling), pollution and potential deep-sea mining (DSM) are threats. Indeed the Treaty came at the same time as global growth of interest in the blue economy in a context where, despite SIDS' disproportionate reliance on ocean resources, larger and wealthier nations were better placed to exploit opportunities, for example by subsidising large-scale fisheries and investing in marine biotechnology. The Treaty was anticipated to enable and support indigenous knowledge and practices at different scales, and provide some opportunities for SIDS to achieve greater ocean equity (Caldeira, Sekinairai, & Vierros, 2025; Hurley, Wilkinson, Weaver, & Kasanawaqa, 2025). Emphasis was placed on the role of indigenous peoples—representing 5% of humanity but protecting 80% of the earth's biodiversity—hence 'indigenous-led conservation models must become the norm this decade if we are to take real action on biodiversity' (Weston & Greenfield, 2022): a striking shift in orientation but without much optimism that this would prove effective.

Progress on biodiversity protection has proved elusive since the first Earth Summit in 1992. Efforts to prevent biodiversity losses experienced funding issues analogous to those concerning climate change. In 2021, a separate Conference of the Parties (COP), hosted by China in Kunming, sought to support the UN Convention on Biological Diversity international treaty. Governments debated measures to protect the diversity and richness of plant and animal species; discussions were centred on the widely supported 30x30 target, alongside the provision of greater financial support to low-income countries to preserve biodiversity. By 2022, frustrations with financial support through the UN Convention intensified alongside opposition to the consumption habits of wealthy and middle-income nations, many of which faced possible recession and sought to prioritise economic recovery. SIDS are often home to areas rich in biodiversity but have fewer means to conserve them. In the Pacific there is a greater irony where some distinct marine and terrestrial species—mainly invertebrates—are still being discovered on high islands. The subsequent COP15 in 2022 sought to ensure that at least 30% of degraded terrestrial, inland water, coastal and marine ecosystems would be under effective restoration by 2030. It also sought 'improved connectivity of biodiversity through green and blue spaces in urban areas', shifting the focus somewhat from rural areas.

Almost 200 nations were signatories to the earlier Kunming-Montreal Global Biodiversity Framework, whose success relied on domestic targets and actions, reflected in National Biodiversity Strategies and Action Plans. However, less than a quarter of countries developed plans that were intended to be committed to the two 30x30 targets. None produced a plan that met these requirements and, of the countries that submitted a plan, only Tonga was a SIDS (Bell-James & Watson, 2025). The Global Biodiversity Framework agreement (part of the Convention on Biological Security) that eventually emerged was not legally binding. Specific targets for tackling the drivers of biodiversity loss were absent. A further objective was for governments to ensure that large transnational corporations disclosed their 'risks, dependencies and impacts on biodiversity' alongside growing pressure for the development of rules for sustainable sourcing of various agricultural products such as rubber and palm oil, but they were not mandated to do so. International cooperation and support from large corporations were not guaranteed even in the face of common threats. Formal processes to retain biodiversity were still unable to overcome the tragedy of the commons.

2.7 Coasts in Crisis?

By the late twentieth century, it was apparent that the world's climate was changing and warming with the repercussions for SIDS involving more hazard events and the potential, through sea-level rise, for severe coastal erosion and flooding. In extreme cases, in the atoll states, that meant the possibility that some of the most recently populated islands in the world would simply disappear, leaving subtle legal questions about sovereignty over empty space. Minimally, it suggested that some populations would require resettlement.

All broadly environmental issues, most influenced by human factors, are intensified on coasts, most obviously through cyclones and floods, and loss of biodiversity. Especially since the 1970s, the coastal zones of SIDS have experienced land use change, from being primarily agricultural to increasingly urban and tourism orientated. The resultant 'coastal squeeze' with the concentration of towns, people, transport and energy infrastructure—and relative wealth—in low-lying coastal areas, is prone to diverse hazards, notably flooding, whose impacts are exacerbated by climate change, vegetation clearance and privatisation of land. Excluding PNG, more than half of Pacific populations live within a kilometre of the coast (Andrew, Bright, de la Rua, et al., 2019), and most of those live on the coast. In 12 Pacific SIDS (including Niue and Cook Islands) as much as 57% of all built infrastructure is within 500 m of coasts (Kumar & Taylor, 2015). International airports are often at risk—intermittently flooded as in Majuro (Marshall Islands), and more exceptionally in Port Vila (Vanuatu). Coastal hardening (concrete construction) continues to replace natural marine habitats and productive ecosystems with impenetrable structures and greater potential for flooding (Floerl, Atalah, Bugnot, et al., 2021). Socially vulnerable communities—those least able to prepare for disasters or recover afterwards—tend to be concentrated in areas that are more susceptible to flooding, one version of environmental injustice. Yet even token integrated coastal planning and management is rare, partly a function of rapid unplanned change (Chap. 6).

Degradation and clearance of reefs, mangroves and seagrass meadows have reduced the viability, productivity and shelter of coastal ecosystems, including coastal lagoons where pollution can be serious (Chap. 5). In marine environments, these pressures include intensive fishing and pollution from both land and marine sources (Bennett et al., 2023). Commercial agriculture is also concentrated in coastal regions. The low-lying Fijian towns of Nadi and Ba have often been badly flooded and, in the Rewa river delta, increased flooding has resulted in mangroves suffering dieback, reducing riverine productivity (Kinch, Vitukawalu, Nalasi, et al., 2019). Many SIDS coastlines have been affected by deforestation, tidal and wave action, storms, sand mining, mangrove clearance and housing, wharf and causeway construction. Sand mining alone, in both regions, particularly from beaches, dunes and riverbeds, has had significant environmental and social impacts, impacting ecosystems and threatening coastal communities. The need for, and shortage of, valuable construction resources has led to habitat destruction, erosion, increased vulnerability to storms and rising sea levels, and potential health hazards, with SIDS typifying the 'global sand crisis' most evident on atolls, at some environmental cost (Babinard, Bennett, Hatziolos, et al., 2014; Lamb, 2023). Human impacts have tended to have a greater influence on localised erosion and coastline changes than more natural processes, with erosion greatest in urbanised areas. Typically, in Barbados, with an estimated 44% of the population living in urbanised areas, most in the capital, Bridgetown, 'flooding in Barbados as a whole, and in Bridgetown specifically, is attributable to anthropogenic pressures such as built development arising from urbanization as well as rising sea-levels and storm surges resulting from climate change' (Mycoo, Robinson, Nguyen, et al., 2021). Honiara, Apia and Suva, amongst other SIDS capitals, have experienced the same issues.

Coastal tourism has encouraged sensitivity to environmental change (Scott, Hall, & Stefan, 2012; see Chap. 4). Beach erosion, coral bleaching, salinisation of drinking water, intense tropical storms and greater heat can deter tourism, and any decline in the tourism industry would have multiplier effects on poverty in most SIDS, even including those with more diverse economies. Despite the need for a pristine environment, the tourist economy itself has occasionally conspired in degradation through damage to coastal habitats (mangroves, seagrass beds, wetlands) and thus diminished biodiversity, over-consumption and drawdown of freshwater aquifers, water pollution and subsequent impacts on marine species and coral reefs, increased erosion and over-fishing (Scott, Hall, & Stefan, 2012). More directly, in terms of sea-level rise, in Caribbean SIDS and territories, 266 (29%) of 906 major coastal resort properties were found to be at risk of partial or full inundation by a 1-m rise in sea level (Scott, Simpson, & Sim, 2012). A similar situation is probable in the Pacific. Inundation would not be uniform, but, among the Caribbean Community (CARICOM) countries, inundation losses would be greatest in St. Kitts and Nevis (and three British territories: Anguilla, British Virgin Islands and Turks and Caicos Islands) and substantial in Barbados, the Bahamas and Trinidad and Tobago.

Most SIDS have policies and organisations set up specifically to deal with coastal management: Jamaica has an Ocean and Coastal Zone Management Policy and Barbados a Coastal Zone Management Unit. They typically have some regulatory functions and manage data collection, monitoring of coral reef and other coastal resources and consultation and coordination with internal and external stakeholders. Here, as elsewhere, capacity and retention of skilled workers is a huge challenge, affecting surveying, monitoring and enforcement, and ultimately effectiveness, while often challenging established vested interests. Coastal management takes time, which is scarce, and necessitates inputs and expertise at different scales. The need for mitigation—often costly—is greatest in the more affected states, which are also the more impoverished SIDS.

2.8 Climate Perceptions and Experiences

Since hazards have always been present, islanders devised ways of responding and coping long before colonialism and well before the contemporary climate crisis (Hofman et al., 2021). Anthropocene climate change has its source almost entirely beyond the SIDS, but island states have some scope for mitigation and adaptation, partly derived from and influenced by interpretations and explanations of climate change and its impacts, and from historical experience of environmental change. Islanders, almost inevitably, perceive climate change and its impacts—or what are assumed to be manifestations of climate change—as harmful, whether storms, erosion, floods, droughts or migratory fish stocks: a combination of causality, circularity and certainty, and are prone to attribute most recent physical changes to climate change (Connell, 2003). Erosion, and the loss of land, is usually the most demonstrable change.

Knowledge is contested and facts are not necessarily the basis of policy formation. Islanders are often sceptical of 'outside experts' and 'placeless' knowledge, whether of governments, NGOs or scientists, whereas local expertise comes from connections to place and long experience of it, as in the Bahamas (Petzold, Ratter, & Holdschlag, 2018). However, explanations are both local and global, evident in diverse Marshallese perceptions of, and explanations for, climate change that balanced complex and ever-changing beliefs and direct experience of physical landscape changes ('visibilism') with (poorly understood) scientific data and predictions, familiar biblical prophecies and indigenous knowledge (Rudiak-Gould, 2014; cf. Lauer, Atger, Holbrook, & Rassweiler, 2022; Connell, 2016).

Where Christianity has an important role, climate change is often linked to religious belief and to failings of belief and action. Recurrent hazards (and ill-health) too were often associated with human flaws and the displeasure and anger, sometimes inexplicable, of the spirits, or the wrath of God, as in Fiji, PNG and several Polynesian SIDS, all versions of moral retribution (e.g. Beyerl, Mieg, & Weber, 2019; Connell & Lutkehaus, 2017; Granderson, 2017; Jackson & Piggott-McKellar, 2021). An i-Kiribati man took a common position in this regard, relating climate change to the 'end times':

> I was born as a Kiribati, so I will die as a Kiribati. I don't care if I would die poor on my own land … if it's God's will to wash Kiribati away, so be it. We do not know the hour and the day for the end of time to arrive … We have to blame ourselves because of the rumours of climate change that have been ongoing; it is a sign that we have changed a lot. Climate is changing because humans are changing; we are modifying the natural resources from God and living rebelliously. (quoted in Gagaeolo, Hemstock, & Price, 2020: 4)

As historic spirits were partly subsumed by a contemporary Christian god, various admonitory gods and malign spirits might be involved. Human failings were sometimes blamed independent of spirituality; in Tuvalu droughts could be the outcome of poor secular leadership (Paeniu, Holland, Miller, & Anderson, 2017). The legacy of diverse supernatural causality remains in place in Melanesia (and is critical to some interpretations and responses). Thus, in northern Malaita (Solomon Islands), more than 90% of farmers attributed the impacts of climate change to God's punishment, because of their own infractions or the fulfilment of a prophecy; 40% of farmers believed that sorcery and breaking taboos had the same effect (Bird, Wairiu, des Combes, & Iese, 2021). Such proportions, while high, are not atypical of other island contexts, where 'traditional' spiritual beliefs combined with 'modern' Christian beliefs, further complicated where fundamentalist biblical values posed a challenge to adaptation and relocation planning (Fache & Fair, 2020; Simonelli & Novalski, 2023) and even, in extremis, where people are able to watch films about themselves and their decision-making (Moesinger, 2019a). Ubiquitously, people vary in their awareness and attitudes to the significance and urgency of climate change (Barbara, Mudaliar, Leach, et al., 2023; Thomas & Benjamin, 2020). Climate change is a scientific idea, a visible phenomenon and a product of western discourse, received, translated, interpreted and negotiated locally, and in a variety of ways (Cox, Finau, Kant, et al., 2018; Kempf, 2020).

Multiple changes have occurred in the physical landscapes of the PICs for many reasons. Those who live there, especially in rural areas, are enormously sensitive to recent and ongoing changes. Coastal erosion is widespread with losses of land, trees, houses, open spaces and graveyards widely reported, as in Fiji (e.g. Martin, Nunn, Leon, & Tindale, 2018; Nichols, 2019), Vanuatu (McNamara & Prasad, 2014; Ruehr, 2022), FSM (McNamara, 2013), Kiribati (Bach, 2017; Jackson & Piggott-McKellar, 2021), and elsewhere. Older people reminisce about playing on long-gone sand flats; lost beaches are widespread markers of better times. In the coastal Fijian village of Togoru, people could readily point to exposed roots or fallen coastal trees, encroaching high tide lines and erosion of burial sites (Yee, Piggott-McKellar, McMichael, & McNamara, 2022). Since coastlines are constantly changing, as they have done for centuries, recognition of coastal change is entirely predictable, but focused on erosion rather than deposition. Riverbank erosion and flooding are evident on high islands, as rivers contain more sediment, agriculture is closer to the banks, nearby houses are at greater risk, and riverine fish content is reduced in number and size (Currenti et al., 2019: 73). People in the Rewa Delta (Fiji) recognised such floods, riverbank erosion and groundwater salinity as among their most serious environmental challenges (Lata & Nunn, 2012) and a consequence of climate change.

Not all environmental changes are visible, but they may nonetheless be experienced. Hotter temperatures and drier climates have been reported. Higher temperatures are more likely to be perceived in densely populated urban heat islands but also between countries: changes in temperature were more likely to be reported in Tuvalu and Samoa than in relatively cooler Tonga (Beyerl et al., 2019). 'More intense rain and sun' were both experienced in Dominica (Cooper, Cloos, Abraham, et al., 2025) and higher temperatures and stronger winds in Efate (Vanuatu) and Antigua (Blair & Momtaz, 2018). Water stresses have been identified in many places, with salinisation of wells, even on high islands (Perumal, 2018). Fijian villagers have pointed to more lengthy droughts (Nichols, 2019; Pearce, Currenti, Mateiwai, & Doran, 2018) and new seasonal patterns (Currenti et al., 2019). However, some advance different explanations of water shortages; in Druadrua village (Fiji) older villagers speculated that burning and clearing of forests for farming and the introduction of goats and pine tree plantations played a part alongside the increased village population (Dumaru, 2010). In Tabiteuea (Kiribati), Tongoa and elsewhere in Vanuatu too, seasons are now said to be blurred and less distinguishable (Alston, Fuller, & Kwarney, 2025; Bach, 2017; Granderson,

2017), so making it harder to forecast and plan. Many claimed that both seasons and fruiting times are changing, with a greater prevalence of pests, and that these have undermined traditional knowledge, which has thus become less valuable, since different seasonal patterns have resulted in lost confidence in planting schedules and the loss of some valuable medicinal species (Cassinat, Cassinat, Segi, et al., 2022; Currenti et al., 2019; Kilroy, McNamara, Witt, & Raynes, 2025). Perceptions were gendered; in Samoa men experienced higher marine temperatures while women perceived agricultural changes (Alston et al., 2025; Cassinat et al., 2022). Differences in market integration and education were probably more significant variables (Pondorfer, 2019). Greater frequency and magnitude of cyclones and storms have been reported by local people in several places, including Fiji (Martin et al., 2018) and Samoa (Flores-Palacios, 2019). Most people on Yadua island (Fiji) stated that climate change had meant higher temperatures and increased frequency and magnitude of heavy rainfall events (Martin et al., 2018). In Samoa, higher temperatures could discourage work and affect productivity (Flores-Palacios, 2019). Almost half of all Malaitan (Solomon Islands) farmers perceived increased temperatures, but a majority had not (Bird, Iese, Combes, et al., 2022), whereas almost all Samoans in different locations had experienced changes in temperatures and seasonality (Latai-Niusulu, Binns, & Nel, 2020). However, even Samoan children claim to have experienced hotter temperatures, but, as elsewhere, that 'experience' is certainly shaped by classroom learning, radio, television and social media (Latai, Taulealo, & Connell, 2024), emphasising that much perception and reception is second-hand and dependent on context.

Some claim to have witnessed greater salinity of soils (Martin et al., 2018) and less productive agricultural systems, notably of root crops (Bird et al., 2022). In Togoru (Fiji) residents pointed to greater saltwater intrusion and the greater difficulty in growing some plants in more saline soils (Yee et al., 2022), even more so in the abandoned taro pits of Abaiang (Kiribati) (Jackson & Piggott-McKellar, 2021). Changing fisheries are as likely to be attributed to physical explanations, as in Samoa with more sediments in run-off (Flores-Palacios, 2019), or perceived changes in wave behaviour (Cooper et al., 2025). In the Fijian village of Navuniivi, residents also pointed to declining ocean fish stocks, but this could be seen as the outcome of ocean acidification from climate change, overfishing or polluted water draining from the village into the nearby ocean (Nichols, 2019). In many places fishers observed that reef-based fish harvests were losing their former diversity and that was most likely so where climate change impacts were negative, emanating from sea temperature changes, lower ocean pH levels, changes in rainfall that affect estuarine fisheries (and the spawning grounds of juvenile fish), bleaching and changes in ocean circulation, which alter habitats and change the distribution and mix of species, resulting in fishers having to travel further (Badjeck, Allison, Hall, & Dulvy, 2010; Blair & Momtaz, 2018). That has been accentuated by overfishing, resulting in the depletion of some vulnerable subsistence and commercial marine species. Characteristically, such losses have been more apparent close to urban areas, with growing populations and less likelihood of effective customary management, even near small towns such as Gizo (Solomon Islands) but evident also in isolated atolls such as Kiritimati (Connell, 2013a; Watson, Claar, & Baum, 2016).

While residents in coastal villages were likely to attribute most such changes to climate change, so discounting local responsibility, some recognised their own contributions. In Tabiteuea (Kiribati), where fishing was crucial to sustainability and food security, most villagers reported fewer fish, smaller fish and fewer fish species, and conceded that this followed their becoming less selective over which fish to catch, taking smaller fish, and disregarding regulations that protected certain species and certain areas (Bach, 2017). By contrast, inland villages, especially in Melanesia, have quite different experiences from coastal residents, face fewer climate change issues and are more concerned with socio-economic issues (e.g. Currenti et al., 2019). In Samoa, most people, wherever they are located, are relatively well informed about climate change, and children too are conscious of changes that have occurred and what their impacts have been, in terms of weather events and agricultural productivity (Latai et al., 2024). As with many changes that might or might not be linked to climate change, few anywhere had the technical knowledge, capacity or data to make more complex attributions. A gap remains between scientific measures and local perceptions. Although that gap is slowly closing, variable explanations and responses remain.

Regardless of cause, such changes on land and sea can be depressing and dramatic as expressed by ni-Vanuatu villagers:

> Nowadays because the coral reefs are dying, the fish have gone … Our crops are producing less yield because of weather changes. Root crops are rotting before they become ready to harvest because of the unusually large amounts of rain that we experience.
>
> We have to spend more money on food now than ever we did in the past. In the future I do not expect for there to be anything left in the waters or in the bush. (quoted in McNamara, Clissold, & Westoby, 2022)

Vulnerability and resilience are intricately combined. It could be useful to have external causes to blame.

In most islands, like Tabiteuea, islanders recognise that they too were complicit in environmental change and degradation, albeit where physical processes, that may be more serious factors, simultaneously occurred. In Tuvalu and Samoa, people referred to a comprehensive range of negative behaviour:

> Over-use, abuse or unwise use of resources: irresponsible and selfish behaviour … valuing money more than the consequences of such behaviours, economic activities, greed and modernisation. Changed conservation and consumption patterns, societal changes, bad manners of youth, and new religious denominations … [and] population growth. (Beyerl et al., 2019: 158)

In Vanuatu people recognised that local activities, including deforestation and sand mining to make cement bricks, contributed to coastal erosion, while growing populations and reduced fallow periods, because of land shortages, reduced crop productivity (Granderson, 2018). In Emau Island (Vanuatu) too, physical degradation was regarded as either resulting from climate change and/or population growth (Ruehr, 2022). In Tuvalu, and more widely, climate change is just one of many factors affecting livelihoods and biodiversity such as economic hardship, food insecurity and overcrowding (McCubbin et al., 2015). People were generally well aware of multiple influences on environmental change, however seemingly erroneous or peripheral some of these might seem.

Many of what are regarded as recent indicators of change have also been reported much earlier but may now be more intense. In Vanuatu at least drought was no new phenomenon (Ruehr, 2022). In Kiribati it was a regular occurrence that had necessitated twentieth-century resettlement. Ecologies were often in flux, especially as new technologies enabled more efficient harvesting of the sea. Flooding of the interior of the main island of Funafuti (Tuvalu) was evident in the 1890s, but contemporary flooding was exacerbated by the construction of a permanent, sealed airstrip occupying as much as a quarter of the island, sand mining, vegetation removal, depletion of the water lens (and thus compaction), causeway construction and other human-induced changes (Yamano et al., 2007; Woodroffe, 2008). On similar atolls both in Woleai (FSM) and Rongelap (Marshall Islands), islanders were conscious that erosion was part of a 'normal process' and had long taken steps to modify and defend eroded areas and secure land expansion elsewhere. Seawalls were built on several atolls, such as Woleai (where erosion of coastal tracks was clearly visible at least by 1990) and in Rongelap by the 1970s (Bridges & McClatchey, 2014). There was no moment or tipping point when the influence of climate change became a dominant factor in environmental change. Not surprisingly, perceptions of the role and significance of climate change and its impacts differed from scientific assessments. There is no means of determining the balance between climate change—in its many manifestations—and human interventions—in their many variations. Consensus over causation and action is rarely achieved (somewhat similarly replicated in opinions about the genesis of, and responses to, COVID-19). Policymakers and especially international agencies draw on more 'conventional' discourses of scientific knowledge, with these perspectives attributed supposed objectivity, independence and rationality. At a quite different scale, national and international policy, supported by mainstream climate science, thus tends to envisage a positive future of 'ecological modernisation' where capitalist economic growth can be sustained through technical solutions and international agreements, and local epistemologies are marginalised.

2.9 Adaptation

Adaptation to the effects of climate change is a continuous process of trial, failure and reflection, influenced by the resources (natural, financial, institutional or human, and including access to ecosystems, information, expertise and social networks) available for adaptation, as well as the ability or capacity to use resources effectively (Brooks & Adger, 2004). Adaptive capacity also involves social, cultural and psychological dimensions (Kuruppu & Liverman, 2011); experience of past environmental hazards; and assumption and expectations over future changes. In every place these factors vary.

Beyond crucial differences in beliefs and understanding, responses necessarily vary between places according to specific problems, opportunities, support, harmony and access to financial and social capital. Responses to real and perceived threats vary in terms of policy and practice at every scale, from designing and building resilient houses, to high-level delegations at global conferences, and from movement away from affected areas (see below), to different planting, fishing, rainwater harvesting and house construction strategies. Increasingly, such strategies and outcomes have become politicised in the context of climate change justice, loss and damage and reparations policies (alongside vague perennial themes of consciousness raising and capacity building). However, mitigation has not always worked. Island shorelines have been 'armoured' through engineered structures such as seawalls, but such hard protection often simply redistributes and sometimes exacerbates shoreline erosion, and fails to protect property, land and food production (Duvat, 2013; Klöck, Duvat, & Nunn, 2022; Nunn, Klöck, & Duvat, 2021). Effective capacity building has been limited. Land reclamation offers at least a short-term solution and has been used in Kiribati and Tuvalu.

Nature-based solutions, centred on planting mangroves that strengthen coasts and are a bulwark against storms, can protect coastal settlements (while also benefiting ecosystem services and contributing to carbon sequestration). In many sites in every PIC, as an alternative to 'hard solutions', mangroves have been planted as barriers. However, while dense mangrove root systems absorb wave energy and provide a buffer against wind, waves and erosion, their ability to defend coasts against high waves is limited, although in some cases they can reduce flood risks to homes and infrastructure more than 400 m inland (Soanes, Pike, Armstrong, et al., 2021). In Kiribati, villagers have used analogous woven fences, made from local materials, to capture sediment and rebuild beaches (Yarina & Takemoto, 2017). Climate adaptation strategies may combine traditional practices and science, to build resilience of coastal communities and ecosystems, including implementing networks of marine protected areas (MPAs), experimenting with salt and drought-tolerant crops, revegetating coastlines with native salt-tolerant plants, revitalising traditional wells and fish traps, and implementing climate-smart development plans (Granderson, 2017; Laalaai-Tuasa, 2022; Mcleod, Bruton-Adams, Förster, et al., 2019; Mercer, Kelman, Alfthan, & Kurvits, 2012; Rarai, Parsons, Nursey-Bray, & Crease, 2022). Such responses are gendered and, as women and the poor are more affected by hazards than others, require considerations of social equity to be built into comprehensive responses (Attzs, 2009; Dulal, Shah, & Ahmad, 2009; Ha'apio, Gonzalez, & Wairiu, 2019). Combining knowledge contributes to local development priorities and the SDGs but is unlikely to ever be entirely adequate, while logistics, disagreements over land and marine tenure, and finance may prove constraints. Cost-effective protection strategies are likely to exceed local ability or willingness to pay and thus require external assistance, but that too can swamp local capacity and even reproduce structural inequalities (Dean, Green, & Nunn, 2017; Donner & Webber, 2014; Klepp & Fünfgeld, 2022; Wrighton & Overton, 2012). Moreover, although growing interest is attached to nature-based solutions to facilitate the continuation of dignified and productive lives, there is growing evidence that these can no longer make a significant contribution to adaptation on atolls, as they did in earlier times (Barnett, Jarillo, Swearer, Lovelock, et al., 2022).

Adaptation is not necessarily successful, as in several Fijian cases where land losses and the destruction of burial grounds resulted in a reduction in family wellbeing, security and harmony, since in situ adaptation was impossible and even new locations nearby were inadequate to provide solace (Nand, Bardsley, & Suh, 2024). Psychological stress is common, pointing to the significance of local values and how loss and damage are immeasurable. Social networks and resources, particularly to access land and housing, are essential, alongside a sense of equity and access to such services as education and health care, some flexibility of choice, past experience and universal involvement (Basel, Goby, & Johnson, 2020; Cinner, Adger, Allison, et al., 2018; Cooper et al., 2025; Hidalgo, Nunn, & Beazley, 2021; Smith, 2025). While community-based adaptation has been portrayed as a panacea, perceiving communities as geographically bounded, temporally fixed and harmonious units, that 'fuzzy, romantic notion' is rarely if ever so (Foale, 2001; Westoby, Clissold, & McNamara, 2021) and divisions may be exacerbated in difficult times. Time and patience are invaluable but often absent. Since the relatively poor are usually at greatest risk, with the least resilient housing, low incomes and weak social networks, they are likely to be most disadvantaged, least included in plans and strategies and with less agency to be aware of options and make informed decisions. Adaptation is most effective when those facilitating it are from the same culture, speak the same language and understand local customs (Dumaru, 2010) but that is not often the case. Even then ecologies vary and resilience is complex and context-specific. Indeed, more isolated villages occasionally fared better since isolation contributes to community autonomy and inclusivity, more diverse local livelihoods and traditional coping mechanisms are more likely to be retained, and social capital remains strong (McNamara, 2013; McMillen, Ticktin, Friedlander, et al., 2014; Korovulavula, Nunn, Kumar, & Fong, 2020; Malherbe et al., 2020). At local scale adaptations primarily restore versions of the 'old order'—a preference for continuity and a basic place attachment—rather than providing opportunities for 'transformational' change and new possibilities, despite some degree of flexibility being invaluable (Rhiney & Baptiste, 2019; Schwarz, Béné, Bennett, et al., 2011; Thomas, 2025). Where assistance is from distant sources, external values and needs can undervalue and neglect local community and individual knowledge, cultural institutions and relational ethics (Perry, 2024), often from governments with their own predilections. Certainly no 'one-size-fits-all' meets the needs of diverse places, and that is further complicated where migration has occurred (see below), necessitating recognition of the inherent inequities and vulnerabilities that are associated with any development trajectory.

At a different scale, almost every SIDS has a policy or plan for climate action, some linked to migration and mobility, but few have proposals for financing what is necessary. Initiatives are limited and under-resourced (Galaitsi, Corbin, Cox, et al., 2024). However, Fiji has developed a National Climate Finance Strategy, aligned to a Climate Change Act, that declares a climate emergency and commits the country to decarbonise by 2050. It identified 12 economic sectors that need adaptation and decarbonisation, from agriculture to health, water and sanitation, and addresses strategies for relocation that are linked to Fiji's Climate Relocation Communities Trust Fund, begun in 2019. Fiji has also been particularly active in supporting and managing resettlement programmes.

In some vulnerable areas, relocation is already being considered and implemented as an adaptation option. Vanuatu developed a National Climate Change and Disaster-Induced Displacement Policy (Vanuatu National Disaster Management Office, 2018) that recognised the probable need for future relocation from some coastal areas, but it faced land tenure challenges (McDonnell, 2021). At a different scale, in 2023, Australia and Tuvalu signed an international treaty, the Falepili Union, that recognised the existential climate-related threats faced by the atoll state of Tuvalu and supported innovatory arrangements for Tuvalu citizens to permanently relocate to Australia (see Chap. 8). Tuvalu has accepted that under a high-emission scenario, much of the area of its atolls (including the capital island, Funafuti) will be flooded by daily tides by 2050 and the country be almost totally submerged by 2100. Its Tuvalu Coastal Adaptation Project (TCAP) in 2024 was shoring up infrastructure around three of the nine atolls (including Funafuti) to prevent such flooding, supported by the Green Climate Fund (GCF) and bilateral partners such as Australia. In the longer term, Tuvalu has developed a Long-Term Adaptation Plan (LTAP) with regional and global partners to sustain its resilience to flooding risk. The cost of LTAP, for just three atolls, is likely to be more than US$1billion, a sum unlikely to be found or donated, indicating that a 'Plan B' will be required for two thirds of the nation's islands and that must centre on migration. Tuvalu has even contemplated a future in the metaverse (Rothe, Boas, Farbotko, & Kitara, 2024). Large islands and nations will be unlikely to disappear in the next few decades but further small-scale migration will continue to be necessary (see Chap. 9). Tuvalu is the only country to have contemplated, and begun to plan for, its own disappearance.

2.10 The Global Threat: The COPs ... and Robbers?

No SIDS are spared the impact of climate change and no SIDS benefit from it: the impacts of climate change are negative and universal, yet the SIDS themselves have made almost no contribution to global warming (being without mines, heavy industries, traffic and even ruminants). That emphasises their limited ability to solve resultant problems through their own efforts so that measures to reduce global warming must be taken at a 'higher' international and global level, with pressure from the SIDS who have taken a disproportionate lead in political deliberations and global policy formation.

As global warming was increasingly regarded as at least partly Anthropocene, primarily associated with the burning of fossil fuels in large developed countries, smaller SIDS pointed to their own innocence in this matter, demanded that developed countries mitigate and end this activity, and organised themselves into groups to press for global change, alongside more militant indigenous groups and NGOs such as Greenpeace. The richest countries, home to 16% of the world population, emit approximately 40% of CO_2 emissions. Island states have effectively grouped themselves into combined international organisations focused on climate change, notably and initially in AOSIS.

Demand from SIDS for action on climate change has increased and accelerated, driven by concerns that islands will be inundated and even that homelands will disappear and that the increased strength of cyclones causes deaths, injuries and the loss of livelihoods, as ecological systems and commercial systems are devastated. Loss and damage became constant themes, complicated by the enduring impacts of colonial dispossession and environmental degradation, where 'non-economic' losses such as burial grounds and sacred sites, especially in the Pacific, were as significant to many as short-term economic losses (McDonnell, 2021). While aid donors have provided funds for climate mitigation activities, disaster relief and consciousness raising, the rate of warming is continuing to increase, possibly even exponentially (Papalexiou & Montanari, 2019). The harsh realities of climate change and the need for action to protect the most vulnerable countries brought early institutional focuses, the establishment of the IPCC, and the Conference of the Parties to the United Nations Framework Convention on Climate Change (COP) as SIDS pressures increased and global concern responded.

At global policy level climate change impacts are assessed through annual United Nations Climate Change Conferences under the United Nations Framework Convention on Climate Change (UNFCCC). First undertaken in 1995, the Conference of the Parties (COPs) constitute the formal meeting of the countries that have agreed to participate in UNFCCC, to examine progress in dealing with climate change and to negotiate the Kyoto Protocol to establish legally binding obligations for developed countries to reduce their greenhouse gas emissions. Alongside the UNFCCC COP, two other environmental COPs exist: the UN Convention on Biological Diversity and the UN Convention to Combat Desertification.

The Paris Agreement, produced at COP21 in Paris in 2015, was the first legally binding global treaty on climate change, with almost all countries in the world committing to keeping the rise in global average temperature to 'well below' 2 °C, and ideally below 1.5 °C, above pre-industrial levels; strengthening the ability to adapt to climate change and build resilience; and aligning finance flows with 'a pathway towards low greenhouse gas emissions and climate-resilient development'. Participating party countries individually decide by how much they will reduce their national emissions each year. Parties communicate these targets to the UNFCCC Secretariat in the form of 'nationally determined contributions' (NDCs), which are revised every 5 years. Initially, NDCs often focused on large-scale infrastructure projects but later emphasised softer, more holistic approaches like ecosystem-based adaptation and community engagement (Mohan, 2025a). Countries are legally bound to publish an NDC, but the Paris Agreement leaves the contents entirely up to them.

SIDS have always played a critical role within COP meetings to secure changes by the major emitting countries directed to reducing the use of fossil fuels. As Mia Mottley, the Barbados Prime Minister, stressed 'small states continue to remain dispensable to the global community. We refuse to be relegated to the footnotes of history, and to be collateral damage to the greed of others. For we have contributed less than 1% of greenhouse gas emissions' (quoted in White, 2021). Like other SIDS, Barbados was experiencing intersecting crises as COVID-19 had decimated the tourism sector, which usually accounted for about 40% of GDP. Nonetheless, Barbados had impressive emission reductions targets and was seeking to be the first entirely green and fossil-free island state in the world. For that, like other SIDS, it required global recognition and support.

Most SIDS participate in COP conferences, as a significant component of foreign policy, and have taken several ministers to ensure coverage of parallel sessions and enable collective teamwork with other SIDS to ensure their critical voices are heard (Klöck, Castro, Chin-Yee, et al., 2025). Thus, the agreed text of the 2023 COP28 in Dubai (a venue criticised for its link to fossil fuel production) was criticised for its wording not including explicit calls to 'phase out' or 'phase down' the use of coal. John Silk, head of the Marshall Islands' delegation, called it a 'canoe with a weak and leaky hull' but one that 'needed to be sailed' because there was no other option. Henry Puna, then Secretary-General of the Pacific Islands Forum, said COP28 did not deliver the outcome the region needed: 'COP28 was a disappointment because of our prioritising 1.5 degrees as our absolute priority with climate change. We will never give up; we have to continue pushing on that 1.5 degrees'.

While denial of the link between greenhouse gas emissions and global heating has largely been discredited, its space has been occupied by a more effective and insidious problem, climate delay, where spurious arguments and disinformation are used to justify not reducing emissions at all, or reducing them at as slow a rate as possible. In Australia, to the dismay of Pacific island SIDS, an approach of 'hollow techno-optimism' in 2021 (Joshi, 2021), where yet uninvented technology was anticipated to solve the problem in the future, barely disguised the lack of effective action. COPs offer only limited optimism for SIDS, despite small states having, in theory at least, as much influence as larger states, since they require support from states (and companies) with at least some degree of self-interest. With the presence of SIDS, resource companies, governments and NGOs, COPs became simultaneously a theatre of climate colonialism and 'a site of decolonial, anti-colonial, anti-racist, and feminist politics' (Sultana, 2022: 119). In the same vein, whether new sources of finance could realistically address extreme vulnerability to climate crisis was challenged (Perry, 2024). At COP, climate politics was both performed and contested, a struggle between competing narratives that could distract from the key issues for SIDS, a reduction in the use of fossil fuels and the need for climate finance.

2.11 Loss, Damage, Finance, Reparations and Equity

Over time, COP conferences increasingly focused on financial mechanisms whereby wealthy countries would provide funding to help poorer countries with clean energy transitions. Likewise, over time, the COP meetings provided an opportunity to press the private sector to play a greater role in driving investment in the renewable energy transition. At the 2009 Copenhagen COP15, developed countries committed to collectively providing US$100 billion a year for climate finance, but the target was not met until 2022. Difficult questions remained. Who should and would provide finance and who receive it? Who was responsible for loss and damage and over what time scale? And would wealthy countries make good on their commitments? Complications over loss and damage have been multiple. Initial disagreement concerned which countries should contribute to the fund, with calls for larger and richer middle-income developing economies with high emissions, such as China, India and the Gulf states, and small but high-income countries, such as Singapore and Brunei, to contribute. Designated contributor countries were reluctant to propose a share, especially while the list of contributor countries was still contentious, while there was no agreed definition of what qualified as climate finance, and ongoing fears that Official Development Assistance aid would be diverted into climate finance, with donors merely reformulating aid projects as climate finance. The predominance of loans over grants was a source of discontent, with most climate finance being loans, thus adding to the financial stress of already debt-burdened developing countries. Many developing countries, and particularly the SIDS, pointed to the difficulty of accessing climate finance, with Australia proposing a 'friends of access' group, whose task would be to find ways for smaller and less-resourced entities, including less developed countries (LDCs) and SIDS, to gain access to funding. Quite simply, the prospect of an increase in overall funding attracted opposition and delays from contributor countries (Ryan, 2024).

A Green Climate Fund (GCF), set up in 2010, also proved difficult to access by SIDS despite their being severely impacted by climate change and thus designated as a priority for adaptation finance. Access was a complex and slow process over several years, when urgent adaptations required solutions, so posing a problem for SIDS with scarce skilled human resources. The Pacific Island Forum's (PIF) programme adviser pointed out that accessing the GCF involved 'the whole ecosystem of governments [where] there's a big turnover in all ministries' (Karlos Moresi, quoted in Naidu, 2024:12). The relative income of SIDS made them ineligible for cheaper finance set aside for low-income countries. Consequently, SIDS have been proportionally less successful than other vulnerable country groups in accessing GCF climate funding, compounding the injustice of climate change for SIDS, that had contributed little to the problem yet struggled to receive meaningful levels of assistance (Treichel, Robertson, Wilkinson, & Corbett, 2024). The Copenhagen Accord from COP15 stipulated that developed countries would provide increased climate finance to developing countries, as part of their sustainable development programmes. The Group of Seven (G7) and other rich countries both agreed to provide 'scaled-up' climate finance that was 'new and additional' to existing overseas development aid. However, rich countries simply took most of the public climate finance from ongoing development aid budgets so that only 6% of climate finance provided between 2011 and 2018 was new and additional. Both Australia and Aotearoa New Zealand (hereafter simply New Zealand) provided small amounts of additional climate finance, marginally ahead of the United States (CARE, 2022). Most finance was therefore merely diverted towards climate change action, resulting in less support for other activities including health, education, women's rights and poverty alleviation.

Vulnerable SIDS regularly argued that losses and damage from climate-induced events occurred 'beyond adaptation' and required compensation from carbon-emitting nations, circumstances particularly true of Tuvalu (Calliari, Serdeczny, & Vanhala, 2020; Calliari & Vanhala, 2022). The difficulty of getting resources to SIDS to meet local mitigation needs and fill the deficits in allocation led to a widening acceptance of the need for global financial support for SIDS (Ciplet, Falzon, Uri, et al., 2022; Giglio, Kelly, & Stroebel, 2021). Donor finance following the Copenhagen COP in 2009 was more likely to go to SIDS and to countries most affected by climate change but not departing far from previous patterns (Weiler, Klöck, & Dornan, 2018). A decade later that resulted in the key outcome from COP29 in Dubai being the establishment of a Loss and Damage Fund to cover unavoidable loss and damage experienced by states as a result of climate impacts. The G77 (a coalition of 134 developing countries) and China wanted financing for a 'loss and damage facility' to be put on the official agenda at the COP27 conference in 2022, but the United States and the European Union—two of the largest greenhouse gas emitters—opposed the idea. Such divisions and divergences were seen by some as the outcome of a deeply entrenched colonial racism that had created and maintained climate injustice, as rich white countries continued to distance themselves from any language of compensation or reparation for both historic and contemporary emissions (Fitz-Henry, 2022). Central to the statements of SIDS leaders, notably Mia Mottley, was that providing finance to mitigate and adapt to climate change was not simply an economic transaction but a means of delivering environmental justice in response to the tribulations and outcomes of colonial economic history. This brought debate over who received what (distributional justice), whose knowledge counts (epistemic justice), who decides (procedural justice) and, ultimately, who is left behind (recognition justice) (Huff & Naess, 2022; Schlosberg & Collins, 2014). Such complexities and legal subtleties meant that claims related to environmental justice became increasingly complex as rich-world countries contested notions of complicity and debated (or ignored) the form of, and responsibility for, any reparations that were bound up in the complexities of historical inequities, uneven power relations and disproportionate contributions to harm. Such 'complex geographies of injustice' (Huff & Naess, 2022: 2) were particularly vivid in the SIDS, whose contributions to climate change were negligible, where colonial histories were recent and not easily erased, and decolonisation was in practice incomplete (Banivanua Mar, 2019; Bordner, Ferguson, & Ortolano, 2020; Gahman, Thongs, & Greenidge, 2021; Moulton, 2024; Perry, 2024). However, no approach could come close to assessing the impacts of colonialism and how they might be remedied. Such challenges became 'super wicked problems': urgent yet without ready solutions where national interests were at stake (Cross & Congreve, 2020) and where the same countries that had caused the problems (and continued to cause them) were expected to provide solutions that were not purely responses to contemporary scientific evidence but value-based and shaped by power structures.

Global frustration with UNFCC and financial support brought the emergence of a new grouping, announced by Antigua and Barbuda with Tuvalu, the Commission of Small Island States on Climate Change and International Law (COSIS), formed to help SIDS seek reparations from, and bring legal action against, major carbon-emitting countries. AOSIS, the negotiating group of SIDS, had first raised the issue of loss and damage (that the SIDS might have preferred to have been seen as euphemisms for liability and compensation, and that would more directly imply reparations). Three decades later, at COP26, any formal and binding mechanism to deliver loss and damage financing remained absent. Kausea Natano, then Prime Minister of Tuvalu, stated that COSIS would be 'a platform for small island states to channel their grievances on the impact of climate change to legal bodies' (quoted in White, 2022). The SIDS argued that the principles of due diligence, that a state must not allow its territory to cause harm to other states, and 'polluter pays', both having been applied for decades in international law, should also be applied in the context of greenhouse gas emissions. The outcome would be 'climate finance' for mitigation (reducing greenhouse gas emissions), adaptation (building resilience), and loss and damage (compensation for harm). As AOSIS' efforts demonstrated, the concept had been repeatedly rejected by wealthy countries that were opposed to accepting blame and liability and providing compensation.

The 2024 COP29 in Azerbaijan (another venue criticised for its link to oil production and with multiple participants from OPEC and oil-producing companies) was regarded by many SIDS as being where the richest nations abandoned their legal and moral obligations, in failing to deliver adequate finance while stalling on crucial issues like phasing out fossil fuels and honouring loss and damage obligations. Pledged finance relied heavily on loans rather than grants, potentially pushing developing nations further into debt while ignoring the growing costs of adaptation and mitigation. The Vanuatu Climate Action Network coordinator, Trevor Williams, said that developed nations systematically dismantled the principles of equity enshrined in the Paris Agreement: 'Their unwillingness to contribute sufficient finance, phase out fossil fuels, or strengthen their NDCs demonstrates a deliberate attempt to evade responsibility. COP29 has taught us that if optionality exists, developed countries will exploit it to stall progress'. Likewise, Grenada's Prime Minister, Dickon Mitchell, speaking at COP29 soon after major damage from Hurricane Beryl, expressed his disappointment.

> We cannot wait any longer for empty pledges to become meaningful action … Wealthy polluting countries are seeking to shift responsibility onto elusive private capital instead of ensuring decisive government action to reduce and disincentivize emissions by introducing a global carbon tax to fund a clean energy revolution that is quintessential to saving our planet. To those who bear the greatest responsibility for climate change, I say this, the time for moral responsibility is now! The time for increased climate ambitions is now!

The Tuvalu Climate Action Network's executive director Richard Gokrun stated that the 'outcome is personal. Every fraction of a degree in warming translates into lost lives, cultures and homelands. Yet, the calls of the Pacific and other vulnerable nations were silenced in Baku. From the weakened Loss and Damage fund to the rollback on Just Transition principles, this COP has failed to deliver justice on any front' (RNZ, 25 November 2024).

The US$700 million committed to the Loss and Damage fund by 2024 was far short of what was required and, as climate change proceeded, was certain to increase over time while implementation raised all the previous issues related to financing. Publication of the IPCC's Working Group III report in 2022 evoked a strong reaction from the United Nations' Secretary-General António Guterres: 'The report is a litany of broken climate promises. It is a file of shame, cataloguing the empty pledges that put us firmly on track towards an unliveable world'. At COP26, the developed countries had collectively pledged a modest US$100 billion for the Adaptation Fund at the same time as the Stockholm International Peace Research Institute (SIPRI) pointed out that world military spending surpassed $2 trillion in 2021. The five largest spenders—the United States, China, India, the UK and Russia—accounted for 62% of this military expenditure (alongside clear evidence that militarism is a major driver of greenhouse gas emissions). By contrast, just four OECD countries (Norway, France, Sweden and Japan) provided what was regarded as a fair share of the global climate finance goal for 2023, with the United States amongst the bottom three. While Australia was not a high performer, a very high proportion of its finance went to SIDS (Pettinotti, Colenbrander, Kamninga, et al., 2025).

The Adaptation Fund ('Committed to Leaving No One Behind') was designed to finance climate change adaptation projects and related programmes based on the priorities of eligible developing countries. Its primary funding came from a 2% share of proceeds of the Certified Emission Reductions issued by the Kyoto Protocol's Clean Development Mechanism. Approved projects in 2023 included US$8 million to build resilience of coastal fisheries and aquaculture in Nauru, and US$10 million in PNG to improve food security through adaptation of small-scale agriculture. Finance was however not easy to obtain, even in the wake of more intense cyclones, that could be seen as outcomes of climate change. In the aftermath of Hurricane Beryl in the Caribbean in mid-2024, which was regarded as being 10–30% more intense than previous hurricanes, access to finance for restoration was no better. Little of the limited previous finance had gone into building resilience—whether in coastal protection, education, health care or housing—that might have mitigated hurricanes (Bardouille, 2024), and despite the role of women in responding to hazards and an intended response to gender implications, no evidence existed, at least in FSM and Tuvalu, that climate finance reached or responded to their needs (Samuwai, Fihaki, & Underhill-Sem, 2020). Coping with hazards and energy transition requirements was scarcely any easier. Despite the Paris Agreement providing a framework for financial support to SIDS and stressing that developed countries should take the lead in providing assistance, the finance required for energy transitions far outweighed commitments. The share of public climate finance from developed countries to SIDS remained below 2% of their financial support for developing countries over the period 2016–2019 in both the Caribbean and Pacific regions, was concentrated more in the LIDS and not always aligned to domestic development priorities (Atteridge & Canales, 2017; Mohan, 2023). It was increasingly evident that no single mechanism was likely to redress the multiple concerns of the SIDS and deliver climate justice, without being specifically directed to their distinctive needs (Lai, Robinson, Salas, et al., 2022).

Some SIDS invested in climate insurance policies, such as the Caribbean Catastrophe Risk Insurance Facility (CCRIF), a regional catastrophe fund established in 2007 aimed at managing risk and the financial impacts of hazards. Payouts have been made to several countries, including Grenada (US$42.4 million after Hurricane Beryl) and Antigua and Barbuda (US$2.9 after Tropical Storm Philippe) culminating in Jamaica receiving US$91 million, the highest ever payout, following Hurricane Melissa in 2025. Some Caribbean SIDS also purchased catastrophe bonds (CAT bonds) through the World Bank, which transferred country risks to the capital markets. Following the passage of Hurricane Melissa, Jamaica received US$150 million from the World Bank, the first island state in the world to independently access the CAT bond market, demonstrating a successful disaster financing tool for climate-vulnerable nations. But even with these proactive insurance measures, SIDS still face enormous financial exposure and risks, while such insurance schemes usually do not cover the full costs of damage, and many households, often the most vulnerable, do not benefit directly from them.

Their middle-income status made most SIDS ineligible for development assistance and international debt relief, and few qualify for concessional borrowing from the World Bank. Additionally, 'Caribbean small islands have restricted fiscal space and inadequate domestic finance to satisfactorily respond to climate change impacts, and are vulnerable to external shocks in particular natural [sic] disasters and require funds for relief and reconstruction, which significantly increases government borrowing and debt which restricts their capacity to borrow' (Mohan & Strobl, 2025). Unsustainable debt burdens continued to plague Caribbean SIDS (Gahman et al., 2021). One outcome of these limitations was the establishment in 2025 of a Pacific Resilience Facility to support projects in two areas: Climate and Disaster Resilience (climate adaptation, disaster preparedness and disaster response) and Social and Community Resilience (community resilience projects to address systemic risks and strengthen regional SDG performance). It was one component of the grand 2050 Strategy for the Blue Pacific Continent (Chap. 7), with financial support promised from Australia (A$100 million), China (US$ 500,000), Saudi Arabia (US$50 million) and the United States (US$ 5 million), and was argued to be the first Pacific-led, member-owned and managed and people-centred climate and disaster resilience financing facility.

Disappointment with the slow process of gaining adequate reparations through the loss and damage fund resulted in SIDS seeking other means of redress. As Ralph Regenvanu, then the Vanuatu Special Envoy for Climate Change and Environment, argued in late 2024 at the Commonwealth Heads of Government Meeting (CHOGM), as finance was only slowly accruing: 'Historical polluters, specifically developed States whose development was achieved through industrialization and colonization, are responsible for the majority of cumulative anthropogenic GHG emissions. This reality should inform both mitigation ambition and contributions to the loss and damage fund'. However, the emerging governmentality of loss and damage obscured the causes and centred on technocratic interventions (Jackson, N'Guetta, De Rosa, et al., 2023). Such orientations and frustration resulted in Vanuatu seeking an advisory opinion from the International Court of Justice (ICJ) in 2023 on states' obligations over transboundary climate change, an initiative developed by University of the South Pacific (USP) law students (Rikimani, 2024): the world's biggest problem was taken to the world's biggest court. Previous efforts along similar lines by Palau and the Marshall Islands in 2011 had been blocked by larger states including the United States.

After its longest ever case, the ICJ provided its advisory opinion in mid-2025, stating that climate change posed 'an urgent and existential threat of planetary proportions' and that all states (and by extension individuals and companies within those countries) had obligations under international law to mitigate the impacts of carbon emissions and limit global temperature increase to 1.5 °C beyond pre-industrial levels, in accordance with the Paris Agreement, and to limit fossil fuel production and consumption, including granting exploration licences. It emphasised that a healthy environment was part of human rights. The Court argued that obligations went beyond existing treaties and that governments were legally bound to compensate vulnerable nations for the harm they had caused, so offering hope for climate-vulnerable states and their people, and to those on the front lines of climate-induced disasters, such as Indigenous and socially disadvantaged people (McDonnell, 2025). While advisory, the Court's ruling seemed likely to result in a new understanding of climate change obligations in international law, in the contexts of treaty-making, state-to-state relations and domestic litigation. It potentially reframed the annual negotiations under the UNFCCC and the Paris Agreement in terms of what was legally required, rather than what was politically expedient or even morally appropriate.

This innovatory approach, 'one of the most remarkable developments in the history of international law is the leadership of small island States' (Akhavan & Holst, 2024), was taken up by COSIS, resulting in 91 states and international organisations participating in the ICJ proceedings. For several SIDS, this was their first appearance before the Court. Lack of human resources, especially in a highly specialised area of international law, prevented some small Caribbean states from participating, but Caribbean and Pacific states worked together on country presentations (Sobion, 2025). The trend towards increased participation was visible in other forums, such as the International Tribunal for the Law of the Sea, where a 2022 request for climate change advisory opinion was also made by COSIS.

Typically, the obligation of states regarding mitigation was a core issue in the International Court. Not surprisingly, developed countries that advocated for the Paris Agreement on climate change as the primary source of the applicable law, took the position that the obligations on mitigation were procedural ones, and compliance with their NDCs would suffice. By contrast, the vast majority of states from the Global South, such as Antigua and Barbuda, disagreed, arguing that other sources of international law, including customary law, applied (Oral, 2025). The ICJ advisory opinion reshaped this process by clarifying the existing legally binding obligations of major polluters to avert catastrophic harm caused by climate change and to make reparations for breaches of such obligations.

Vanuatu and two other PICs (Samoa and Fiji) went further to encourage legal opinion through the International Criminal Court (ICC) to recognise ecocide—the destruction of nature—as an international crime. As Ralph Regenvanu stated in 2024, 'Vanuatu considers it imperative that the international community takes this conversation seriously, and we warmly invite all member states to engage. Legal recognition of severe and widespread environmental harm holds significant potential to ensure justice and, crucially, to deter further destruction' (quoted in Conley, 2024). Environmental destruction was absent from the statutes of the ICC so that success was unlikely, but it drew further attention to the significance, scale and origins of environmental change and to the increased militancy of its victims.

2.12 Displacement and Resettlement

In several places, adaptation to environmental change has brought displacement and resettlement of populations, usually following the erosion of coastal land and infrastructure, and often after cyclones. This displays continuities with past resettlement schemes that followed high population densities, volcanic eruptions, mining, weapons training and nuclear testing (Connell, 2012; McAdam, 2014). Since 2015, cyclones in Vanuatu and Fiji, volcanic eruptions in Manam (PNG), Ambae and Ambrym (Vanuatu), tsunamis in the Solomon Islands in 2007 and Samoa in 2009 have all provoked despondency at responding to frequent hazards (with agricultural systems taking minimally a year to revive) with every cyclone a catalyst to some abandonment and migration. Nonetheless, islanders are universally reluctant to leave their home islands and villages, even where damage has been substantial (Bratrud, 2024; Perumal, 2018), and many island populations have remained or returned, even after their demise was predicted. After every home and structure on Ragged Island (Bahamas) was destroyed by Hurricane Irma in 2017, and all the residents (about 70 people) were evacuated, the government recommended relocation but few were in favour, and 6 months later almost all had returned to rebuild their houses but without government services (Thomas & Benjamin, 2020). The same process of return occurred on Manam Island (PNG) through the 2010s, after the volcanic eruption had subsided, despite government refusal to restore services.

More successful small-scale moves, in numbers and distances, have followed cyclone damage, tsunamis, coastal erosion and flooding, as in Fiji, Samoa and Vanuatu (e.g. Latai-Niusulu, Tsujita, & Neef, 2023), and offer precedents for further moves in future years. Preferred moves are local, where land is owned, with movement uphill and from smaller islands to larger ones, where that is possible, as in several cases in Solomon Islands, Vanuatu and Fiji (Ha'apio et al., 2019; Human Rights Watch, 2025; McMichael, Katonivualiku, & Powell, 2019; McMichael, Powell, Piggott-McKellar, & Yee, 2025; Piggott-McKellar, Pearson, McNamara, & Nunn, 2020; Ruehr, 2022; Yee et al., 2022; Yee, Piggott-McKellar, McMichael, & McNamara, 2025) and from Mango (Tonga), emphasising increasing centralisation of population and the decline of outer islands, but that may also be a function of political issues, evident in PNG and Barbuda.

On several islands in Fiji, shoreline recession and erosion have displaced villages from established sites (Charan, Kaur, & Singh, 2018; McMichael et al., 2019; Neef et al., 2018). Migration from the atolls of Bougainville (PNG) and small islands off Choiseul (Solomon Islands) has occurred in advance of substantial climate change, in the former case because of high population pressure on scarce land areas and diminished livelihoods (Connell, 2016), and in the latter through tectonic changes (Albert et al., 2018). Similar tectonic changes have resulted in migration away from small islands in Buka (Bougainville), the Torres Islands (Vanuatu) and Ranongga (Solomon Islands) (Connell, 2015). Where island populations were quite small, adaptation in place is particularly difficult so that smaller islands, such as those off Choiseul and Buka, and Ragged Island, are most likely to experience migration and resettlement.

Resettlement has proved difficult even where people move over short distances and within the same language group. Past resettlement schemes provide unpromising precedents. Resettlement programmes have existed at least since the 1900s when people from some FSM atolls were formally relocated on Pohnpei after cyclones. In the 1930s people from Kiribati were resettled in Solomon Islands and Fiji; these international movements, of people experiencing drought and reduced livelihoods, occurred within and between British possessions in colonial times. While the immediate post-war resettlement of Banabans in Fiji was relatively successful, since an 'empty' island without local claimants was available and at the time involved no more than about 1000 people, uncertainty still reigned over land tenure 70 years later (Connell & Tabucanon, 2016). Otherwise, most resettlement after environmental hazards, mainly volcanic eruptions and tsunamis (and forced migration from military and mine sites), has been problematic with migrants seeking to return home, land disputes intensifying, friction occurring with nearby people and challenges in adapting to different ecologies and cultures. In the Caribbean, Montserrat islanders found it easier to go to the UK than establish new lives in the north of the island or in nearby Antigua (Griffen & Robinson, 2023). Resettlement of Manam and Carteret Islanders in PNG was thwarted by 'host' landowners, the impossibility of gaining adequate access to land and land rights, government inactivity, even after decades of attempts, and, in the case of Manam, urgent needs: 'social relationships are written on the ground and editing or removing the writing is almost impossible, establishing a geometry of power that absolutely marginalises potential settlers' (Connell & Lutkehaus, 2017: 92). Ironically, without access to land, or finance, Carteret Islanders moving from their intermittently flooded atoll, mainly live in the most flood-prone part of urban Buka, their main destination.

Similar disputes and conflicts have occurred in Solomon Islands, Fiji and elsewhere. Such relative failures in even short-distance resettlement raised questions of national identity, citizenship, ethics and morality. Landowners have been increasingly reluctant to cede land to others, however moral and worthy their claims, even when they shared kinship ties or exchange relationships, especially as their own populations and needs had grown, and land become more valuable. Neighbours are not necessarily friends. Settlers have been perceived as outsiders and rival claimants to valuable coastal resources, yet it is valuable for familiar livelihoods to be negotiated. Without some significant political and cultural changes, there is little likelihood of effective resettlement within the Pacific region, but international moves tend to be costly and risky, and require advance planning, political acquiescence and acceptability. The problems experienced by quite small populations moving short distances in similar cultural contexts are indicative of the potential future problems facing environmental (or other) migrants. Although Kiribati has purchased a small island in Fiji, for possible resettlement, the 'welcome' that might be attached to migration of i-Kiribati is likely to be ambiguous. Multilevel and integrated governance is necessary but usually absent for acquiring tenure to land and for planning, enabling, establishing and supporting new settlements (Connell, 2011; Giardino, Nederhoff, & Vousdoukas, 2018). This is likely to be essential in future, but more difficult where power inequalities have worsened, the agency of the most affected has declined and where urban areas are often more prone to flash flooding during extreme weather events.

Response to flood events, notably the Mataniko river flood in Honiara, Solomon Islands in 2014 and the flooding of Ba town (Fiji) in 2018, reflects the problems and costs of resettlement, including land, infrastructure, and credit; lack of formal education and skills; and limited flexible livelihood options (Ha'apio et al., 2019). Nadi, the second city of Fiji, on the edge of a coastal delta, is flood-prone—the outcome of combined delta subsidence and sea-level rise, in the 2020s about 4 mm/year, accentuated by cyclones—yet as tourism development has expanded nearby it 'seems clear that the nature and pace of coastal development is incompatible with the accelerating rate of climate change' (Nunn, Smith, & Elrick-Barr, 2021: 6). Its increasingly frequent flooding exemplified the problems emanating from coastal squeeze in combination with climate change, to increase the risk of environmental hazards, water contamination and reduced agricultural and fisheries productivity (Bennett, Talouli, & Dixit, 2024; Varea, Varea, Kant, & Farrelly, 2022). Typically, as in the 2009 floods around Nadi, poorer farmers with smaller landholdings and houses on lower ground were the most affected (Lal, Kinch, & Wickham, 2009). In a single inland village in Fiji, households with access to tangible assets such as land, equipment and resources to utilise it, along with intangible assets such as motivation and knowledge, were better positioned to adapt to extreme events such as cyclones and flooding (Currenti et al., 2019). Those better positioned have capacity, flexibility and diverse strategies not available to all.

A 'cognitive map' developed from Nadi residents' diverse perceptions of these processes and the relationships between them, as they changed over time, demonstrated the complexity of human and physical influences on flooding and therefore the tasks of integrated management and planning (Chandra & Gaganis, 2016). Adaptive measures intended to mitigate and minimise flooding, such as clearing drainage channels and rebuilding infrastructure, were limited to improving practices, being minimalist, low-cost and incremental while mangrove clearances continued. Such incremental adaptation takes the place of necessary transformational shifts (such as the relocation of Palau's capital from Koror to Babeldaob in 2006), a function of limited finance and scientific knowledge: an intensifying path dependency (Nunn, Klöck, & Duvat, 2021). Meanwhile floods have become increasingly costly and are expected to become more so (Brown, Daigneault, & Gawith, 2017), and mitigation is more difficult as urban populations grow and hard infrastructure extends further, with the urban poor most at risk and resettlement more crucial and less probable.

Communal and household adaptive strategies can differ widely, even in localised cultural contexts, due to diverse individual and household assessments of livelihood opportunities and climate-related (and other) risks. All households have access to social capital through extended family networks, which lessens the burden of hazards. Households with access to tangible assets such as land and the equipment and resources to use it, together with intangible assets like motivation and knowledge, are better positioned to respond and adapt. Decisions to relocate from 'risky environments' are influenced by combinations of past experience, local power relations, attachment to social space and the provision of external assistance (Neef et al., 2018). That is complicated by changes and limitations to governance, where 'the combined interactions exceed the current knowledge, resources, traditional coping mechanisms and institutional capacity of stakeholders to plan for adaptation measures' (Chandra & Gaganis, 2016: 265), even were such stakeholders to have choices and options and be involved in a consultative planning process, and where some policies and practices are necessarily at a level and scale where local inputs are slight. Vulnerability may be place-specific and contextual, but it is influenced by variables at multiple scales.

Hazards have gender implications. People experience hazards and consequent migration processes according to gender-specific inequalities, vulnerabilities, and access to resources and opportunities. The impact of floods and other hazards is greatest for those living in marginal environments, such as riverbanks, who tend to be relatively poor, with poor access to services, and who become 'environmentally crippled' where their livelihoods also disappear, while burdens on women are greater where their household chores and livelihood activities increase (Chandra & Gaganis, 2016). Hurricane Dorian affected women more than men, with LGBTI people and migrant Haitian women and girls even more disadvantaged in the face of multiple forms of discrimination and a heightened risk of gender-based violence. Emergency shelters lacked gender-responsive planning and management and Haitian migrants with irregular status avoided government-run shelters. Tropical Storm Erika and Hurricane Maria in Dominica most affected indigenous Kalinago Carib women and girls. Many Kalinago women, previously employed in the agriculture and tourism sectors, were unable to earn sufficient income to recover economically from these disaster events. Erika resulted in higher displacement rates and longer stays in emergency shelters for female-headed households. Incidents of family separation and gender-based violence, along with inadequate privacy and security measures at these shelters, were reported following both Erika and Maria. Few residents returned home within a year, and some were displaced as far as Guadeloupe (Bleeker, Escribano, Gonzales, et al., 2021; Cooper et al., 2025). Disparities in land ownership and access to financial services increase the economic insecurity of women involved in agriculture and their vulnerability to climate change. That could be worsened in single-parent households and by rising economic insecurity, degradation of rural livelihoods, and where men are employed overseas.

Ironically, migrants are also often most affected by environmental hazards. In Solomon Islands migrants on flood-prone land in Honiara and, in Fiji, migrants to the Ba flood plain and the Rewa delta have been most affected by floods. In Solomon Islands, migrants from Kiribati, resettled on coastal land, were most affected by a tsunami, being unfamiliar with such an event. On the island of Abaco (The Bahamas), many illegal migrants living in marginal settlements (The Mudd and Pigeon Peas—the former whose name came from its propensity to flood in heavy rain) failed to move before the arrival of Hurricane Dorian, since they had no kin or financial support elsewhere in the country, and so became the main victims of the hurricane and of subsequent structural violence and spatial injustice (Benett, 2023; Cooper, 2020). In Fiji, rural Indo-Fijians responded differently from indigenous Fijians to the consequences of Tropical Cyclone Winston in 2016, owing to relatively limited access to traditional knowledge of climate responses, lack of communal labour sharing, and weaker intra- and inter-community networks, pointing to the necessity to implement more inclusive disaster risk reduction policies and practices (Nakamura & Kanemasu, 2022). Such differentiated impacts and responses emphasised how the impact of hazards is not arbitrary but shaped by colonial histories that left geographical and structural inequalities. In turn, these contributed to land tenure issues, and pressure on coastal land, that make the future of displacement and resettlement problematic.

Few welcome resettlement, much preferring local adaptation. When the Marshall Islands rejected resettlement in 2013, its lead climate negotiator (an elegant example of a new moral terminology) Tony de Brum commented that 'the prospect of planned displacement of communities and peoples is repugnant … an admittance of defeat before we are actually defeated' but conceded that 'it is prudent for leaders to keep it in mind as a matter of survival' (quoted in Stege, 2018: 381). He had also stated:

> People do not want to be detached from their homelands. It's their soul. If you were to take the Marshallese community as it is now and say, we're going to move you some place else, that's the end of a people and a culture and a tradition … destroying the soul of a society. (quoted in Jarillo & Barnett, 2022: 854)

At the same time, Marshallese were migrating to the United States at unprecedented rates to the extent that the national population was declining faster than in any other SIDS. If nothing else, people wanted the ability to shape their own lives, livelihoods and futures.

2.13 Hazards and Hopes

Both groups of SIDS have experienced a long history of hazards—some dramatic both in the past and present—but the context has shifted from historic 'natural' hazards to the recognition that, in this Anthropocene era, human impacts are an increasingly significant influence so that discussing and mitigating hazards have become increasingly political and multi-scalar as much as social and physical. That has become less easy, and more costly for health, food security and reconstruction, with cyclones and other hazards more damaging. Choices in response have become greater—but never unconstrained—from restoration to migration, to an adjacent hillside or a metropolitan city. Thus far relocation has occurred only in exceptional circumstances, often quite localised, but more dramatic change is envisaged. As Hilda Heine, the President of the Marshall Islands, pointed out at COP25 in 2019:

> The most vulnerable – atoll nations like my country – already face death row. Water covers much of our land at one or other point of the year as we fight rising tides … It's a fight to the death for anyone not prepared to flee. As a nation we refuse to flee. But we also refuse to die … Climate is the most serious human rights issue we have ever faced. (quoted in Rikimani, 2024: 374–5)

Climate change has provoked an existential crisis, especially in the atoll states and on smaller islands where migration issues are already daunting: whether, why, when and where to go? Yet climate change is not the primary factor in migration.

With sustained hazards and SIDS able to exert their authority and agency, new organisations have emerged to press the case for adequate loss and damage payments. SIDS have limited capacity to mitigate and respond to climate change themselves, but since 1988, they have been at the vanguard of climate action. Malta initiated the first United Nations recognition of climate change as a common concern of humankind, followed by the adoption of the Male [Maldives] Declaration on Global Warming and Sea Level Rise in 1989. By then Pacific island leaders had declared global warming and sea-level rise to be serious threats and taken a lead 'from the frontline' in shaping climate change diplomacy (Morgan, Carter, & Manoa, 2024). That has taken several forms, primarily through formal diplomatic representation at global and regional forums, and through more dramatic representations of vulnerability, again at different scales.

While floods, erosion, resettlement and lost biodiversity in several SIDS suggest major problems, SIDS have managed to establish a common front in the vital struggle against climate change and for adequate loss and damage restitution: an obvious, necessary and growing refusal to be passive bystanders. Working through AOSIS over decades, and fighting in various international contexts as far as the International Court of Justice, has demonstrated the ability of SIDS individually and collectively to achieve positive changes, increasingly by focusing on the role of equity and environmental justice and by developing and collaborating on physical changes. SIDS pressure brought the Green Climate Fund, a critical element of the historic Paris Agreement. The Bridgetown Initiative (named after the Barbados capital) provided an action plan to reform the global financial system so that the world could better respond to crises. The Marshall Islands, having played a key role in negotiating the Paris Agreement, instigated the High Ambition Coalition (Corbett, 2015). From its start in 2021, the initial nine SIDS in COSIS made history by initiating and litigating the first ever inter-state advisory proceeding on climate change and international law, combining the existential threat to their survival, with irrefutable scientific evidence and legal arguments. Ultimately, for all their disappointments with COPs, the SIDS achieved a considerable degree of success over a quarter of a century by working together and making a strong moral case, despite their small delegations being at a negotiating disadvantage but offset by the 'competent performance of vulnerability' (Benwell, 2011; Corbett & Connell, 2015; Corbett, Xu, & Weller, 2019; Panke, 2012; Sobion, 2025). A concentrated vision and focus on the existential threat of climate change proved advantageous.

SIDS also acted much less formally, stressing both resilience and external support. In Kiribati, for example, elites, like other officials elsewhere, were somewhat perversely required to enact the travails of climate change, what Sophie Webber (2013) has described as 'performative vulnerability', drawing on images of sinking islands and climate refugees, literally performing the problems in order to secure adaptation finance. Mike Goldsmith (2015) has argued that the Tuvalu government deliberately promoted the nation as an icon of vulnerability to increase aid and even increase 'last-chance tourism'—'see it before it disappears'—a variant on 'wishful sinking' (Farbotko, 2010a, 2010b). That was emphasised when, prior to COP26, Tuvalu's Minister of Foreign Affairs, Simon Kofe, famously stood knee-deep in Tuvalu's lagoon, dressed in business suit and tie, to proclaim 'We are sinking'. Such vivid encounters, 'saturated with power' (Webber, 2013: 2722) and manifestations of smallness at least temporarily silenced 'alternative identities that emphasise more empowering qualities of resilience and resourcefulness' (Farbotko, 2005: 289). The concept of vulnerability, never easily defined, became a weapon to be deployed strategically by those at risk, by the 'development industry', and by NGOs (and academic researchers) in their quest and need for relevance. A 'sinking islands' narrative could be attributed to opportunism from government agencies, donors, NGOs and local people, in what has been described as 'climate change bandwagoning' where people strategically 'play along' to obtain support, yet in so doing foster aid-dependency, clientelism and more generally a hand-out mentality (Cox, 2009; Foale, 2001; Jinnah, 2011) that 'represents a strategic leveraging of resources' (Ruehr, 2022: 173) or even a 'strategy of ignoring scientific knowledge for political gain' (Lauer et al., 2022: 67). As the *New York Times* proclaimed after COP27, 'Tiny Vanuatu Uses its "Unimportance" to Launch Big Climate Ideas' (Sengupta, 2022). Such strategies became a 'microstate craft' where small nations like Tuvalu leveraged their international position as independent, yet threatened, states (Yarina & Takemoto, 2017). Supplication for funds however risked inverse sovereignty and some loss of control and authority over aid direction and expenditure, and its concentration purely on climate finance (Chap. 8). What amounts to a form of environmental determinism has suited the needs of local people, national governments, NGOs, journalists (and researchers) in search of an issue and a story: a neat conjuncture of media interest, alongside local, national and international interest.

Such national activism (and inadequate response) was accompanied, notably in the Pacific, by accelerated regional activism. The Pacific Climate Warriors, for example, used the slogan, 'not drowning but fighting' to oppose the narrative of the 'sinking island states' to emphasise and enhance resilience. The Warriors challenged the belittlement of the Pacific Islands region to present a vision of Oceanic power whose activism, amongst other things, involved blockading the world's largest coal port in Newcastle (Australia) to present a vision of Oceanic power (Steiner, 2015; McNamara & Farbotko, 2017; Fair, 2020; Farbotko & Kitara, 2021). That brought more women to the forefront, and a growing focus on equity and environmental justice, centred on homelands rather than on migration (Mcleod et al., 2019). The previous absence of women 'as female experts on climate change issues' (Chandra & Gaganis, 2016: 266) had resulted in technical approaches and responses rather than broader social engagement, whilst politicisation stressed that islands were not sinking, but were being sunk, as increasingly argued in global forums and law courts. Vulnerability to climate change was not just an accidental product of sea-level rise but of a colonial and global history of harmful extractive industries and practices that produced fossil fuels and undermined local adaptation practices and knowledge systems.

Climate change brought both the need for effective environmental management, that in places resulted in a revival of older knowledge, and for economic transition to more heat-resistant plants and work practices. Further adaptations and partnerships between communities and governments, and initiatives that promote Indigenous knowledge alongside scientific expertise are invaluable for relevant nature-based solutions. So too management; even in larger SIDS such as Fiji, 'regulations on urban planning, land use and EIA are weakly implemented by the government. These laws require enforcement and review, particularly in the land use, agriculture, infrastructure and construction sectors' (Chandra & Gaganis, 2016: 266; Scobie, 2016), hitherto resulting in technical approaches and responses rather than broader social and environmental engagement.

Too often combatting climate change has been a lethargic and regressive process. Two countries with climate sceptics as presidents, the United States and Argentina, left the Paris Agreement in 2025 (with President Trump encouraging 'drill baby drill'), at odds with climate science, damage and risks, even in their own countries. Regression occurred in the same week that scientists declared 2024 the hottest year on record. Scarcely surprising then that it was strongly criticised in the SIDS. Vanuatu's Attorney-General, Arnold Loughman, said the US withdrawal was an 'undeniable setback' and a 'troubling precedent' since 'the Paris Agreement remains key to the world's efforts to combat climate change and respond to its effects, and the participation of major economies like the US is crucial. Ultimately, the whole world stands to lose if the international legal framework is allowed to erode': a troubling potential precedent given many countries' continued reliance on fossil fuels. In the short term, the SIDS can expect no real changes to the environmental threats they face. Climate change poses security risks, through social responses to changing environmental conditions or in anticipation of them. While local participation and local solutions are crucial, global responses are even more crucial as losses and damage mount, without adequate reparations. At every scale action is required to manage environmental degradation, reduce structural inequities and shape evolving island economies. The effects of climate change on agriculture and fisheries are likely to drive up the reliance on imported foods of low nutritional quality, leading to a rising prevalence of obesity, and economic and social issues that are the subject of the following chapters. Meanwhile deleterious environmental degradation also stems from direct practices and policies detached from global warming. As we show in the next chapter the SIDS offer multiple examples of diverse experiences, responses, adaptations and expectations.

Bibliography

Akhavan, P., & Holst, R. (2024). *What are the legally binding obligations of states in respect of climate change?* IISD SDG Knowledge Hub.

Albert, S., Leon, J., Grinham, A., et al. (2016). Interactions between sea-level rise and wave exposure on reef island dynamics in the Solomon Islands. *Environmental Research Letters, 11*(5), 054011.

Albert, S., Bronen, R., Tooler, N., et al. (2018). Heading for the hills: Climate-driven community relocations in the Solomon Islands and Alaska provide insight for a 1.5°C future. *Regional Environmental Change, 18*(8), 2261–2272.

Alofa, P., & Costello, C. (2025). Reshaping reality: Climate change's impact on Pasifika girls' lives. *ANU DevPolicyBlog*, 20 February.

Alston, M., Fuller, S., & Kwarney, N. (2025). Women and climate change in Vanuatu, Pacific Islands Region. *Gender, Place & Culture, 32*(1), 83–104.

Andrew, N., Bright, P., & de la Rua, L., et al. (2019). Coastal proximity of populations in 22 Pacific Island Countries and Territories. *PLoS One, 14*(9), e0223249.

Atteridge, A., & Canales, N. (2017). *Climate finance in the Pacific: An overview of flows to the region's small island developing states.* Stockholm Environmental Institute Working Paper 2017-04.

Attzs, M. (2009). Natural disasters and remittances: Poverty, gender and disaster vulnerability in Caribbean SIDS. In W. Naudé, A. Santos-Paulino, & M. McGillivray (Eds.), *Vulnerability in developing countries* (pp. 249–261). United Nations University Press.

Babinard, J., Bennett, C., Hatziolos, M., et al. (2014). Sustainably managing natural resources and the need for construction materials in Pacific island countries: The example of South Tarawa, Kiribati. *Natural Resources Forum, 38*(1), 58–66.

Bach, J. (2017). *Perceptions of environmental change: Nikutoru, Tabiteuea Maiaki, Kiribati.* PhD thesis, University of Montana.

Badjeck, M., Allison, E., Hall, A., & Dulvy, N. (2010). Impacts of climate variability and change on fishery-based livelihoods. *Marine Policy, 34*, 375–383.

Ballard, C., McDonnell, S., & Calandra, M. (2020). Confronting the naturalness of disaster in the Pacific. *Anthropological Forum, 30*(1–2), 1–14.

Banivanua Mar, T. (2019). *Decolonisation and the Pacific: Indigenous globalisation and the ends of empire*. Cambridge University Press.

Barbara, J., Mudaliar, C., Leach, M., et al. (2023). Popular attitudes to climate change in the Pacific: The case of Samoa. *Regional Environmental Change, 23*, 81.

Bardouille, P. (2024, July 31). Disaster response isn't keeping up with the climate crisis. *Project Syndicate*.

Barnett, J. (2017). The dilemmas of normalising losses from climate change: Towards hope for Pacific atoll countries. *Asia Pacific Viewpoint, 58*(1), 3–13.

Barnett, J., Jarillo, S., Swearer, S., Lovelock, C., et al. (2022). Nature-based solutions for atoll habitability. *Philosophical Transactions of the Royal Society B, 377*, 20210124.

Basel, B., Goby, G., & Johnson, J. (2020). Community-based adaptation to climate change in villages of Western Province. *Solomon Islands. Marine Pollution Bulletin, 156*, 111266.

Bell-James, J., & Watson, J. (2025). Ambitions in national plans do not yet match bold international protection and restoration commitments. *Nature Ecology and Evolution, 9*, 417–424.

Benett, I. (2023). Space unmade: Hurricane Dorian, autoethnography and dislocation. *Caribbean Studies, 51*(1), 39–75.

Bennett, N., Alava, A., Ferguson, C., Blythe, J., et al. (2023). Environmental (in)justice in the Anthropocene ocean. *Marine Policy, 147*, 105383.

Bennett, S., Talouli, A., & Dixit, V. (2024). Towards a circular logistics partnership: Regional trade, waste, and inter-sectoral cooperation. In A. Dansie, H. Alleway, & B. Böer (Eds.), *The water, energy, and food security nexus in Asia and the Pacific: The Pacific* (pp. 197–199). Springer.

Benwell, R. (2011). The Canaries in the Coalmine: Small states as climate change champions. *The Round Table, 100*(413), 199–211.

Beyerl, K., Mieg, H., & Weber, E. (2019). Comparing perceptions of climate-related environmental changes for Tuvalu, Samoa and Tonga. In C. Klöck & M. Fink (Eds.), *Dealing with climate change on Small Islands: Towards effective and sustainable adaptation?* (pp. 143–174). Göttingen University Press.

Bird, Z., Wairiu, M., des Combes, H., & Iese, V. (2021). Religious and cultural-spiritual attributions of climate-driven changes on food production: A case study from North Malaita, Solomon Islands. In J. Luetz & P. Nunn (Eds.), *Beyond belief: Opportunities for faith-engaged approaches to climate-change adaptation in the Pacific Islands* (pp. 39–56). Springer Nature.

Bird, Z., Iese, V., des Combes, H., et al. (2022). Assessing the impacts of climate change on domestic crop production: Experience and perceptions of local farmers in North Malaita, Solomon Islands. *Pacific Dynamics, 6*(1), 77–98.

Blair, A., & Momtaz, S. (2018). Climate change perception and response: Case studies of fishers from Antigua and Efate. *Ocean and Coastal Management, 157*, 86–94.

Bleeker, A., Escribano, P., Gonzales, C., et al. (2021). *Advancing gender equality in environmental migration and disaster displacement in the Caribbean.* ECLAC and IOM.

Bordner, A., Ferguson, C., & Ortolano, L. (2020). Colonial dynamics limit climate adaptation in Oceania: Perspectives from the Marshall Islands. *Global Environmental Change, 61*, 10205.

Bove, C., Mudge, L., & Bruno, J. (2022). A century of warming on Caribbean reefs. *PLOS Climate, 1*(3), e0000002.

Bratrud, T. (2024). Climate change in the Pacific and the question of relocation. *Human Organization, 83*(3), 209–215.

Bridges, K., & McClatchey, W. (2014). Living on the margin: Ethnoecological insights from Marshall Islanders at Rongelap Atoll. *Global Environmental Change, 19*(2), 140–146.

Brooks, N., & Adger, N. (2004). Assessing and enhancing adaptive capacity. In B. Lim & E. Spanger-Siegfried (Eds.), *Adaptation policy framework for climate change* (pp. 165–182). Cambridge University Press.

Brown, P., Daigneault, A., & Gawith, D. (2017). Climate change and the economic impacts of flooding on Fiji. *Climate and Development, 9*(6), 493–504.

Caldeira, M., Sekinairai, A., & Vierros, M. (2025). Weaving science and traditional knowledge: Toward sustainable solutions for ocean management. *Marine Policy, 174*, 106591.

Calliari, E., & Vanhala, L. (2022). The 'national turn' in climate change loss and damage governance research: Constructing the L&D policy landscape in Tuvalu. *Climate Policy, 22*(2), 184–197.

Calliari, E., Serdeczny, O., & Vanhala, L. (2020). Making sense of the politics in the climate change loss and damage debate. *Global Environmental Change, 64*, 102133.

CARE. (2022). *That's not new money.* CARE Denmark.

Cassinat, J., Cassinat, K., Segi, T., et al. (2022). Traditional village roles and gender shape Samoan perceptions of climate change. *Current Research in Environmental Sustainability, 4*, 100173.

Cavallo, E., & Noy, I. (2009). *The economics of natural disasters: A survey*. The Inter-American Development Bank Working Paper No. 124.

Chandra, A., & Gaganis, P. (2016). Deconstructing vulnerability and adaptation in a coastal river basin ecosystem: A participatory analysis of flood risk in Nadi, Fiji Islands. *Climate and Development, 8*(3), 256–269.

Chandra, A., McNamara, K., Clissold, R., et al. (2023). Climate-induced non-economic loss and damage: Understanding policy responses, challenges and future directions in Pacific small island developing states. *Climate, 11*, 74.

Charan, D., Kaur, M., & Singh, P. (2018). Customary land and climate change induced relocation - A case study of Vunidogoloa Village, Vanua Levu, Fiji. In W. Filho & J. Nalau (Eds.), *Limits to climate change adaptation* (pp. 345–358). Springer.

Cinner, J., Adger, N., Allison, E., et al. (2018). Building adaptive capacity to climate change in tropical coastal communities. *Nature Climate Change, 8*(2), 117–123.

Ciplet, D., Falzon, D., Uri, I., et al. (2022). The unequal geographies of climate finance: Climate injustice and dependency in the world system. *Political Geography, 99*, 102769.

Cloos, P., Belloiseau, M., McPherson, N., et al. (2023). Discussing linkages between climate change, human mobility and health in the Caribbean: The case of Dominica. A qualitative study. *The Journal of Climate Change and Health, 11*, 100237.

Conley, J. (2024, September 10). Pacific Islands Push ICC to Recognise Ecocide as International Crime. *Common Dreams*.

Connell, J. (2003). Losing ground? Tuvalu, the greenhouse effect and the garbage can. *Asia Pacific Viewpoint, 44*(2), 89–107.

Connell, J. (2011). Elephants in the Pacific? Pacific urbanisation and its discontents. *Asia Pacific Viewpoint, 52*(2), 121–135.

Connell, J. (2012). Population resettlement in the Pacific: Lessons from a hazardous history. *Australian Geographer, 43*(2), 127–142.

Connell, J. (2013a). *Islands at risk: Environments, economies and contemporary change*. Edward Elgar.

Connell, J. (2013b). Ants, rats, snakes and taro blight: Contemporary invasions in the Pacific Islands. In S. Larrue (Ed.), *Biodiversity and societies in the Pacific Islands* (pp. 51–70). PUP.

Connell, J. (2015). Vulnerable islands: Climate change, tectonic change, and changing livelihoods in the Western Pacific. *The Contemporary Pacific, 27*(1), 1–36.

Connell, J. (2016). Last days in the Carteret Islands? Climate change, livelihoods and migration on coral atolls. *Asia Pacific Viewpoint, 57*(1), 3–15.

Connell, J. (2018). Nothing there Atoll? "Farewell to the Carteret Islands". In T. Crook & P. Rudiak-Gould (Eds.), *Pacific climate cultures* (pp. 73–87). DeGruyter.

Connell, J. (2023). *Climate change and health: A security challenge in the Pacific Islands*. Toda Peace Institute Policy Brief No. 166.

Connell, J., & Lutkehaus, N. (2017). Escaping Zaria's fire? The volcano resettlement problem of Manam Island, Papua New Guinea. *Asia Pacific Viewpoint, 58*(1), 14–26.

Connell, J., & Tabucanon, G. (2016). From Banaba to Rabi: a Pacific model for resettlement? In S. Price & J. Singer (Eds.), *Global Implications of Development, Disasters and Climate Change* (pp. 142–158). Routledge.

Connell, J., & Taulealo, T. (2021). Island tourism and COVID-19 in Vanuatu and Samoa: An unfolding crisis. *Small States and Territories, 4*(1), 105–124.

Cooper, T. (2020). Hurricane Dorian: A case for building comprehensive climate change resilience frameworks for small island developing states in the Caribbean. In S. Rolle, J. Minnis, & I. Bethell-Bennett (Eds.), *Tourism development, governance and sustainability in the Bahamas* (pp. 201–215). Routledge.

Cooper, S., Cloos, P., Abraham, C., et al. (2025). What can be said about risks, vulnerabilities, and adaptation to climate change in Caribbean small island developing states (SIDS)? The case of Dominica. A qualitative study. *PLOS Climate, 4*(4), e0000275.

Corbett, J. (2015). "Everybody knows everybody": Practising politics in the Pacific Islands. *Democratization, 22*(1), 51–72.

Corbett, J., & Connell, J. (2015). All the world is a stage: Global governance, human resources and the 'problem' of smallness. *The Pacific Review, 28*(3), 435–459.

Corbett, J., Xu, Y., & Weller, P. (2019). Norm entrepreneurship and diffusion 'from below' in international organisations: How the competent performance of vulnerability generates benefits for small states. *Review of International Studies, 45*(4), 647–668.

Cox, J. (2009). Active citizenship or passive clientelism? Accountability and development in Solomon Islands. *Development in Practice, 19*(8), 964–980.

Cox, J., Finau, G., Kant, R., et al. (2018). Disaster, divine judgement, and original sin: Christian interpretations of tropical cyclone Winston and climate change in Fiji. *The Contemporary Pacific, 30*(2), 380–410.

Cross, I., & Congreve, A. (2020). Teaching (Super) wicked problems: Authentic learning about climate change. *Journal of Geography in Higher Education, 45*(4), 491–516.

Currenti, R., Pearce, T., Salabogi, T., Vuli, L., et al. (2019). Adaptation to climate change in an interior Pacific Island village: A case study of Nawairuku, Ra, Fiji. *Human Ecology, 47*, 65–80.

Dean, A., Green, D., & Nunn, P. (2017). Too much sail for a small craft? Donor requirements, scale and capacity discourses in Kiribati. In E. Stratford (Ed.), *Island geographies: Essays and conversations* (pp. 54–77). Routledge.

Donner, S., & Webber, S. (2014). Obstacles to climate change adaptation decisions: A case-study of sea-level rise and coastal protection measures in Kiribati. *Sustainability Science, 9*(3), 331–345.

Dulal, H., Shah, K., & Ahmad, N. (2009). Social equity considerations in the implementation of Caribbean climate change adaptation policies. *Sustainability, 1*, 363–383.

Dumaru, P. (2010). Community-based adaptation: Enhancing community adaptive capacity in Drua`drua Island, Fiji. *WIRES Climate Change, 1*, 751–763.

Duvat, V. (2013). Coastal protection structures in Tarawa Atoll, Republic of Kiribati. *Sustainability Science, 8*(3), 363–379.

Edmonds, C., & Noy, I. (2018). The economics of disaster risks and impacts in the Pacific. *Disaster Prevention and Management, 27*(5), 478–494.

ESCAP. (2024). *People and planet: Addressing the interlinked challenges of climate change, poverty and hunger in Asia and the Pacific*. United Nations.

Fache, E., & Fair, H. (2020). Turning away from wicked ways: Christian climate change politics in the Pacific Island Region. *Anthropological Forum, 30*(3), 233–253.

Fair, H. (2020). Their sea of islands? Pacific climate warriors, oceanic identities, and world enlargement. *The Contemporary Pacific, 32*(2), 341–369.

Farbotko, C. (2005). Tuvalu and climate change: Constructions of environmental displacements in the Sydney Morning Herald. *Geografiska Annaler B, 7*(4), 279–293.

Farbotko, C. (2010a). 'The global warming clock is ticking so see these places while you can': Voyeuristic Tourism and Model Environmental Citizens on Tuvalu's Disappearing Islands. *Singapore Journal of Tropical Geography, 31*, 224–238.

Farbotko, C. (2010b). Wishful sinking: Disappearing islands, climate refugees and cosmopolitan experimentation. *Asia Pacific Viewpoint, 51*(1), 47–60.

Farbotko, C., & Campbell, J. (2022). Who defines atoll 'uninhabitability'? *Environmental Science and Policy, 138*, 182–190.

Farbotko, C., & Kitara, K. (2021). The political ecology and visual culture of the pacific climate warriors. In T. Demos, E. Scott, & S. Banerjee (Eds.), *The Routledge companion to contemporary art, visual culture, and climate change* (pp. 399–408). Routledge.

Fitz-Henry, E. (2022, July 12). Climate change is white colonisation of the atmosphere. It's time to tackle this entrenched racism. *The Conversation.*

Floerl, O., Atalah, J., Bugnot, A., et al. (2021). A global model to forecast coastal hardening and mitigate associated socioecological risks. *Nature Sustainability, 4*, 1060–1067.

Flores-Palacios, X. (2019). Climate-induced migration in Lotofaga village in Samoa. In C. Klöck & M. Fink (Eds.), *Dealing with climate change on small islands: Towards effective and sustainable adaptation?* (pp. 313–337). Göttingen University Press.

Foale, S. (2001). 'Where's our Development?' Landowner aspirations and environmentalist agendas in Western Solomon Islands. *Asia Pacific Journal of Anthropology, 2*, 44–67.

Fonua, F. (2023, January 12). Tonga volcano eruption: Mango and 'Atata Islanders resettle nearly a year on from disaster. *Radio New Zealand.*

Gagaeolo, F., Hemstock, S., & Price, C. (2020). Christianity, Kiribati and climate change-induced migration. In F. Gagaeolo, S. Hemstock, & C. Price (Eds.), *Science, faith and the climate crisis* (pp. 1–14). Emerald.

Gahman, L., & Thongs, G. (2020). Development justice, a proposal: Reckoning with disaster, catastrophe, and climate change in the Caribbean. *Transactions of the Institute of British Geographers, 45*(4), 763–778.

Gahman, L., Thongs, G., & Greenidge, A. (2021). Disaster, debt, and 'underdevelopment': The cunning of Colonial-Capitalism in the Caribbean. *Development, 64*(1–2), 112–118.

Galaitsi, S., Corbin, C., Cox, C., et al. (2024). Balancing climate resilience and adaptation for Caribbean small island developing states (SIDS): Building institutional capacity. *Integrated Environmental Assessment and Management, 20*(5), 1237–1255.

George, E. (2024, May 29). Barbuda's unending disaster. *Caribbean Investigative Journalism Network.*

Giardino, A., Nederhoff, K., & Vousdoukas, M. (2018). Coastal hazard risk assessment for small islands: Assessing the impact of climate change and disaster risk reduction measures on Ebeye (Marshall Islands). *Regional Environmental Change, 18*, 2237–2248.

Gibson, K., Barnett, J., Haslam, N., & Kaplan, I. (2020). The mental health impacts of climate change: Findings from a Pacific Island atoll nation. *Journal of Anxiety Disorders, 73*, 102237.

Giglio, S., Kelly, B., & Stroebel, J. (2021). Climate finance. *Annual Review of Financial Economics, 13*, 15–36.

Gilbert, N. (2022, April 28). Climate change will force new animal encounters - and boost viral outbreaks. *Nature.*

Goldsmith, M. (2015). The big smallness of Tuvalu. *Global Environment, 8*(1), 134–151.

Gordon-Strachan, G. P., S, Harewood, H., et al. (2025). The 2024 small island developing states report of the *Lancet* Countdown on health and climate change. *The Lancet Global Health, 13*(1), e146–e166.

Granderson, A. (2017). The role of traditional knowledge in building adaptive capacity for climate change: Perspectives from Vanuatu. *Weather Climate and Society, 9*(3), 545–561.

Granderson, A. (2018). Value conflicts and the politics of risk: Challenges in assessing climate change impacts and risk priorities in rural Vanuatu. *Climate and Development, 10*(6), 481–494.

Griffen, S., & Robinson, S. (2023). (Un)just post-disaster mobilities in small island developing states: Revisiting the patterns and outcomes of three major environmental disasters in the Caribbean. *International Journal of Disaster Risk Reduction, 97*, 104029.

Ha'apio, M., Gonzalez, R., & Wairiu, M. (2019). Is there any chance for the poor to cope with extreme environmental events? Two case studies in the Solomon Islands. *World Development, 122*, 514–524.

Hallegatte, S., Bangalore, M., Bonzanigo, L., et al. (2016). *Shock waves: Managing the impacts of climate change on poverty.* World Bank.

Handmer, J., Monson, R., & Schinko, T. (2025). Addressing the diversity of loss and damage in Pacific Island countries to foster a just transition towards a climate-resilient future. *Climate and Development, 17*(7), 657–669.

Heffernan, O. (2024). *The high seas.* Greystone.

Hein, J., Klepp, S., & Bercht, A. (2024). Bringing a justice lens to ocean access. *Maritime Studies, 23*, 45.

Hidalgo, D., Nunn, P., & Beazley, H. (2021). Uncovering multilayered vulnerability and resilience in rural villages in the Pacific: A case study of Ono Island, Fiji. *Ecology and Society, 26*(1), 26.

Hofman, C., Stancioff, C., Richards, A., et al. (2021). Resilient Caribbean communities: A long-term perspective on sustainability and social adaptability to natural hazards in the Lesser Antilles. *Sustainability, 13*(17), 9807.

Howden, M. (2022, July 5). Floods will continue to intensify so we must act. *Sydney Morning Herald*.

Huff, A., & Naess, L. (2022). Introduction: Reframing climate and environmental justice. *IDS Bulletin, 53*(4), 1–11.

Hughes, T., Barnes, M., Bellwood, D., et al. (2017). Coral reefs in the anthropocene. *Nature, 546*(7656), 82–90.

Human Rights Watch. (2025). *'There's just no more land'. Community-led planned relocation as last-resort adaptation to sea-level rise in Solomon Islands*. Human Rights Watch.

Hurley, G., Wilkinson, E., Weaver, J., & Kasanawaqa, V. (2025). *Turning the tide. Enhancing ocean equity for small island developing states*. ODI Global.

Intergovernmental Panel on Climate Change [IPCC]. (2022). *Climate change 2022: Impacts, adaptation and vulnerability (Working Group II contribution to the IPCC's Sixth Assessment Report)*. Cambridge University Press.

Jackson, G., & Piggott-McKellar, A. (2021). Gods, spirits and natural hazards: Ontologies and epistemologies of natural hazards and climate change in Kiribati and Papua New Guinea. In J. Luetz & P. Nunn (Eds.), *Beyond belief: Opportunities for faith-engaged approaches to climate-change adaptation in the Pacific Islands* (pp. 81–98). Springer Nature.

Jackson, G., N'Guetta, A., De Rosa, S., et al. (2023). An emerging governmentality of climate change loss and damage. *Progress in Environmental Geography, 2*(1–2), 33–57.

Jarillo, S., & Barnett, J. (2022). Repositioning the (is)land: Climate change adaptation and the atoll assemblage. *Antipode, 54*(3), 872–892.

Jinnah, S. (2011). Climate change bandwagoning: The impacts of strategic linkages on regime design, maintenance and death. *Global Environmental Politics, 11*(2), 1–9.

Joshi, K. (2021, December 29). In Australia's climate wars, delay and deception are the new denial. *Guardian*.

Kempf, W. (2020). Introduction: Climate change and Pacific Christianities. *Anthropological Forum, 30*(3), 215–232.

Kench, P., Ford, M., & Owen, S. (2018). Patterns of island change and persistence offer alternate adaptation pathways for atoll nations. *Nature Communications, 9*, 605.

Kench, P., Sengupta, M., Ford, M., & Owen, S. (2024). Island change framework defines dominant modes of atoll island dynamics in response to environmental change. *Communications Earth & Environment, 5*, 585.

Kilroy, J., McNamara, K., Witt, B., & Raynes, S. (2025). 'I don't know anything else besides gardening': Exploring sensitivities to climate change in Vanuatu. *Asia Pacific Viewpoint, 66*(3), 328–340.

Kinch, J., Vitukawalu, B., Nalasi, U., et al. (2019). Socioeconomic aspects of oyster harvesting in the Rewa River delta area, Fiji. *SPC Fisheries Newsletter, 159*, 45–54.

Klepp, S., & Fünfgeld, H. (2022). Tackling knowledge and power: An environmental justice perspective on climate change adaptation in Kiribati. *Climate and Development, 14*(8), 757–769.

Klöck, C., Duvat, V., & Nunn, P. (2022). Maladaptive diffusion? The spread of hard protection to adapt to coastal erosion and flooding along island coasts in the Pacific and Indian Ocean. *Regional Environmental Change, 22*, 136.

Klöck, C., Castro, P., Chin-Yee, S., et al. (2025). Beyond AOSIS: Small island states' presence and participation at COP27. *Climate and Development*, 1–15.

Korovulavula, I., Nunn, P., Kumar, R., & Fong, T. (2020). Peripherality as key to understanding opportunities and needs for effective and sustainable climate-change adaptation: A case study from Viti Levu Island, Fiji. *Climate and Development, 12*(10), 888–898.

Kumar, L., & Taylor, S. (2015). Exposure of coastal built assets in the South Pacific to climate risks. *Nature Climate Change, 5*(11), 992–996.

Kuruppu, N., & Liverman, D. (2011). Mental preparation for climate adaptation: The role of cognition and culture in enhancing adaptive capacity of water management in Kiribati. *Global Environmental Change, 21*(2), 657–669.

Laalaai-Tuasa, C. (2022, November 15). Climate change is a harsh reality in Sāmoa – and the world's leaders need to face it. *The Spin-Off.*

Laffoley, D., Baxter, J., Amon, D., Claudet, J., et al. (2022). The forgotten ocean: Why COP26 must call for vastly greater ambition and urgency to address ocean change. *Aquatic Conservation: Marine and Freshwater Ecosystems, 32*(1), 217–228.

Lai, M., Robinson, S., Salas, E., et al. (2022). Climate justice for small island developing states: Identifying appropriate international financing mechanisms for loss and damage. *Climate Policy, 22*(9–10), 1213–1224.

Lal, P., Kinch, J., & Wickham, F. (2009). *Review of economic and livelihood impact assessments of, and adaptation to, climate change in Melanesia*. SPREP.

Lamb, V. (2023). Constructing the global sand crisis: Four reasons to interrogate crisis and scarcity in narrating extraction. *The Extractive Industries and Society, 15*, 101282.

Lata, S., & Nunn, P. (2012). Misperceptions of climate-change risk as barriers to climate-change adaptation: A case study from the Rewa Delta, Fiji. *Climatic Change*, 169–186.

Latai-Niusulu, A., Binns, T., & Nel, E. (2020). Climate change and community resilience in Samoa. *Singapore Journal of Tropical Geography, 41*(1), 40–60.

Latai-Niusulu, A., Tsujita, M., & Neef, A. (2023). Climate micro-mobilities as adaptation practice in the Pacific: The case of Samoa. *Philosophical Transactions of the Royal Society B, 378*, 20220392.

Latai, A., Taulealo, A., & Connell, J. (2024). 'Burn like Hot Stones': Children's perceptions of climate change in Samoa. *Island Studies Journal, 19*(1).

Lauer, M. (2023). *Sensing disaster: Local knowledge and vulnerability in Oceania*. University of California Press.

Lauer, M., Atger, T., Holbrook, S., & Rassweiler, A. (2022). Who is perturbed by ecological perturbations? Marine scientists' and Polynesian fishers' understandings of a crown-of-thorns starfish outbreak. In S. Hoffman, T. Eriksen, & P. Mendes (Eds.), *Cooling down: Local responses to global climate change* (pp. 65–89). Berghahn.

Lenzen, M., Moran, D., Kanemoto, K., et al. (2012). International trade drives biodiversity threats in developing nations. *Nature, 486*(7401), 109–112.

Llana, S. (2019, December 17). Rebuilding the Bahamas: How a hurricane blows up social divides. *Christian Science Monitor*.

Lowe, R., Gasparrini, A., Van Meerbeeck, C., Lippi, C., et al. (2018). Nonlinear and delayed impacts of climate on dengue risk in Barbados: A modelling study. *PLoS medicine, 15*(7), e1002613.

Malherbe, W., Sauer, W., & Aswani, S. (2020). Social capital reduces vulnerability in rural coastal communities of Solomon Islands. *Ocean & Coastal Management*, 191, 105186.

McAdam, J. (2014). Historical cross-border relocations in the Pacific: Lessons for planned relocations in the context of climate change. *Journal of Pacific History, 49*, 301–332.

McAdoo, B., Moore, A., & Baumwoll, J. (2009). Indigenous knowledge and the near-field population response during the 2007 Solomon Islands tsunami. *Natural Hazards, 48*(1), 73–82.

McCubbin, S., Smit, B., & Pearce, T. (2015). Where does climate fit? Vulnerability to climate change in the context of multiple stressors in Funafuti, Tuvalu. *Global Environmental Change, 30*, 43–55.

McDonnell, S. (2021). The importance of attention to customary tenure solutions: Slow onset risks and the limits of Vanuatu's climate change and resettlement policy. *Current Opinion in Environmental Sustainability, 50*, 281–288.

McDonnell, S. (2025). Renewed hope: Why the ICJ decision matters to us all. *ANU DevPolicyBlog*, 25 July.

Mcleod, E., Bruton-Adams, M., Förster, J., et al. (2019). Lessons from the Pacific Islands – Adapting to climate change by supporting social and ecological resilience. *Frontiers in Marine Science, 6*, 289.

McMichael, C., Katonivualiku, M., & Powell, T. (2019). Planned relocation and everyday agency in low-lying coastal villages in Fiji. *Geographical Journal, 185*(3), 325–337.

McMichael, C., Powell, T., Piggott-McKellar, A., & Yee, M. (2025). Climate change and the planned relocation of people: A longitudinal analysis of Vunidogoloa. *Fiji, Ambio, 54*, 1043–1056.

McMillen, H., Ticktin, T., Friedlander, A., et al. (2014). Small islands, valuable insights: Systems of customary resource use and resilience to climate change in the Pacific. *Ecology and Society, 19*(4), 44.

McNamara, K. E. (2013). Taking stock of community-based climate-change adaptation projects in the Pacific. *Asia Pacific Viewpoint, 54*(3), 398–405.

McNamara, K., & Farbotko, C. (2017). Resisting a 'doomed' fate: An analysis of the Pacific Climate Warriors. *Australian Geographer, 48*(1), 17–26.

McNamara, K., & Prasad, S. (2014). Coping with extreme weather: Communities in Fiji and Vanuatu share their experiences and knowledge. *Climate Change, 123*, 121–132.

McNamara, K., Clissold, R., & Westoby, R. (2022). Teaching our children from books, not the sea': How climate change is eroding human right in Vanuatu. *The Conversation.*

Martin, P., Nunn, P., Leon, J., & Tindale, N. (2018). How sea-level and climate change affect island people: The example of Yadua Island, Fiji. *Climate Risk Management, 21*, 7–15.

Mercer, J., Kelman, I., Alfthan, B., & Kurvits, T. (2012). Ecosystem-based adaptation to climate change in Caribbean small island developing states: Integrating local and external knowledge. *Sustainability, 4*, 1908–1932.

Moesinger, A. (2019a). Influence of socio-economic stressors on interpretations of climate change on Takuu Atoll, Papua New Guinea. *Journal de le Société des Océanistes, 149*, 223–234.

Moesinger, A. (2019b). *Modifications to natural resource use in response to perceptions of changing weather conditions on Takuu Atoll, Papua New Guinea*. SPC Traditional Marine Resource Management and Knowledge Information Bulletin No. 40.

Mohan, P. (2023). Financing needs to achieve nationally determined contributions under the Paris Agreement in Caribbean small island developing states. *Mitigation and Adaptation Strategies for Global Change, 28*, 26.

Mohan, P. (2025a). Climate change adaptation in small island developing states: Evidence from the nationally determined contributions of Caribbean States. *Regional Environmental Change, 25*(1), 32.

Mohan, P. (2025b). Climate reparations for a just response to climate change: A review of historical responsibility and future implication. *Wiley Interdisciplinary Reviews: Climate Change, 16*, e70007.

Mohan, P., & Strobl, E. (2025). The macroeconomic impact of disasters: A review of hurricanes in Caribbean small island developing states. *Climate and Development, 17*(7), 602–615.

Morgan, W., Carter, S., & Manoa, F. (2024). Leading from the frontline: A history of Pacific climate diplomacy. *Journal of Pacific History, 59*(3), 353–374.

Moulton, A. (2024). Modernity's Antillean ecologies: Dispossession, disasters, justice and repair across the Caribbean archipelago. *Progress in Environmental Geography, 3*(1), 3–16.

Moulton, A., & Machado, M. (2019). Bouncing forward after Irma and Maria: Acknowledging colonialism, problematizing resilience and thinking climate justice. *Journal of Extreme Events, 6*(1), 1940003.

Mycoo, M., Robinson, S.-A., Nguyen, C., et al. (2021). Human adaptation to coastal hazards in Greater Bridgetown, Barbados. *Frontiers in Environmental Science, 9*, 647788.

Naidu, R. (2024). Dropouts declining, but education challenges remain. *Islands Business, 50*(2), 25–26.

Nakamura, N., & Kanemasu, Y. (2022). A minority group's response to a severe climatic event: A case study of rural Indo-Fijians after Tropical Cyclone Winston in 2016. *Disasters, 46*(1), 1–20.

Nand, M., Clissold, R., McNamara, K., & Yee, M. (2024). Guiding our responses to climate change by what people value: Insights from Fiji. *Geoforum, 155*, 104080.

Nand, M., Bardsley, D., & Suh, J. (2024). Climate change loss and damage from droughts: Key insights from Fiji's sugar industry. *Local Environment, 28*(6), 768–783.

Neef, A., Benge, L., Boruff, B., Pauli, N., Weber, E., & Varea, R. (2018). Climate adaptation strategies in Fiji: The role of social norms. *World Development, 107*, 125–137.

Ngcamu, B. (2023). Climate change effects on vulnerable populations in the Global South: A systematic review. *Natural Hazards, 118*, 977–991.

Nichols, A. (2019). Climate change, natural hazards, and relocation: Insights from Nabukadra and Navuniivi villages in Fiji. *Climatic Change, 156*, 255–271.

Nunn, P. (2009). Responding to the challenges of climate change in the Pacific Islands: Management and technological imperatives. *Climate Research, 40*(2-3), 211–231.

Nunn, P., Klöck, C., & Duvat, V. (2021). Seawalls as maladaptations along island coasts. *Ocean and Coastal Management, 205*, 105554.

Nunn, P., Smith, T., & Elrick-Barr, C. (2021). Path dependency and future adaptation of coastal cities: Examples from the Asia-Pacific. *Frontiers in Environmental Science, 9*, 642385.

Oral, L. (2025). *ICJ climate change advisory opinion: Noteworthy trends*. IISD Knowledge Hub.

Paeniu, L., Holland, E., Miller, C., & Anderson, G. (2017). Rainfall trends, drought frequency and La Niña in Tuvalu: A small equatorial island state in the Pacific Ocean. *Journal of Environmental and Analytical Toxicology, 7*, 501.

Panke, D. (2012). Dwarfs in international negotiations: How small states make their voices heard. *Cambridge Review of International Affairs, 25*(3), 313–328.

Papalexiou, S., & Montanari, A. (2019). Global and regional increase of precipitation extremes under global warming. *Water Resources Research, 55*(6), 4901–4914.

Pearce, T., Currenti, R., Mateiwai, A., & Doran, B. (2018). Adaptation to climate change and freshwater resources in Vusama village, Viti Levu, Fiji. *Regional Environmental Change, 18*(2), 501–510.

Perry, K. (2024). Epistemic silences, subversive politics: Post-disaster economic assessments as technologies of persistent coloniality and route to an emancipatory climate justice agenda in the Caribbean. *Climate and Development, 16*(9), 798–810.

Perry, K., & Sealey-Huggins, L. (2023). Racial capitalism and climate justice: White redemptive power and the uneven geographies of eco-imperial crisis. *Geoforum, 145*, 103772.

Perumal, N. (2018). "The place where I live is where I belong": Community perspectives on climate change and climate-related migration in the Pacific island nation of Vanuatu. *Island Studies Journal, 13*(1), 45–64.

Pettinotti, L., Colenbrander, S., Kamninga, T., et al. (2025). *A fair share of climate finance? Assessing quantity, quality and alignment with gender goals*. ODI Global.

Petzold, J., Ratter, B., & Holdschlag, A. (2018). Competing knowledge systems and adaptability to sea-level rise in The Bahamas. *Area, 50*(1), 91–100.

Piggott-McKellar, A., Pearson, J., McNamara, K., & Nunn, P. (2020). A livelihood analysis of resettlement outcomes: Lessons for climate-induced relocations. *Ambio, 49*(9), 1474–1489.

Piguet, E. (2022). Linking climate change, environmental degradation, and migration: An update after 10 years. *Wiley Interdisciplinary Reviews: Climate Change, 13*(1), e746.

Pondorfer, A. (2019). The perception of climate change: Comparative evidence from the small-island societies of Bougainville and Palawan. *Environmental Development, 30*, 21–34.

Prakash, R., & Nakamura, N. (2024). Sugarcane farmers' perceptions of climate change and their "adaptation" methods in Fiji: adaptation for the industry or for their livelihood?. *Climate and Development, 16*(8), 665–672.

Raddatz, C. (2009). *The Wrath of God: Macroeconomic costs of natural disasters*. World Bank Policy Research Working Paper 5039.

Rarai, A., Parsons, M., Nursey-Bray, M., & Crease, R. (2022). Situating climate change adaptation within plural worlds: The role of Indigenous and local knowledge in Pentecost Island, Vanuatu. *Environment and Planning E: Nature and Space, 5*(4), 2240–2282.

Reaser, J., Meyerson, L., Cronk, Q., et al. (2007). Ecological and socioeconomic impacts of invasive alien species in island ecosystems. *Environmental Conservation, 34*, 98–111.

Reed, K., Wehner, M., & Zarzycki, C. (2022). Attribution of 2020 hurricane season extreme rainfall to human-induced climate change. *Nature Communication, 13*, 1905.

Rhiney, K. (2015). Geographies of Caribbean vulnerability in a changing climate: Issues and trends. *Geography Compass, 9*(3), 97–114.

Rhiney, K. (2020). Dispossession, disaster capitalism and the post-hurricane context in the Caribbean. *Political Geography, 78*, 102171.

Rhiney, K., & Baptiste, K. (2019). Adapting to climate change in the Caribbean: Existential threat or development crossroads? *Caribbean Studies, 47*(2), 59–80.

Rikimani, B. (2024). Climate justice and Pacific Island countries – A case study on grassroots advocacy. *The Round Table, 113*(4), 374–384.

Rise, N., Oura, C., & Drewry, J. (2022). Climate change and health in the Caribbean: A review highlighting research gaps and priorities. *The Journal of Climate Change and Health, 8*, 100126.

Rothe, D., Boas, I., Farbotko, C., & Kitara, T. (2024). Digital Tuvalu: State sovereignty in a world of climate loss. *International Affairs, 100*(4), 1491–1509.

Rudiak-Gould, P. (2014). *Climate change and tradition in a small island state: The rising tide*. Routledge.

Ruehr, S. (2022). Beyond the vulnerability/resilience dichotomy: Perceptions of and responses to the climate crisis on Emau. *Vanuatu. Island Studies Journal, 17*(1), 157–176.

Russell, J., & Kueffer, C. (2019). Island biodiversity in the Anthropocene. *Annual Review of Environment and Resources, 44*, 31–60.

Ryan, F. (2024). A quantitative goal with no quantum – three years of climate finance talks. *ANU DevPolicyBlog*, 15 October.

Samuwai, J., Fihaki, E., & Underhill-Sem, Y. (2020). Demystifying climate finance impacts in small island developing states: Pacific women's perspectives from Funafuti and Weno. *Small States and Territories, 3*(2), 283–302.

Schlosberg, D., & Collins, L. (2014). From environmental to climate justice: Climate change and the discourse of environmental justice. *WIREs Climate Change, 5*(3), 359–374.

Schrader, J., Weigelt, J., Cai, L., et al. (2024). Islands are key for protecting the world's plant endemism. *Nature, 634*(8035), 868–874.

Schwarz, A., Béné, C., Bennett, G., et al. (2011). Vulnerability and resilience of remote rural communities to shocks and global changes: Empirical analysis from Solomon Islands. *Global Environmental Change, 21*(3), 1128–1140.

Scobie, M. (2016). Policy coherence in climate governance in Caribbean small island developing states. *Environmental Science & Policy, 58*, 16–28.

Scorlazzi, C., Gingerich, S., van Dongeren, A., Cheriton, O., et al. (2018). Most atolls will be uninhabitable by the mid-21st century because of sea-level rise exacerbating wave-driven flooding. *Science Advances, 4*, eaap9471.

Scott, D., Hall, C., & Stefan, G. (2012). *Tourism and climate change: Impacts, adaptation and mitigation*. Routledge.

Scott, D., Simpson, M., & Sim, R. (2012). The vulnerability of Caribbean coastal tourism to scenarios of climate change related sea level rise. *Journal of Sustainable Tourism, 20*(6), 883–898.

Seebens, H., Meyerson, L., Richardson, D., et al. (2025). Biological invasions: A global assessment of geographic distributions, long-term trends and data gaps. *Biological Reviews*.

Sengupta, M., Ford, M., & Kench, P. (2021). Shoreline changes in coral reef islands of the Federated States of Micronesia since the mid-20th century. *Geomorphology, 377*, 107584.

Sengupta, S. (2022, December 8). Tiny Vanuatu uses its 'unimportance' to launch big climate ideas. *The New York Times*.

Sheller, M. (2025). Caribbean Islands and the coloniality of climate change. In P. Noxolo, K. Rhiney, & R. Cummings (Eds.), *Routledge handbook of Caribbean studies* (pp. 35–46). Routledge.

Shultz, J., Sands, D., Kossin, J., & Galea, S. (2020). Double environmental injustice—climate change, Hurricane Dorian, and the Bahamas. *New England Journal of Medicine, 382*(1), 1–3.

Simonelli, A., & Novalski, K. (2023). Rainbows of comfort in rising seas: How literalist Bible interpretations impact climate change communication in the Marshall Islands. *Weather, Climate and Society, 15*, 773–785.

Smith, R. (2025). Indigenous vulnerability, disaster governance, and environmental justice. Case study of the 2021 La Soufrière Eruptions in St. Vincent and the Grenadines. In P. Noxolo, K. Rhiney, & R. Cummings (Eds.), *Routledge handbook of Caribbean Studies* (pp. 75–98). Routledge.

Smith, R., & Rhiney, K. (2015). Climate change, vulnerability, land and livelihoods: The case of the Black Caribs in northeastern St Vincent. *Geoforum, 73*, 22–31.

Soanes, L., Pike, S., Armstrong, S., et al. (2021). Reducing the vulnerability of coastal communities in the Caribbean through sustainable mangrove management. *Ocean and Coastal Management, 210*, 105702.

Sobion, J. (2025). *ICJ advisory opinion on climate change: The case for the Caribbean*. IISD SDG Knowledge Hub.

Spennemann, D. (1996). Nontraditional settlement patterns and typhoon hazard on contemporary Majuro Atoll, Republic of the Marshall Islands. *Environmental Management, 20*, 337–348.

Stege, M. (2018). Atoll habitability thresholds. In W. Filho & J. Nalau (Eds.), *Limits to climate change adaptation* (pp. 381–399). Springer.

Steibl, S., Kench, P., Young, H., Wegmann, A., et al. (2024). Rethinking atoll futures: Local resilience to global challenges. *Trends in Ecology & Evolution, 39*(3), 258–266.

Steiner, C. (2015). A sea of warriors: Performing an identity of resilience and empowerment in the face of climate change in the Pacific. *The Contemporary Pacific, 27*, 147–180.

Sultana, F. (2022). Critical climate justice. *Geographical Journal, 188*(1), 118–124.

Taylor, M., Clarke, L., Centella, A., et al. (2018). Future Caribbean climates in a world of rising temperatures: The 1.5 vs 2.0 dilemma. *Journal of Climate, 31*(7).

Thomas, A. (2025). Transformational adaptation to climate change in the Caribbean. In P. Noxolo, K. Rhiney, & R. Cummings (Eds.), *Routledge handbook of Caribbean studies* (pp. 47–60). Routledge.

Thomas, A., & Benjamin, L. (2020). Non-economic loss and damage: Lessons from displacement in the Caribbean. *Climate Policy, 20*(6), 715–728.

Tollefson, J. (2020). Why deforestation and extinctions make pandemics more likely. *Nature, 584*, 175–176.

Treichel, P., Robertson, M., Wilkinson, E., & Corbett, J. (2024). Scale and access to the green climate fund: Big challenges for small island developing states. *Global Environmental Change, 89, 102943*.

Trotman, A., Mahon, R., Meerbeeck, C., & Riley, J. (2021). *Advancing drought risk management in the Caribbean: A multi-sectoral perspective*. Caribbean Institute for Meteorology and Hydrology.

Turner, R., McConney, P., & Monnereau, I. (2020). Climate change adaptation and extreme weather in the small-scale fisheries of Dominica. *Coastal Management, 48*(5), 436–455.

United Nations. (2024). *Surging seas in a warming world*. UN.

Urban, M. (2024). Climate change extinction. *Science, 386*, 1123–1128.

van der Velde, M., Green, S., Vanclooster, M., & Clothier, B. (2007). Sustainable development in small island developing states: Agricultural intensification, economic development, and freshwater resources management on the coral atoll of Tongatapu. *Ecological Economics, 61*, 456–468.

Vanuatu National Disaster Management Office. (2018). *National climate change and disaster-induced displacement policy*. Port Vila.

Varea, R., Varea, R., Kant, R., & Farrelly, T. (2022). *Qi no tu i baba ni qwali* (living down by the river): Impacts of flooding and mining on ecosystems and livelihoods. *Frontiers in Marine Science, 9*, 954062.

Vinke, K. (2025). *Redrawing humanity's map*. Springer.

Wairiu, M. (2017). Land degradation and sustainable land management practices in Pacific Island Countries. *Regional Environmental Change, 17*, 1053–1064.

Watson, M., Claar, D., & Baum, J. (2016). Subsistence in isolation: Fishing dependence and perceptions of change on Kiritimati, the world's largest atoll. *Ocean and Coastal Management, 123*, 1–8.

Weatherill, C. (2023). Sinking paradise? Climate change vulnerability and Pacific island extinction narratives. *Geoforum, 145*, 103566.

Weatherill, C. (2025). Colonial fantasies of invulnerability to climate change. *International Feminist Journal of Politics, 27*(1), 34–55.

Webber, S. (2013). Performative vulnerability: Climate change adaptation policies and financing in Kiribati. *Environment and Planning A, 45*(11), 2717–2733.

Weiler, F., Klöck, C., & Dornan, M. (2018). Vulnerability, good governance or donor interests? The allocation of aid for climate change adaptation. *World Development, 104*, 65–77.

Weinberg, H., Bhardwaj, J., Watkins, A., & Kuleshov, Y. (2024). The impact of marine heatwaves on isotherm displacement and tuna distribution in Vanuatu. *Climate, 12*(11), 181.

Westoby, R., Clissold, R., & McNamara, K. (2021). Alternative entry points for adaptation: Examples from Vanuatu. *Weather, Climate and Society, 13*(1), 211–222.

Weston, P., & Greenfield, P. (2022, December 19). 'Crucial' Cop15 deal includes target to protect nature on Earth by 2030. *Guardian.*

White, B. (2021, October 29). Barbados: The small but mighty force for climate action. *Open Democracy.*

White, A. (2022, February 18). Island states meet to discuss suing Global North over climate change. *Open Democracy.*

World Bank. (2024). *Global rapid post-disaster damage estimation (Grade) report.* World Bank.

Woodroffe, C. (2008). Reef-island topography and the vulnerability of atolls to sea-level rise. *Global and Planetary Change, 62*, 77–96.

Wrighton, N., & Overton, J. (2012). Coping with participation in small island states: The case of aid in Tuvalu. *Development in Practice, 22*(2), 244–255.

WWF. (2024). *Living planet report: A system in Peril.* WWF.

Yamano, H., Kayanne, H., Yamaguchi, T., Kuwahara, Y., et al. (2007). Atoll island vulnerability to flooding and inundation revealed by historical reconstruction: Fongafale Islet, Funafuti Atoll, Tuvalu. *Global and Planetary Change, 57*, 407–416.

Yarina, E., & Takemoto, S. (2017). Interrupted atolls: Riskscapes and edge imaginaries in Tuvalu. *The Plan Journal, 2*(2), 461–495.

Yee, M., Piggott-McKellar, A., McMichael, C., & McNamara, K. (2022). Climate change, voluntary immobility, and place-belongingness: Insights from Togoru, Fiji. *Climate, 10*, 46.

Yee, M., Piggott-McKellar, A., McMichael, C., & McNamara, K. (2025). Framing locally led adaptation in a planned relocation in Fiji. *Geoforum, 159*, 104196.

Zittis, G., Zoumides, C., Zemah-Shamir, et al. (2025). Insular ecosystem services in peril: A systematic review on the impacts of climate change and other drivers. *Climatic Change, 178*, 127.

3

Viable Economies, Flexible Livelihoods?

All Caribbean and Pacific SIDS face multiple well-documented constraints to achieving conventional versions of economic development, often involving isolation and remoteness, archipelagic populations, few natural resources, and susceptibility to extreme events and other external shocks, now including climate change. Models from larger continental states work poorly in small states, raising questions about the feasibility of, and need for, alternative structures of development, and whether there might or should be a Pacific Way or a Caribbean Way. Economic growth, where it occurs, is not necessarily compatible with achieving the SDGs. Broad socio-economic vulnerability is the outcome of population size and distribution, small open economies, scarce human resources (especially skilled populations), limited and ageing infrastructure, distance from markets and few economies of scale. Island economies are further constrained by poor connectivity to regional and international markets, high costs of doing business (transport, insurance and energy costs), scarce local skills, problems of matching skills and jobs, and limited domestic capital. SIDS do not experience all these constraints at the same time; none are exclusive to SIDS, and economic development has nonetheless occurred.

J. Connell, Y. Campbell, *Island Challenges*,
https://doi.org/10.1007/978-981-95-8177-1_3

SIDS tend to be distant from financial centres and markets—including labour and tourism markets—and beyond major shipping lanes. Nonetheless, certain financial services such as, ironically, registration of international cargo ships are less closely linked to geographic location. Small domestic markets pose barriers to the development of a dynamic private sector and a competitive economy. Entrepreneurialism can be stifled by a lack of skills, capital and contacts but also by scarce opportunities. Basic infrastructure often needs to be duplicated if populations are dispersed, as they are in the archipelagic SIDS.

Transport infrastructure and transport routes disadvantage SIDS. Infrastructure to facilitate mobility—roads, wharves, airports (and electricity pylons)—is costly to build and maintain. Funding shortfalls, engineering complexities and maintenance offer routine challenges while cyclones, landslides and floods disrupt mobility and communication. In several SIDS, only those living in urban communities enjoy full 'infrastructural citizenship' in access to adequate transport and the internet. Smaller SIDS export little and have substantial trade deficits; hence, container ships that deliver imports on their outbound journey have little for the return, raising import costs. Fuel costs and disrupted supply chains have proved costly, especially where several Pacific SIDS are more than 3000 kilometres from the nearest continent: a distance and thus cost structure unlike that of Caribbean islands. The Pacific is the most heavily dependent world region for imported fossil fuel, with 95% of its supplies being imported, yet ocean connectivity, transport and trade are critical. Because of the great distances, the majority (70%) of the Pacific's fossil fuel usage is concentrated in maritime transport (Newell, Nuttall, Prasad, & Veitayaki, 2017). Higher costs have caused already small domestic demand for goods to diminish, shrinking shipping demand so that most shipping companies cannot justify the high financial and time costs to travel to the region. Such transport issues are beyond SIDS control or influence: 'Islanders in the Pacific by nature must travel significant distances by air or by sea to reach one another. Increased cost of shipping has made their economies, their development, and their integration unaffordable' (Gustav Aitaro, Minister of State, Palau, quoted in Labh, 2024).

After the Russian invasion of Ukraine in 2022, Pacific SIDS faced significant increases in fuel and commodity prices, and conflict in the

Middle East a year later continued these trends. By 2024, the cost of global freight shipping had reached its highest level on record, increasing the cost of food and basic commodities across the Pacific. A lack of significant commercial presence in the region makes basic goods and services less accessible, and infrequent and unreliable shipping services increase the costs of development projects (Asian Development Bank, 2024; Labh, 2024). The volatility of fuel and shipping prices has exacerbated existing economic vulnerability. By contrast, the Caribbean's closeness to North America has enabled Trinidad to be a major hub for energy and the Bahamas for logistics.

Such cumulative disadvantages have emerged from and resulted in narrow export specialisation and market orientation so that export prices and foreign exchange earnings may fluctuate considerably. Manufacturing is conspicuous by its absence and with minimal backward linkages to agriculture or fisheries. Tourism has boomed in most SIDS, belatedly and unevenly in the Pacific, but above all in the Caribbean, being effectively a monoculture in the Bahamas (and in smaller politically dependent territories) and the basis of economic growth, despite widespread lack of local control and ownership. So significant is the past, present and future of tourism that we consider it separately in Chap. 4.

Most SIDS have experienced a continued downturn in local production in this century. Some Caribbean SIDS are almost entirely without commercial agriculture and with only residual fisheries. The breakdown in supply chains (and closure of some local markets) during COVID-19 led to considerations of the need for and possibility of greater self-reliance, centred on food security and the blue economy, to support and encourage local producers, enable improved nutrition, and resile from uneven and inequitable trade relationships (Campbell & Connell, 2011). However, exemplified in the tourist industry, moving towards greater self-reliance consistently proved difficult and the illusion never lasted long. Commodity production has fared better but in only a few SIDS and for a few products. By contrast, in the Pacific at least, the effective collective regulation of tuna fishing by a group of SIDS has offered a model for success, albeit a partial and unusual model (Hanich, Jung, McDonald, et al., 2021; Karcher, Fache, Breckwoldt, Guilloux, et al., 2020; Seto, Miller, Young, & Hanich, 2022). Relatively generous overseas aid, an unusual positive small country bias (Chap. 8) and remittances (Chap. 9) have partly compensated for disadvantage.

Compounding these economic challenges, development has usually been conceptualised in standard Western, 'modern' terms that have had little to do with any local alternative perceptions or with the environmental footprint of that development, or much to do with many of the SDGs. Vanuatu briefly achieved fame when it topped the Happy Planet Index in 2016 that sought to measure subjective wellbeing and led to Vanuatu pioneering an 'Alternative Indicators of Wellbeing for Melanesia' initiative that centred around traditional knowledge and practice, and community vitality, offering a holistic approach more akin to the broad objectives of the SDGs. These have brought a greater focus on more immeasurable environmental matters, equity, gender and security, but no lesser perspective on economic growth. However, SIDS have struggled to achieve the SDGs. Poor monitoring systems and lack of data have made it difficult to track progress, for both the Caribbean and Pacific SIDS, especially in relation to inequalities and access to infrastructure and basic services (Persaud & Dagher, 2021).

3.1 Towards a Blue Economy?

The notion of a blue economy emerged in the early 2010s, following the Rio+20 conference, as both rhetorical device and development practice, often vague but which hinted at new directions with marine connotations, and a focus more consonant with equitable development and the SDGs (Hurley, Wilkinson, Weaver, & Kasanawaqa, 2025). It referred to the sustainable use of ocean resources for economic growth, improved livelihoods and jobs while preserving the health of the ocean ecosystem. It therefore included established sectors, such as fisheries, marine transport and marine-based tourism, while expanding to include emerging sectors, such as aquaculture, marine biotechnology, seabed mining, blue carbon and marine renewable energy. However, competing interpretations of what the blue economy might mean complicated SIDS' global engagement and resulted in differences within CARICOM (Hassanali, 2020; Chap. 7). A 'blue-green economy' found a place in some discourses that focused on sustainability but without any restored priority for agriculture. Different meanings of the term were used to support different agendas and viewpoints (Dornan, Morgan, Cain, & Tarte, 2018). The blue economy was also adopted by some corporate entities, notably mining companies, to more elegantly refer to what others saw as 'seabed grabbing' or 'ocean grabbing' (Barbesgaard, 2018; Fache, Le Meur, & Rodary, 2021; Mallin, 2018). Where there was some departure from the past, most evident in the Pacific, a strategic element became more evident with suggestions, at the 2017 Pacific Islands Forum, that the oceans could be seen as a new contested and productive frontier and that the Pacific especially might act as a 'blue continent', and thus perhaps as a bastion and defence against external intervention (Keen, Schwarz, & Wini-Simeon, 2018; Louey, 2022). The 'blue continent' became a theme in the *Boe Declaration on Regional Security* in the following year, was later taken up more controversially in the Partners in the Blue Pacific and was more central to regional agencies as both narrative and strategy, as a rejoinder to the Indo-Pacific strategy associated with the United States and to a lesser extent Australia, and to China's maritime Silk Road in the Pacific, a variation on its Belt and Road Initiative (Kabutaulaka, 2021, Chap. 8). The Samoan Prime Minister, Tuilaepa Malielegaoi, declared in 2018 that 'under the flagship of our Blue Pacific identity, we are building a collective voice amidst the geopolitical din', where the Blue Pacific narrative 'represents our recognition that as a region we are large, connected and strategically important' (quoted in Waqavakatoga, Habru, & Koro, 2025: 212). By then the notion of a blue economy was simply used by many institutions and individuals to indicate that they had moved with the times: 'a buzzword among policy makers across the globe' (Bhattacharya & Dash, 2021; Patil, Virdin, & Colgan, 2023) but of no analytical or policy value. In most SIDS the 'old order' was stubbornly persistent.

3.2 The Death of Agriculture?

The commercial economies of SIDS began with agriculture and fishing (and trade and migration). Some SIDS have changed little, but they nevertheless vary considerably. On atolls and where tourism is dominant, there is minimal interest in agriculture. Commodity exports are key sources of national and household revenue in some SIDS, especially in the larger Pacific SIDS, but export incomes have declined over time with deteriorating terms of trade while plantations are largely memories of the past. Most Pacific SIDS combine subsistence and cash economies. More than two thirds of Melanesians, in the PNG islands, Vanuatu and Solomon Islands, live on their traditional lands, growing food crops, and harvesting forest and marine resources for personal consumption, exchange and market sales. The cash economy is present everywhere and critically important for earning income for education, transport and household needs. Agriculture is valuable, providing income, employment and livelihoods, and playing a crucial role in ensuring food security in SIDS. However, the narrow economic base and small markets in most SIDS states have resulted in a limited range of commodities, notably cocoa, coffee and copra in the Pacific, the 'traditional' plantation commodities, and more recently oil palm, as in Solomon Islands.

The inexorable decline of agriculture has been a slow process in the Caribbean, starting in the 1960s. A legacy of slavery and colonialism and loss of soil fertility first accounted for the decline; competing economic activities and displacement of workers as well as frustrations over (and so independence from) foreign market price fluctuations brought further problems. Agriculture became vulnerable to changing oil prices, having followed 'policies of neoliberalism and trade liberation' along with 'heightened environmental vulnerabilities' (Barker, 2014: 43). This emphasised the role of 'shocks' and 'stresses' such as hurricanes, shifts in seasonal rainfall patterns, poor marketing and changes in global commodity prices. Decline in the Pacific came much later, as alternatives were fewer. But, as in the Caribbean, agriculture was increasingly tainted by some as demeaning, due to its association with colonialism, and a poor source of employment, at best providing a 'last-resort' career (FAO, 2014: 1). That created problems as SIDS sought to revive agriculture and achieve greater food security and better livelihoods. However, in this century, agriculture's contribution to GDP and employment has declined almost universally. Agriculture accounts for about 27% of GDP in both Vanuatu and FSM, 17% of GDP in Dominica but only 4% in Barbados and a mere 0.5% in Trinidad and Tobago.

Subsistence agriculture is particularly important in Melanesia, based on a diminishing range of root crops, dominated by sweet potato and complemented by tree crops, such as breadfruit and almonds, bananas and pawpaws. Beyond ubiquitous chickens and pigs throughout Melanesia, where they have a valuable social role, livestock are of little significance. Cattle remain in some coastal areas, especially in Vanuatu, where beef is a major export, but have proved too demanding of land to be successful in most SIDS. Food security is primarily achieved through subsistence farming and, to a lesser extent, fishing and market trading. As food security has shifted from mostly locally produced food to a mix of local and imported food, the ability to purchase is crucial. Increased dependence on imported food, notably rice, sugar, noodles and flour, brought a new kind of food insecurity and exposed SIDS to the vagaries of global food production and to NCDs (Chap. 6).

Copra prices have meant that traditional coconut plantations that once dominated all PICs have disappeared. Many 'plantation' products of earlier centuries, such as bananas, sugar, rice and copra, have faded away, sometimes victims of economies of scale, loss of protection and competition from elsewhere. Coconut plantations, which played a key role in 'opening up' much of the Pacific to capitalism and once dominated Pacific SIDS, have virtually disappeared. Poor competitiveness was the result of intervening opportunities, low profit margins due to transport distances, while quality requirements for copra increased with Pacific producers unable to adapt to meet new requirements. The once-near-universal copra trade now offers a negative 'model' of change: low prices and hard work. In south Bougainville (PNG), where a cash-crop economy is centred on cocoa, copra is essentially a standby crop, only worth harvesting and processing when prices are unusually high. SIDS play no part in influencing such prices. Low prices have contributed to the loss of commercial agriculture in outer islands and especially atolls, such as Ujae (Marshall Islands) (Rudiak-Gould, 2012), as transport costs make production not feasible. A 'transport stranglehold' has worsened with containerisation bypassing smaller islands, a process that had begun in the 1960s. Typically, between 1982 and 2003, the number of ships visiting the Fijian island of Kadavu declined from 82 to 67 and, rather than visiting several island ports, stopped only at one (Sofer, 2018). Only subsidies have kept production possible in such SIDS as Kiribati and Marshall Islands, in efforts to support livelihoods in the outer islands.

Competition from Latin America and the end of preferential access to European markets under the Lomé convention of 1975 ended the productive Caribbean banana export sector (Connell, 2013a; Grossman, 1998). Sugar production, once the mainstay of Fiji, has declined since global trade regulation meant it was no longer subsidised; factories are ageing and renewed land leases difficult to negotiate. With the expiry of preferential sugar price arrangements in 2017, coming much later than their expiry in the Caribbean, the industry was no longer competitive but still provided a livelihood for about 15% of Fiji's population. However, younger people were no longer interested in sugar and most cane farmers were aged between 50 and 70 (*Fiji Times*, 23 February 2024). Trinidad closed its last sugar factory in 2007, and without EU support, St. Kitts later lost the entirety of its own rather smaller sugar industry (Richardson-Ngwenya, 2010). By contrast, oil palm development has expanded in this century, especially in Solomon Islands, but smallholders are only marginally involved in oil palm production, and environmental degradation has followed large-scale land clearing and production.

Coffee and cocoa are major export crops, especially in the PNG islands, but have been less successful elsewhere. Bula Coffee in Fiji and Tanna Coffee in Vanuatu have gained only small markets, despite considerable potential. Cocoa is an important export for Solomon Islands and Vanuatu. In Vanuatu, a quarter of rural households are involved in cocoa production, and in Solomon Islands, cocoa is the biggest agricultural export earner. With rising global demand, especially in Asia, and higher market prices, Solomon Islands cocoa production has become increasingly valuable. However, poor-quality beans (especially smoky beans), lack of market knowledge and limited access to finance hinder development. Cocoa production in Vanuatu faces related challenges including skill training, financing and adequate infrastructure for quality improvements and export production. Both coffee and cocoa have been set back by cyclone damage. In Samoa and Tonga, both coffee and cocoa have foundered, but in the Caribbean, notably in St. Lucia, cocoa has expanded with the revival of high prices, partly replacing the banana industry. Further development requires high-yielding varieties, superior management processes, transport infrastructure and effective marketing, inherently difficult in even the larger SIDS (Griswold, 2021). Thus, Bougainville (PNG) produces valuable chocolate on a small scale, but it is almost unknown even within PNG. Chocolate niches face marketing issues, but cocoa and coffee have unrealised potential.

Attempts to develop niche commodities, as comparative advantage for other crops declined (Grynberg & Razzaque, 2004), have been limited over time and space. Grenada has effectively been dominated by niches, with a long history of regional and global markets for its locally grown spices, such as nutmeg, cinnamon and cloves, and has sought to be the only Caribbean country exporting soursop. Dominica has had various plantation crops—sugar, cocoa, limes, then bananas—all poorly suited to the country's terrain and all failed, wiped out by cyclones, diseases and global food price rises (Wilkinson, 2023). Initial success stories, such as ginger in Fiji and vanilla and squash in Tonga (van der Grijp, 1997), have failed to survive and foundered on marketing, taste, quality assurance, consistency of supply, environmental stresses and especially intervening opportunities. Prospects for bamboo and other non-traditional exports have been affected by supply challenges, inadequate investments and international regulations that often discourage new entrants and small farmers. Niches have come and gone. In villages near Suva cut flowers have become a small-scale commodity. Pineapples have succeeded in Fiji. Citrus, spices, cucumber, sweet potatoes, dasheen and yams are occasional exports in the Caribbean. Vanilla has been an intermittent success story, with Tonga, Vanuatu and Fiji producing small quantities. However, Mexico, Indonesia and Madagascar produce much larger quantities so that their intervening opportunities dictate market prices. Although Pacific vanilla growers and exporters continue to face the global challenges of price volatility and market access, optimistically 'the Pacific can continue to play an important role in the global vanilla trade, offering a touch of Pacific flavour to the tapestry of this plant's bittersweet history' (*Pacific Trade Invest*, 23 October 2024). The global context of vanilla is much like that restricting other niche products.

In the Pacific, kava, sometimes referred to as 'green gold', boomed at the end of the last century (Murray, 2000) until it was banned in its main European market, officially on health grounds, but revived as international bans ended. Opening of the Australian market in 2022, hitherto closed to kava as an illegal drug, further boosted exports, benefiting Fiji, Vanuatu and Tonga (Aporosa, 2019). In Vanuatu, between 2017 and 2019, exports of kava increased by more than 130%, and in most areas of the country new plantings were expanding (Bertana & Gaytán, 2025; Nef, Kotra, Stauffacher, et al., 2022). However, the end of controls in the United States brought competitive production there, and the need for PICs to secure Geographical Indication protection. That was further complicated by concerns over how best to protect and preserve the cultural and economic value of kava; hence, Tonga has sought restrictions that it be consumed only in its traditional, drinkable form—not processed into extracts for capsules, powders or supplements abroad—whereas Fiji and Vanuatu have sought expanded trade. Tonga argued that processing encourages abuse and commodification, distancing the product from its sacred ceremonial roots (*Vanuatu Daily Post*, 28 May 2025). Marijuana has found a clandestine domestic and international economic niche in several Pacific and Caribbean SIDS, notably following the opening of the market in North America. The 2015 amendment of the Dangerous Drug Act in Jamaica, decriminalising marijuana for medical, scientific and therapeutic purposes, boosted production and later export. St. Vincent decriminalised production in 2018, but its Rastafarian producers, a marginalised group, faced competition from Canada for regional export markets (Duncan & Mctair, 2025). Barbados subsequently developed a small export industry. Serious consideration has been given in the Bahamas to both establishing a small export industry and even creating marijuana agritourism (Rolle, 2020).

Significant production of food crops for flourishing urban and rural markets occurs in a few fertile coastal areas, such as the Guadalcanal plains (Solomon Islands) and the Sigatoka Valley (Fiji) with good access to markets. Other than Tuvalu and Nauru, all Pacific SIDS have at least one fresh food market. In PNG, the most valuable locally marketed commodity is betel nut. Fresh food prices are often higher than imported foods, such as rice, partly a function of limited local supply again posing a problem for food security and nutrition, yet local markets have kept a diverse agricultural sector alive. While marketing empowers many female market vendors, who constitute the great majority of traders, many are disadvantaged, as in Vanuatu, by a combination of heavy workloads, poor infrastructure, transport costs and a reduction in traditional knowledge (Kilroy, McNamara, Witt, & Raynes, 2025).

In most SIDS every facet of agriculture has been threatened by shifting work preferences, migration, food imports and pressure on coastal land, contributing to reduced diversity. Land tenure, once regarded as a constraint to commercial agriculture (notably by the World Bank that constantly sought tenure reform), has increasingly been seen as no real brake on development but a source of stability. Land certification has not discouraged agricultural investment (Huntington & Shenoy, 2021). Land tenure can affect adaptability to climate change as insecurity of tenure also extends to the use of land as collateral for credit, reluctance of farmers to move during natural disasters for fear of losing their land and family disputes (Mitchell & McEvoy, 2019). Moreover, insecure land tenure can hinder adaptation to climate change and limited recognition of customary land rights may increase vulnerability.

Beyond market gardening near urban areas, new sustainable success stories are few. Agriculture has generally failed to develop linkages with the tourism industry because of uncertain irregular production (and, frustratingly, visitors' distaste for 'island nights' of local food). At least 60% of food used in the tourism sector in eastern Caribbean SIDS is imported (Jansen, Stren, & Weiss, 2015; see Chap. 4). In Vanuatu, even eggs must be imported, despite some creative success stories such as the Farm to Hospital project that created jobs and increased food production in peri-urban Santo, reduced food costs in the hospital and improved nutrition (Silas, Makikon, & Kinsella, 2022). Creativity is invariably at a small scale. By contrast, one outcome of guestworker migration from the Pacific SIDS over the last 15 years, for work in agriculture in New Zealand and Australia, and from the Caribbean to Canada, has been at the expense of agriculture in the SIDS, as return migrants with capital seek new non-agricultural activities, disdaining agriculture and losing their connection to land (Chap. 9). Agriculturalists also face food system challenges and structural barriers where, for example, women farmers and food producers in eastern Caribbean SIDS, notably Grenada, St. Lucia and St. Vincent, are inextricably linked into a form of path dependency derived from persistent colonial-plantation relations, patriarchal social norms and liberal-capitalist logic (Barry & Gahman, 2020, 2021; Perry, 2024) that are less binding and inbuilt in the Pacific. Nonetheless, as attention shifts towards a blue economy, and frustration with agriculture contributes to this, almost everywhere that livelihoods are dominated by fishing, they are also accompanied by agriculture. Productive agricultural systems are crucial to sustainable development and food security, if not seemingly as exciting or dominant as a new blue economy.

3.3 All at Sea?

Recent focus, however rhetorical or political, on a blue economy, has suggested that the old 'green agriculture' order had limitations and the future economies of the SIDS were more likely to be marine but as much out of frustration with the old rather than by discovery of a dynamic new economy. Nonetheless, beyond agriculture, fisheries, tourism and deep-sea mining have increasingly been seen as key components in an emerging and more diverse 'blue economy'. While regional organisations have touted the flourishing of a future blue economy and claimed the Pacific as the future 'blue continent', unless fishing delivers even greater rewards or mining seabed minerals becomes economically and environmentally feasible, the future economies of the SIDS will not be greatly different, and certainly not visually different, from the present ones. In an ocean covering a third of the world's surface, this is where a blue economy should be. Globally fishing occupies four times the area of agriculture (Kroodsma, Mayorsga, Hochberg, Miller, & Park, 2018). The intensification of the relationships between land and sea, extending maritime commodity frontiers into a Pacific Ocean that is not simply a place of trade routes but a site of value, has contributed to its becoming a strategic frontier. As both the EEZs and the deep ocean become of greater value, and marine resources decline elsewhere, the Pacific is becoming a more significant locus of economic activity, evident in the expanding role of China's fishing fleet that belatedly 'mirrors the pelagic imperialism of the Western powers' (Campling & Colás, 2021: 210; see Chap. 8). 'Blue growth' reproduces earlier imperialist dynamics of order and appropriation. As fishing has over-exploited fisheries elsewhere, including the Caribbean, attention has turned to the last productive ocean fishery.

But for so many, in the Caribbean too, most of the ocean is out of sight—beyond the horizon and easily ignored: so vast that it can easily be assumed static. A whole new continent in the south-western Pacific—Zealandia—has been discovered in this century (Mortimer, Campbell, Tulloch, et al., 2017): one measure of how little we know about oceans. However, satellites, drones and sonar technology are assisting in discovery and surveillance. Just a single reference to the ocean appeared in the Paris Agreement on climate change. Oceans are subject to a patchwork of laws and agreements, where enforcement is difficult and objectives can be divergent and conflicting. Beyond the EEZs, the high seas remain a global commons beyond defined commodity rights but where tuna offer opportunities. Caution is warranted over optimistic notions of 'blue growth' through bringing new territory—ocean space—under a more effective domain of capitalist commodity production (Mallin & Barbesgaard, 2020; Brent, Barbesgaard, & Pedersen, 2020), evident in a deepening of capitalist relations in fishing and the marginalisation of small-scale fishers, or, alternatively, where a sustainable blue economy might be a much-needed panacea for terrestrial environmental degradation.

3.3.1 Local Fisheries Under Pressure

Multiple forms of fisheries exist in the Caribbean and the Pacific (although freshwater fishing and aquaculture are almost absent). Small-scale fishing is invaluable for many households, for consumption and possible marketing, while the regulation and management of deepwater fisheries in the Pacific at least have contributed major economic gains where the Caribbean has struggled. The main economic sectors and most livelihoods in SIDS (whether in tourism, fisheries and agriculture) are dependent on the natural resource base of coastal and marine environments and its vulnerability to environmental change (Robinson & Wren, 2020). Declines in Caribbean fish stocks and catches have occurred, in tandem with declines in coral reef cover and health, so that nearshore resources and most oceanic resources (tuna and billfish) are either fully exploited or over-exploited (Jackson, Donovan, Cramer, & Lam, 2014; McConney, Cox, & Parsram, 2020). However, the dominance of artisanal, small-scale fisheries in scattered and remote locations has made generalisations over regional trends unusually difficult (Campbell, Steenbergen, Li, et al., 2024; Dunn, Stewart, Bjorkland, et al., 2010).

At local level, marine resources provide daily subsistence for coastal dwellers who may catch fish, harvest marine resources from reefs and the sea, and sell surpluses at markets. At another scale, small fishing boats catch fish offshore for local markets, while some small- and medium-sized fisheries-related industries, such as seaweed harvesting and pearl cultivation and production, are highly localised. In subsistence economies regulation of marine resources by traditional management practices supported sustainability but came under growing pressure as populations grew. Commercialisation of marine resources further challenged sustainability. An early casualty in the Pacific was bêche-de-mer (sea cucumber), harvested in unsustainable quantities to satisfy the Chinese and Asian market (see below). The aquarium trade has been seen as a negative influence on reef depletion, with the 2003 film *Finding Nemo* said to have led to over-exploitation of clownfish on Pacific reefs. In practice, demand for clownfish only marginally increased and the furore over possible exploitation may have contributed to greater reef conservation, at least where tourism was significant (Militz & Foale, 2017). Other coastal species are under pressure, and increasingly, some tuna species are at risk in deeper waters.

Coastal fisheries face typical coastal squeeze challenges. Competing claims to ownership, control and access to coastal land threaten access to reefs and their resources as well as lagoons and inland waters. Pollution threatens marine environments, while burgeoning populations, increased building development and poor management combine to endanger the ecology of coral reefs and mangrove beds. Sustainability is constantly threatened by overfishing in densely populated coastal regions, especially in the Caribbean, where fishing is a fraction of its former importance, and the region is one of the most overexploited in the world. Overfishing, illegal fishing and illegal fishing techniques damage ecosystems. Damage to and bleaching of coral reefs and coral bleaching are widespread. By the start of this century, 5% of marine bony shorefish in the broader Caribbean were 'threatened due to overfishing, invasive lionfish predation and the degradation of coral reefs and estuaries, which provide habitats and feeding grounds for many species'; the most vulnerable species are the red snapper and the now endangered Atlantic bluefin tuna. The removal of large parrotfish has also affected the benthic dynamics of coral reefs (Shantz, Ladd, & Burkepile, 2020). Conch fish in the Bahamas are similarly under threat from overfishing and lack of enforcement and management of illegal fishing, in a country where it is a cultural symbol, since its popularity and economic value especially in the out islands conflicts with the need for ecological, environmental and economic management, and a lack of agreement on how to achieve this and who should be responsible (Bomhauer-Beins, de Guttry, & Ratter, 2019). On Curaçao, indicative of changes in Caribbean SIDS, in response to technological and environmental changes since 1905, fishers targeted species increasingly further from shore after nearby fish had become rare. Large predatory reef fishes living close to shore had virtually disappeared from catches around the mid-twentieth century. After 1969, the decline in catches in coral-dominated reef communities suggested that habitat loss, in addition to fishing, had led to the decline in reef fish abundance (Vermeij, Latijnhouwers, Dilrosun, et al., 2019) and the virtual cessation of effective fishing in several places. Poaching from Central American countries, such as Nicaragua and Honduras, remains a problem, exacerbated by limited capacity to police international laws that protect SIDS territorial waters. Since 2011, the Bahamas has made it illegal for foreign-owned vessels to use its EEZ; Barbados likewise does not allow foreign commercial fishing. Little is left of a fishing industry in many islands, and its retention often primarily suits tourism objectives.

Since so many SIDS people live in coastal areas and depend on fishing, its success is crucial for lives and livelihoods, which provides the problem. Many inshore fisheries have experienced unsustainable pressures, local knowledge has declined (and been ignored), new technologies have proved too efficient (or destructive) and management ineffective. Species diversity and availability are reduced and ecosystems stressed throughout the SIDS (e.g. Christensen, 2011; Rhodes, Warren-Rhodes, Sweet, et al.,

2015). That has been especially so around urban areas where commercialisation is greatest (e.g. Aswani, Basurto, Ferse, et al., 2018). Yet, similar declines have occurred in villages and islands at a distance from rapidly growing urban centres, as in the Bahamas (Casola, Oren, Register, et al., 2021; Hidalgo, Witten, Nunn, et al., 2020). By the 1990s, on remote atolls like Woleai (Yap, FSM), more efficient fishing techniques, aided by the increased accessibility that outboard motors on fibreglass boats enabled, had reduced catches and sustainability. A diversity of more than 50 past techniques had declined by more than half (Smith & Dalzell, 1991). There as elsewhere communal fishing and canoes had given way to motorboats and more individualistic techniques, including night-time spear fishing, both contributing to uneven development, since not all could afford them, and a disregard for conservation. The larger hauls that new technology permitted resulted in local sales rather than more traditional patterns of social distribution, as in Woleai and Nanumea (Tuvalu) (Chambers & Chambers, 2018). That was not always so. The same new technology on the remote island of Ono-i-Lau (Fiji) resulted in a reduction in fishing effort rather than an increased yield since remoteness prevented commercialisation. Yet although the population was declining, some reef species were still overfished (Kuster, Vuki, & Zann, 2005). Pacific trends followed those of the Caribbean and, even where populations were declining, challenged food security.

The newer technology, of outboard motors and night-time spearfishing, initially excluded women who no longer needed to fish. However, female fishers have made significant contributions, mainly handlining and collecting from reefs and in shallow water. Many fishing methods used by women involve minimal technology, making them more sustainable and less vulnerable to external socio-economic impacts. Among policymakers and fishery managers, a longstanding assumption has been that women fish for household consumption while men fish for income and livelihood support. However, on Kadavu (Fiji) and elsewhere, women also sold fish locally, emphasising the important role of female fishers in terms of food security for their own households and beyond (Gotschall, 2021; Marre & Hunt, 2021; Thomas, Baptiste, Martyr-Koller, et al., 2020). Not marginalising women increases the overall effectiveness of small-scale fishing. However, small-scale women fishers have rarely been held in high regard, as in Jamaica where they have mainly been involved in post-catch activities (Wade, 2025; Pena, McConney, Perch, & Phillips, 2022), despite women fishers demonstrating their value in contributing to food security for their own households and the wider community, one of several instances where expanding women's livelihood opportunities would be more generally beneficial for food and wider security (Ferguson Irlanda et al., 2025; Thomas, Baptiste, et al., 2020; Purcell, Tagliafico, Cullis, & Cocks, 2024; Batalofo, Pollard, Ride, et al., 2024) and also to equity.

Traditional marine management practices have come under intense pressure through commercialisation, local population growth and changed social relationships, including more mobile local populations with less commitment to local matters. Most conservation actions have been at a small-scale, lacking a broader spatial context, where management issues differ (Keppel, Morrison, Watling, et al., 2012; Mills, Pressey, Weeks, Foale, & Ban, 2010). Thus, attempts to transfer relatively successful models from Fiji to the Solomon Islands and Vanuatu have proved almost impossible since land and marine tenure systems and related political systems in the three countries are quite different. Indeed, coastal fisheries in adjoining Vanuatu and Kiribati are characterised by a diversity of contexts, fishing practices and fisher participation, which makes developing appropriate management regimes unusually difficult (Campbell et al., 2024). Thus, transferring the more centralised tenure and political system of Fiji to the decentralised and 'politically eclectic' Vanuatu and Solomon Islands proved unsuccessful: 'one size does not fit all' (Aswani, Albert, & Love, 2017; Foale, 2001). Most marine protected areas (MPAs) are small-scale, linked to small clusters of villages or areas of sea and lagoon, but even then management functions poorly since social dynamics operate at a much smaller scale (e.g. Buggy & McNamara, 2015). Communities, insofar as they exist, cut across and through villages, and are not homogeneous but constitute a range of clans and factions, alongside particular interest groups with various values, images and principles (Voyer et al., 2015) between which relationships can be challenging and contested, if usually muted to avoid confrontation (D'Arcy, 2006; Lauer, 2016; Lutz, 1988). Successful MPAs, as in the Bahamas, were the outcome of trust in management by a relatively educated population with access to income other than fishing (Casola, Rehnberg, Peterson, et al., 2022). In Kiribati, for example, customary marine tenure has declined with laws passed by island councils not enforced through lack of chiefly authority and weak traditional sanctions (Thomas & Tonganibeia, 2007). However, processes affecting MPAs cannot be contained at the local scale. In some places the decline of marine species has produced a 'race to fish' that simply enhanced the rate of degradation of stocks (Rhodes, Warren-Rhodes, Houk, Cuetos-Bueno, & Fong, 2011), exemplifying the tragedy of the commons where individual resource users deplete and degrade common resources. Local conflicts can occur where efforts have been made to establish MPAs to reverse resource depletion, but where fishing livelihoods are perceived to be at stake. Management is more difficult when indigenous ecological knowledge varies and local people cannot easily comprehend the scale, direction and causes of change in marine environments (Connell, 2013a; Lauer & Aswani, 2010; O'Garra, 2012). Conversely, a disconnect often occurs between the urgency of local voices and delayed response in national plans (Scobie, Gopinathan, Kelman, & Myhre, 2023). Without government support and trust in government, adaptation to changing ecological and environmental circumstances can be difficult. In the small town of Whitehouse (Jamaica), depleted fish stocks were the outcome of pollution, habitat destruction, overfishing and illegal fishing (from two LIDS), but fishers were unable to readily adapt to this, at the same time as climate change (where dry periods were more evident), because of a lack of capital, no adequate land tenure (hence no collateral for loans), weak human capital and lack of alternative employment. In Soufriere (St. Lucia), a similar decline in both fisheries and agriculture was partly absorbed by an expanding tourism industry, but many people also faced problems of adjustment, without land or capital and limited education. The collapse of institutional support for copra and banana exports at the end of the twentieth century had reduced trust in government and slowed adaptation (Hogarth & Wójcik, 2016a, 2016b). In these parallel circumstances, longstanding constraints, in part the lack of land ownership that was a legacy of slavery, poor education and weak government support, dampened the resilience and social capital that existed and made response to environmental change ineffective.

As tensions between fisheries and tourism emphasise, and diversification into agriculture fails, external interventions in particular sectors tended to be technical and managerial, at best 'ecological modernization', raising questions about efficiency, economics, ethics and even aesthetics, and wider questions about governance, commitment, involvement, comprehension and scale: above all acute differences and divergences exist between local and distant knowledge and objectives and local capacities for resilience. Multiple case studies have pointed to social-ecological traps, where poverty exists and local institutions are weak, so that overfishing with destructive gear with disregard for seasons and species can push reef and lagoon socio-ecological systems past key thresholds. In Caribbean SIDS, small-mesh nets that typically scrape the ocean floor are used illegally. Fishing techniques recommended by governments to be banned because of their destructive effect on stocks, including SCUBA spearfishing, blast fishing, gillnetting and the use of poisons (Rhodes et al., 2011), remain in use. That can be a function of poverty, which may be the outcome of lack of land or marine tenure or inequality between local social groups (Fabinyi & Barclay, 2022). Yet repeatedly it has become evident that small-scale fisheries are remarkably efficient at enabling access to food and income for those who need them most, in contrast to industrial fleets, which waste an estimated 20% of fish caught as bycatch (Hendriks, 2022).

Illustrative of pressure on marine resources and the need for income in remote islands have been the fluctuations in bêche-de-mer production (complicated by its slow reproduction rate). Bêche-de-mer has gone through boom-and-bust cycles, responding to high prices since the 1980s. Intermittent closure of the Chinese market has hampered development, but overfishing has occurred when markets have been open. Ease of fishing makes bêche-de-mer fisheries difficult to manage sustainably and regular moratoriums have been necessary (Hair, Foale, Kinch, et al., 2016; Watt, 2024). It has intermittently provided good incomes for more remote islands, such as Ontong Java (Solomon Islands), where other commercial opportunities are scarce (Christensen, 2011). The tenuous survival of the bêche-de-mer trade, centred on Pacific SIDS for luxury markets in Asia, epitomises fisheries management problems. In Palau, and Pohnpei and Yap (FSM), despite short-term financial incentives to export bêche-de-mer, various groups and civil society organisations coordinated to ban the trade at its peak, using public protests and court battles, to maintain moratoriums and enable sustainability. Similar fluctuations in production and overproduction exist in the valuable sea urchin fisheries of Barbados and St. Lucia, with the additional threat of illegal night-time fishing supported by 'rogue' social networks (McConney et al., 2020). In both these fisheries, the poverty of fishers has put intermittent extreme pressure on their resources. The valuable spiny lobster fishery of the Bahamas has been regulated more effectively, despite some resistance (Doerr, 2021), suggesting that management can be effective where adequate human resources and alternative income sources exist.

Other coastal marine possibilities offer some potential. Seaweed farming has disappointed in Kiribati, without more weather-resistant varieties, indicating the difficulty of innovation in risk-averse, low-income, subsistence environments (Thomas & Tonganibeia, 2007). It offers possibilities in Solomon Islands and elsewhere, where agriculture is limited by land and water availability and climate variability. Seaweed offers a resilient alternative that does not require land, fresh water or fertilisers and contributes to the health of marine ecosystems by improving water quality and providing habitats for marine life. It is, however, labour intensive and market access is poor. While global demand for seaweed products is growing—driven by the food, cosmetics, pharmaceutical and biofuel industries—potential small-scale farmers have limited access to market information.

Aquaculture, long undeveloped, has made progress in this century, notably in Fiji where women and community groups play a significant role on about 500 tilapia farms, offering opportunities where coastal fisheries experienced declining yields (Harohau, Allison, & Teoh, 2025; Waqairatu-Waqainabete & Kaumaitotoya, 2021). In Solomon Islands it proved unsuccessful in contributing to incomes or food security, since the tilapia were rarely eaten or sold. Generally, it has rarely been successful unless subsidised but may prove a partial replacement for declining coastal fisheries, especially in the Melanesian SIDS, if the utility of tilapia can be improved, and constraints over tenure and biosecurity can be resolved (Gillett & Fong, 2023; Harohau et al., 2025; Harohau, Blythe, Sheaves, & Diedrich, 2020). Pearls have been grown successfully in some

PICs but on a small scale and, as in Tonga and Fiji, with government subsidies. The pearl industry has had a chequered history, and few ventures have achieved long-term success, despite its potential for providing employment for men and women and diversifying rural livelihoods in a culturally compatible and rewarding way (Mikhailovich, Mackenzie, & Smith, 2023). Several countries—most recently the Marshall Islands—have sought to encourage its development in a globally competitive niche industry.

Fisheries reforms and technological advances in offshore aquaculture (mariculture) could increase production, but the existing and potential mobility of fisheries stocks will increase and challenge this process, if ocean temperatures continue to increase significantly. The mobility of marine species has meant that fisheries have been particularly likely to be affected by ongoing climate change but especially through the movement of tuna species (Bell, 2011; see below). A similar process has already happened in Barbados, with flying fish migrating to Trinidad, prompting arguments over ownership and maritime boundaries (Kwiatkowska, 2007). Fish eat more in warmer waters; hence, climate change potentially alters marine ecosystems. Increased ocean temperatures and acidification, changing patterns of nutrient supply and greater storms and sea-level rise affecting reefs and coastal habitats, all reduce fisheries potential. Around Australia, more than 100 fish species, most of commercial value, have migrated south to cooler waters, something of an inequitable 'donation' from nearby SIDS.

The emergence of markets for high-value seafood products has produced a predictable pattern of steady depletion of resources and social disruption in coastal communities: a 'tragedy of the commodity' (Ferguson, Bennett, Kostka, et al., 2022). Remote islands were increasingly affected, with many communities concerned about the impact of industrial fishing on the depletion of fish in their coastal waters (James, Tidd, & Kaitu, 2018). Before the end of the last century, fishers from Woleai were finding fewer and fewer tuna on their own fishing trips outside the lagoon as the incursions of Distant Water Fishing Nation (DWFN) vessels had depleted the stock. Small local fisheries frequently face a significant competitive disadvantage against highly subsidised and industrialised foreign fleets (Hurley et al., 2025). Like the tragedy of the commons, Indigenous peoples and local communities can organise to resist the tragedy of the commodity, by asserting Indigenous values, rights and institutions, 'recommonizing' the resource and preventing fisheries collapse (Ostrom, 2000). The continued bêche-de-mer trade challenges both the inevitability of the 'tragedy of the commodity' and the narrative of poor fishers as vulnerable and disempowered, alongside the value of reasserting traditional values and knowledge. But implementing regulations that demand some reduced degree of fishing effort is difficult and has often been unsuccessful (Fabinyi & Barclay, 2022). It has been no easier in the Caribbean. Globally three pathways have contributed to greater stability: revitalising Indigenous knowledge and stewardship, fostering community-based management and empowering women (Eriksson, Sulu, Blythe, et al., 2020), but each is difficult partly because they reverse the scale of change by (re)turning to the local.

Conservation efforts can also be frustrated by pandemics or any form of economic crisis. Pressure on marine resources increased during the onset of COVID-19 partly as people returned to a more subsistence lifestyle, other sources of food were more costly and agriculture was unable to be boosted quickly (Ride, Eriksson, Tutuo, et al., 2021). In turn, that contributed to overfishing and disputes over local boundaries, marine tenure and fishing techniques with some collapses of MPAs. It also coincided with limitations on fishing effort because of fuel price rises and marketing costs with negative impacts on incomes for almost all small-scale fishers in coastal communities. People consequently spent more time on agriculture to generate income (Bennett, Finkheiner, Ban, et al., 2020; Marre & Imhof, 2021). Women were further disadvantaged by acquiring extra domestic tasks limiting their availability to fish while those who returned to their home villages often lacked (or had forgotten) fisheries skills and techniques (Garcia, Makhoul, Fox, & Tuxson, 2022; Ride et al., 2021). COVID thus pointed to the difficult tasks involved in regulating, managing and supporting small-scale fisheries at different scales, tasks as evident in the Caribbean as elsewhere (Cumberbatch, Compton, & McConney, 2023) and which recurred in different forms in the context of 'meso-fishing' with larger boats, sometimes national fleets, in deeper waters.

Where inshore fisheries have been depleted and modern technology (initially basic outboard motors) has enabled fishing further from shore, other new techniques have emerged including the use of fisheries aggregation devices (FADs) that have become more important in the 2010s, enabling small-scale fishers in Pacific SIDS to economically access the considerable tuna resources, being less costly (in terms of fuel and other expenditure). In the Caribbean, FADs have provided opportunities for fishing further from overfished reefs, improved catch rates for fishers, better access to more pelagic seafood, especially tuna, as in Kadavu (Fiji) and northern Malaita (Solomon Islands), although it has proved more difficult to support and manage effectively, and incidentally reinforced traditional gender roles (Labuinao, 2020; Marre & Hunt, 2021). In this century coastal fisheries have become unusually complex and diversified but increasingly critical for development.

3.3.2 In Deeper Waters

Several Pacific SIDS have vast EEZs, unlike those of Caribbean SIDS, immediately suggesting the potential of deep-sea fisheries. The countries with large EEZs also tend to be smaller in land area and population, indicating that their ability to benefit from deep-sea resources might contribute not just to income generation but to greater regional equity. Yet, only relatively recently have SIDS been able to take advantage of this, with 'traditional' fisheries rarely venturing as far as 20 kilometres from shore. That advantage has come from the leasing of these waters to DWFNs rather than from national fisheries. The Pacific Ocean has become more valuable in this century as oceans elsewhere have been depleted, constituting a bulwark against global overfishing, and the consequent disruption of marine ecosystems and loss of sustainability (Graham, 2024). Here the Pacific SIDS have a very considerable advantage over the Caribbean SIDS that now lack a real tuna fishery.

Around 58% of global fish stocks are already decimated (including, famously, the North American cod fishery), partly because global populations have grown but also through unsustainable practices, pollution and habitat loss, and further threatened by both climate change and illegal fishing, the latter accounting for about 15% of globally caught fish (Graycar, 2022). These trends have also occurred in the SIDS, with vessels of other nations illegally fishing in the Pacific, the last fisheries frontier. Response to global pressures and problems demanded more careful management, the core of which has come from several SIDS working together through the Parties to the Nauru Agreement (PNA) to regulate access to the Pacific EEZs: probably the greatest example of effective regional cooperation in the Pacific, and one that has provided substantial economic benefits.

Despite their combined EEZs constituting the world's largest tuna fishing area, at the turn of the century, Pacific SIDS caught just $200 million worth of tuna from these fisheries while foreign nations fishing in the same waters caught over $1 billion. That discrepancy was the instigation for the PNA, as the resource became more obviously valuable. Commercial tuna exploitation first began in the early 1950s, due largely to the development of purse seining for skipjack and yellowfin tuna, with the Pacific now providing a third of the world's total tuna supply. Most tuna are close to the equator, benefiting those countries with EEZs in that area. Eight SIDS—FSM, Kiribati, the Marshall Islands, Nauru, Palau, PNG, Solomon Islands and Tuvalu (with the Cook Islands and Tokelau)—constitute the PNA. Kiribati, with much the largest EEZ, covering three island groups and most accessible to US fleets, has been by far the greatest beneficiary, accounting for a third of the regional catch. Ironically, Nauru, which is on the equator and hosted the meeting from which the agreement got its name, has a very small EEZ, although the revenue is of enormous significance to its economy. The overall revenue from tuna has been so great that the eight SIDS have been described as 'tuna-dependent' (Bell et al., 2021; cf. Tarte, 2006). The impact of climate change, where tuna will migrate to more distant waters in the high seas beyond Pacific EEZs, has been estimated to be so substantial that in 2025 the PNA SIDS received US$107 million from the Green Climate Fund to enable their economies to adapt to that contingency.

First established in 1982, under the Pacific Islands Forum Fisheries Agency (FFA), the PNA became a separate entity based in the Marshall Islands in 2010. In that year, after 18 years of effort, it began its increasingly effective control of tuna purse-seine fishing in approximately 4.5

million square kilometres of the Pacific by licensing vessels from DWFNs to fish in their combined EEZs. Income comes from the licence fees, through the PNA's Vessel Day Scheme (VDS). The total annual purse-seine fishing effort within the combined EEZs of the PNA partners is set at around 45,000 fishing days a year, allocated to the participating parties according to their individual EEZs and their previous decade of fishing effort. The VDS offers a trading mechanism among the SIDS, allowing them to respond to climate fluctuations within the region, where the best catches vary over the 4000 kilometres from west to east according to the nature of ENSO events, evening out previous variability in access fees, so helping to stabilise government revenue (Bell, Senina, Adams, et al., 2021). Since the fees paid by vessel owners are based on the number of days fished, and not on catch, the incentive to mis-declare catch disappeared.

Fisheries income has proved invaluable for PICs, notably the northern SIDS—Marshall Islands, Nauru, FSM, Tuvalu and Kiribati—where minimal alternative local income generation exists. In Kiribati that income has been used to subsidise the price of copra from outer islands (2019). Of the more southern SIDS who did not join the PNA, only Fiji has a substantial EEZ, but it is less productive than those of the PNA states. Fiji has focused on developing a marine infrastructure of servicing and processing, airfreighting tuna to the United States and Japan, and retaining a small proportion for the less valuable cannery industry.

The SIDS themselves have increased their participation in the skipjack purse-seine fishery, with more than 50% of purse-seine vessels flagged to, or chartered by, Pacific SIDS. Taking part initially proved difficult. Early ventures failed through lack of experience and knowledge of commercial fishing and of the fisheries, acquiring vessels with the wrong specifications and excess downtime in ports (Johnson, 2021; Schurman, 1998). Fishing has become more technical and sophisticated, with onboard helicopters seeking out shoals of fish and monitoring catches. After the success of VDS, SIDS were more easily able to participate by using their days for their own vessels (Johnson, 2021). More successful participation also enabled local workers to secure employment on national ships, and men and women to be employed in land-based processing activities, or as statutory observers and scientists, even on foreign vessels. In 2022, eight women were celebrated as the first all-women deck crew on a tuna long-line fishing vessel in Fiji and perhaps in the world. Observers on vessels have dual roles in ensuring compliance and collecting scientific data, contributing to a detailed body of knowledge across a complex socio-ecological system and enabling management at several scales (Syddall, Fisher, & Thrush, 2022; Syddall, Thrush, & Fisher, 2021).

The PNA states have also gained from securing land-based activities, directly related to processing and canning, and indirectly in servicing the vessels, so generating substantial employment. Some DWFNs have invested in domestic processing capacity and constructed factories, as in PNG and Solomon Islands, or simply engaged in transhipment to containers, as in Tuvalu, employing a few hundred workers. Transhipment is irregular, depending on the location of tuna schools, that is a function of ENSO-La Niña, and competition between ports. In Madang (PNG) women made up 90% of the 2800 cannery employees, a considerable proportion of all urban employment (Havice & Reed, 2012). The tuna fishing and processing centre at Noro (Solomon Islands) has long been the largest component of the private sector in the country, employing around 2400 people in 2018, and even buying fresh vegetables for its canteen from as many as 500 local farmers (Fabinyi & Barclay, 2022). Such integration and linkages are however rare. Women valued the regularity of wages—'fast money' versus the 'slow money' of agricultural sales—and the greater autonomy that followed (Havice & Reed, 2012). After the lease fees, employment proved particularly beneficial, much being factory employment for women, albeit with some social costs and inadequate pay, sometimes leaving women open to exploitation (Rajan, 2005; Syddall et al., 2021), with others drawn into sex work (Havice & Reed, 2012). Likewise, men who worked onboard Chinese and Taiwanese tuna vessels were exposed to difficult working conditions, with a high risk of injury, and were more likely to become HIV-AIDS victims (Connell & Negin, 2012). Bycatch (of local use and value) increased employment in processing and other land-based activities. Careful management of the VDS scheme, the rescue of one key tuna species—bigeye—from overfishing, and a reduction in bycatch emphasised the contribution of tuna fishing to food security, sustainability and development in the PNA states. Its main architect, Transform Aqorau, has aptly stated: 'The salaries of health workers, teachers, and civil servants depend on the revenue generated from the sale of days from the VDS' (Aqorau, 2019: 127).

Differences between states exist in opportunities for marketing marine resources, depending on market access and supporting infrastructure, including adequate water resources for fish canneries, affordable power for refrigeration plants, and sufficiently rapid transport links between catch landing points and markets. The Marshall Islands has particularly benefitted from developing land-based infrastructure and has gone from being a bystander, selling a few licences to DWFN fleets that reaped the benefits, and with no other source of national economic strength, to playing a key role in the PNA. With less fish in its EEZ, it effectively became a transhipment and processing centre (and a lucrative shipping registry) with relatively good air transport connections to Hawai'i and Japan to satisfy the sashimi market. Majuro eventually became the busiest tuna fishing port in the world. Since 2010, the Marshall Islands Marine Resources Authority has seen commercial tuna fisheries revenue increase ten-fold (with the purse-seine fishery going from US$4 million a year to US$30 million) while implementing a conservation and management regime that, together with its partners in the PNA, maintained catch levels for all four tuna species at sustainable levels (Johnson, 2021). Marshall Islands' success, however, has made it more difficult for other SIDS, notably neighbouring Kiribati, to compete for transhipment and processing activities, or benefit from offloading tuna and bycatch that might be marketed locally and contribute to food security (Pilling et al., 2020). Even with basic transhipment, as in Tuvalu, the smallest SIDS within PNA, the country has benefited from port entry and mooring charges, inspection fees, bycatch sales, crew expenditure on land (which in 2015 alone was estimated at A$750,000 equivalent to 2.2% of the country's GDP) and some support infrastructure such as repair facilities, but not without social costs, including prostitution and substance abuse, and environmental problems associated with spillages, all curiously described as 'industrial tourism' (James et al., 2018).

Sustainability is crucial; hence, the need for careful observing and monitoring made more difficult as global fisheries depletion (including in the Caribbean) had forced deep-sea fishers to concentrate increasingly in the Pacific. In 2020, the four key tropical tuna stocks were all in a relatively healthy state, classified as not overfished and not experiencing overfishing, with sustainably high catches being taken. Efforts to reduce bycatch have had mixed success, with some shark species overfished. However, the Western and Central Pacific remains the only oceanic region in the world where tropical tuna resources are being fished sustainably but with constant concern that DWFNs wish to maximise the current value of tuna resources rather than ensure the long-term sustainability crucial to the sector and to the Pacific SIDS.

3.3.3 Illegality and Regulation

The decline of fisheries elsewhere has placed considerable pressure on Pacific fisheries—some perhaps inevitable—such as various forms of overfishing and illegal fishing. That may increase where there are conflicts between fishing nations in overlapping fishing areas where borders are not finalised (Nomura, Woodill, Sweeney, Harte, & Watson, 2024). More than a third of global fish stocks are fished at unsustainable levels, degrading biodiversity, and threatening the future of fisheries and fishers. Despite this, many governments encourage overfishing, through fisheries subsidies, such as fuel subsidies and tax exemption programmes that contribute to overcapacity and destructive fishing practices—with several large states including China, the United States, South Korea, Thailand and Japan, the biggest offenders. Attempts to eliminate this grand-scale tragedy of the commodity started at the turn of the century when it was put on the agenda of the World Trade Organization (WTO). In 2015, world leaders recognised the damage that harmful subsidies were causing to fish stocks and the marine environment, adopted the SDGs and finally reached agreement at the WTO in 2022 to eliminate harmful subsidies.

Illegal fishing remains a problem. At every scale fisheries face the problem of illegality, from small-scale pilfering and avoiding regulation in MPAs to substantial depredations within EEZs where monitoring is still difficult. Globally about 20% of fish are estimated to be taken illegally, notably in the South China Sea where boundaries are contested and marine productivity has declined substantially. Since Chinese ships catch a third of the world's fish—mainly in the Pacific—and the quantity has been increasing, as national demand and the national fleet have increased, legislation and regulation are crucial, now involving the onboard monitoring of legal operations within the PNA SIDS. The main countries at risk from illegal fishing by China are all SIDS: Kiribati, Solomon Islands, Vanuatu, FSM and PNG, where illegal, unreported and unregulated (IUU) fishing adds to fishing under formal agreements and lacks adequate onboard observer coverage (Massola, 2020). Prior to PNA the SIDS had limited capacity for monitoring and management, but surveillance vessels have been a valuable component of overseas aid, satellite imagery indicates where ships are fishing, 'reasonable agreements' regulate who might fish, reducing illegal fishing, and constraining formerly rapacious fishing from Japan and the United States. Surveillance and enforcement by member states and their former colonial allies coordinated by the regional Forum Fisheries Agency (FFA) have been extremely effective, indicative even of a further model emanating from the Pacific but perhaps too late for other fisheries regions.

Nevertheless, the total value of illegally harvested or transhipped tuna in the Pacific region has been estimated at about US$616 million a year, equating to more than 12% of the overall value of fish taken from the region and a significant loss to fishing nations with limited resources. Illegal, unreported and unregulated (IUU) fishing in the Caribbean is responsible for about 30% of the total catch (Neufville, 2022). Fleets licensed to operate in the Pacific accounted for over 95% of the total volume and value of IUU activity, indicating the need to improve monitoring and traceability throughout the supply chain, strengthen incentives to foster compliance and increase deterrents to non-compliance. The depletion of tuna stocks—whether through IUU fishing, overfishing, ocean warming or a combination of these—puts the health of the ocean at risk, jeopardises the long-term economic security of island nations and threatens sustainability in the last fisheries frontier (Pew Charitable Trusts, 2016). Because of ocean warming, reef bleaching and acidification, as few as 10% of reefs in Micronesia may remain in 2050 so that extinction rates of some reef species could be as high as 70% by the end of the century, and 40% in other tropical seas, with SIDS most likely to experience scarcer resources to the extent that conflicts over fish resources may occur (Asch, Cheung, & Reygondeau, 2018; Heck, Beck,

Reguero, et al., 2023; Lam, Allison, Bell, et al., 2020). Despite the High Seas Treaty, maintaining effective regulation, compliance, security and sustainability over a vast ocean remains difficult.

3.3.4 Climate Change

As ocean temperatures increase, fish, fisheries and fishing all change. Fishers have always had to adapt to the vagaries of weather and climate, but current rates of change are unprecedented in recent history and are accompanied by complex socio-economic changes. Climate change will influence activities already characterised by mismanagement and conflict, revealing the fragility or resilience of certain livelihood systems. Inshore fisheries will increasingly be affected, at some economic cost to SIDS, because of the impact of hazards such as cyclones on fishing and on infrastructure, already substantial in Dominica and Jamaica (Campbell & Lee, 2021; Dey, Gosh, Valmonte-Santos, et al., 2016; Galappaththi, Susarla, Loutet, et al., 2022; Hanich, Wabnitz, Ota, Amos, et al., 2018; Monnereau, Mahon, McConney, Nurse, et al., 2021; Turner, McConney, & Monnereau, 2020). Tuna in the deep oceans are more likely to be affected by, and more mobile in response to, climate change and evolving ocean temperatures than lagoon and reef species; hence, climate-driven redistribution of tuna has the potential to disrupt the economies of Pacific SIDS and the sustainable management of the tuna fishery. However, coral reef fisheries could decrease by 20% by 2050 and coastal aquaculture be less efficient (Bell, Ganachaud, Gehrke, et al., 2013). Eastwards and northwards, fish migration has already reduced the supply of albacore tuna in Fiji by around 30%, leading the Levuka factory of the Pacific Fishing Company (PAFCO) to reduce its operations to a four-day work week to maintain economic viability. That has benefited Tuvalu and Kiribati, whereas Samoa, Tonga and Fiji 'have very little' left (*Seafood Source*, 11 December 2019). By 2025, Fiji was seeking a deal that would enable access to the EEZs of Kiribati and Tuvalu.

Under a high greenhouse gas emissions scenario, tuna in the waters of the eight Pacific PNA SIDS could experience a 20% decline by 2050, with a similar reduction in access fees, estimated at an annual US$90 million for the SIDS, and thus substantial government revenue, hence the need for loss and damage reparations. To avoid such losses, the SIDS have sought to lock in the present system in perpetuity (Bell, Latu, et al., 2021). As fisheries extend eastwards, monitoring, control and surveillance to combat illegal and unregulated fishing would also become more difficult, threatening sustainable management. This suggests future development issues since the region accounts for 86% of jobs in the global fishery and aquaculture industries. Six Pacific SIDS—Kiribati, FSM, Solomon Islands, Marshall Islands, Tuvalu and Palau—are among the 14 countries most at-risk to marine fisheries loss (Thiault, Mora, Cinner, et al., 2019). Moreover, fishing in the high seas is surprisingly unprofitable because of transhipment costs (and losses from illegal transhipment) and the complex vertical integration of fisheries. Predicted change, although relatively slow, will reduce the potential of the fishery significantly, despite the obvious mobility of fish, but uncertainties exist over future climate change, ocean temperatures, and any impacts of acidification and pollution. Climate change strikes at the core of island economies since little else offers the potential of fisheries. It is vital that it survive climate change, overfishing and illegality.

3.4 Food Security

The shift away from agriculture, and to a much lesser extent fisheries, has culminated in Caribbean SIDS like the Bahamas, and in the Pacific, Nauru, having little more than a few token backyard gardens. Only exceptionally is urban food gardening of much significance despite a brief COVID-inspired revival. Increasing proportions and volumes of food are imported and purchased, with a consequent decline in nutritional status in most SIDS. On coral atolls and other low islands, lack of fertile land constrains opportunities to diversify food sources, alongside coastal land losses, but much is influenced by convenience, tastes and fashion, prices and preferences. Greater pressure on land, seas and forests has contributed to a downward spiral in production and productivity. Although food insecurity in both regions was worsening in the early 2020s, the Caribbean fared better than the Pacific, where Nauru and Tuvalu were particularly disadvantaged (Atzori, Sonneveld, Alfarra, et al., 2024; FAO, 2024; Stanberry & Fletcher-Paul, 2024).

Food security is a key determinant of health and an indicator of household economic stability. A resilient food system is where food is accessible, affordable, healthy and desirable, that is produced and distributed sustainably (including sufficient local production to support farmer livelihoods, so contributing to the economy). All SIDS remain susceptible to food insecurity, with high levels of food imports, demonstrably problematic during COVID-19 when transport was disrupted. Sharp increases in food insecurity, associated with rising food costs, followed both the pandemic and the war in Ukraine (affecting grain exports), while climate change has worsened food security, disrupting agricultural and fishing strategies, and contributing to deficiencies in micronutrients, particularly amongst children and other vulnerable groups. High imports, amounting to 60–80% of food consumed in the Caribbean, are driven by domestic consumption and the tourist industry. The disadvantaged urban poor rely on cheap imports since price is usually a factor, and prepackaged, processed foods usually provide cheap options. But urban middle-income groups consume more expensive food imports. Individual choices have been reinforced by governments' cheap food policies that have exploited cost economies from global food sources but disadvantaged local producers (Mohammadi, Singh, McCordic, & Pittman, 2022). Accompanying this are oligopolistic food distribution tendencies, low investment in the domestic food market, the shift towards a corporate food regime, inefficient marketing and institutional policies supporting export regimes, as, for example, in St. Lucia (Saint Ville, Hickey, & Phillip, 2017). Small-scale fishing faces the same 'power asymmetries' that make combining local knowledge and practices with market needs difficult (Fache & Pauwels, 2020).

Both agriculture and fishing have been threatened by 'more sophisticated resource extraction technologies' without more sophisticated management mechanisms (Fazey, Pettorelli, Kenter, et al., 2011). Complex human ecologies have become more fragile with greater individualism, resulting in less social cohesion and more disputes over land and marine tenure. Changing production practices and the coastal squeeze have intensified pressure on remaining agricultural land, with production per capita falling rapidly in the late twentieth century. In extreme cases, in settlements such as Sisiak (Madang, PNG), residents simply lacked the space to grow food and the cash to purchase it and were often hungry (Gewertz & Errington, 2021). Only exceptionally in atoll SIDS, over place and time, where intensive aid projects have existed, as in Kiribati and Marshall Islands, have such circumstances been reversed and a greater degree of food security achieved (Cheng, Tseng, Iwaki, & Huang, 2021), but that has often been illusory or temporary. In parts of the Caribbean, such as the Bahamas, urbanisation and population migration have meant the neglect of regions that might have contributed to food security (Mercer, Kelman, Alfthan, & Kurvits, 2012). Food security and food sovereignty have worsened.

Food security, once primarily achieved through subsistence farming, fishing and small-scale market trade and exchange, now involves a mix of local and imported foods; hence, the ability to purchase adequate food has become crucial, especially in Caribbean SIDS. As local food systems changed with new cultural and taste preferences, status, and time-short convenience, urban wages and incomes have brought movement away from a degree of monotony but increased dependence on store-bought food. The recent explosion of 'fast foods' has introduced a new kind of food insecurity and undernutrition and exposed SIDS both to the uncertainties of global food production, supply and cost, and to NCDs (e.g. Guell, Brown, Iese, et al., 2021; Guell, Brown, Navunicagi, et al., 2022; Guell, Saint Ville, Anderson, et al., 2024; Halliday, Morrissey, Saint Ville, et al., 2023). 'Two-minute' noodles have much to answer for (Errington, Fujikura, & Gewertz, 2013). Diversity has declined as local varieties, often attuned to local conditions, have disappeared; in Fiji more than half the varieties of coconuts disappeared as the focus shifted to producing only those varieties suitable for coconut plantations and copra production (Shah, Moroca, & Bhat, 2018). Many breadfruit, taro and banana varieties have similarly disappeared (e.g. Kagy, Wong, Vandenbroucke, et al., 2016) as have intensive agricultural techniques such as mulching and terracing. A narrowing of plant species and diets benefits efficiency but neither nutrition nor sustainability. Only in the Melanesian SIDS is food consumption more likely to be dominated by local foods, and a greater resilience retained, evident in several Solomon

Islands and Vanuatu Islands (Allen, 2020; Georgeou, Hawksley, Wali, Lountain, et al., 2022; Nef et al., 2022; Vogliano, Raneri, Coad, et al., 2021; Wilson, 2013), pointing to the value of retaining diverse agricultural and fisheries systems (that also cope better with hazards). More often, contemporary work in food gardens, as in St. Vincent, Fiji and peri-urban Vanuatu, provides nostalgic images as much as significant production (Guell et al., 2021; Savage, Bambrick, & Gallegos, 2021a). Gathered wild foods are similarly declining as components of diets.

COVID-19 resulted in disrupted supply chains, higher market costs and reduced food purchases (e.g. Iese, Kiem, Mariner, Malsale, et al., 2021; Lau & Sutcliffe, 2021) but without local food production increases. That pushed international food prices to record levels, with food particularly expensive in SIDS, where supply chains were badly affected. In atoll states where a high proportion of foods were ordinarily imported, consumer price indices rose alarmingly (Forbes, 2021). Demand for high-cost market food, including fruit, livestock and fish, tended to fall as incomes fell. Many displaced workers and their families, not only from the tourism sector (see Chap. 4), had to return to agricultural livelihoods. Not all displaced workers could reclaim and revitalise rural lives, so putting a burden on local ecologies and environments and on women who were the primary producers there. Tensions could occur over land and sea tenure, as people returned from lost urban work and sought to claim land they barely remembered or was already in use. Since agriculture could not immediately be productive and output consumable or marketable, the transition was difficult. Similar outcomes occurred in inshore fishing but with increased pressure on resources since, unlike agriculture, immediate returns were possible (Steenbergen, Neihapi, Koran, et al., 2020). Where rural population densities were high, as on most Tuvalu atolls, established residents resented returnees, who soon remembered that trying to develop agriculture on coral 'soils' was one reason they had moved away in the first place (Connell, 2021). Indeed, prior to the arrival of COVID-19, urban women, including those in the impoverished urban settlement of Sisiak (PNG), had no wish 'to acquire the skills necessary for village life. Women did not look forward to paddling a canoe, setting fishing nets, cutting firewood and smoking fish' (Gewertz & Errington, 2021: 229).

Many returned to urban areas as soon as it became feasible. Attitudes to agriculture have hindered food security.

Backyard agriculture intensified in urban areas, as in Fiji and Vanuatu, while public spaces were claimed for production. Households in urban areas engaged in 'pop-up' agriculture, cultivating new plots and resurrecting older, sometimes-abandoned techniques of food production, storage and preparation, as they faced an enforced self-reliance. Disrupted supply chains and their effects on nutrition as food supplies declined and prices of food increased, resulted in households in many places eating fewer meals and less food (Lau & Sutcliffe, 2021; Randin & Turagabaleti, 2021; Ride et al., 2021). By default, more sustainable agricultural practices were briefly more affordable, provided more nutritious food and enhanced biodiversity (Davila et al., 2021). That was also linked to building resilience to natural disasters. Reversing the trend away from agriculture was scarcely by choice. Local food production could not immediately compensate for the lost incomes that enabled food purchases. Theft of kava plants, trees, livestock and root crops was reported mainly in peri-urban areas of Pacific SIDS. Re-establishing agricultural livelihoods proved challenging. In several SIDS, including Samoa, Tonga, Tuvalu and Fiji, governments introduced programmes to train local youth in agricultural techniques and distributed seeds and planting material of fast-growing crops such as sweet potato and vegetables, increasing self-reliance if sometimes adding to the domestic labour burdens of women (Iese et al., 2021; Vainikolo, 2021). Rural communities that had retained more traditional practices around food production were better able to cope. Sago production revived in parts of Melanesia as did once-abandoned food preservation techniques in Micronesia.

In 2020, local food production and food sharing practices across 199 coastal villages in seven SIDS (FSM, Fiji, Palau, PNG, Solomon Islands, Tonga and Tuvalu) conferred some degree of resilience. Communities in countries more reliant on imports were almost twice as likely to report food insecurity compared to those least reliant; in Tuvalu and FSM shortages of imported food in local stores were critical problems, and food sharing was less likely. Food security was also less likely in villages experiencing in-migration, with greater market integration and significant imports, and on unproductive atoll soils (Ferguson, Tuxson, Manghubai, et al., 2022).

While that suggested a plausible way to a more sustainable future, many 'new agriculturalists' returned to urban centres in due course (abandoning the hard work entailed in agriculture), demonstrating that increased food security would not be easily achieved and certainly not in-country. Yet imports meant lengthy 'food miles' and carbon footprints while moving fruit and vegetables in refrigerated containers has a particularly large footprint.

Climate change worsens prospects of food security throughout the SIDS both directly and indirectly, mainly by exacerbating existing trends, notably the increase in significant hazard events (with direct impacts on crops, reefs and fish, and indirect impacts on infrastructure, greater mobility away from former agricultural areas and a dependence on imported foods) with lack of water security likely to become a problem (Hidalgo et al., 2020). Climate change has however been blamed where population pressures, overfishing and other factors are more probable explanations (Savage, Bambrick, & Gallegos, 2021b). The food security of relatively poor population groups in coastal locations is most likely to be negatively affected (Lenderking, Robinson, & Carlson, 2021; Petzold & Scheffran, 2024). In some Pacific SIDS, local people have responded to the increased propensity for cyclones and droughts by changing planting strategies, both spatially and temporally, with several smaller gardens in varied locations and accumulated seed banks, and by sharing practices (McNamara, Clissold, & Westoby, 2020; Pickering, Pearce, Manuel, Doran, & Smith, 2023). Such strategies aligned with Melanesian agricultural systems that existed more than a century earlier. Some SIDS have attempted to incorporate climate-resilient crops, but these have proved difficult to develop where ecological range is limited on small islands, just as high-yielding varieties have had little relevance to SIDS. Taking out crop insurance is prohibitively expensive and entirely absent in the Pacific (Carballo & Dos Reis, 2018).

High levels of food imports (not all of which are detrimental to health) and changing tastes and practices have accentuated the movement away from food security, which has long exasperated Pacific islanders themselves. As the Tongan anthropologist, 'Epeli Hau'ofa, pointed out:

> Once you are educated, once your mind is expanded, subsistence on a remote little island is simply unacceptable … Travel has changed our material values and expectations, drained us of talent … and transformed eating habits to the point that most of our protein and flour is imported … Our imports are six times higher than our exports, yet one-third of that import bill is for food we have to buy only because our tastes are swinging away from what we are able to grow for ourselves. (quoted in Dyson, 1982: 120–121)

For over half a century such trends have been hard to reverse, whether policies were directed at economic issues related to trade and taxation, nutrition or agricultural development. Various remedies have failed: islanders are used to eating rice and noodles, cultivating traditional food crops such as taro is demanding, market prices for local foods are high and food aid has focused on quick solutions rather than on local foods, with reversing the trend particularly difficult on atolls (Ahlgren, Yamada, & Wong, 2014; Connell, 2015; Tui & Fakhruddin, 2022; World Food Programme, 2025; Chap. 6). Taxing imports of harmful foods is against WTO regulations. Not only has food insecurity been costly, but it has been a primary contributor to the increased significance of such NCDs as cancers, cardiovascular diseases and diabetes that indirectly have been a substantial drain on household and national lives and economies (Chap. 6).

3.5 Timber: A Renewable Resource?

Other than in PNG, only the larger islands of Fiji and Solomon Islands have significant timber industries, the former based on renewable species (made up of introduced pine and mahogany plantings, and a small 'modern' timber industry even producing a small amount of furniture), but the latter has been exploited by outside interests, at unsustainable rates, with damaging economic and environmental consequences. Otherwise, SIDS have limited forest cover, and, independent of forestry, this is often declining because of growing pressures for commercial agriculture and the use of timber for construction and firewood (and copra or cocoa drying) (e.g. Garonna et al., 2009). PNG is China's single largest supplier of tropical logs and sends almost all its felled timber to China. The economic benefits have been widely perceived as being uneven, and a basis for corruption, and far less than the social and environmental costs, with landowners gaining little (West, 2016).

In Solomon Islands, logging has created ongoing problems, being often undertaken without adequate local comprehension or compensation and with severe environmental consequences (Kabutaulaka, 2006), disadvantaging local landowners in remote areas with limited access to alternative income-earning opportunities and other forms of support. Despite at times contributing more than 60% of exports by value, the industry has failed to generate lasting local benefits; unsustainable logging has undermined subsistence livelihoods, including fisheries, and contributed to social conflict outlasting the operations themselves, sexual exploitation, excessive alcohol use, and reinforced gender disparities through structurally excluding women from decision-making and benefit-sharing (Minter & van der Ploeg, 2023). Logging continues at unsustainable levels, with support from politicians (some of whom own logging companies) and traditional landowners (or claimants to land) paid off to secure rights, and with the timber destined for China (Anderson, 2023). Forest exploitation continues, both legally and illegally. Although some of the worst excesses in Kolombangara and New Georgia islands have been curtailed, many of the larger forests have been effectively worked out.

Vanuatu minimised environmental problems by banning log exports, assisted by the fortuitous geographical fragmentation of its forests. Vanuatu learned from Solomon Islands and closed two major Asian logging companies in the 1980s and 1990s on Malakula Island after their operations antagonised local people. Vanuatu once possessed valuable stands of sandalwood that traders removed in early colonial times, but efforts are being made to redevelop them and have been successful on some islands such as Aniwa, based on high national and international prices and global demand for health and beauty-care products (Thomson, 2020; Wilson, 2013). Potential exists for developing niche forest product value chains, including small-scale producers in the large Melanesian islands, as in Vanuatu, but difficulties inherent in sourcing from smallholders, whether for sandalwood and Canarium almonds, include small production volumes in dispersed plots with difficult access, high transport costs, low technical know-how, difficulties in accessing inputs and finance, unreliability of supply in terms of quality and quantity, market access, and lack of reliable information about the resource base and potential management (Blumfield, Reverchon, & Vigulu, 2018; Vega, Page, & Ota, 2023). Canarium almonds (galip) have a significant local role in Melanesian markets and have unrealised commercial potential (FAO, 2024). Unlike nuts, sandalwood is not perishable so is particularly useful for small remote islands with poor transport connections (Thomson, Bush, & Lesubula, 2020). It is endemic to Fiji and Tonga, but projects to develop production in both countries have been challenged by the need for broad forestry technical skills offset by a large turnover and migration of professional and technical staff in both countries (see Chap. 9).

The Caribbean, a biodiversity hotspot, has experienced a high rate of deforestation and forest degradation (Maharaj, Asmath, Ali, Agard, et al., 2019), but compared to extensive deforestation in the early twentieth century, forest cover has increased in several Caribbean SIDS since independence as agricultural lands have been gradually abandoned and left to naturally regenerate or be replanted: albeit a revival from a very small base. Increased forest area in St. Vincent followed the establishment of small plantations, primarily for water, soil and coastal protection, and some timber and fuelwood production. Whereas in Dominica, a continued decline followed agricultural expansion, although deforestation was slowed by mountainous terrain and inaccessibility (Peteru, Regan, & Klak, 2010). Otherwise in the Caribbean SIDS deforestation has largely 'bottomed out' with increased areas of forest gradually becoming more common, contributing to greater resilience, but as the outcome of declining agricultural expansion rather than adoption of forest conservation practices.

With few exceptions, forests are being cleared rather than planted; hence, developing adequate forestry management practices is difficult, despite environmental damage. Soil erosion and the siltation of rivers and coastal waters have reduced water quality and damaged some coral reefs and coastal fisheries. Deforestation has degraded river habitats; legislation to regulate forestry operations is unwieldy and provides little compensation to landowners, while governments have had neither capacity nor will to enforce the legislation (Connell, 2013a). Removal of mangroves has produced numerous environmental costs; the coastal squeeze and mangrove sales for firewood have brought erosion and loss of juvenile fish habitats. In several SIDS, different versions of deforestation, riddled with promise and corruption, have brought short-term benefits and long-term disappointment.

3.6 The Lure of Mining

Mining has a long history in the Caribbean, and it was the lure of gold (King Solomon's mine) that first brought Europeans to the Pacific, successfully in the PNG islands, where it marked the start of 'modern' history but less so in Solomon Islands. Mining history in the Caribbean can be traced back to the pre-independence period but was established in the post-war period as bauxite extraction expanded. Mineral deposits exist in the wider Caribbean region but are almost absent from the island SIDS. In the Pacific, too, they exist on just a few islands.

Bauxite is mined in the Cockpit Country in Jamaica, contributing about 4% to Jamaica's GDP, but it has been a controversial development, owing to its extractive and exploitative nature, where aluminium is smelted elsewhere and profits made and transferred overseas. The bauxite industry has also created environmental challenges due to waste disposal, alongside early problems of displacement that have persisted despite new regulations, and complications over both the wellbeing of the local Maroon population and mining in a conservation area (Cowell, 1987; Sprague, 2015). Cockpit Country is the largest remaining forest in Jamaica, the source of six rivers and about 40% of Jamaica's water needs. Ensuring its preservation is key to promoting climate resilience. The challenge of mining without causing irreversible environmental damage, and with any plausible degree of sustainability, is an issue throughout the SIDS, especially where mining, as on Lihir (PNG), is coastal.

Oil is invaluable in Trinidad and Tobago, arguably the world's oldest site of commercial oil extraction, but natural gas has gradually become more valuable. Trinidad is the largest producer of natural gas in CARICOM and the second largest producer of oil after Guyana. Its economy (and ethnic structure) is thus very different from smaller SIDS, being heavily dependent upon these resources. Oil and gas account for about 40% of GDP and 80% of exports but only 5% of employment. The electricity sector is fuelled entirely by natural gas, and Trinidad is the world's largest exporter of ammonia. Like other mineral producers, the national economy is negatively affected by fluctuating oil and gas prices, and Trinidad has sought economic transformation through diversification, whether a growing tourist industry or the creative industries, and has been able to ward off 'Dutch disease', the resource curse, where other components of the economy are in disarray (Hosein, 2021). Only if such diversification proceeds will Trinidad have much resemblance to other island SIDS.

Mining has had a long history in PNG, with alluvial gold mining opening up parts of New Guinea at the end of the nineteenth century. It boomed in PNG after the construction of the Panguna copper and gold mine on Bougainville Island shortly before independence, which then provided more than half the nation's exports. Fifteen years later, disputes over environmental impacts and the distribution of profits and royalties brought a violent end to the mine, and a civil war, at the end of the 1980s. Two decades later disputes have continued to centre on the ownership and role of any new mine (especially its role were Bougainville to become independent), remediating past environmental damage, and compensation for affected communities, complicated by unresolved disputes between different landowner groups over the mine's future (Laslett, 2014; Regan, 2017). Only the smallest mines have been without similar problems but less violently. Matthew Allen has pointed out that 'one would be hard-pressed to name a single extractive project in Melanesia that has not produced violence in some form' (2018: 6). The larger PNG mines, notably Ok Tedi and Porgera, have both intermittently closed over versions of the problems that plagued Panguna. The Lihir mine is a rare example of a mine that has both proceeded without substantial disruptions (despite, or perhaps because of, waste being dumped at sea) and is in a relatively small island (Bainton, 2010). Gold Ridge mine, the only mine in Solomon Islands, has gone through similar phases of disputes and unrest over environmental damage and land tenure, being closed down since 2014, despite its potential.

Ninety per cent of the entire nation of Nauru has experienced environmental devastation from a century of phosphate mining that briefly turned the country into the second wealthiest country in the world. When mining largely ended at the turn of the century, bad capital investments had left a degraded Nauru impoverished, the most globally significant example of the resource curse (Morris, 2023). Mining in Fiji has been as long-established as in PNG, but no new mines have been developed in recent years, and it remains largely a residual economic activity. The Vatukoula gold mine and a small mining town began in the 1930s, but its subsequent history has been fraught with uncertainties, especially for the local population dependent on this single industry for their livelihood. Over the years, much as in PNG, concerns and periodic conflicts covered wages, safety practices, housing conditions, gender inequities, associated social problems and the mine's environmental impact. It briefly closed in 2007 but has continued to produce smaller quantities of gold that have drawn attention to dependence on global prices and the tasks of achieving sustainability. On the other side of the main Fijian island, Viti Levu, the Namosi area has promised copper mining for 50 years, but prices have never been high enough to compensate for the low-grade deposit. Similarly, manganese in Vanuatu has suggested possibilities, but they have remained unrealised while Gold Ridge remains mired in disputes. Mining has never quite delivered economic and social success. Too often it has generated social and economic costs. It employs few (and numbers of workers tend to fall rather than increase), has often foundered on land, environment and compensation issues, and has contributed to resource curses. It has, however, generated substantial wealth for large multinational companies and for national governments, as opposed to landowners, and it has not deterred several SIDS, including Nauru, from considering the possibilities of deep-sea mining.

3.6.1 Scraping the Bottom? Deep-Sea Mining

The possibility of deep-sea mining (DSM), much like tuna fishing, has globally centred almost exclusively on the Pacific, initially around the New Guinea islands but later in the eastern Pacific. Two thirds of the oceans are beyond national jurisdictions and constitute the high seas; hence, an International Seabed Authority (ISA) was created in 1994 through the United Nations Convention on the Law of the Sea (UNCLOS), to regulate mining and other sea-floor activities. Despite the ISA being in Jamaica, only Jamaica in the Caribbean has expressed any interest in DSM, because of the costs and uncertainties of involvement, and the absence of any known seabed minerals there (Davis-Mattis, 2023). Mining has been anticipated since the 1970s but has been delayed by technological, economic, environmental, legal and political complications. Multiple research cruises have identified promising mineral resources, mainly manganese nodules and cobalt, where Kiribati, the Marshall Islands, the Cook Islands and the Clarion-Clipperton Zone (CCZ) appear to have the greatest potential (Waiti & Lorrenij, 2018). Although assessments only hint at content and volume, in a global context where demand for minerals critical to the clean energy transition is considerable, some regard DSM as now 'imperative' (Wang, Shu, Zhou, et al., 2025). Others see broad 'socio-ecological injustices' (Murphy & Gard, 2025), environmental costs and few employment opportunities.

An estimated 8.9 billion tons of nodules are scattered over the ocean floor, worth a possible US$20 trillion: the world's largest and richest known resource of polymetallic nodules, and a substantial proportion of the world's currently known cobalt resources. The nodules are rich in the four essential metals needed for batteries: cobalt, nickel, copper and manganese (Bainton & Skrzypek, 2022). As land resources are mined out and technology evolves, the polymetallic nodules of the eastern Pacific will become increasingly attractive, hence the growing interest of various countries, including China (Zhang, 2018). This interest comes at a time of heightened geostrategic competition, with China seeking influence and resources, including timber, fish and minerals (Chap. 8). China holds several exploration permits, has invested significantly in DSM machinery and is a world leader in both submersible technology and in manufacturing solar panels, batteries and other technology that require critical minerals.

Hundreds of exploration licences have been issued to Pacific SIDS by ISA, but no mining activities have started anywhere in the world, with technology and strategies of investment still evolving, and doubts attached to economic viability. SIDS, conscious that technology was in its infancy, have adopted a precautionary approach. By mid-2022, at least seven Pacific SIDS, including Palau, Fiji, Samoa, PNG, Tuvalu, Vanuatu and FSM, had either declared their total opposition to mining or called for a moratorium on exploration licences and any actual mining licences (RNZ, 29 June 2022). At the United Nations Ocean Conference in Lisbon in June 2022, the then Vanuatu opposition leader, Ralph Regenvanu, called for a regional ban:

> I appeal to the Republic of Nauru, the Cook Islands, the Republic of Kiribati and the Kingdom of Tonga to follow the lead of Tuvalu and draw back from the brink ... deep-sea mining will disturb the largest and most stable carbon sink we have, and in the event that mining accidentally hits a methane seep on the ocean floor, scientists warn of 'doomsday climate scenarios'. [We are] at the edge of a great abyss. A critical juncture in human endeavour, a time of complex and interlinked crises—the climate crisis, the biodiversity crisis, the pandemic, and the economic crisis strangling our countries under the weight of debt. These interlinked crises are crises of our own making, a spectacular failure of the dominant value system that underpins our economic structures. Yet stories of mountains of gold persist, and now we witness another wave of greed to exploit the ocean floor for minerals. (PacNews, 29 June 2022)

Concern was greatest where tourism was particularly significant, as in Fiji (Folkersen, Fleming, & Hasan, 2018) and Palau, where the President, Surangel Whipps, stated in 2024 that 'the deep seabed ... is a treasure to be held in trust for future generations, not a commodity to be exploited by any nation state or corporation' (quoted in Murphy & Gard, 2025: 11). SIDS lacked the regulatory framework and human resources to deal with a new and rapidly evolving form of development, where land and marine tenure took on a new dimension: a commodity frontier within the blue economy (Bourrel, Swaddling, Atalifo, & Tawake, 2018). Over half a century, divisions over DSM have intensified.

Not all SIDS were concerned, and Nauru, Tonga, Kiribati (and the Cook Islands) have supported DSM, with Nauru, Tonga and Kiribati issued mining licences by the ISA. Nauru, despite its environmental and economic devastation from terrestrial mining, has constantly sought economic diversity through DSM. Its Ambassador to the United Nations and the ISA, Margo Deiye, has said: 'We need to diversify, we have more ocean than land'. Mining would however be in international waters. Through Nauru Ocean Resources Inc. (NORI), Nauru gained a 15-year exploration contract from ISA and has sought to begin mining near Clipperton, contingent on ISA approval which required analysis of who would be the beneficiaries (Jaeckel et al., 2023). Nauru resisted opposition from other PICs, who sought a moratorium due to environmental concerns, and the lack of scientific data related to these concerns, and remained the strongest proponent of DSM, whereas Kiribati lost interest in 2025, recognising high costs of exploration, continued economic uncertainty and the lack of companies outside China with the technical capacity to begin mining.

Proponents of DSM claimed that it was scarcely mining at all, since nodules would simply be vacuumed up by robots from the seabed and then pumped to the surface, with minimal impact on the seabed or the sea. Sediment would not be churned up nor carbon be released (despite oceans being a carbon sink). Opponents argued that DSM amounts to scraping the seabed and would create sediment plumes that threaten ecosystems, and especially fish stocks, where PICs depend on fisheries for revenue, nutrition and livelihoods. Damage would be almost invisible by being centred hundreds of metres below sea level and distant from human populations. Over half a century, as the possibility of mining increased and interest expanded, the environmental pendulum swung against deep-sea mining and even exploration.

More environmental research, in one of the least-known ecosystems on earth, intensified opposition to mining. DSM could negatively impact biodiversity to a greater extent than terrestrial mining, and restoring diversity would be particularly difficult in a context where little is still known about the marine fauna and flora, or seabed regeneration, and where the deep sea operates on a slower time scale than the land. Seafood organisations have called for a moratorium on the grounds that mining plumes and noise may impact the migration of tuna species, presently moving eastwards towards the CCZ, following climate change (Amon, Palacios-Abrantes, Drazen, et al., 2023). Analysis of a very small area of the Pacific, a year after a two-hour-long test excavation, showed that the density of fish and shrimp had declined by about half in areas directly affected and in adjoining areas (Washburn, Simon-Lledó, Soong, & Suzuki, 2023). Excavation over a longer period and over a greater area would have greater impact. In 2023 scientists reported that they had identified 5578 species of mainly benthic invertebrate fauna in the CCZ, 92% of which were previously unknown to science, with the probability that mining would have a negative impact on this biodiversity (Rabone, Wiethase, Simon-Lledo, Emery, et al., 2023). Such discoveries may hinder decisions over mining. DSM retains an uncertain future but pressures for mining in the 'earth's final frontier' will intensify, exemplified by President Trump's desire to ignore the ISA and fast-track ocean mining while the Pacific SIDS remained divided on the issue.

3.7 Making Things: A Role for Manufacturing?

Manufacturing is at best a small-scale urban phenomenon in most SIDS, where it typifies import-substitution industrialisation in its early phases, being concentrated on light manufacturing (such as packaging) and food production (such as bread and biscuits). Aspirations to being Newly Industrialising Countries (NICs) have long gone. What was regarded as excessive rural-urban migration in the Caribbean in the mid-1970s was 'not doing so to fuel a new wave of industrialisation. It is too late for that' (Cross, 1979: 149). Manufacturing is most evident in the relatively low-value and high-weight food and drink industries (where a significant component is water); hence, many SIDS have soft drink and beer industries. Even here, the costs of importing machinery and bottles and even printing labels can be prohibitive. In 2022, Paradise Beverages, the manufacturers of Vailima beer (advertised as 'The Taste of Samoa'), chose to close its brewery operations in Samoa and produce the beer in Fiji, with the loss of around 50 jobs. Its commercial viability had declined because of ageing machinery, a 2019 measles outbreak and the disruption caused by COVID-19 (RNZ, 1 July 2022). This typified a general pattern of centralisation, and the problems of scale, skills and technological development, but especially transport costs for component imports and possible exports.

Geography (including human and physical resource endowments), culture and politics hinder industrial development. Limited natural resources, small domestic markets, scarce skills, high energy, water and insurance costs, inadequate access to technology (including spare parts) and investment capital and the lack of economies of scale have hampered even limited development. Conventional disadvantages for export production, notably transport and labour costs, are accentuated for smaller SIDS. In Samoa and Fiji agricultural processing firms operate at less than full capacity because of small markets and the seasonal supply of agricultural produce. Small businesses struggle with limited access to finance, inadequate supportive business development services, lack of entrepreneurial skills (including lack of strategic planning), copying already existing businesses and challenges in adopting digital technologies (Bevacqua, Hunter, Deshpande, & Morgan, 2025). Beyond this, businesses in Vanuatu are disadvantaged by high electricity costs, insurance costs, cyclone damage (and, more recently, earthquake damage) and weak government support because of political instability (Vanuatu Chamber of Commerce, 2023). Higher wages than in Asia and Latin America, open economies and consumer preferences for overseas products, including food and drink, and lack of market knowledge have further restricted industrialisation. The only significant exceptions are the fish processing and packing industries of some Pacific SIDS.

In most SIDS manufacturing wages remain low (and unions rare) compared with public service employment. Footloose industries benefited from such suppliant regimes and were often outposts of transnational corporations that were able to migrate easily if conditions, such as 'tax holidays', changed. Several countries established export processing zones, with special land and legal regimes, financial incentives and attractive taxation policies. At the peak of Caribbean industrialisation in the late 1990s, St. Lucia had 70 factories employing over 3500 workers—almost 10% of the formal workforce. Most workers were women 'explained by their willingness to work for lower wages and to be more tolerant than men', but they could not survive and raise families on manufacturing wages without additional income-earning opportunities (Potter, Barker, Conway, & Klak, 2004: 340). In the 1980s an electronics industry was successful in Barbados, St. Lucia and St. Kitts, and exports briefly thrived. Even in the heyday of manufacturing in St. Lucia, Tonga and probably elsewhere, factory workers in such low-wage circumstances saw the principal goal of factory employment as that of obtaining skills valuable for subsequent emigration (Connell, 2013a). Certainly, as elsewhere, an inefficient and even inappropriate industrialisation, dependent on imports of capital and intermediate goods, was sheltered behind tariff walls and barriers.

Where industry was relatively successful, as in Barbados, the factors that contributed to that were good labour relations, a skilled and pliable labour force (usually female), low (usually minimum) wages and weak unions, financial incentives, political stability, suitable real estate (for expatriate owners and managers) and ease of access to foreign markets (Connell, 2013a). However, steady decline in Barbados' manufacturing industry from the 1970s, when the industry contributed significantly to national economic activity, representing around 20% of GDP, had become just 4% by 2014 (Drakes & Moore, 2015). Island industries were less able to compete in global markets since they were too small (in terms of capital and workforces), and unable to achieve basic economies of scale and productivity, so were assisted with subsidised infrastructure, the ability to expatriate profits and avoid exchange regulations, and concessionary trade agreements. Without subsidies, de-industrialisation continues, as in Tonga, where an infant industrial estate, protected by tariff walls, once had 21 firms which exported footballs, woollen garments and watches; little remains beyond a brewery and a small fish-exporting business. In such circumstances, with tax breaks and other concessions offered to industries, national benefits were relatively few, and tightened global trade restrictions effectively ended them. A car components factory in Samoa, employing 950 people in 2012, that had survived through special amendments to Australian trading regulations on 'rules of origin', closed in 2017 after the car industry wound down in Australia and Australian subsidies ended: an example of the flow-on effects of changes in global supply chains and how the neoliberal global economy could undermine the livelihoods of workers. The company had predominantly employed young early school leavers, contributing to some degree of empowerment for less powerful members of Samoan society who otherwise had very limited prospects to develop confidence in themselves and earn money and status in their families, their villages and the workplace (Levi, 2019). Manufacturing employment might be limited, but it could be invaluable.

In this century, only Fiji has had a relatively successful manufacturing industry but mainly confined to the garment industry. At its peak towards the end of the 1990s the clothing industry employed about 19,000 people (and garment exports constituted a third of the value of all exports in 2000), but factories have closed, wages have stagnated and by the mid-2000s employment had fallen to around 8000 people (Connell, 2013a). By early 2025 that number had halved to around 4000 while between 2015 and 2024 the value of exports had also halved. Higher wages, and a new minimum wage, suggested further decline as the industry struggled to remain competitive in the global market, due to increasing operational costs, a shortage of skilled labour and inadequate training opportunities. Concern followed the inability to retain workers when superior opportunities attracted workers to migrate to the Australian PALM scheme (see Chap. 9; RNZ 20 March 2025). Small clothing factories still exist, mainly employing women on low wages, but employment in manufacturing is falling rather than growing.

The establishment of export-oriented manufacturing largely occurred through protective trade agreements and widespread local concessions, and collapsed when these conditions disappeared. The optimism that once prevailed over industrialisation in Fiji, and in rather larger Trinidad, that sometimes made them seem models for SIDS, has been vanquished in more liberal trading conditions. Niche successes, such as the global marketing of FIJI Water (Connell, 2006; Jones, Murray, & Overton, 2017), have been threatened by concerns over environmental impacts, rival products and changes in fashion. Likewise import-substitution industries have been constrained by various factors, not least transport costs, small local markets and absent economies of scale. Manufacturing, particularly for export, has rarely been a viable option in the SIDS, and diversification has not ensued.

3.8 Services New and Old

The absence of significant manufacturing has meant that the service sector makes a much greater proportionate contribution to SIDS economies and employment than in most larger states. After the public service, the largest components of the service sector are tourism (Chap. 4), followed by retail services. A few SIDS developed finance sectors, although business development has been constrained by many of the same factors that have limited manufacturing: government attitudes, limited technical and management skills, small markets, and lack of capital. In rural areas especially, small businesses, especially stores, are often not centred on profitability and have supported communities rather than contributed to a more extensive commercial development. Where the public service, finance and tourism dominate, and agriculture and fishing are trivial, as in the Bahamas, employment in services involves over 90% of the workforce. In smaller SIDS similar percentages prevail but are often biased towards the public service. Universally outer islands have become 'government islands', with employment almost exclusively in the public services, mainly education and health, with public sector employment eagerly sought after as stable, relatively well-paid, undemanding and in pleasant surroundings.

3.8.1 Markets

With scarce formal employment opportunities, the informal sector is a major source of urban livelihoods (Chaps. 4 and 5). Governments have been slow to recognise its role in employment creation and poverty reduction, and have actively and regularly opposed its development despite the visibly vibrant parts of all towns and cities being the markets, and especially the central markets of the capital cities. Outside the atoll states, multiple markets employ hundreds of workers—selling and transporting goods, vendors and patrons, and maintaining the more formal urban markets. Larger markets incorporate agricultural products, fish and other marine products, cooked food, handicrafts and newer traded and imported goods such as batteries, clothes and SIM cards. While markets are often regarded as the core of the informal sector, many market traders work regular hours, pay for market access and effectively run small businesses: neither informal nor marginal to mainstream development. Street vendors are more likely to be informal, although not necessarily intermittent, but often perceived as inimical to the modern city.

In the large Melanesian capitals especially, markets provide much of the fresh fruit, vegetables and fish for urban residents and offer a basis for food security. Capital cities like Suva and Honiara have several formal markets with hundreds of vendors (about 90% of whom are women in both cities) and various informal markets, most operating at least six days a week, and each with up to around a hundred vendors (Dewey, 2011). The informal Honiara markets mainly sold local produce, including baked goods, but also betel nuts, cigarettes and alcohol. Honiara market vendors were mostly middle-aged women working long hours five or six times a week, selling vegetables; men sold fish. In most SIDS, markets have expanded in size, number, scale and frequency in both urban and rural areas. By the 1970s, imported and manufactured goods were increasingly sold in some marketplaces, and now sometimes represent a high proportion of goods sold. At the same time, where remittance economies were emerging, as in Tonga, overseas Tongans were sending back goods to their kin at home to be sold in markets that often became dominated by such goods (Brown & Connell, 1993). Broadly similar situations exist in Caribbean SIDS, where women purchase produce from farmers and sell in urban markets, but local markets are growing as farmers prefer to sell their own produce more directly. No regulated marketing systems exist, but vendors usually rent their stalls from urban governments, which has made marketing a 'model' informal sector activity, generating linkages and elaborate domestic supply chains between islands (Beckford & Campbell, 2013; Keen & Ride, 2019), despite gender norms restricting the role of women in society and reducing market capacity and the possibility of transformative change (Barbara & Baker, 2020), but nonetheless making a valuable contribution to nutrition and women's employment and income, with minimal national support.

3.8.2 Finance

From the late 1960s, when international capital became more mobile and rapid global telecommunications were established, several SIDS became offshore financial centres of various kinds. Tax havens such as the Bahamas had begun in the 1930s, and, at one point, Grenada had over 30,000 registered banks. By contrast, Pacific tax havens have only ever had a modest and belated presence, entering the scene only after the boom years of the 1970s. By the end of the 1990s, the Caribbean had become the core of the industry, but, in this century, intense regulation of tax havens greatly reduced their presence, profits and operations. Russian oligarch and mafia money, reputedly laundered in Nauru and elsewhere, became better controlled. Under pressure from Australia, Vanuatu opted to end its tax haven status in 2008. Nauru, Samoa and the Marshall Islands also closed their offshore financial centres. By 2023, the Marshall Islands had been blacklisted twice by the EU for financial irregularities. At its peak in the 1980s, the Vanuatu tax haven had employed over 300 people, a substantial component of the urban economy (Connell & Pritchard, 1990). Once common in the Caribbean, only the Bahamas remains a significant SIDS participant. The loss of offshore finance capacity, however illegal, has weakened the economies of several SIDS, but economic growth came at a cost. Illegal financial behaviour sometimes 'spilled over' into other spheres, such as tourism in the Bahamas (Block & Krausner, 1987); tax havens did nothing for equity but attracted highly skilled workers 'in chronically short supply' from other sectors of the economy where the national benefits were greater (Sharman, 2011: 59). Politically dependent Caribbean territories have fared better.

3.8.3 Service Diversity

Other service activities have made some contribution, sometimes short-term, to local economies, again especially in the Caribbean. A fast-growing industry in the 1980s in several SIDS was data processing (where transport costs were irrelevant). At its peak the Barbados industry had several hundred workers, almost all women, processing airline tickets and medical records, but it failed to survive. Deregulation and electronic technology were unable to transform SIDS economies and advances in AI may further limit development. Even in services where electronic communication is central, locational disadvantages, a shortage of skilled labour, a different trade regime and greater scrutiny have restricted growth, beyond small-scale retailing and transport services. Services such as banking and insurance are often foreign-owned, even regionally; Trinidad owns most of these services in the eastern Caribbean. The Pacific has experienced 'debanking': the loss of correspondent banking services in some SIDS with the withdrawal of major banks from the region, resulting in aid donor subsidies to major banks (notably ANZ) to remain in the region and avert financial crisis.

Other service industries have had some success. The Marshall Islands is the third largest shipping nation in the world (after Liberia and Panama) through providing flags of convenience, with the Bahamas and Vanuatu some way behind. Countries that prefer to sail under an 'open register'—not that of their home country—may do so for tax avoidance, to avoid national labour and environmental regulations or the ability to hire crews from lower-wage countries. Such shipping registers bordered on the illegal. Both the Marshall Islands and Vanuatu were criticised for deriving income from registering ships and also drilling rigs, involved in producing oil and gas (Srinivasan, 2023).

Several Caribbean SIDS have sought economic diversification through medical tourism, partly building on the tourist-oriented economies. The Bahamas, Barbados and also the Cayman Islands sought different strategies for medical tourism, in order to generate foreign exchange and new employment, and reduce costs from overseas referrals. Barbados is well advanced in 'fertility tourism' offering IVF, egg freezing, with the Barbados Fertility Centre (and its offshoot in Trinidad) viewed as successes. The industry was developed by overseas corporations and oriented to a US market and employed many islanders. Business principles rather than health care dominated the development strategies of emerging transnational medical corporations and raised ethical issues. The Bahamas has similarly sought to develop medical tourism, based on its proximity to North America and the existing tourism infrastructure, but faced strong competition from Mexico. Jamaica too has tried to market itself at least to the North American diaspora but without success (Connell, 2013b; Johnston, Crooks, & Ormond, 2015).

Various SIDS, mainly in the Caribbean, where islands have good air connections, have encouraged the growth of a 'second home' industry, with expatriates purchasing homes—so providing immediate revenue—and subsequent related services, stimulating some local employment. COVID-19 brought creativity and new responses to 'working from home' that contributed to residing for long periods in those SIDS with superior internet connectivity (and time zones like North America) with the emergence of special work visas to facilitate the residence of 'digital nomads' who might contribute to local economies. Thus, Barbados introduced a 12-Month Welcome Stamp (which 3000 people had applied for in 6 months) and Dominica a Work in Nature Program. Dominica's policy encouraged individuals, families and business teams to relocate there and 'live, work and rejuvenate' for up to 18 months. However, COVID-19 was not propitious for such programmes, potential nomads had families, children and possessions, and few took advantage of such schemes.

Disappointments with relatively conventional strategies have brought ever more unusual development strategies, several of which quickly flickered and burned. Smaller SIDS especially have had unusual strategies for generating revenue. Grenada has hosted offshore gambling sites; Antigua battled with the United States to establish an offshore gambling industry and remarkably won the battle of structural asymmetry, if only temporarily, despite American non-compliance with the legal outcome (Cooper, 2023). Hosting 'reality' television shows has brought in occasional wealth (and tourist publicity). Fiji earned about US$20 million from staging Celebrity Love Island in 2005, and Vanuatu and Samoa have both hosted versions of Survivor and other series. Tuvalu rents out its internet domain name.tv for a valuable and perhaps unique income; early in this century it received US$5 million per year or approximately 7% of its GDP. For many years the sale of stamps was a development strategy for several SIDS. Some multi-island states, such as Tonga and St. Vincent, brought out stamps for different islands while Antigua produced stamps for unpopulated Redonda Island. However, the propensity of several states to bring out too many stamps of limited aesthetic value or local relevance and the global decline of philately brought this near an end. Stamp, coin and passport sales, being a tax haven or providing flags of convenience, were disproportionately more attractive to the smallest states, despite some social costs, as even the small revenue they generated was significant.

By the start of this century, Vanuatu, Tonga and several other SIDS had begun to sell 'diplomatic' passports, mainly to Chinese entrepreneurs but the crime-free status of some purchasers was not always carefully tested. Passport sales were usually linked to 'Citizenship by Investment' (CBI) schemes that were a means of more autonomous development. In 2011, Dominica and St. Kitts came under pressure for selling passports under an Economic Citizenship Program to Iranian nationals. Other SIDS found themselves assisting international criminals, although such 'golden passports' did not necessarily guarantee residence. In Vanuatu they did allow foreign nationals to purchase citizenship and therefore visa-free access to the EU and the UK. Many of Vanuatu's newest citizens included senior political figures from around the world as well as several convicted and alleged criminals. In 2020, just over half the 2200 passports that were issued went to Chinese, followed by Nigerians, Russians and Lebanese. Consequently, in 2024 Vanuatu became the first country to lose its visa-free access to the EU and the UK, and a year later its citizens were denied entry to the United States. The Marshall Islands also sold passports, mainly to Chinese nationals, until the United States indicated that such passports did not guarantee entry to the United States. By 2024 Nauru was offering a passport costing US$140,000, giving visa-free access to 89 countries. The programme was optimistically described as enabling conscientious investors ('climate citizens') to acquire an additional citizenship with the revenue contributing to climate adaptation projects in Nauru. Several Caribbean SIDS introduced CBI programmes. To be qualified for citizenship under this programme in Antigua and Barbuda, Dominica, Grenada, St. Kitts and Nevis or St. Lucia, the value of an investment or a donation had to be at least US$100,000. Purchasing a house there came with a free passport, but residence was not necessary. By 2024, over 80,000 visas had been sold, and by 2025, for the 5 eastern

Caribbean SIDS, passport sales were estimated to account for 10–30% of GDP. Leaders of Caribbean countries that did not issue passports criticised the sale of citizenship and identity. Visa sales were regarded negatively by metropolitan countries for their impacts on security, as one cause of the infiltration of organised crime, money laundering, tax evasion and corruption (McNeill & Walton, 2025).

Caribbean countries' creative solutions proved to complicate international relations. The smaller eastern Caribbean SIDS faced EU pressure to revisit this policy when it became a threat to their own visa-free arrangements, with the UK cancelling visa-free travel for Dominica, owing to 'clear and evident abuse of the scheme, including the granting of citizenship to individuals known to pose a risk to the UK' (Jolly & O'Carroll, 2023). Dominica was particularly reliant on selling citizenship to fund its public services, with this income accounting for about half of all government revenue, effectively funding national development projects including its ambitious geothermal power plant (Chap. 6). A surge in CBI revenue early in 2024 significantly contributed to Grenada's economic health, resulting in a large budget surplus, increased government deposits and lower public debt. St. Kitts became over-reliant on visa sales, some of which were illegal, so that it too was threatened with losing its visa-free access to the UK and the EU and had to abandon the 'reckless exploitation' and mismanagement of the scheme (*Jamaica Observer*, 18 October 2024). Nauru, seeking to benefit from 'investment migration', introduced in 2024 a grandiosely entitled Nauru Economic and Climate Resilience Citizenship Program. Palau created a law that allowed it to grant digital residency, an identification that did not require physical residence in Palau but provided functional residence. In 2022 a venture capitalist and billionaire became the first digital resident of Palau, receiving his new identity as the founding digital resident, and proclaiming: 'I think it's the beginning of something extraordinary, that is going to spread around the world and is as innovative as this government' (quoted in Reklai, 2022). The benefits to Palau were not obvious. Here too, in the search for niches, policies and practices came close, sometimes too close, to illegality but were occasional useful sources of foreign exchange.

3.9 Leasing

Many SIDS have leased out land or entire islands—effectively selling their sovereignty—but often under pressure before, during or after decolonisation. Global superpower rivalry has enabled a few SIDS to 'benefit' from strategic land rentals. The Marshall Islands receives substantial income from the rent of Kwajalein atoll for a US missile range, and some islanders have gained substantial compensation payments for the damage and loss of their atolls for nuclear testing, but the long-term social costs of such economic benefits have been immeasurable. China had constructed a satellite-tracking station in Kiribati in 1997. Such activities, and the possibility of China constructing ports in the region, mooted for Solomon Islands and Vanuatu, became increasingly sensitive and controversial (Chap. 8).

From the start of the century Nauru has provided a service to Australia by leasing land for a detention centre: a contemporary version of placing political exiles and lepers on remote islands. Nauru has earned a substantial income, when little else was possible, through housing one of the three Australian island refugee processing centres, that contributed more than a quarter of its GDP and directly employed 100 people, indicating both the challenges of achieving development in small SIDS and how unusual strategic shifts can change national values and incomes. In 2025, it took on another group of unwanted refugees (Chap. 8). Such distinctive 'development strategies' provide little sustainability, although the income has boosted local incomes (from employment and land leasing), enabled the revival of a national airline and supported several small-business operations (Morris, 2021, 2023). By 2024 Nauru was regarded as able to graduate from the list of SIDS eligible for development assistance.

A particular form of land lease has been associated with numerous fanciful libertarian schemes, usually in the Pacific, more recently involving bitcoin and other variants of cryptocurrency, pushed onto SIDS (if not always resisted) by outsiders, dating back to attempted libertarian ventures before independence in Vanuatu and for the Minerva Reef (Tonga). These have persisted in various forms such as the intended sustainable community of Tribewanted.com in Fiji (Keene, 2008) reflecting the persistent allure of private islands to outsiders and attempts to establish fantasy micronations (Connell & Aldrich, 2021). In 2021 a British investor sought to transform his 'private island' of Latoro (off Santo, Vanuatu), by encouraging 21,000 cryptocurrency investors to move there and form a regulation-free utopian 'blockchain-based democracy' and 'crypto capital of the world' (Neate, 2022), one of a series of schemes aimed at bringing cryptocurrency enthusiasts into real-world communities in small island states. None offered obvious benefits to any SIDS that might become involved. Plans to establish 'Cryptoland' on a Fijian island, where investors were told they would 'enjoy a first-class crypto lifestyle', collapsed when the promoters failed to buy the island. For a short time, Vanuatu let people buy citizenship with bitcoin. Nauru has sought to become the 'crypto powerhouse' of the Pacific (RNZ, 28 July 2025). The Marshall Islands previously sought to make bitcoin legal tender there, but it never eventuated, and links to actual residence faded. SIDS islands were no longer for sale or lease. With the exception of long-term land leases for military or quasi-military facilities (for refugees) that generate a considerable revenue—but well-documented social costs (as at Kwajalein atoll where Marshall Islanders are not allowed on two main islands, other than for transit and employment, and confined to one tiny atoll island)—most such ventures have been short-term, speculative and unable to generate any significant income.

3.10 The Travails of Trade

SIDS experience significant trade deficits, made possible because of aid and remittances, with imports substantially greater than exports. Productive island economies have been weakened through recent trade liberalisation, multinationalisation of production and capital deregulation, reducing protection for traditional commodities, such as bananas and sugar, hastening the demise of plantation-like systems. More problematically, external economic prescriptions presented to SIDS, primarily in the Caribbean, have 'significantly contributed to the cycle of indebtedness, austerity and the dismemberment of social systems' (Dagher, 2019: 1041). Since the 1970s, the rate of growth of exports from small states has been inferior to that of larger states, and open economies mean that even small trade shocks have a large impact because of narrow export concentrations. SIDS characteristically trade with distant metropolitan states rather than with similar neighbours. Trade is constrained by various factors including capacity, limited diversification and trade barriers. Hitherto successful export industries in both regions, such as sugar production and garment manufacturing in Fiji, have barely survived, unable to compete with better-placed countries with lower wages.

Regional integration was intended to address this problem, but challenges of economic integration have prevented cooperation. Lack of complementarity, and greater competition rather than cooperation between states in a more deregulated context, has meant that regional trade is minimal, even resulting in localised conflicts such as the 'kava-biscuit war' between Vanuatu and Fiji (Connell, 2007). Regional trade is further constrained by biosecurity regulations covering various agricultural products such as kava. Diversification of exports is rarely an option, due to narrow resource bases, diseconomies in investment and high transport costs. Even domestically, as in the Marshall Islands, moving people and goods within the countries is 'expensive, time consuming, inconvenient, irregular and unreliable' (McMurray & Smith, 2001: 1320). That has resulted in many SIDS focusing on tourism and on creative niches, where competition is limited and where 'local branding', as with FIJI Water, adds value. It has been suggested that one area that SIDS might develop is the cosmetics sector since 'cosmetics are not ordinarily thought of as a major sector for economic growth and trade in developing countries' while other opportunities existed in processing sargassum and sea-moss (UN OHRLLS, 2024: 31). At much the same time it was pointed out that the CARICOM region has 'an almost negligible number of highly sophisticated exports' (Satnarine-Singh, Hosein, & Saridakis, 2023: 1163) as does the Pacific. Both regions have yet to establish significant creative ventures, but film, through digital media and traditional crafts, suggests unrealised opportunities.

More rapid growth and development, derived from free market reform, have been widely touted as justification for reducing trade barriers, but little empirical evidence exists of real advantages to SIDS from a more open economy, and SIDS are particularly vulnerable to fluctuating international trade regimes. The elimination of trade barriers has not fostered economic growth, despite this being a basic tenet of the World Trade Organisation and international financial institutions, because trading environments differ between developed states and island states, and geographical distance matters; hence, Pacific SIDS are particularly disadvantaged. Similarity of island products deters regional trade, although regional trade agreements are more beneficial than broader and more general trade agreements through the WTO (Carmichael & Jia, 2022).

Free trade, centred on notions of comparative advantage, increasingly threatens countries primarily producing agricultural commodities, particularly where diversification from traditional export staples is difficult. Protection policies have had to be abandoned; hence, Samoa's accession to the World Trade Organisation (WTO) in 2011 meant that it was no longer able to impose a ban on turkey tails and other problematic imports, despite negative nutritional and health consequences (Snowdon, Moodie, Schultz, & Swinburn, 2011; Thow, Swinburn, Colagiuri, et al., 2010). Trade concessions were costly, further entrenching the image of SIDS as remittance and aid-dependent economies (Wallis, 2010). In the Pacific, two regional agreements, the Pacific Island Countries Trade Agreement (PICTA) and the Pacific Agreement on Closer Economic Relations (PACER), have attempted to use economic integration to overcome systematic challenges (Morgan, Carter, & Manoa, 2024). The assumptions that Pacific SIDS can produce tradeable goods of value elsewhere, and that this will outweigh the costs in terms of sovereign control over economic affairs, have however yet to be demonstrated. SIDS can rarely compete on a level-playing field. Trade has effectively emphasised asymmetry and uneven development. One consequence of limited trade in commodities is that SIDS have increasingly focused on trade in services, and especially on tourism, where a degree of similarity with neighbours is not necessarily problematic. We turn to this in the next chapter.

3.11 The Diversification Paradigm: Towards Development and the SDGs?

It is widely accepted that livelihood diversification is valuable for rural development, usually involving some combination of agriculture and fisheries, and national development. Yet, in a blue economy, agriculture has been relatively neglected, or policies have favoured urban elites, despite substantial scope for diversification of livelihoods in many coastal communities and consequent movement towards achieving the SDGs (Basurto, Gutierrez, Franz, et al., 2025; Roscher, Eriksson, Harohau, et al., 2022). At national level, the resource curse has meant that economies dominated by resource extraction find necessary diversification difficult, while distant places are disadvantaged by resource limitations, transport and skill deficits, despite having similar needs but higher costs. Meanwhile all SIDS have sought to go beyond the established combination of fisheries and agriculture in the search for 'something better', but manufacturing and services too often have proved something of a blind and intermittently illegal alley.

In the small-scale fisheries and agricultural sectors in Pacific SIDS, livelihood diversification has been promoted by donors, governments and civil society organisations as a means of alleviating poverty, improving food insecurity and reducing pressure on marine resources. Coastal fisheries development programmes have encouraged alternative livelihoods, such as ecotourism (e.g. Diedrich & Aswani, 2016) or handicraft production, but, in practice, livelihood diversification has often been achieved through migration, but that may not necessarily increase domestic incomes or enable superior food security (Sulu, Eriksson, Schwarz, et al., 2015; Eriksson, Blythe, Österblom, & Olsson, 2021; Li, 2009; Chap. 9). Even limited diversification has declined over time. At local and national levels, achieving economic growth and welfare provision that would enable the SIDS to come closer to achieving the SDGs has remained frustrating. Alternatives to prevailing economic orthodoxy, enabling a more sustained, inclusive and equitable development, have proved more rhetorical than real and generated no new directions within a blue or green economy, but future regional economic security will most likely be derived from the oceans.

We have shown that there have been at least two 'success stories' both based on natural resources. The success of the PNA in regulating access to deepwater tuna in the Pacific high seas took nearly two decades to achieve but demonstrated that Pacific islanders were able to manage and regulate their resources in the right circumstances, just as they have achieved successes within international organisations. Few obvious potential parallels exist at this scale, but it is at least plausible that regulation, management and involvement in deep-sea mining might eventually take a similar form. For any variant of a blue economy, much depends on securing maritime security (Fabinyi, Cvitanovic, Barclay, et al., 2025). At a quite different scale and in a different cultural context, the emerging success of kava in Vanuatu, Fiji and Tonga, with expanding continental markets, suggests a development possibility for other SIDS and for other niche crops, despite intervening opportunities. By contrast, 'legacy' crops, notably coffee and cocoa, that also have niche potential, have been ignored for too long in both regions. Yet PICs have complained regularly about various legal prohibitions on kava, and tariffs potentially hit Fiji hard (notably for FIJI Water), indicating that commodity exports can never be entirely secure. Geography suggests that Caribbean SIDS are better placed to generate agricultural niches, but one size does not fit all. While it is valuable for creativity to be involved, as an economically valuable means of sustaining resilience and identity (Frauts, 2019), that has rarely had enduring resonance. It seems unlikely that cosmetics will fill a niche. It is undeniable that more creative solutions are required to enable a better diversification; hence, it is no surprise that what Nicholas Thomas (2014: 82) has called 'the textbook commodities of early Pacific trade'—bêche-de-mer and sandalwood—linger on hopefully in many small islands with few new opportunities to replace them.

Once tourism was an innovative and creative possibility. Now, as the ADB has pointed out, 'the temptation is to let tourism do the heavy lifting, but now is the time to take advantage of the opportunity a more balanced economy can bring' (quoted in *Islands Business*, January 2013: 12). It is almost certain that tourism will continue to provide one near universal development strategy, being ubiquitously regarded as a lucrative option that generates income, provides formal and informal employment, is not environmentally damaging and is superior to a future based on natural resource exploitation. After all, whatever its flaws, it has been the one substantial contemporary economic success story. Few options—and neither tuna nor kava—appear as glamorous; however, much tourism depends on external circumstances. We now turn to the tourist industry.

Bibliography

Ahlgren, I., Yamada, S., & Wong, A. (2014). Rising oceans, climate change, food aid, and human rights in the Marshall Islands. *Health and Human Rights Journal, 16*(1), 69–80.

Allen, M. (2018). *Resource extraction and contentious states: Mining and the politics of scale in the Pacific Islands*. Springer.

Allen, M. (2020). Framing food security in the Pacific: Resilience in Malo, Vanuatu. In J. Connell & K. Lowitt (Eds.), *Food security in small island states* (pp. 91–110). Springer.

Amon, D., Palacios-Abrantes, J., Drazen, J., et al. (2023). Climate change to drive increasing overlap between Pacific tuna fisheries and emerging deep-sea mining industry. *npj Ocean Sustainability, 2*, 9.

Anderson, B. (2023). The costs of logging in Solomon Islands. *ANU DevPolicyBlog*, 3 August.

Aporosa, S. (2019). De-mythologizing and re-branding of kava as the new 'world drug' of choice. *Drug Science, Policy and Law, 5*, 1–13.

Aqorau, T. (2019). *Fishing for success: Lessons in Pacific regionalism*. ANU Department of Pacific Affairs.

Asch, R., Cheung, W., & Reygondeau, G. (2018). Future marine ecosystem drivers, biodiversity and fisheries maximum catch potential in Pacific Island countries and territories under climate change. *Marine Policy, 88*, 285–294.

Asian Development Bank. (2024). *Assessing and addressing risks in Pacific construction supply chains*. ADB.

Aswani, S., Albert, S., & Love, M. (2017). One size does not fit all: Critical insights for effective community based resource management in Melanesia. *Marine Policy, 81*, 381–391.

Aswani, S., Basurto, X., Ferse, S., et al. (2018). Marine resource management and conservation in the Anthropocene. *Environmental Conservation, 48*(2), 192–202.

Atzori, D., Sonneveld, B., Alfarra, A., et al. (2024). Nutrition fragility in isolation: Food insecurity in small island developing states. *Food Security., 16*, 437–453.

Bainton, N. (2010). *The Lihir destiny. Cultural responses to mining in Melanesia.* ANU Press.

Bainton, N., & Skrzypek, E. (2022, August 3). Pacific nations are extraordinarily rich in critical minerals. But mining them may take a terrible toll. *The Conversation.*

Barbara, J., & Baker, K. (2020). Addressing collective action problems in Melanesia: The Northern Islands Market Vendors' Association in Vanuatu. *Development in Practice, 30*, 994–1004.

Barbesgaard, M. (2018). Blue growth: Savior or ocean grabbing? *Journal of Peasant Studies, 45*(1), 130–149.

Barker, D. (2014). Caribbean agriculture in a period of global change: Vulnerabilities and opportunities. *Caribbean Studies, 40*(2), 41–61.

Barry, T., & Gahman, L. (2020). Agrarian struggle and food system injustice in the Anglo-Caribbean: Centering social reproduction by (re)turning to Creft and Fanon. *Human Geography, 13*(2), 174–178.

Barry, T., & Gahman, L. (2021). Food system and social reproduction realities for women in agriculture across the Caribbean: Evidence from Grenada, St. Lucia and St. Vincent and the Grenadines. *Journal of Agrarian Change, 21*, 815–833.

Basurto, X., Gutierrez, N., Franz, N., et al. (2025). Illuminating the multidimensional contributions of small-scale fisheries. *Nature, 637*, 875–884.

Batalofo, M., Pollard, A., Ride, A., et al. (2024). What can the experiences of rural women in Solomon Islands teach us about innovation in aquatic food systems? *Asia Pacific Viewpoint, 65*(1), 96–110.

Beckford, C., & Campbell, D. (2013). *Domestic food production and food security in the Caribbean.* Palgrave Macmillan.

Bell, C., Latu, C., Na'ati, E., Snowdon, W., Moodie, M., & Waqa, G. (2021). Barriers and facilitators to the introduction of import duties designed to prevent noncommunicable disease in Tonga: A case study. *Globalization and Health, 17*, 136.

Bell, J., Ganachaud, A., Gehrke, P., et al. (2013). Mixed responses of tropical Pacific fisheries and aquaculture to climate change. *Nature Climate Change, 3*, 591–599.

Bell, J, Reid, C, Batty, Allison, E., et al. (2011). Implications of climate change for contributions by fisheries and aquaculture to Pacific Island economies and communities. In J. Bell, J. Johnson & A. Hobday (Eds.), *Vulnerability of tropical Pacific fisheries and aquaculture to Pacific Island economies and communities* (pp.733–801), SPC.

Bell, J., Senina, I., Adams, T., et al. (2021). Pathways to sustaining tuna- dependent Pacific Island economies during climate change. *Nature Sustainability, 4*, 900–910.

Bennett, N., Finkheiner, E., Ban, N., et al. (2020). The COVID-19 pandemic, small-scale fisheries and coastal fishing communities. *Coastal Management, 48*(4), 336–347.

Bertana, A., & Gaytán, M. (2025). Racialized downgrading and upgrading: Dis/articulation and the Fijian kava commodity chain. *Environment and Planning A: Economy and Space, 57*(1), 22–39.

Bevacqua, R., Hunter, S., Deshpande, S., & Morgan, P. (2025). *Supporting small-scale entrepreneurs in the Pacific Region*. ADB Institute Policy Brief No. 2025-10.

Bhattacharya, P., & Dash, A. (2021). Determinants of blue economy in Asia-Pacific island countries: A study of tourism and fisheries sector. *Ocean and Coastal Management, 211*, 105774.

Block, A., & Krausner, P. (1987). Masters of Paradise Island: Organized crime, neo-colonialism and the Bahamas. *Dialectical Anthropology, 12*, 85–102.

Blumfield, T., Reverchon, F., & Vigulu, V. (2018). The importance of market access for timber growers in small island developing states. *Land Use Policy, 77*, 598–602.

Bomhauer-Beins, L., de Guttry, C., & Ratter, B. (2019). When culture materializes: Societal dynamics in resilience of social-ecological systems in the case of conch management on Abaco, the Bahamas. *Sustainability, 11*(4), 1080.

Bourrel, M., Swaddling, A., Atalifo, V., & Tawake, A. (2018). Building in-country capacity and expertise to ensure good governance of the deep sea mineral industry within the Pacific region. *Marine Policy, 95*, 372–379.

Brent, Z., Barbesgaard, M., & Pedersen, C. (2020). The blue fix: What's driving blue growth? *Sustainability Science, 15*, 31–43.

Brown, R., & Connell, J. (1993). The global flea market: Migration, remittances and the informal economy in Tonga. *Development and Change, 24*(4), 611–647.

Buggy, L., & McNamara, K. (2015). The need to reinterpret "community" for climate change adaptation: a case study of Pele Island, Vanuatu. *Climate and Development, 8*(3), 1–11.

Campbell, Y., & Connell, J. (Eds.). (2011). *COVID in the Islands: A comparative perspective on the Caribbean and the Pacific*. Palgrave Macmillan.

Campbell, D., & Lee, S. (2021). Vulnerability of Jamaica's South coast fishing communities to coastal erosion and flooding. In S. Moncada et al. (Eds.), *Small island developing states: Vulnerability and resilience under climate change* (pp. 211–231). Springer.

Campbell, B., Steenbergen, D., Li, O., et al. (2024). Characterising a diversity of coastal community fisheries in Kiribati and Vanuatu. *Fish and Fisheries, 25*, 837–857.

Campling, L., & Colás, A. (2021). *Capitalism and the sea: The maritime factor in the making of the modern world*. Verso.

Carballo, D., & Dos Reis, L. (2018). *The agricultural insurance market in the Caribbean*. World Bank.

Carmichael, C., & Jia, G. (2022). Economic growth through trade liberalization for small island developing states in the Pacific: Regionalism versus globalization. In J. Randall (Ed.), *Annual report on Global Islands 2021* (pp. 135–148). Island Studies Press.

Casola, W., Oren, J., Register, M., et al. (2021). Modernization of artisanal fishing communities on Andros Island, The Bahamas, as a treadmill of production. *Ocean & Coastal Management, 201*, 105487.

Casola, W., Rehnberg, M., Peterson, M., et al. (2022). Drivers of long-term support for marine protected areas in The Bahamas. *Ocean & Coastal Management, 217*, 106000.

Chambers, A., & Chambers, K. (2018). Engaging the world: Four decades of intensifying change in Tuvalu. In J. Connell & H. Lee (Eds.), *Change and continuity in the Pacific: Revisiting the region* (pp. 17–36). Routledge.

Cheng, Y., Tseng, Y., Iwaki, Y., & Huang, M. (2021). Sustainable food security in small island developing states (SIDS): A case study of horticulture project in Marshall Islands. *Marine Policy, 128*, 104378.

Christensen, A. (2011). Marine gold and atoll livelihoods: The rise and fall of the bêche-de-mer trade on Ontong Java, Solomon Islands. *Natural Resources Forum, 35*, 9–20.

Connell, J. (2006). "The taste of paradise": Selling Fiji and FIJI Water. *Asia Pacific Viewpoint, 47*, 342–350.

Connell, J. (2007). Towards free trade in the Pacific? The genesis of the 'Kava-Biscuit War' between Fiji and Vanuatu. *Geographical Research, 45*, 1–12.

Connell, J. (2013a). *Islands at risk: Environments, economies and contemporary change*. Edward Elgar.

Connell, J. (2013b). Medical tourism in the Caribbean islands: A cure for economies in crisis? *Island Studies Journal, 8*(1), 115–130.

Connell, J. (2015). Food security in the island Pacific: Is Micronesia as far away as ever? *Regional Environmental Change, 15*(7), 1299–1311.

Connell, J. (2021). Double jeopardy: Distance and decentralisation in Tuvalu. In Y. Campbell & J. Connell (Eds.), *COVID in the Islands: A comparative perspective on the Caribbean and the Pacific* (pp. 163–175). Palgrave Macmillan.

Connell, J., & Aldrich, R. (2021). *The ends of empire: The last colonies revisited.* Palgrave Macmillan.

Connell, J., & Negin, J. (2012). The malaise of modernity: Urbanization, mobility and HIV in the Pacific. *Pacific Studies, 35*(1–2), 1–35.

Connell, J., & Pritchard, W. (1990). Tax havens and global capitalism: Vanuatu and the Australian Connection. *Australian Geographical Studies, 28*, 38–50.

Cooper, A. (2023). A small state taking on a Giant: Antigua and Barbuda 'getting the better' of the United States without overturning structural asymmetry. In G. Baldacchino (Ed.), *The success of small states in international relations* (pp. 96–17). Routledge.

Cowell, N. (1987). The impact of bauxite mining on peasant and community relations in Jamaica. *Social and Economic Studies*, 171–216.

Cross, M. (1979). *Urbanization and urban growth in the Caribbean.* Cambridge University Press.

Cumberbatch, J., Compton, S., & McConney, P. (2023). Managing change to implement the small-scale fisheries guidelines: Perspectives from a Caribbean fisheries policy process. *Maritime Studies, 22*, 23.

Dagher, R. (2019). Policy space under a constraining combination- open economies, austerity and small island states. *Third World Quarterly, 40*(6), 1040–1063.

D'Arcy, P. (2006). *The people of the Sea: Environment, identity and history in Oceania.* University of Hawai'i Press.

Davila, F., Bourke, R., McWilliam, A., Crimp, S., et al. (2021). COVID-19 and food systems in Pacific Island Countries, Papua New Guinea, and Timor-Leste: Opportunities for actions towards the sustainable development goals. *Agricultural Systems, 191*, 103137.

Davis-Mattis, L. (2023). The future of deep-seabed minerals and marine genetic resources in blue economies. In P. Clegg, R. Mahon, P. McConney, & H. Oxenford (Eds.), *The Caribbean blue economy* (pp. 182–194). Routledge.

Dewey, S. (2011). Markets and women's market trading in the Pacific Islands: An overview of social contexts and ongoing challenges. *Asian Women, 27*(3), 1–23.

Dey, M., Gosh, K., Valmonte-Santos, R., et al. (2016). Economic impact of climate change and climate change adaptation strategies for fisheries sector in Solomon Islands: Implication for food security. *Marine Policy, 67*, 171–178.

Diedrich, A., & Aswani, S. (2016). Exploring the potential impacts of tourism development on social and ecological change in the Solomon Islands. *Ambio, 45*, 808–818.

Doerr, A. (2021). Regulating for social and environmental sustainability: A stakeholder perspective from the Bahamian spiny lobster fishery. *Marine Policy, 124*, 104366.

Dornan, M., Morgan, W., Cain, T., & Tarte, S. (2018). What's in a term? 'Green growth' and the 'blue-green' economy in the Pacific islands. *Asia and the Pacific Policy Studies, 5*, 408–425.

Drakes, C., & Moore, W. (2015). The success factors of manufacturing firms in Caribbean small island states: A look at Barbados. *Central Bank of Barbados, 15*(10).

Duncan, N., & Mctair, D. (2025, January 5). The Rastafarians opening up Caribbean cannabis farms to promote the benefits of marijuana. *Guardian*.

Dunn, D., Stewart, K., Bjorkland, R., et al. (2010). A regional analysis of coastal and domestic fishing effort in the wider Caribbean. *Fisheries Research, 102*(1-2), 60–68.

Dyson, J. (1982). *The south seas dream. Adventures in paradise*. Heinemann.

Eriksson, H., Sulu, R., Blythe, J., et al. (2020). Intangible links between household livelihoods and food security in Solomon Islands: Implications for rural development. *Ecology and Society, 25*(4), 18.

Eriksson, H., Blythe, J., Österblom, H., & Olsson, P. (2021). Beyond social-ecological traps: Fostering transformations towards sustainability. *Ecology and Society, 26*(1), 13.

Errington, F., Fujikura, T., & Gewertz, D. (2013). *The noodle narratives*. University of California Press.

Fabinyi, M., & Barclay, K. (2022). *Asia-Pacific fishing livelihoods*. Palgrave Macmillan.

Fabinyi, M., Cvitanovic, C., Barclay, K., et al. (2025). Rethinking maritime security from the bottom up: Four principles to broaden perspectives and centre humans and ecosystems. *npj Ocean Sustainability, 4*(1), 29.

Fache, E., & Pauwels, S. (2020). Tackling coastal "overfishing" in Fiji: Advocating for indigenous worldview, knowledge, and values to be the backbone of fisheries management strategies. *Maritime Studies, 19*(1), 41–52.

Fache, E., Le Meur, P.-Y., & Rodary, E. (2021). The new scramble for the Pacific: A frontier approach. *Pacific Affairs, 94*(1), 57–75.

FAO. (2014). *Youth and agriculture: Key challenges and concrete solutions*. FAO.

FAO. (2024). *The state of food security and nutrition in the world 2024*. FAO.

Fazey, I., Pettorelli, N., Kenter, J., et al. (2011). Maladaptive trajectories of change in Makira, Solomon Islands. *Global Environmental Change, 21*, 1275–1289.

Ferguson, C., Bennett, N., Kostka, W., et al. (2022). The tragedy of the commodity is not inevitable: Indigenous resistance prevents high-value fisheries collapse in the Pacific islands. *Global Environmental Change, 73*, 102477.

Ferguson, C., Tuxson, T., Manghubai, S., et al. (2022). Local practices and production confer resilience to rural Pacific food systems during the COVID-19 pandemic. *Marine Policy, 137*, 104954.

Ferguson Irlanda, C., Mangubhai, S., Waqa, E., Govan, H., et al. (2025). Insights on the roles of women in effective and procedurally just environmental governance from coastal fisheries management in Fiji. *Conservation Biology, 39*(6), e70121.

Foale, S. (2001). 'Where's our Development?' Landowner aspirations and environmentalist agendas in Western Solomon Islands. *Asia Pacific Journal of Anthropology, 2*, 44–67.

Folkersen, M., Fleming, C., & Hasan, S. (2018). Deep sea mining's future effects on Fiji's tourism industry: A contingent behaviour study. *Marine Policy, 96*, 81–89.

Forbes, K. (2021, April 27). Caribbean food security during COVID-19 can only be ensured through debt relief. *The Conversation*.

Frauts, M. (2019). Resilience and the creative economy in Kingston, Jamaica. *Cultural Studies, 33*(3), 391–413.

Galappaththi, E., Susarla, V., Loutet, S., et al. (2022). Climate change adaptation in fisheries. *Fish and Fisheries, 23*, 4–21.

Garcia, C., Makhoul, N., Fox, M., & Tuxson, T. (2022). Looking at the impacts of COVID- 19 on coastal communities in the Pacific using a gender and social inclusion lens. *SPC Women in Fisheries Information Bulletin, 35*, 21–25.

Garonna, I., Fazey, I., Brown, M., & Pettorelli, N. (2009). Rapid primary productivity changes in one of the last coastal rainforests: the case of Kahua, Solomon Islands. *Environmental Conservation, 36*(3), 253–260.

Georgeou, N., Hawksley, C., Wali, N., Lountain, S., et al. (2022). Food security and small holder farming in Pacific Island countries and territories: A scoping review. *PLoS Sustainability and Transformation, 1*(4), e0000009.

Gewertz, D., & Errington, F. (2021). Stingy galitarianism: Precarity and Jealousy at the Sisiak Settlement, Madang, Papua New Guinea. In N. Bainton, D. McDougall, K. Alexeyeff, & J. Cox (Eds.), *Unequal lives: Gender, race and class in the Western Pacific* (pp. 213–235). ANU Press.

Gillett, R., & Fong, M. (2023). *Fisheries in the economies of Pacific Island countries and territories.* Pacific Community.

Gotschall, I. (2021). An overview of I-Kiribati women in fisheries. *SPC Women in Fisheries Information Bulletin, 33*, 31–36.

Graham, R. (2024). The marine litter issue in the Windward Islands- a pathway to responses using the DPSIR framework. *Frontiers in Environmental Science, 11*, 1150722.

Graycar, A. (2022, March 16). Sustainable catch: Navigating global fisheries crime. *Interpreter.*

Griswold, D. (2021). Growing hope: Island agriculture and refusing catastrophe in climate change adaptation in Fiji. *Island Studies Journal, 16*(2), 136–155.

Grossman, L. (1998). *The political ecology of bananas.* University of North Carolina Press.

Grynberg, R., & Razzaque, M. (2004). *The trade performance of small states.* Commonwealth Secretariat.

Guell, C., Brown, C., Iese, V., et al. (2021). "We used to get food from the garden." Understanding changing practices of local food production and consumption in small island states. *Social Science & Medicine, 284*, 114214.

Guell, C., Brown, C., Navunicagi, O., et al. (2022). Perspectives on strengthening local food systems in small island developing states. *Food Security, 14*(5), 1227–1240.

Guell, C., Saint Ville, A., Anderson, S., et al. (2024). Small island developing states: Addressing the intersecting challenges of non-communicable diseases, food insecurity, and climate change. *The Lancet Diabetes & Endocrinology, 12*(6), 422–432.

Hair, C., Foale, S., Kinch, J., et al. (2016). Beyond boom, bust and ban: The sandfish (*Holothuria scabra*) fishery in the Tigak Island, Papua New Guineas. *Regional Studies in Marine Science, 5*, 69–79.

Halliday, C., Morrissey, K., Saint Ville, A., et al. (2023). Trends in food supply, diet, and the risk of non-communicable diseases in three small island developing states: Implications for policy and research. *Frontiers in Sustainable Food Systems, 7*, 1058540.

Hanich, Q., Wabnitz, C., Ota, Y., Amos, M., et al. (2018). Small-scale fisheries under climate change in the Pacific Islands region. *Marine Policy, 88*, 279–284.

Hanich, Q., Jung, M., McDonald, A., et al. (2021). Tuna fisheries conservation and management in the Pacific Islands region: Implications for Korean distant water fisheries. *Asia-Pacific Journal of Ocean Law and Policy, 6*(2), 192–220.

Harohau, D., Blythe, J., Sheaves, M., & Diedrich, A. (2020). Limits of Tilapia aquaculture for rural livelihoods in Solomon Islands. *Sustainability, 12*, 4592.

Harohau, D., Allison, E., & Teoh, S. (2025). Nature-based solutions to aquaculture: A way forward for aquaculture development in Pacific Island countries. *SPC Fisheries Newsletter, 176*, 50–60.

Hassanali, K. (2020). CARICOM and the blue economy- Multiple understandings and their implications for global engagement. *Marine Policy, 120*, 104137.

Havice, E., & Reed, K. (2012). Fishing for development? Tuna resource access and industrial change in Papua New Guinea. *Journal of Agrarian Change, 12*(2–3), 413–435.

Heck, N., Beck, M., Reguero, B., et al. (2023). Global climate change risk to fisheries – A multi-risk assessment. *Marine Policy, 148*, 105489.

Hendriks, S. (2022). Sustainable small-scale fisheries can help people and the planet. *Nature, 606*(7915), 650–652.

Hidalgo, D., Witten, I., Nunn, P., et al. (2020). Sustaining healthy diets in times of change: Linking climate hazards, food systems and nutrition security in rural communities of the Fiji Islands. *Regional Environmental Change, 20*, 73.

Hogarth, J., & Wójcik, D. (2016a). An evolutionary approach to adaptive capacity assessment: A case study of Whitehouse, Jamaica. *Journal of Rural Studies, 43*, 248–259.

Hogarth, J., & Wójcik, D. (2016b). An evolutionary approach to adaptive capacity assessment: A case study of Soufriere, Saint Lucia. *Sustainability, 8*, 228.

Hosein, R. (2021). *Oil and gas in Trinidad and Tobago managing the resource curse in a small petroleum-exporting economy*. Palgrave Macmillan.

Huntington, H., & Shenoy, A. (2021). Does insecure land tenure deter investment? Evidence from a randomized controlled trial. *Journal of Development Economics, 150*, 102632.

Hurley, G., Wilkinson, E., Weaver, J., & Kasanawaqa, V. (2025). *Turning the tide. Enhancing ocean equity for small island developing states*. ODI Global.

Iese, V., Kiem, A., Mariner, A., Malsale, P., et al. (2021). Historical and future drought impacts in the Pacific islands and atolls. *Climatic Change, 116*, 19.

Jackson, J., Donovan, M., Cramer, K., & Lam, V. (Eds.). (2014). *Status and trends of Caribbean Coral Reefs: 1970–2012*. IUCN.

Jaeckel, A., Harden-Davies, H., Amon, D., van der Grient, J., et al. (2023). Deep seabed mining lacks social legitimacy. *npj Ocean Sustainability, 2*(1), 1.

James, P., Tidd, A., & Kaitu, L. (2018). The impact of industrial tuna fishing on small-scale fishers and economies in the Pacific. *Marine Policy, 95*, 189–198.

Jansen, H., Stren, A., & Weiss, E. (2015). *Linking farmers and agroprocessors to the tourism industry in the Eastern Caribbean*. World Bank.

Johnson, G. (2021). *Our ocean's promise: From aspirations to inspirations: The Marshall Islands fishing story*. Create Space.

Johnston, R., Crooks, V., & Ormond, M. (2015). Policy implications of medical tourism development in destination countries: Revisiting and revising an existing framework by examining the case of Jamaica. *Global Health, 11*, 29.

Jolly, J., & O'Carroll, L. (2023, October 20). EU raises security concerns as it reveals five Caribbean states have sold 88,000 'golden passports'. *The Guardian*.

Jones, C., Murray, W., & Overton, J. (2017). FIJI Water, water everywhere: Global brands and democratic and social injustice. *Asia Pacific Viewpoint, 58*(1), 112–123.

Kabutaulaka, T. (2006). Global capital and local ownership in Solomon Islands' Forestry Industry. In S. Firth (Ed.), *Globalisation and governance in the Pacific Islands* (pp. 239–257). ANU Press.

Kabutaulaka, T. (2021). Mapping the blue Pacific in a changing regional order. In G. Smith & T. Wesley-Smith (Eds.), *The China alternative. Changing regional order in the Pacific Islands* (pp. 41–70). ANU Press.

Kagy, V., Wong, M., Vandenbroucke, H., et al. (2016). Traditional banana diversity in Oceania: An endangered heritage. *PloS One, 11*(3), e0151208.

Karcher, D., Fache, E., Breckwoldt, A., Guilloux, B., et al. (2020). Trends in South Pacific fisheries management. *Marine Policy, 118*, 104021.

Keen, M., & Ride, A. (2019). Trading places: Inclusive cities and market vending in the Pacific Islands. *Asia Pacific Viewpoint, 60*(3), 239–251.

Keen, M., Schwarz, A., & Wini-Simeon, L. (2018). Towards defining the blue economy: Practical lessons from Pacific ocean governance. *Marine Policy, 88*, 333–341.

Keene, B. (2008). *Tribe wanted: My adventure on paradise or bust*. Ebury Press.

Keppel, G., Morrison, C., Watling, D., et al. (2012). Conservation in tropical Pacific Island countries: Why most current approaches are failing. *Conservation Letters, 5*(4), 256–265.

Kilroy, J., McNamara, K., Witt, B., & Raynes, S. (2025). 'I don't know anything else besides gardening': Exploring sensitivities to climate change in Vanuatu. *Asia Pacific Viewpoint, 66*(1) (in press).

Kroodsma, D., Mayorsga, J., Hochberg, T., Miller, N., & Park, J. (2018). Tracking the global footprint of fisheries. *Science, 359*(6378), 904–908.

Kuster, C., Vuki, V., & Zann, L. (2005). Long-term trends in subsistence fishing patterns and coral reef fisheries yield from a remote Fijian island. *Fisheries Research, 76*(2), 221–228.

Kwiatkowska, B. (2007). The 2006 Barbados/Trinidad and Tobago Award: A landmark in compulsory jurisdiction and equitable maritime boundary delimitation. *International Journal of Marine and Coastal Law, 22*(1), 7–60.

Labh, N. (2024, May 18). Connectivity in crisis: The problem with shipping in the Pacific. *Visions and Voices*.

Labuinao, E. (2020). Kastom, gender and economic development: The case of the fish aggregating device in To'abaita, Solomon Islands. *SPC Women in Fisheries Information Bulletin, 32*, 13–18.

Lam, V., Allison, E., Bell, J., et al. (2020). Climate change, tropical fisheries and prospects for sustainable development. *Nature Reviews Earth and Environment, 1*(9), 440–454.

Laslett, K. (2014). *State crimes on the margins of empire: Rio Tinto, the war on Bougainville and resistance to mining*. Pluto Press.

Lau, J., & Sutcliffe, S. (2021). Of isolation and atolls: Coping with COVID-19 in Manus, Papua New Guinea. In Y. Campbell & J. Connell (Eds.), *COVID in the Islands: A comparative perspective on the Caribbean and the Pacific* (pp. 353–366). Palgrave Macmillan.

Lauer, M. (2016). Governing uncertainty: Resilience, dwelling and flexible resource management in Oceania. *Conservation and Society, 14*(1), 34–47.

Lauer, M., & Aswani, S. (2010). Indigenous knowledge and long-term ecological change: Detection, interpretation, and responses to changing ecological conditions in Pacific Island communities. *Environmental Management, 45*, 985–997.

Lenderking, H., Robinson, S., & Carlson, G. (2021). Climate change and food security in Caribbean small island developing states: Challenges and strategies. *International Journal of Sustainable Development & World Ecology, 28*(3), 238–245.

Levi, M. (2019). Employment of the weak: The role of a multinational factory in the life trajectory of early school leavers in Sāmoa. *The Contemporary Pacific, 31*(2), 417–446.

Li, T. (2009). Exit from agriculture: A step forward or step backward for the rural poor. *Journal of Peasant Studies, 36*(3), 629–636.

Louey, P. (2022). The Pacific blue economy: An instrument of political maneuver. *Marine Policy, 135*, 104880.

Lutz, C. (1988). *Unnatural emotions: Everyday sentiments on a Micronesian Atoll.* University of Chicago Press.

McConney, P., Cox, S.-A., & Parsram, K. (2020). Fisheries governance and food security in the Eastern Caribbean. In J. Connell & K. Lowitt (Eds.), *Food security in small island states* (pp. 199–218). Springer.

McMurray, C., & Smith, R. (2001). *Diseases of globalization. Socioeconomic transitions and health*. Earthscan.

McNamara, K., Clissold, R., & Westoby, R. (2020). Women's capabilities in disaster recovery and resilience must be acknowledged, utilized and supported. *Journal of Gender Studies, 30*(1), 119–125.

McNeill, H., & Walton, G. (2025). Passports, prosperity and problems: (Counter)geopolitical narratives of Vanuatu's citizenship by investment scheme. *Geopolitics, 30*(4), 1795–1825.

Maharaj, S., Asmath, H., Ali, S., Agard, J., et al. (2019). Assessing protected area effectiveness within the Caribbean under changing climate conditions: A case study of the small island, Trinidad. *Land Use Policy, 81*, 185–193.

Mallin, F. (2018). From sea-level rise to seabed grabbing: The political economy of climate change in Kiribati. *Marine Policy, 97*, 244–252.

Mallin, F., & Barbesgaard, M. (2020). Awash with contradiction: Capital, ocean space and the logics of the Blue Economy Paradigm. *Geoforum, 113*, 121–132.

Marre, J.-B., & Imhof, C. (2021). An assessment of socio-economic impacts due to COVID-19 on coastal fisheries in Tonga. *SPC Fisheries Newsletter, 165*, 28–32.

Marre, J.-B., & Hunt, A. (2021). Coastal fisheries and FAD monitoring project in Kadavu, Fiji: An assessment of data collected from 2017-2020. *SPC Fisheries Newsletter, 166*, 29–37.

Massola, J. (2020, July 13). Tension on the high seas. *Sydney Morning Herald*, 14–15.

Mercer, J., Kelman, I., Alfthan, B., & Kurvits, T. (2012). Ecosystem-based adaptation to climate change in Caribbean small island developing states: Integrating local and external knowledge. *Sustainability, 4*, 1908–1932.

Mikhailovich, K., Mackenzie, E., & Smith, D. (2023). Enhancing coastal livelihoods through half-pearl aquaculture and handicrafts in Tonga and Fiji. *Development in Practice, 33*(1), 110–121.

Militz, T., & Foale, S. (2017). The "Nemo Effect": Perception and reality of Finding Nemo's impact on marine aquarium fisheries. *Fish and Fisheries, 18*(3), 596–606.

Mills, M., Pressey, R., Weeks, R., Foale, S., & Ban, N. (2010). A mismatch of scales: Challenges in planning for implementation of marine protected areas in the Coral Triangle. *Conservation Letters, 3*, 291–303.

Minter, T., & van der Ploeg, J. (2023). 'Our happy hour became a hungry hour': Logging, subsistence and social relations in Solomon Islands. *International Forestry Review, 25*(1), 113–135.

Mitchell, J., & McEvoy, D. (2019). *Land tenure and climate vulnerability*. UN Habitat.

Mohammadi, E., Singh, S., McCordic, C., & Pittman, J. (2022). Food security challenges and options in the Caribbean: Insights from a scoping review. *Anthropocene Science, 1*(1), 91–108.

Monnereau, I., Mahon, R., McConney, P., Nurse, L., et al. (2021). Fisheries sector vulnerabilities to climate change in small island developing states. In S. Moncada et al. (Eds.), *Small island developing states: Vulnerability and resilience under climate change* (pp. 233–255). Springer.

Morgan, W., Carter, S., & Manoa, F. (2024). Leading from the frontline: A history of Pacific climate diplomacy. *Journal of Pacific History, 59*(3), 353–374.

Morris, J. (2021). Colonial afterlives of infrastructure: From phosphate to refugee processing in the Republic of Nauru. *Mobilities, 16*(5), 688–706.

Morris, J. (2023). *Asylum and extraction in the Republic of Nauru*. Cornell University Press.

Mortimer, N., Campbell, H., Tulloch, A., et al. (2017). Zealandia: Earth's hidden continent. *GSA Today, 27*(3), 27–35.

Murphy, J., & Gard, R. (2025). The deep-sea mining commodity frontier: Blue economy agenda in the Pacific Ocean. *Academia Environmental Sciences and Sustainability, 2*(1).

Murray, W. (2000). Neoliberal globalisation, "exotic" agro-exports and local change in the Pacific islands: A study of the Fijian kava sector. *Singapore Journal of Tropical Geography, 21*, 355–373.

Neate, R. (2022, February 12). Bitcoin paradise? Briton creates 'crypto utopia' in South Pacific. *Guardian*.

Nef, D., Kotra, K., Stauffacher, M., et al. (2022). Vulnerability to food insecurity in a telecoupled world: Insights from Vanuatu. *Frontiers in Sustainable Food Systems, 6*, 855272.

Neufville, Z. (2022). *Overharvesting and poaching devastate Jamaica's fisheries*. InSight Crime Investigation Report.

Newell, A., Nuttall, P., Prasad, B., & Veitayaki, J. (2017). Turning the tide: The need for sustainable sea transport in the Pacific. *Marine Policy, 75*, 249–259.

Nomura, K., Woodill, A., Sweeney, J., Harte, M., & Watson, J. (2024). Emergent geopolitical risks from fishing activities and past conflicts in the Pacific Ocean. *Marine Policy, 166*, 106234.

O'Garra, T. (2012). Economic valuation of a traditional fishing ground on the coral coast of Fiji. *Ocean and Coastal Management, 56*, 44–55.

Ostrom, E. (2000). Collective action and the evolution of social norms. *Journal of Economic Perspectives, 14*(3), 137–158.

Patil, P., Virdin, J., & Colgan, C. (2023). The blue economy as a global initiative. In P. Clegg, R. Mahon, P. McConney, & H. Oxenford (Eds.), *The Caribbean blue economy* (pp. 1–8). Routledge.

Pena, M., McConney, P., Perch, L., & Phillips, T. (2022). Capacities and gaps in Caribbean fisherfolk organization leadership. *Gender, Technology and Development, 26*(3), 503–521.

Perry, K. (2024). Epistemic silences, subversive politics: Post-disaster economic assessments as technologies of persistent coloniality and route to an emancipatory climate justice agenda in the Caribbean. *Climate and Development, 16*(9), 798–810.

Persaud, N., & Dagher, R. (2021). *The role of monitoring and evaluation in the UN 2030 SDGs agenda*. Palgrave Macmillan.

Peteru, S., Regan, S., & Klak, T. (2010). Local vibrancy in a globalizing world: Evidence from Dominica. *Focus, 53*(4), 125–133.

Petzold, J., & Scheffran, J. (2024). Climate change and human security in coastal regions. *Cambridge Prisms: Coastal Futures, 2*(e5), 1–14.

Pew Charitable Trusts. (2016). *Netting billions. A global valuation of Tuna*. Pew Charitable Trusts.

Pickering, K., Pearce, T., Manuel, L., Doran, B., & Smith, T. (2023). Socio-ecological challenges and food security in the 'salad bowl' of Fiji, Sigatoka Valley. *Regional Environmental Change, 23*(2), 61.

Pilling, G., Harley, S. J., Nicol, S., Williams, P., & Hampton, J. (2020). Can the tropical Western and Central Pacific tuna purse seine fishery contribute to Pacific Island population food security?. In J. Connell & K. Lowitt (Eds.), *Food Security in Small Island States*. Springer.

Potter, R., Barker, D., Conway, D., & Klak, T. (2004). *The contemporary Caribbean*. Pearson.

Purcell, S., Tagliafico, A., Cullis, B., & Cocks, N. (2024). Wicked problem of improving fishery livelihoods through capacity building. *Marine Policy, 163*, 106108.

Rabone, M., Wiethase, J., Simon-Lledo, E., Emery, A., et al. (2023). How many metazoan species live in the world's largest mineral exploration region? *Current Biology, 33*, 2383–2396.

Rajan, J. (2005). Gilt-edged packet or economic straight jacket? A case study of cannery workers in Levuka, Fiji Islands. In I. Novaczek, J. Mitchell, & J. Veitayaki (Eds.), *Pacific voices: Equity and sustainability in Pacific Island fisheries* (pp. 153–166). Institute of Pacific Studies, USP.

Randin, G., & Turagabaleti, P. (2021). COVID-19 and food security in Fiji's urban areas: The rise of urban farming and home gardening as a 'new way of life'. In Y. Campbell & J. Connell (Eds.), *COVID in the islands: A comparative perspective on the Caribbean and the Pacific* (pp. 353–366). Palgrave Macmillan.

Regan, A. (2017). Bougainville: Origins of the conflict, and debating the future of large-scale mining. In C. Filer & P.-Y. Le Meur (Eds.), *Large scale mines and local-level politics. Between new Caledonia and Papua New Guinea* (pp. 353–414). ANU Press.

Reklai, L. (2022, August 17). Palau inaugurates its first digital residence. *Island Times*.

Rhodes, K., Warren-Rhodes, K., Houk, P., Cuetos-Bueno, J., & Fong, Q. (2011). *An interdisciplinary study of market forces and nearshore fisheries management in Micronesia*. Nature Conservancy Asia Pacific Conservation Region Marine Program Report No 6/11.

Rhodes, K., Warren-Rhodes, K., Sweet, S., et al. (2015). Marine ecological footprint indicates unsustainability of the Pohnpei (Micronesia) coral reef fishery. *Environmental Conservation, 42*(2), 182–190.

Richardson-Ngwenya, P. (2010). The EU sugar reform and the responses of Caribbean sugar producers. *Geography, 95*(2), 70–79.

Ride, A., Eriksson, H., Tutuo, J., et al. (2021). Catching fish in COVID-19 currents: Food security and governance in rural communities in Solomon Islands. In Y. Campbell & J. Connell (Eds.), *COVID in the islands: A comparative perspective on the Caribbean and the Pacific* (pp. 367–384). Palgrave Macmillan.

Robinson, S.-A., & Wren, C. (2020). Geographies of vulnerability: A research note on human system adaptations to climate change in the Caribbean. *Geografisk Tidsskrift-Danish Journal of Geography, 120*(1), 79–86.

Rolle, S. (2020). Marijuana agro-tourism habitat. In S. Rolle, J. Minnis, & I. Bethell-Bennett (Eds.), *Tourism development, governance and sustainability in the Bahamas* (pp. 177–188). Routledge.

Roscher, M., Eriksson, H., Harohau, D., et al. (2022). Unpacking pathways to diversified livelihoods from projects in Pacific Island coastal fisheries. *Ambio, 51*(10), 2107–2117.

Rudiak-Gould, P. (2012). *Surviving paradise. One year on a disappearing Island.* Sterling.

Saint Ville, A., Hickey, G., & Phillip, L. (2017). Institutional analysis of food and agriculture policy in the Caribbean: The case of Saint Lucia. *Journal of Rural Studies, 51*, 198–210.

Satnarine-Singh, N., Hosein, R., & Saridakis, G. (2023). Structural change and export diversification: A comparison of CARICOM's position. *The Journal of International Trade & Economic Development, 33*(6), 1163–1189.

Savage, A., Bambrick, H., & Gallegos, D. (2021a). From garden to store: Local perspectives of changing food and nutrition security in a Pacific island country. *Food Security, 13*, 819–847.

Savage, A., Bambrick, H., & Gallegos, D. (2021b). Climate change and socio-economic determinants are structural constraints to agency in diet-related non-communicable disease prevention in Vanuatu: A qualitative study. *BMC Public Health, 21*, 1231.

Schurman, R. (1998). Tuna dreams: Resource nationalism and the Pacific Islands. *Development and Change, 29*, 107–136.

Scobie, M., Gopinathan, U., Kelman, I., & Myhre, S. (2023). The agency of community groups in health and climate change adaptation governance and policy in SIDS: The case of Toco, Trinidad and Tobago. *Environmental Science & Policy, 147*, 116–125.

Seto, K., Miller, N., Young, M., & Hanich, Q. (2022). Toward transparent governance of transboundary fisheries: The case of Pacific tuna transshipment. *Marine Policy, 136*, 104200.

Shah, S., Moroca, A., & Bhat, J. (2018). Neo-traditional approaches for ensuring food security in Fiji Islands. *Environmental Development, 28*, 83–100.

Shantz, A., Ladd, M., & Burkepile, D. (2020). Overfishing and the ecological impacts of extirpating large parrotfish from Caribbean coral reefs. *Ecological Monographs, 90*(2), e01403.

Sharman, J. (2011). *The money laundry. Regulating criminal finance in the global economy*. Cornell University Press.

Silas, E., Makikon, J., & Kinsella, S. (2022). Service delivery reform- lessons from Vanuatu. *ANU DevPolicyBlog*, 15 March.

Smith, A., & Dalzell, P. (1991). *Fisheries resources and management investigations in Woleai Atoll, Yap State*. Federated States of Micronesia, Noumea: South Pacific Commission.

Snowdon, W., Moodie, M., Schultz, J., & Swinburn, B. (2011). Modelling of potential food policy interventions in Fiji and Tonga and their impacts on noncommunicable disease mortality. *Food Policy, 36*, 597–605.

Sofer, M. (2018). Revisiting the Fijian periphery: A 30-year perspective on Kadavu Island. In J. Connell & H. Lee (Eds.), *Change and continuity in the Pacific: Revisiting the Region* (pp. 166–186). Routledge.

Sprague, J. (2015). From international to transnational mining: The industry's shifting political economy and the Caribbean. *Caribbean Studies*, 73–112.

Srinivasan, P. (2023, October 3). 'End the licences': Vanuatu oil rig registry sparks concern amid climate advocacy. *Guardian.*

Stanberry, D., & Fletcher-Paul, L. (2024). *Food security in the Caribbean.* Palgrave Macmillan.

Steenbergen, D., Neihapi, P., Koran, D., et al. (2020). COVID-19 restrictions amidst cyclones and volcanoes: A rapid assessment of early impacts on livelihoods and food security in coastal communities in Vanuatu. *Marine Policy, 121*, 104199.

Sulu, R., Eriksson, H., Schwarz, A., et al. (2015). Livelihoods and fisheries governance in a contemporary Pacific Island setting. *PLoS One, 10*(11), e143516.

Syddall, V., Thrush, S., & Fisher, K. (2021). Transdisciplinary analysis of Pacific tuna fisheries: A research framework for understanding and governing oceans as social-ecological systems. *Marine Policy, 134*, 104783.

Syddall, V., Fisher, K., & Thrush, S. (2022). Collaboration a solution for small island developing states to address food security and economic development in the face of climate change. *Ocean and Coastal Management, 221*, 106132.

Tarte, S. (2006). Managing Tuna fisheries in the Pacific. In J. Bryant-Tokalau & I. Fraser (Eds.), *Redefining the Pacific? Regionalism, past, present and future* (pp. 89–99). Ashgate.

Thiault, L., Mora, C., Cinner, J., et al. (2019). Escaping the perfect storm of simultaneous climate change impacts on agriculture and marine fisheries. *Science Advances, 5*(11), eaaw9976.

Thomas, F., & Tonganibeia, K. (2007). Pacific island rural development: Challenges and prospects in Kiribati. In J. Connell & E. Waddell (Eds.), *Environment, Development and Change in Rural Asia-Pacific: Between local and global* (pp. 38–55). Routledge.

Thomas, N. (2014). The age of empire in the Pacific. In D. Armitage & A. Bashford (Eds.), *Pacific histories: Ocean, land, people* (pp. 75–96). Palgrave Macmillan.

Thomas, A., Baptiste, A., Martyr-Koller, R., et al. (2020). Climate change and small island developing states. *Annual Review of Environment and Resources, 45*, 1–27.

Thomson, L. (2020). Looking ahead – Global sandalwood production and markets in 2040, and implications for Pacific Island producers. *Australian Forestry, 83*(4), 245–254.

Thomson, L., Bush, D., & Lesubula, M. (2020). Participatory value chain study for yasi sandalwood (*Santalum yasi*) in Fiji. *Australian Forestry, 83*(4), 227–237.

Thow, A.-M., Swinburn, B., Colagiuri, S., et al. (2010). Trade and food policy: Case studies from three Pacific Island countries. *Food Policy, 35*, 556–564.

Tui, S., & Fakhruddin, B. (2022). Food for thought: Climate change risk and food (in) security in Tuvalu. *Progress in Disaster Science, 16*, 100255.

Turner, R., McConney, P., & Monnereau, I. (2020). Climate change adaptation and extreme weather in the small-scale fisheries of Dominica. *Coastal Management, 48*(5), 436–455.

UN OHRLLS. (2024). *Promising sectors for economic transformation of small island developing states*. United Nations.

Vainikolo, L. (2021). Small and isolated Vava'u, Tonga: From weakness to strength during COVID-19. In Y. Campbell & J. Connell (Eds.), *COVID in the Islands: A comparative perspective on the Caribbean and the Pacific* (pp. 207–218). Palgrave Macmillan.

van der Grijp, P. (1997). Leaders in squash export: Entrepreneurship and the introduction of a new cash crop in Tonga. *Pacific Studies, 20*(1), 29–62.

Vanuatu Chamber of Commerce. (2023). *Vanuatu private sector economic update*. Chamber of Commerce and Industry.

Vega, D., Page, T., & Ota, L. (2023). Challenges and opportunities for inclusive value chains of niche forest products in small island developing states: Canarium nuts, sandalwood, and whitewood in Vanuatu. *Journal of Rural Studies, 100*, 103036.

Vermeij, M., Latijnhouwers, K., Dilrosun, F., et al. (2019). Historical changes (1905-present) in catch size and composition reflect altering fisheries practices on a small Caribbean island. *PLoS One, 14*(6), e0217589.

Vogliano, C., Raneri, J., Coad, J., et al. (2021). Dietary agrobiodiversity for improved nutrition and health outcomes within a transitioning indigenous Solomon Island food system. *Food Security, 13*, 819–847.

Voyer, M., Gollan, N., Barclay, K., & Gladstone, W. (2015). 'It's part of me'; understanding the values, images and principles of coastal users and their influence on the social acceptability of MPAs. *Marine Policy, 52*, 93–102.

Wade, E. (2025). Fishers' responses to future catch declines in Jamaican small-scale fisheries. *ICES Journal of Marine Science, 82*(5), fsaf070.

Waiti, D., & Lorrenij, R. (2018). Sustainable management of deep sea mineral activities: A case study of the development of national regulatory frameworks for the Republic of the Marshall Islands. *Marine Policy, 95*, 388–393.

Wallis, J. (2010). 'Friendly islands' in an unfriendly system: Examining the process of Tonga's WTO accession. *Asia Pacific Viewpoint, 51*, 262–277.

Wang, C., Shu, X., Zhou, S., et al. (2025). Embracing a new era of deep-sea mining: Research progress and prospects. *Marine Policy, 180*, 106778.

Waqairatu-Waqainabete, S., & Kaumaitotoya, L. (2021). Women and the business of aquaculture: A case for women tilapia farmers in Fiji. *SPC Women in Fisheries Information Bulletin, 34*, 4–8.

Waqavakatoga, W., Habru, P., & Koro, M. (2025). How do Pacific Island countries respond to China's statecraft? In J. Wallis, H. McNeill, M. Rose, & A. Tidwell (Eds.), *Power and influence in the Pacific Islands. Understanding statecraftiness* (pp. 201–218). Routledge.

Washburn, T., Simon-Lledó, E., Soong, G., & Suzuki, A. (2023). Seamount mining test provides evidence of ecological impacts beyond deposition. *Current Biology, 33*(14), 3065–3071.

Watt, L. (2024). Relational resource geographies of beche-de-mer under moratorium. *Asia Pacific Viewpoint, 65*(1), 40–54.

West, P. (2016). *Dispossession and the environment: Rhetoric and inequality in Papua New Guinea*. Columbia University Press.

Wilkinson, E. (2023). How a small Caribbean island is trying to become hurricane-proof. *The Conversation,* 29 November.

Wilson, K. (2013). Wan Laki Aelan? Diverse development strategies on Aniwa, Vanuatu. *Asia Pacific Viewpoint, 54*(2), 246–263.

World Food Programme. (2025). *Cost of the diet analysis in Samoa*. WFP.

Zhang, D. (2018). *China's growing interest in deep sea mining in the Pacific*. ANU Department of Pacific Affairs InBrief 2018/11.

4

Tourism: From the Land to the Sea

Tourism is of obvious and established economic and critical importance in most SIDS, sometimes by default or even reluctant choice, when other sectors have floundered but now more often regarded as the primary source of economic growth and development. In some respects, this chapter is the converse of the previous one where diverse development strategies were more evident in the Pacific, since tourism is more prevalent in the Caribbean. Most Caribbean countries, because of significant entry barriers to other markets and their natural resource endowments, had gravitated towards land- and sea-based tourism as a development option by the 1970s (Hayle, Singh, & Wright, 2010; Payne & Sutton, 2007). Typically, as for example in St. Vincent and St. Lucia, the end of preferential access to metropolitan markets for agriculture brought the decline of the primary sector (Bishop, 2010). It took little time to recognise that for SIDS, 'tourism offers the best chance for development in terms of creating growth and employment, generating foreign exchange and reducing poverty' (Croes, 2006: 455). It offered comparative advantages, rare in most SIDS, being closely linked to sea, sun and sand. It proved a considerable source of employment and grew rapidly as cheaper airlines, higher incomes, longer vacations, cruise ships and improved

J. Connell, Y. Campbell, *Island Challenges*,
https://doi.org/10.1007/978-981-95-8177-1_4

facilities all facilitated and encouraged travel. As the Pacific industry was beginning, overtourism was emerging in parts of the Caribbean, where cruise ship tourism, initially hindered and then helped by the COVID-19 pandemic, dominated previous land-based tourism.

By the late twentieth century, as commodity production declined, tourism had become the dominant source of income in many SIDS, generating US$96 billion per year and employing 40,000 people (Gayle & Goodrich, 1993). For the overall Caribbean, mass tourism had taken over from a small, exclusive resort sector and, by the 1980s, had displaced the sugar industry as the largest source of foreign exchange (Duval, 2004). By the late 2010s its share of export earnings was more than 75% in six Caribbean SIDS (Antigua, Bahamas, Dominica, Grenada, St. Lucia and St. Vincent) compared with just one Pacific SIDS (Palau) (Armstrong & Read, 2021). That suggested the need for caution over creating a monoculture, brought home by its abrupt disappearance during COVID-19, and the need for resilience.

Tourism is labour intensive, creating many jobs (though few are well-paid, with promotion prospects, hence many are taken by women, for whom alternative employment possibilities could be scarce), and generates considerable activity in the service sector. Integration of agriculture with the tourist industry has been disappointing, usually because of low volumes, uncertain supplies, seasonality, land alienation and inaccessibility (Rhiney, 2011; Scheyvens & Russell, 2012). Much coastal land has been lost to tourism, agricultural production has declined through the withdrawal of land and labour, mangroves have been removed and coastal erosion has sometimes followed: major elements of the coastal squeeze and too often in fragile, aesthetic sites (Lewis-Cameron, 2021). In Vanuatu as much as 90% of coastal land has been alienated by foreigners, much of that for tourism with reduced access to the sea for local people: epitomising 'accumulation by dispossession'. Demands on services, especially water, are often considerable. However, since tourists favour clean environments, localised environmental benefits exist. In most SIDS, a challenge has been to make a transition from tourism economies and infrastructure that deplete and alter natural landscapes and resources, to low impact, nature-based ecotourism, more difficult to market, manage and attract tourists.

Benefits from tourism have been queried. After more than half a century, large-scale tourism in the Caribbean had scarcely contributed to poverty alleviation, failed to reduce wealth inequalities, but played a part in increasing inequality, without significant linkages to stimulate sustainable small businesses. Such concerns emerged later in the Pacific where there was less rush to enter the market, as states feared harmful and unmanageable social change (Macnaught, 1982), but the Pacific was more distant from likely markets and disadvantaged by intervening opportunities. Tourism is often highly centralised around the main urban areas, as in Grenada and Vanuatu. Though cruise ships reach more isolated areas, their impact is localised. Economic benefits are less than in larger states because of disproportionate imports of foods and other materials, and overseas ownership of tourism infrastructure, but tourism has generated more income and more foreign exchange than commodity exports in almost all SIDS, comprehensively so in the Caribbean. However, tourists increasingly spend relatively little in the SIDS compared with pre-travel spending, notably on travel and accommodation. Many SIDS are highly dependent on a monoculture of tourism, yet it is vulnerable to perceptions of political unrest and violence, and to environmental degradation and climate change.

Tourism is vulnerable to external shocks (and internal shocks such as coups or other forms of political violence). Tourism ended abruptly in both regions after the arrival of COVID-19 in early 2020. Its collapse focused attention on the lack of local ownership and the possibility of developing a smaller-scale, more sustainable locally based industry. However, in the rush to restart, restructuring never happened as vested interests prevailed. The COVID-19 collapse raised questions about both the viability of 'monocultural' tourist economies and the possibility of attracting niche tourism industries, such as whale watching.

The growing focus on tourism, although rarely central to discourses of the blue economy, reflects an accelerating transition in most SIDS from the productive to the service sector. Islands were regarded and marketed as having comparative advantages in terms of clean beaches, unpolluted seas, warm weather and water, and distinct cultures. Only some Pacific SIDS have remained on the tourist periphery, distant from conventional tourist circuits and markets, without distinctive cultural or natural attractions. Few visitors reach such SIDS as Nauru, Tuvalu and Kiribati, while within archipelagic states tourists are less likely to visit outer islands, knowing little about them, preferring to stay put, or without enough time or money. Nauru is the only SIDS that intentionally discourages tourists by making access to visas unusually difficult. Concentration of infrastructure enables some local economies of scale and makes it more difficult for remote places to develop tourism even within the SIDS (e.g. Mohan, 2022a). More than almost any other economic activity—except mining—tourism has an unusually uneven impact.

4.1 Being Small: Transnational Tourism

Tourism typifies in exaggerated form how a single sector within the SIDS exemplifies, at different scales, tensions between society, ecology, environment and economics. It increases pressures on energy, water resources, coral reefs and coastal lagoons. An obvious relationship usually exists between tourist numbers and environmental degradation. That endpoint is yet to occur in Caribbean and Pacific SIDS, despite occasional sharp conflicts between the development of the tourism industry and sustainability. Much hangs on the complex relationships between governance, regulation, corporate social responsibility and numbers, especially on cruise ships. Mass tourism in some Caribbean islands has exceeded any notion of tourism carrying capacity. It is no accident that Butler's (1980) tourism life cycle model worked well in the Caribbean (where, at a certain point, and without modifications, tourism numbers stabilise and possibly decline as environmental pressures, and simply overcrowding, intensify). That life cycle sometimes paralleled a different socio-economic life cycle where local owners were bought out by overseas owners once success was assured, changing the nature and impact of the industry.

Under the guise and banner of sustainability many SIDS, including Jamaica, Barbados, Grenada, Antigua and Barbuda, adopted policies and practices, such as the Caribbean Sustainable Tourism Policy Framework, the Caribbean Regional Sustainable Tourism Development Programme, Barbados' Tourism Master Plan, Grenada National Sustainable Tourism Plan, the Organisation of Eastern Caribbean States (OECS) Common

Tourism Policy. They covered everything from tourism numbers (usually for particularly sensitive sites), vehicle hire, restrictions on some activities (mainly in marine areas), to artificial beach nourishment and coral restoration. The Caribbean Tourism Organization (CTO), the region's main tourism development agency, took on the rather obvious 'responsibility to secure the sustainability and growth of Caribbean tourism, and to market and promote the region'. The Pacific Tourism Organisation followed. The ability to advertise green credentials and sustainability has resulted in the 'marketization of environmental responsibility' (Grydehøj & Kelman, 2017). Fortunately, quite often, carefully managed resorts have been outposts, islands and even models of sustainability. Regional challenges to sustainability have included climate vulnerability and disaster risk management (Becken & Hay, 2007; Hall & Higham, 2005; Reddy & Wilkes, 2012, Gössling & Higham, 2021, McLeod, Dodds, & Butler, 2024). Capacity constraints, and a wider regional capability to manage the sustainability of tourism, involve problems of coordination, ownership, creating climate resilient infrastructure and waste management. Not all such challenges are easily overcome.

In small states valuable components of the tourist industry are owned by overseas interests. Where tourists are almost exclusively from elsewhere (and are hopeful of some degree of hedonism), and many of the key economic interests (hotel chains, hire car fleets, airlines and, above all, cruise ships) are mostly externally owned, local direction and regulation of the tourist industry is particularly difficult. Tourism in SIDS is thus characterised by the substantial overseas leakage of expenditure, much being made prior to travel, to real and virtual travel agencies (such as booking.com), rental car chains, big corporations (especially hotel chains and cruise ships) such as Disney and foreign airlines. Much of the tourism industry in SIDS is centred on resorts and cruise ships. Luxury resorts attract higher spending tourists, who consume more resources per capita, especially water and energy, and generate more waste than do residents (Connell, 2013; Kelman, 2021). Some Caribbean resorts are becoming linked to marinas and enclaves of private residences, including second homes, a growing trend that minimally benefits SIDS.

The trend towards all-inclusive resorts, begun in the Caribbean and followed in the Pacific, has resulted in expenditure being concentrated there. In Negril (Jamaica) resort development was encouraged by subsidies and infrastructure provision by the national government, supported by USAID and Japanese aid, leading to expenditure being oriented to larger overseas-owned hotels and companies. Several resorts discouraged tourists from venturing beyond, further diminishing local expenditure especially on food, crafts and transport (Rhiney, 2012). Little of the income generated by tourism filtered into the nearby towns of Montego Bay and Falmouth, which experienced uneven development (Dodman, 2009). Much the same is true of 'leakage' of expenditure from Denarau island resorts in Fiji to the city of Nadi, just ten kilometres away. In the resort towns of Ocho Rios and Montego Bay (Jamaica), illegal drugs were sold to tourists, squatter settlements appeared, local beaches were privatised and real estate prices soared, creating tourist dual cities (Brooks, 2016; Mycoo, 2022). More generally, resorts have been able to capture the bulk of tourist expenditure with limited interaction with local culture (beyond highly choreographed tours) and are marketed as 'sea, sun and sand' destinations, with local people as exotic bystanders and staff.

In similar vein, cruise ship tourism adds even less to local economies relative to other types of tourism, since passengers sleep, eat, drink and are entertained on the cruise ship, and have ample opportunities for onboard 'retail therapy' (MacCarthy, 2023). That has occurred even where there is evidence that tourists would welcome something more—at least when they are specifically asked that question (e.g. Graci & van Vliet, 2020; Grilli, Tyllianakis, Luisetti, et al., 2021; Tyllianakis, Grilli, Gibson, et al., 2019). Ironically, the most exclusive resorts in the world are on some of the most isolated islands, such as Turtle Island and Wakaya in Fiji, Petit St. Vincent in the Grenadines, and Calivigny in Grenada. All obvious needs are satisfied within such resorts, almost all payments are made outside the country, and at best local people gain only from some lease payments and employment: versions of 'zero-dollar tourism' that minimise local expenditure and may offer few links to local cultures with the erosion and exclusion of history. Leisure, hedonism, heritage, culture and sustainability are not often complementary. Even changes in tourism philosophies, that have sought to value local culture, do not necessarily result in changes to established resorts (Spennemann, 2021). Path dependency has usually prevailed. Consequently, resort development has tended to clash with notions of sustainable development where that involves more subtle and complex links with local people and cultures and with marine and terrestrial ecosystems.

Ecotourism provides an alternative, emphasising nature and 'traditional values' and which is ethically managed to be low impact and contribute to conservation. This partial and supposed antithesis of resort tourism has been widely touted but has rarely been very successful—and predictably then only on a small scale—because it is difficult to market internationally, and sometimes poorly accessible, alongside the disinterest and lack of time and knowledge of tourists (and overseas travel agents), and skill shortages, so that many such ventures have been transient (Prentice, Kundra, Alam, et al., 2023). Success has only been exceptionally achieved where networks of civil society organisations have persisted in putting pressure on local governments and companies and been able to escape top-down directions (Lewis-Cameron & Brown-Williams, 2023). Agritourism and farm stays have been discouraged by a similar range of constraints including business regulations, lack of financial support, absent entrepreneurial skills, a motivated limited labour force, and a fear of damage from extreme weather events related to climate change (Pratt, Magbalot-Fernandez, & Ohe, 2022). Where ecotourism ventures are successful, they are usually targeted at domestic visitors and educational experiences.

Many small-scale entrepreneurs and entire villages have sought to establish a niche presence in the industry, with the lure of 'authenticity' and 'ecotourism', building small guesthouses in the hope that some tourists might stay or drift away from resorts, all too often a forlorn hope. Scattered through Vanuatu, Fiji and elsewhere, such ecotourist ventures have largely disappeared as too few tourists sought to get off the beaten track. In Tanna (Vanuatu) particularly, villagers established such possibilities as a 'cannibal trail' and other 'traditional' practices, but even where tourists included such ventures in tight itineraries, disputes over the distribution and use of the revenue brought their closure (Robinson & Connell, 2008, Lindstrom, 2018). Ventures that have survived have tended to be close to urban centres and able to market themselves as one transitory component of an island tour.

Community interests are typically overwhelmed by the power, influence and prestige of outside tourism capital, and communities are often divided, as at Soufriere in St. Lucia, where tourism took precedence over the aspirations of local fishers (Pugh, 2005). Slow tourism (Timms & Conway, 2012; Walker, Lee, & Li, 2021) has also been seen as an alternative to mass tourism and a form of 'mobility justice'. Yet tourism, especially in the Caribbean, remains 'inequitable and unsustainable' being grounded in coloniality, racial inequity, white supremacy and 'extractive neoliberal development' (Sheller, 2021). Too often, capital-, water- and energy-intensive hotels, with high import requirements, have contributed little beyond poorly paid employment to the local economy. Without positive local involvement, social and ecological disparities, a diminishing economic contribution, concerns about environmental pollution and destruction, the loss of quality of life and income equality, plus overcrowding, have led to a growing animosity towards tourism and further tourism growth (Peterson, DiPietro, & Harrill, 2020). Reconstruction or diminution of the industry has not followed, while cruise ships (see below) brought other problems and more opposition.

4.2 Linkages

Tourism stimulates development in other economic sectors, notably transport, entertainment, agriculture, fisheries and handicrafts, but multiplier effects are least in smaller islands, where goods are more likely to be imported (including such 'tourism artefacts' as swimwear, surfboards, snorkels and scuba gear), as in cruise ship tourism and enclave resort tourism. Tourism has created few agricultural linkages on islands and sometimes resulted in a decline of local production as land and labour move into tourism (Addinsall, Weiler, Scherrer, & Glencross, 2017; Connell, 2013). The transport sector has fared better than most, mainly owing to taxis and car rentals. International air transportation linkages are with metropolitan countries; hence, 'island hopping' is difficult to achieve (and expensive) so that tourists rarely visit more than a single country, emphasising competition rather than cooperation between countries.

Agriculture and fisheries are only poorly linked to tourism, because of infrastructure problems, resorts' need for regular and substantial supplies, while tourists prefer familiar food. The lack of linkages is no new problem for either region. A reliance on imports emerged from the mono-crop plantation economies, and dependence on imports further increased in the post-war period, with prevailing neoliberal economic policies further reducing prospects for linkages by focusing almost exclusively on promoting foreign investment for development and hoping that linkages would occur spontaneously through 'rational' market forces (Timms, 2006). International hotel chains are least likely to purchase locally, through concerns about delivery, price and quality. Demand may also be limited where children—conservative in their tastes—are prominent, yet 'island nights', with 'authentic' local foods, have not always tempted tourists to try them. Ironically, incorporating local produce into mainstream resort menus has also proved difficult for chefs (Scheyvens & Laeis, 2021). In Fiji three quarters of the imported foods consumed in the main tourism hotels were meat, seafood and dairy products (International Finance Corporation, 2018), all less easily produced locally on a consistent basis.

Some successful attempts have been made to encourage more local participation. In St. Lucia, a local produce purchasing policy in place since 2002 with three Sandals hotels has incorporated more than 50 local farmers supplying weekly orders, with training for purchasing clerks and receivers, regular visits to farms by chefs and kitchen staff, and discussions with farmers and suppliers on issues of traceability. Likewise in Jamaica, the Sandals Resort Farmers Programme grew from 10 farmers supplying 2 hotels to 80 farmers supplying hotels across the island in 2004. Farmers' sales increased over 55 times in 3 years, from US$60,000 to US$3.3 million. Long-term commitment and coordination succeeded, with the participation of cooperatives (and wholesalers used by large hotels), enabling institutional stability, and the hiring of an agricultural extension officer to coordinate and manage supply and demand. Larger SIDS have more chance of being able to develop such local supply chains, if not counteracted by excessive land alienation. Occasional events and festivals, such as at the weekly Oistins Fish Fry in Barbados, where fishermen, fish sellers, farmers and local chefs sell local vegetables and fish, may be more successful in stimulating and sustaining links between tourism and agriculture, fisheries and handicrafts. Frequently, tourism has fallen short of its local value-adding potential, the outcome of both local and international factors, emphasising the overseas leakage of income.

4.3 Employment

In many SIDS, total visitor expenditure represents more than half the gross domestic product, and employment in the tourism sector can absorb more than a third of the formal working population and generate as many indirect jobs. In 2021, Caribbean tourism accounted for 15% of total employment with 2.75 million people employed in the sector, much more than in the Pacific. However, a high proportion of formal employment is labour intensive, is poorly paid, lacks a career path and requires no high educational status (including wait staff, gardeners and cleaners). Although it is substantial, tourism work can be precarious (Robinson, Martins, Solnet, & Baum, 2019). Managerial roles can often be taken by expatriates. Employment is also indirect in the tourism-linked informal sector of markets, transport, construction, retailing and entertainment. Much of that employment is of women, usually concentrated in less-skilled areas. Transport is almost always a male arena and a rare part of the industry where men outnumber women. Formal employment opportunities for women have enabled some empowerment even where, as in Fiji, that employment is routine and menial (Kanemasu, 2008). In Samoa, in 2019, 80% of workers in the formal tourist industry were women, and the proportion of women in the related informal sector was probably even higher. In the Caribbean region, tourism accounts for 12% of female employment and 6% of male employment, and up to 70% of employment in accommodation and food services is of women. Women are strongly represented in the hotel and tourism sector, due to its characteristics: less emphasis on formal education and qualifications, greater importance attached to communication and hospitality skills, flexible working opportunities and more diversity in opportunities for entrepreneurship that do not require heavy start-up financing (UNFPA, 2025). Despite limited prospects, tourism has offered and provided women with new opportunities for social mobility and greater control over household incomes and, in some contexts, a break from patriarchal society. Social costs have sometimes ensued, with uneven development and conflicts over land and incomes, and persistent gendered inequality, but alongside recognition that cherished values, centred on kinship relations, were usually resilient and tourism enabled superior lifestyles.

4.4 Cruise Ships

The Caribbean Sea has become the main global hub for cruise tourism. With around 40% of the world market, it surpasses the Mediterranean by a wide margin, with most of the global total of 12.8 million tourists in 2023 being from North America. By then numbers had passed pre-COVID-19 levels. Although the main destination countries are in Central America, Bahamas received 5.4 million passengers (more than 10 times the resident population and 3 times the number of air arrivals), and around 500,000 passengers in St. Kitts also dominated the resident population. The Pacific had fewer numbers in 2023, collectively about 680,000 passengers (a total that included several dependent territories) with Vanuatu having 260,000. It had still to surpass pre-COVID levels.

In half a century cruise ships of ever-increasing sizes have added a new dimension to tourism, placing some distant and isolated islands firmly on tourist maps for the first time, and contributing to maritime versions of overtourism. Whilst cruise tourism also has an urban bias, with large ports increasingly necessary, it extended the reach of tourism and brought unusual cash incomes to some relatively remote islands, such as Dravuni (Fiji) and 'Mystery Island' (Vanuatu). However, stays are longer and opportunities for expenditure greater in larger port cities so that, in Fiji, cruise ship tourists spent an average of F$104 (US$52) in the capital Suva but only F$3 in the small island of Dravuni (International Finance Corporation, 2019). In Vanuatu, in 2014, cruise passengers spent around US$125 whereas other tourists brought US$1514 during a nine-day stay (Tolkach, 2023). That ratio was almost identical to that in Jamaica where longer-stay tourists had 12 times the expenditure of cruise ship passengers (*Jamaica Gleaner*, 4 September 2019). In the Caribbean, where most tourists are American, individual spending of cruise ship passengers in 2015 was somewhat less than in the Pacific: US$78 in Barbados and US$83 in the Bahamas (Jessop, 2019). That may reflect ethnicity, affluence, the availability of goods or, most likely, less time spent on land. Similar patterns of expenditure may also be true of land-based tourists, who arrive by air and remain in one location for longer, so spending more than cruise passengers.

Since cruise ships bring their own accommodation, entertainment and dining facilities, and stay only briefly, the 'capacity' of the islands to absorb tourist numbers has increased substantially, but the economic benefits have been disproportionately fewer and the environmental costs greater. Cruise ship tourism differs from land-based tourism in creating much less local employment, with more short-term economic opportunities, for the local provision of goods (particularly café food and handicrafts) and services (especially cleaning, transport and tours). However, the sheer numbers of passengers, with individual cruise ships disgorging over a thousand visitors, ensure that economic benefits are still considerable, especially in stores. Thus, Vanuatu received 143,512 cruise ship visitors in 2019, contributing an estimated A$18.7 million to the country's economy, with the average passenger spending A$85 there (Pacific Private Sector Development Initiative, 2021): a significant decline in five years. Cruise ship passengers in Fiji, where annual arrivals had increased fourfold between 2010 and 2018, spent an average of US$44 there, mostly on pre-booked tours, but would have liked to spend more; instead, they faced 'unmet spending opportunities' (International Finance Corporation, 2018). At least the costs of marketing were all external. Yet, as one small-scale Fijian tourism operator observed:

> In the year or so before COVID, we watched the massive cruise ships arrive in Savusavu Bay, dwarfing the town. Even then, a cruise ship in town created tension. As hundreds of tourists spilled onto the streets for just a few hours, small business owners almost came to blows in their efforts to capture some of the trade. Ironically, there wasn't very much money available anyway – most of the tourist dollars were siphoned off by players higher up the chain long before the visitors disembarked. We learned that our own

> tour, for which we charged F$25 (around A$16.50), was being sold on board for up to A$110. (Moorhead, 2021)

The newest cruise ships offer an increasing range of onboard attractions and services, enabling cruise companies to increase revenue, and run land-based excursions, reducing local income possibilities (Hoarau-Heemstra, Wigger, Olsen, & James, 2023). Beginning in the Caribbean, as cruise ship tourism boomed in the wake of *The Love Boat*, mass market cruise ships—the vast majority of those in the SIDS—marketed Eurocentric understandings of Caribbean cultures, homogenising and sanitising the people and countries to produce 'tropical Disneylands'. Caribbean cruise companies have also sought to buy or lease particular islands, often uninhabited, or enclosed and privatised beaches, to concentrate all land activities (and expenditure) in their own facilities, turning places into 'sanitised enclaves' (Lallani, 2024). Royal Caribbean owns private islands in the Bahamas—the Royal Beach Club and Perfect Day in CocoCay—and other land is being purchased, at an additional environmental cost, as protective mangroves are destroyed, a new 'accumulation by dispossession'. Local anti-cruise protests in the Bahamas have opposed more cruise ports, the alienation of small islands, and the consequent pollution, environmental damage and loss of income. The Caribbean 'model' was subsequently replicated in the Pacific. In 2019, following Caribbean precedents, the Royal Caribbean line was seeking to develop what they described as the 'world's first carbon neutral cruise destination' on the private island of Lelepa (Vanuatu), officially in partnership with the local people, but wider economic benefits were likely to be small. COVID-19 brought at least a temporary halt to the proposal.

Cruise ships pose different environmental issues, related to carbon emissions and waste disposal. By the end of the 2010s, the cruise ship industry was increasingly being viewed as less beneficial to destinations than previously assumed, despite the increased numbers of ships, and with unanticipated environmental costs. Ships place pressure on fresh water supplies and generate considerable waste streams, including fumes, grey water and sewage, with their impact on marine pollution. Water is required for laundries, kitchens, showers, wellness centres and swimming pools, and cleaning, with only a small part being recycled. Waste disposal, intended to occur in international waters, is difficult to monitor. Since most cruise ships operate under flags of convenience—two thirds being registered in just three countries: Bahamas, Panama and Bermuda—and so without 'genuine' homes, they were poorly regulated for workplace health and safety, and environmental management (Mahoney & Collins, 2020). Islands and their inhabitants were being reduced to mere colourful backdrops for an industry based elsewhere, with cruises focused on entertainment onboard rather than on the destinations, culminating in 'cruises to nowhere' that visited no ports.

4.5 Space, Place and Equity

Cruise ships bringing tourists to often quite remote and isolated islands, especially in the Pacific, drew attention to uneven tourism development, as some islands benefited while others missed out. Even in the smallest SIDS, tourism is usually concentrated in a few areas: a process that began early—with concentration in particularly pleasant, accessible and serviced sites and which then brought sequential development there—as in St. Vincent in the 1970s—the outcome of the industry usually being controlled by the metropolitan travel industry rather than islanders (Britton, 1977; Britton, 1980). With partial exceptions, such as Fiji, that concentration was a variant on urban bias, as infrastructure provision made concentration cheaper. Enclaves of luxury surrounded by poverty produced deleterious social and cultural effects. Even where facilities existed, tourists were less likely to venture off main islands, and social media have not reduced the resultant uneven development.

Exceptions to urban bias exist, but few enough to distract from the way infrastructure provision (roads, water and electrification) benefited urban tourist regions. Fiji is one of very few countries where islands other than the main island are marketed in key source countries and where many islands have experienced tourism (Connell, 2015). Cruise tourism produced more exceptions as a few small islands, that previously had few formal income-generating opportunities, became part of cruise ship itineraries. More rarely, regular tourism and incomes even reversed rural-urban migration flows by supporting economic needs and diversifying limited existing livelihoods, as in the Yasawas (Fiji), Aneityum/Mystery Island (Vanuatu), the Grenadines (St Vincent) and Exuma (Bahamas). That could still result in intense competition for a limited market as cruise companies sought some variations to their itineraries.

Even where cruise ships reached remote islands and were relatively frequent, frustrations remained. In the Trobriand Islands (PNG), economic and social relations have been substantially transformed, creating new forms of inequality both between islanders and cruise passengers (on ships perceived as 'markers of privilege and mobility') and within the islands. Optimism ('It will be a like a town here'—that hinted about development preferences) was misplaced, culture was eroded and the limited incomes gained (mainly from cold beer sales) failed to trickle far socially or geographically. What was touted as 'development' beneficial to all, led to 'unequal resources distribution, jealousy and … exclusion from a world of wealth and exchange … clearly visible yet just out of reach' (MacCarthy, 2023: 86). Here and elsewhere, as in the Bahamas, 'the trickledown effect [wa]s yet another myth about tourism development' (Minnis, Rolle, & Bethell-Bennett, 2020: 40). Visible and tangible inequity pointed to localised uneven development and suggested the value of ecotourism, pro-poor tourism and niches based on local attractions and resources, but few have succeeded in the SIDS.

Few ecotourism ventures or successful tourism niches exist, and many have failed. In recent times, however, what has been described as 'Anthropocene tourism', has centred around global change science, where islands have become destinations and laboratories for scientific study and training, as in the Bahamas (Moore, 2019). Fiji and Solomon Islands have promoted surfing, but efforts to achieve diversity through sports have largely failed, as in Barbados, where sports tourism failed to enhance economic growth, despite its hosting major cricket and golf tournaments (Alleyne, Alleyne, Cezair, & Charles-Soverall, 2024). SIDS like Kiribati, much more distant from markets, had even less potential for sport tourism, lacking adequate infrastructure (Molu, Pratt, & Tarte, 2024). However, Fiji has gained from international rugby matches with its national team, the Fijian Drua, creating in 2023 the equivalent of more than 240 full-time jobs, adding F$11.5 million (US$5.1 million) to the economy, with 20% of the spectators being international fans, so contributing to economic growth and community wellbeing (*FBC News*, 24 July 2025). Diaspora or VFR (Visiting Friends and Relatives) tourism has become significant in Caribbean SIDS (Connell & Stanley-Niaah, 2008; Duval, 2003; Mortley, 2011) and increasingly in the Polynesian SIDS. Marine niches have evident potential, including whale watching in Tonga, where small groups swim with the whales (Orams, 2013), surfing in Fiji, diving in Chuuk (FSM) and Palau. Wedding tourism has worked in Fiji and unusual terrestrial attractions include volcanic activity as in Soufriere (St. Lucia) and Tanna (Vanuatu). Local foods may have some limited potential, but few SIDS are renowned for their cuisine. Food tourism has yet to be successful (Berno, 2025; Gibson, 2025). Several Caribbean SIDS have sought economic diversification through medical tourism (Chap. 3). A niche shared by all, but proportionately most effective in Kiribati and Tuvalu, is that of global 'country collectors', ticking off one more place. Some islands have also experimented with tourism that caters to alternative medicine such as 'magic mushroom' and psychedelic therapy retreats. In Jamaica, where this is unregulated, hotels have capitalised on the demand for psilocybin mushrooms.

Equity fails to benefit where 'public' infrastructure provision has favoured tourist regions, and resorts are more likely to alienate land (and sometimes sea) from local uses, as a central part of the 'coastal squeeze'. Islanders seeking to belatedly enter the industry have been frustrated by various factors: a lack of education and familiarity with international tourism, the difficulty of marketing that limits their degree of participation, the inability to raise adequate capital while some may even be marginalised by political patronage as in Tobago (Mohan, 2022b). Established small-scale tourism operators may simply be crowded out. However, the growth in employment and incomes (national and individual), despite overseas ownership, and the lack of obviously successful alternatives providing employment and national income have insulated the industry from rigorous criticism and ensured that it is central to national development.

4.6 Environmental Degradation and Climate Change

Tourism has multiple environmental impacts. Change is inevitable where facilities must be constructed, but they are usually less visually intrusive and damaging to the environment than many new structures. In some cases this has resulted in a subsequently improved environment, where governments, hotels, NGOs, villages (and occasionally tourists) undertake some conservation and management. In 2025 Tourism Fiji introduced 'Loloma Hour', a commitment by 21 resorts to include guests for 1 h in a 'sustainability activity': from guided nature walks to planting coral, learning traditional crafts, helping protect Fiji's wildlife, or being part of beach clean-ups or mangrove restoration. Several island hotels have pioneered environmental conservation ahead of governments when it became evident that this was valuable for the industry. It may not necessarily extend far from the resorts and sometimes be no more than greenwashing and subtle advertising.

Tourism has placed considerable pressure on reefs and coasts, through reconstruction (such as seawall and marina development), reclamation and recreational uses. Golf courses have sometimes erased both agricultural land and history and require chemicals and water in maintenance (Spennemann, 2021). Exceptionally, the correlation between concentrated tourism growth and environmental degradation may be strong, especially with coastal squeeze, as on the Mulinuu peninsula (Samoa) or Erakor lagoon (Vanuatu). But, in the Bahamas, several hotels have sought to pioneer effective waste management (Chap. 6). Cruise ships bring marine pollution and pressure on scarce water supplies, even resulting in legal challenges. Carnival Corporation, one of the largest cruise lines, was charged in 2019 with dumping food mixed with plastic waste in Bahamian waters, falsifying records of environmental compliance plans and illegally releasing over 500,000 gallons (1.9 million litres) of sewage and 11,000 gallons (41,000 litres) of food waste globally.

Overall global tourism has been estimated to produce 8% of greenhouse gases (Lenzen, Sun, Faturay, et al., 2018), a proportion that was still growing before COVID-19 (Higham, Font, & Wu, 2022; Wolf, Filho,

Singh, et al., 2021). Tourism thus contributed to the climate change that will challenge its own future. That is particularly so in the Pacific where reaching destinations involves relatively long-haul flights. Few airlines have sought to reduce emissions and carbon offsets and barely a tenth of tourists are willing to contribute to doing so. Growth in both international and domestic travel, and energy-intensive transportation, has increased the non-renewable energy dependence of tourism. Other environmental issues include excessive water consumption in relation to residential water use, untreated water discharge, waste generation and damage to local terrestrial and marine biodiversity. Environmental degradation deters tourists. At the start of the twenty-first century, 77% of tourists in Barbados stated that they would be unwilling to return for the same price, under a scenario where 'beaches largely disappeared' (Uyarra, Cote, Gill, Tinch, et al., 2005). That is scarcely surprising where tourism involves sea, sun and sand.

Climate change poses a physical threat to island tourism, especially where almost all its infrastructure (from hotels to airports) is on modified and less protected coasts. Island destinations are particularly affected because of their extensive coasts and critical marine ecosystems, and because tourism is such a crucial economic activity. Erosion of beaches will have a substantial economic cost if coastal management practices are not also intensified (Spencer, Strobl, & Campbell, 2022). Increased temperatures and humidity are directly and indirectly problematic but may exacerbate bleaching of coral reefs (a significant drawcard for the industry) and some loss of biodiversity. Such losses might reduce the potential for some niche tourism. Remarkably, Tuvalu had a brief phase of 'last resort tourism', where a global crisis that seemed likely to render the islands uninhabitable was turned into a tourism opportunity (Farbotko, 2010; Huebner, 2012). More intense cyclones, and longer cyclone seasons, may damage tourism infrastructure but with effects on the industry rather than on tourist choices. That might result in higher prices if agricultural and fisheries production declines. More than half of all tourism assets in both regions face some risk, including island airports, almost all of which are coastal. Flight disruptions and airport closure in extreme weather conditions could have major impacts on the transport of goods, the transport of tourists and accessibility for emergency services. The international airports of Vanuatu and the Marshall Islands have both been closed because of flooding. For many tourists, air travel is the main means of access to the islands, but it is likely to become more expensive and less attractive as tourists and airlines try to reduce their carbon footprints.

Policy adjustments in response to climate change are necessary at multiple levels and scales, only some of which involve tourism infrastructure development and land management, but require collaboration between private and public sectors (Holden, Jamal, & Burini, 2022; Mahon, Petrie, Trotman, et al., 2021; Moghal & O'Connell, 2018; Scheyvens & Cheer, 2021; Wolf, Moncada, Surroop, et al., 2024; Wong, Jiang, Klint, et al., 2013). Physically, these are likely to be centred on building construction, water conservation and management, sewage treatment and physical planning, rather than on changing the structure of the industry. As has been pointed out for Barbados, 'what remains to be answered is who will pay and who will benefit from the policies and projects implemented. This matter calls for stakeholder input informed by an awareness of the philosophy of equity, economic efficiency, and environmental sustainability' (Mycoo, 2014: 56). That is difficult enough to conceptualise let alone implement, and it should also involve greater consideration of equity and ethical issues, decarbonisation and the greater role of nature-based tourism, little of which is presently evident. Conceivably, links to the local economy may be even more difficult if climate change affects agricultural and fisheries systems. Eventually, the threat from climate change may well be primarily influenced by changes in policies and practices in tourism source countries that affect ethics, purchasing power and airline technology.

4.7 Beyond the COVID Crash

Early in 2020, steady tourism growth in the SIDS came to a dramatic stop—thus far the ultimate external shock—with the global spread of COVID-19. No global industry was more greatly affected. Within two months, all Pacific and Caribbean ports (and borders) had closed, cruise ship tourism was over and land-based tourism ended almost as quickly.

Some SIDS moved fast despite the significance of tourism. Samoa, with a history of epidemics, was particularly quick to close its borders. Not only were cruise ships banned, but they were perceived in a new and negative light, as floating 'petri dishes' implicated in the global spread of COVID-19. Employment in ports and airports, and in a range of services and suppliers, from cleaning and markets to restaurants and taxis, disappeared. Hotels and resorts gradually closed, alongside restaurants and bars, dive shops, handicraft markets, local travel agents and car hire companies. Staff were laid off; local fresh produce and fish and taxis were no longer needed. Closures were dramatic. Typically, in Samoa, for example, within a month of the border being closed, some 70 hotels, resorts and other businesses closed, at least half the industry, leaving 1000 people without employment, where there was little job security and no social security safety net. Skeleton workforces usually remained, including maintenance staff (such as gardeners and handypersons) but not housekeeping staff (most of whom were women) and wait staff (again mostly women). In Samoa, two thirds of those who lost their jobs were women; many returned to their home villages (Connell & Taulealo, 2021). New unemployment numbers were even worse in the Caribbean. In both Pacific and Caribbean SIDS, that reflected economic losses amounting to around a third of national GDP from the tourist sector alone, a shock of a magnitude virtually impossible to manage for small economies (Coke-Hamilton, 2020).

The absence of tourists had direct multiplier effects. Resorts ended most community engagement activities, including sourcing supplies from local markets and small 'businesses' that provided security, conservation, handyman, gardening and cleaning services so that domestic value chains were no longer relevant (Naupa, Mecartney, Pechan, & Howlett, 2021). Since tourists were absent and domestic travel declined rapidly (especially where public health concerns discouraged travel between islands), national island airlines, often government-subsidised, were unused. Despite subsidies, Air Vanuatu collapsed (to be much later revived) and LIAT Airlines in the eastern Caribbean was liquidated (but it too was eventually re-established). Most SIDS sought to stimulate domestic tourism, by reducing costs and introducing special offers for variants of 'staycations', but a small local market, widespread recession and caution usually only delayed inevitable closures.

COVID-19 emphasised the marginality of women in the tourism industry and the related informal sector, despite their numbers. Women were usually first to be displaced from the formal sector (being cleaners and waitresses, and rarely chefs and accountants) and lost their roles in informal markets, handicraft production and cultural performances. In due course, as the industry began to cautiously reopen, women also took the lead in its revival both in the Caribbean and in the Pacific, as they had also done following cyclones. Resourceful women, often unacknowledged, proved invaluable in response and reconstruction through mobilising capital, working collectively and being innovative small-scale entrepreneurs (Clissold, Westoby, & McNamara, 2020; Connell, 2023; Neef, 2021). In almost every country, even where tourism was trivial, unemployment worsened, most seriously in and around urban areas, where the tourism sector was concentrated and where there was greater dependence on cash incomes. Poverty increased, especially amongst those in the already disadvantaged informal sector, accentuated in both Vanuatu and Samoa, where COVID-19 followed several cyclones in the 2010s.

No economic sector is more sensitive to external shocks than tourism. And no shock was more devastating than COVID-19 because of the abrupt ending of the entirety of international tourism and because 'conventional' shocks (such as coups, bombs, cyclones or even financial crises) are instantaneous rather than of indeterminate length. Uncertainty over when, if ever, normality might return made response particularly difficult. While the pandemic remained, the infrastructure of the industry—airlines (particularly small national airlines), resorts, travel agents, hire cars and cruise ships—gradually folded as debts mounted. The abrupt halt provided unwelcome evidence of the costs of absolute dependence on international tourism, raising questions about its future.

No other sector of SIDS economies was as devastated as the tourism industry. During its closure, numerous organisations (such as chambers of commerce) and governments suggested that this provided the opportunity to review its role. Strong arguments were made that the industry should become more sustainable with more local involvement and control. That was partly related to a trend during COVID-19 where tourism sector workers had welcomed a return to traditional values as they returned to the security of village life (Movono & Scheyvens, 2021). Yet sharing and reciprocity did not always succeed when there was less food available, bartering was never certain and early satisfaction in being able to revive old skills rarely lasted, as in Tuvalu where many were anxious to return to town, white-collar jobs and 'modern' infrastructure (Connell, 2021), or in Fiji as they returned to reliable wage employment. That was a repeat of what had first drawn workers into the industry, as in the Bahamas where 'the allure of less hard work for more money … being a maid or a bartender was easier than working the fields and these offered a more "sustainable" income' (Minnis et al., 2020: 33). Even more attractive was the possibility of living at home and working in the industry, a possibility that turned employment in villages close to coastal Fijian resorts into localised tourism monocultures (Connell & Rugendyke, 2008).

Caribbean SIDS, where tourism had greater significance for national economies than in most Pacific SIDS and that had experienced COVID-19 (and believed themselves capable of managing it), were most anxious to re-establish the industry. Several countries had implemented various schemes to resuscitate the tourism sector during the pandemic, including the implementation of resilient 'corridors' (Jamaica), exclusive use of beaches (Grenada) and hotels offering quarantine services. Eventually, the Caribbean experienced the most rapid global post-pandemic recovery. By 2023 numbers were comparable with pre-pandemic numbers, while in Jamaica and the Bahamas they had passed pre-pandemic levels. In the Pacific, Fiji achieved pre-pandemic levels relatively quickly, but with few exclusive resorts, Vanuatu had not reached pre-pandemic numbers by early 2025 (when a major earthquake caused further disruptions). In both regions, the return to pre-pandemic levels was not simply transient—a 'binge effect' driven by pent-up demand—but was sustained.

Ironically, as fear of COVID-19 faded (if cases did not), and people in lockdown were anxious to travel again, the ships that once seemed to harbour the disease took elaborate precautions, including enhanced medical facilities and health staff, and turned themselves into quarantined spaces. The industry itself claimed that it experienced lower rates of illness and hospitalisation than on land (Scott, 2022). Typically, as a Sydney resident said in 2022:

> COVID is just a part of life now along with any other illness or virus and I'm confident the cruise industry has put the very best practices in place to ensure the safety of passengers. I believe many people are over the hype of COVID and just want to enjoy their lives. (quoted in Scott, 2022: 24)

Organisations such as the Pacific Tourism Organisation and the Pacific Islands Forum Secretariat stressed the need and value of aligning tourism redevelopment with sustainability and with the SDGs (Hughes, 2023) while some SIDS such as Vanuatu drew up new plans that incorporated such directions. Neither in the Pacific nor even more so in the Caribbean where cruising dominated (McLeod, 2024), was there much indication of evident change in direction. Indeed, the previously maligned resorts and cruise ships led the way in reopening. By 2023, barely after its reopening, the continued recovery of tourism in Fiji was the main source of economic growth, with pent-up demand leading to record visitor arrivals despite high inflation and tight monetary policies in key tourism source countries. Of the Pacific SIDS where tourism is significant, only Palau had not come close to pre-COVID levels within two years, as the one Pacific SIDS mainly dependent on Asian tourism, which had been slow to rebound.

Workers in tourism and related activities were generally anxious to return to their previous employment as soon as possible and, like the resorts, had no interest in restructuring. While in Fiji, a third of Fijian tourist workers were wary about staying in the industry and some contemplated the possibility of better jobs in creative industries (Movono & Scheyvens, 2021). Most simply wanted to return to work in what had been a growing industry with reasonable wages and conditions. A small sample of travel agents in Sydney, the major market for most Pacific SIDS, indicated that neither they nor their clients sought anything other than a return of the old order—the old normal—principally as a sign of security, familiarity and continuity. Sunk capital in resorts and cruise ships discouraged interest in experimentation and difference, an unintentional material and cultural path dependence, alongside little SIDS capacity to reformulate directions for an industry substantially overseas-controlled and anxiety to get 'back to normal' and restore employment and income. Cruise ships returned even faster with the old order firmly intact (Newell, Nuttall, & Newell, 2025). SIDS have minimal influence on the cruise ship lines where itineraries could easily be changed if countries imposed arduous conditions. Much earlier, plans for a CARICOM-wide standard tax on cruise ship passengers collapsed after cruise lines threatened to remove destinations from their itineraries.

For some parts of the industry, revival came too late. In Vanuatu, as Air Vanuatu folded, the industry became even more concentrated in the capital city and the main island. Some workers had been able to find alternative jobs or move overseas to become guestworkers, creating a labour shortage. Outer islands had become less accessible. A rhetoric of sustainable and regenerative tourism was divorced from practice with agricultural enterprises too small to be linked to resorts (Connell, 2024). As significant were the aspirations of tourists, many returning to familiar places. No much-touted 'new normal' followed in any of the SIDS as their revivals were largely out of local hands but were dependent on decisions made elsewhere by international resort chains, airlines, cruise ships and other services. Thus, China effectively sealed its borders much longer, delaying revival of the market in Fiji and Palau. The trajectory and structure of tourism were not, however, easy to change.

4.8 Beyond Revival: A New Future?

Although the COVID-19 collapse raised questions about the future of tourism, it was universally assumed that the revival of tourism was invaluable. The hiatus provided an unparalleled pause to reconsider but, as the rush to return took a familiar form, that pause proved inconsequential. There was no sense that some version of small might have become more beautiful. The future of tourism was being shaped by social, psychological and emotional issues as much as by economic factors. As COVID-19 persisted to deny a rebirth of Pacific tourism, hopeful marketing phrases such as 'we're keeping it beautiful for you' (Vanuatu) proclaimed continuity.

Invocations to change sought new directions that would offer not merely a 'new normal' but a more inclusive tourism where local linkages and environmental management were more evident, restructured to benefit local communities and provide social, economic and ecological justice (e.g. Figueroa & Rotarou, 2021; Gibson, 2021; Higgins-Desbiolles, 2020; Movono & Scheyvens, 2021; Sheller, 2021). Yet restarting was everything and the moment passed. A World Bank (2023) report on the Pacific, despite noting 'limited progress towards environmentally sustainable tourism', largely focused on technical changes. Mimi Sheller's suggestion (2021: 11) that 'rather than quick-fix rebuilding', what was required was 'a regenerative approach to building mobility justice within an integrated, inter-regional, reparative framework, into which sustainable tourism could be integrated with a view to supporting resilient food systems and jobs, while preserving biodiversity' proved more than a step too far. Invariably, the main objective was to get underway, revive employment and generate income flows. Finance rather than ecology and equity dominated future directions, most evident at both national and global level in the longstanding and ongoing failure of even slight progress towards climate policy action, globally evident at COP26 (Higham et al., 2022). At every scale restoring income was crucial, and overseas interests were as welcome as ever. When Fiji reopened late in 2021, its 'tentative steps towards sustainable tourism' proved mere tokenism in the rush to recovery (Moorhead 2021).

In some respects, no new directions were necessary. The substantial 'unmet spending opportunities' experienced in Fiji, where a quarter of cruise passengers spent nothing at all on land—circumstances no doubt replicated elsewhere—suggested the potential for increasing sales of such basic goods as 'handicrafts, clothing, tours and excursions, and food and beverage' (International Finance Corporation, 2019). However, Palau, which had implemented a Palau Pledge in 2017, requiring international visitors to sign a pledge stamped in their passports that promised the children of Palau that they would 'tread lightly' and 'preserve and protect' the islands, instigated a new programme (Ol'au Palau) in 2022, where tourists would be offered exclusive experiences based on how they treated the environment and culture, not by how much they spent. Rewards would be given to those who treated the island nation 'gently and respectfully' by making sustainable decisions such as using reef-safe sunscreen, visiting culturally important sites and eating sustainably sourced local food (Galloway, 2022). There was no indication that this had any traction.

While SIDS themselves might well have been interested in more local participation and ownership, that was rarely stated and their control over a footloose transnational industry was limited. External players were seeking nothing other than revival. Officially, at least, many would value some form of local content that respected local cultures and local foods (e.g. Movono & Scheyvens, 2021): small changes but hard to implement. Tourists still eschewed 'island nights' but unsurprisingly preferred to visit sustainably managed destinations, whether in Fiji or more generally in SIDS (e.g. Grilli et al., 2021; Tyllianakis et al., 2019). Doubt however surrounded the extent to which tourists might recognise innovative and sustainably managed destinations.

Tourism has always been uneven and unpredictable, where islands are dependent on decisions and events elsewhere (whether by potential tourists, airlines or cruise companies, or by viruses and cyclones). That is true for both governments and tourists. Oil and energy costs (and therefore air fares), political stability, cyclones and climate change remain uncertain variables. A time to pause and think about the future was not a time to experiment, though it suggested to many that a portfolio of livelihood possibilities was superior to a narrow focus on tourism. Early on, some displaced workers expressed a degree of pride and satisfaction in being able to revive old skills and restore vestiges of earlier subsistence livelihoods, but that rarely lasted (Movono & Hughes, 2022; Movono & Scheyvens, 2021). None of this enthusiasm and revivalism was converted into ecotourism. Employment, regular wages and job security were preferable.

Although no new tourism niches have emerged, creativity suggested diversity. In the days after Julien Alfred won the gold medal at the 2024 Summer Olympics in the 100-metre event, the first-ever Olympic medal for St. Lucia, the island's official tourism board made her the face of its #LetHerInspireYou marketing campaign, coined because St. Lucia is the only country named after a woman, using the attention garnered by the newly proclaimed world's fastest woman to invite visitors to 'explore St

Lucia like an Olympian' and enjoy, as Julien had done, the healing properties of the world-renowned, soothing mud baths at its naturally hot Sulphur Springs. In the same month (July 2024) Fiji set a record for arrivals in a single month, partly attributed to the presence of the World Surf League Championship, which drew attention to its 'world-class waves'. Scope for creativity and diversity existed.

4.9 Plus Ça Change

While tourism generated jobs and incomes for all SIDS, it was never without uncertainties, largely stemming from its dependence on foreign investment and foreign tourists and the extent of income leakage. Yet, by most standards, it has been an unquestioned success story, perhaps especially so in such dry and unproductive islands as the Bahamas where diversity is now implausible, but even where SIDS were reluctant entrants into the tourism economy. Some degree of caution has usually been present, evident in the smaller eastern Caribbean states, such as Antigua and Barbuda's reluctance to be involved in an industry that had contributed to environmental degradation and mismanagement in Jamaica (Carr & Heyman, 2009), but caution and reluctance could be quickly bypassed. Cruising is the fastest-growing category in the leisure travel market, but it is also perhaps the least beneficial to local communities. Ecotourism and adventure tourism are yet to realise their much vaunted 'potential' and few SIDS have attracted 'high-value' tourists who spend more and are aware of their environmental footprint. The uneven development that often followed tourism tended to build on existing inequalities where they relate to power, land tenure and access to resources, with local people often constrained by a lack of economic and social capital and basic understanding of the increasingly global dimensions of the industry, which prevents effective engagement with tourists and their intermediaries (Cheer, Cole, Keir, et al., 2017). COVID-19 provided time to reflect on the possibilities of future development, but little or nothing changed. Nowhere in the SIDS was there any suggestion of a new order or even simply the more effective participation of local people. Nor were there seemingly more than token considerations of a more diversified national economic focus (that could also generate linkages) rather than a continued dependence on externally oriented tourism: income was being generated and filtering down, however unevenly.

COVID-19 demonstrated how dependent most SIDS economies were on tourism. It had provided a shock more devastating than the 2008 global financial crisis where small economies specialising in tourism, and in financial services, were the worst affected. Further specialisation had made them more susceptible to shocks than at any time in the past (Armstrong & Read, 2018). The case for retaining diversity—and a greater degree of food security—was evident, yet analogous to the resource curse that plagued mining economies, a 'tourism curse' was becoming more evident. Yet, tourism has hastened modernity in various guises and generated employment, the economic benefits from which have been used, in much the same way as remittances, on improved consumption, housing, education and technology (Scheyvens & Russell, 2012, Movono, Scheyvens, & Auckram, 2022; Chap. 9). Environments have been degraded but others have been renovated and preserved; coastal zone management has benefited.

Even as a future of long-haul air travel—the fate of the Pacific—and of large polluting cruise ships—the fate of the Caribbean—suggests challenges, for many SIDS tourism has proved nothing less than economic salvation and a panacea for development, especially as plantation economies collapsed and little else seemed profitable. The potential for tourism growth has not been fully exploited due to challenges such as a lack of air connectivity, the high cost of air travel, and absent infrastructure and human capital. Despite its flaws, employment and income reach the otherwise relatively poor informal sector. However exploitative—culturally, environmentally and economically—tourism might be, a future without it is inconceivable. Too much expenditure may have returned overseas, local cultures reshaped and inequalities enhanced, but for governments it has been too good to be true. Tourism is consequently seen by governments, and by international agencies such as the World Bank and the Asian Development Bank (ADB), as a valuable means of enabling both agricultural development and poverty reduction. Sensitive restructuring, however improbable, would make it even more valuable.

Ironically, the economic sector that has more to gain than most by being sustainable, depends on market access and external orientation, often with a considerable carbon footprint, yet it is impossible to envisage tourism in SIDS without air travel. Although the industry is burdened with impressive adjectives—eco, sustainable, slow, alternative, responsible—that challenge overtourism and would shape a more equitable distribution of tourism's benefits, few have resonance in the SIDS. Island tourism especially has yet to respond adequately to the challenges of climate change (Loehr & Becken, 2025). Nonetheless, whatever global environmental and socio-economic shifts occur, islands will probably remain attractive destinations (Scheyvens & Momsen, 2007). Ultimately, the experience of COVID-19 provided a harsh lesson that some greater diversity and self-reliance were essential, within and beyond tourism, an easy conclusion to reach but a challenging one to implement. The abrupt and unpleasant pause eventually simply meant a return to familiar tourism industries and familiar national economies. Even in a sector that would seemingly benefit most from creativity, in every sense and in every place structural changes were slight and slow. We can now extend beyond this and the previous chapter to take a more political perspective on economic change, including tourism, and its implications for employment, poverty and governance.

Bibliography

Addinsall, C., Weiler, B., Scherrer, P., & Glencross, K. (2017). Agroecological tourism: Bridging conservation, food security and tourism goals to enhance smallholders' livelihoods on South Pentecost, Vanuatu. *Journal of Sustainable Tourism, 25*(8), 1100–1116.

Alleyne, R., Alleyne, A., Cezair, N., & Charles-Soverall, W. (2024). Sport tourism in small island developing states: A case study of the Barbados Sport Tourism Industry. In D. Van Rheenen, O. Naria, R. Melo, & C. Sobry (Eds.), *Sport tourism, island territories and sustainable development. A comparative perspective* (pp. 223–238). Springer.

Armstrong, H., & Read, R. (2018). The impact of the 2008 global crisis on small economies in the Caribbean. *Canadian Journal of Latin American and Caribbean Studies, 43*(3), 394–416.

Armstrong, H., & Read, R. (2021). Tourism and sustainable growth in small (island) economies. In J. Roberts et al. (Eds.), *Shaping the future of small Islands* (pp. 93–108). Springer.

Becken, S., & Hay, J. (2007). *Tourism and Climate Change: Risks and opportunities*. Multilingual Matters.

Berno, T. (2025). "Do you want fries with that?". Reflections on the relationship between Pacific cuisine, tourism and sustainability. In E. Fusté-Forné & E. Wolf (Eds.), *Food and beverage tourism on Islands* (pp. 133–146). Edward Elgar.

Bishop, M. (2010). Tourism as a small-state development strategy: Pier pressure in the Eastern Caribbean. *Progress in Development Studies, 10*(2), 99–114.

Britton, R. (1977). Making tourism more supportive of small state development: The case of St. Vincent. *Annals of Tourism Research, 4*(5), 268–278.

Britton, S. (1980). The spatial organisation of tourism in a neocolonial economy: A Fiji case study. *Pacific Viewpoint, 21*(2), 144–165.

Brooks, S. (2016). Informal settlements in Jamaica's tourism space: Urban spatial development in a small island developing state. *Urban Island Studies, 2*, 72–94.

Butler, R. (1980). The concept of a tourist area cycle of evolution: Implications for management of resources. *Canadian Geographer, 24*(1), 5–12.

Carr, L., & Heyman, W. (2009). Jamaica bound? Marine resources and management at a crossroads in Antigua and Barbuda. *Geographical Journal, 175*(1), 17–38.

Cheer, J., Cole, S., Keir, J., et al. (2017). Tourism and islandscapes: Cultural realignment, social-ecological resilience and change. *Shima, 11*(1), 40–54.

Clissold, R., Westoby, R., & McNamara, K. (2020). Women as recovery enablers in the face of disasters in Vanuatu. *Geoforum, 113*, 101–110.

Coke-Hamilton, P. (2020). Impact of COVID-19 on tourism in small island developing states, UNCTAD.

Connell, J. (2008). Niue: Embracing a culture of migration. *Journal of Ethnic and Migration Studies, 34*(6), 1021–1040.

Connell, J. (2013). *Islands at risk: Environments, economies and contemporary change*. Edward Elgar.

Connell, J. (2015). Competing Islands? The Mamanuca and Yasawa Islands, Fiji. In G. Baldacchino (Ed.), *Archipelago Tourism* (pp. 183–197). Ashgate.

Connell, J. (2021). Double jeopardy: Distance and decentralisation in Tuvalu. In Y. Campbell & J. Connell (Eds.), *COVID in the Islands: A comparative perspective on the Caribbean and the Pacific* (pp. 163–175). Palgrave Macmillan.

Connell, J. (2023). COVID-19 and tourism in the island Pacific: Gender tribulations and transformations in different seas. In A. Datta, J. Momsen, & A. Oberhauser (Eds.), *Bridging worlds- Building feminist geographies* (pp. 212–220). Routledge.

Connell, J. (2024). Culture, COVID-19, cyclones and supply chains: The contraction of tourism in Vanuatu. In G. Baldacchino (Ed.), *Archipelago tourism after the pandemic* (pp. 137–152). Routledge.

Connell, J., & Rugendyke, B. (2008). Tourism and local people in the Asia-Pacific region. In J. Connell & B. Rugendyke (Eds.), *Tourism at the grassroots. Villagers and visitors in the Asia-Pacific* (pp. 1–40). Routledge.

Connell, J., & Stanley-Niaah, S. (2008). Remembering the fire: Music tourism and the Bob Marley Heritage in Jamaica. *Caribbean Geography, 15*(1), 76–90.

Connell, J., & Taulealo, T. (2021). Island tourism and COVID-19 in Vanuatu and Samoa: An unfolding crisis. *Small States and Territories, 4*(1), 105–124.

Croes, R. (2006). A paradigm shift to a new strategy for small island economies: Embracing demand side economics for value enhancement and long-term economic stability. *Tourism Management, 27*, 453–465.

Dodman, D. (2009). Globalization, tourism and local living conditions on Jamaica's north coast. *Singapore Journal of Tropical Geography, 30*(2), 204–219.

Duval, D. (2003). When hosts become guests: Return visits and diasporic identities in a Commonwealth Eastern Caribbean Community. *Current Issues in Tourism, 6*(4), 267–308.

Duval, D. (2004). *Tourism in the Caribbean: Trends, development, prospects.* Routledge.

Farbotko, C. (2010). 'The global warming clock is ticking so see these places while you can': Voyeuristic Tourism and Model Environmental Citizens on Tuvalu's Disappearing Islands. *Singapore Journal of Tropical Geography, 31*, 224–238.

Figueroa, E., & Rotarou, E. (2021). Island tourism-based sustainable development at a crossroads: Facing the challenges of the COVID-19 pandemic. *Sustainability, 13*, 10081.

Galloway, L. (2022). *Palau's world-first 'good traveller' incentive.* Retrieved from www.bbc.com/travel/article/20220517

Gayle, D., & Goodrich, J. (1993). Caribbean tourism marketing, management and development strategies. In D. Gayle & J. Goodrich (Eds.), *Tourism marketing and management in the Caribbean* (pp. 1–19). Routledge.

Gibson, C. (2021). Theorising tourism in crisis: Writing and relating to place. *Tourist Studies, 21*(1), 84–95.

Gibson, D. (2025). Island delights: Exploring the culinary tapestry of Fiji through food and beverage tourism. In E. Fusté-Forné & E. Wolf (Eds.), *Food and beverage tourism on Islands* (pp. 4–14). Edward Elgar.

Gössling, S., & Higham, J. (2021). The low carbon imperative: Destination management under urgent climate change. *Journal of Travel Research, 60*(6), 1167–1179.

Graci, S., & van Vliet, L. (2020). Examining stakeholder perceptions towards sustainable tourism in an Island destination: The case of Savusavu, Fiji. *Tourism Planning Development, 17*, 62–81.

Grilli, G., Tyllianakis, E., Luisetti, T., et al. (2021). Prospective tourist preferences for sustainable tourism development in small island developing states. *Tourism Management, 82*, 104178.

Grydehøj, A., & Kelman, I. (2017). The eco-island trap: Climate change mitigation and conspicuous sustainability. *Area, 49*(1), 106–113.

Hall, C., & Higham, J. (Eds.). (2005). *Tourism, recreation and climate change.* Channel View Publications.

Hayle, C., Singh, R., & Wright, A. (2010). Planning for tourism resilience in the Caribbean. *Journal of Eastern Caribbean Studies, 35*(1), 36–59.

Higgins-Desbiolles, F. (2020). Socialising tourism for social and ecological justice after Covid-19. *Tourism Geographies, 22*(3), 610–623.

Higham, J., Font, X., & Wu, J. (2022). Code red for sustainable tourism. *Journal of Sustainable Tourism, 30*(1), 1–13.

Hoarau-Heemstra, H., Wigger, K., Olsen, J., & James, L. (2023). Cruise tourism destinations: Practices, consequences and the road to sustainability. *Journal of Destination Marketing & Management, 30*, 100820.

Holden, A., Jamal, T., & Burini, F. (2022). The future of tourism in the Anthropocene. *Annual Review of Environment and Resources, 47*(1), 423–447.

Huebner, A. (2012). Public perceptions of destination vulnerability to climate change and implications for long-haul travel decisions to small island states. *Journal of Sustainable Tourism, 20*(7), 939–951.

Hughes, E. (2023). Tourism and sustainable development goals in small island states in the Pacific. In M. Stephenson (Ed.), *Routledge handbook on tourism and small states in the Pacific* (pp. 28–43). Routledge.

International Finance Corporation. (2018). *From the farm to the tourist's table: A study of fresh produce demand from Fiji's hotels and resorts.* IFC.

International Finance Corporation. (2019). *Assessment of the economic impact of cruise ship tourism in Fiji*. IFC.

Jessop, D. (2019, September 15). Reassessing the role of cruise tourism. *Tourism Analytics.*

Kanemasu, Y. (2008). Weapons of the weak: Employees in the Fiji hotel scene. In J. Connell & B. Rugendyke (Eds.), *Tourism at the grassroots* (pp. 114–130). Routledge.

Kelman, I. (2021). Words without meaning? Examining sustainable development terminology through small states and territories. *Island Studies Journal, 4*(2), 231–244.

Lallani, S. (2024). Cultural globalization at sea: The rise of the modern Caribbean cruise industry. *Globalizations, 21*(7), 1290–1308.

Lenzen, M., Sun, Y., Faturay, F., et al. (2018). The carbon footprint of global tourism. *Nature Climate Change, 8*, 522–528.

Lewis-Cameron, A. (2021). Understanding small island states and territories. In A. Lewis-Cameron, L.-A. Jordan, & S. Roberts (Eds.), *Managing crises in tourism. Resilience strategies from the Caribbean* (pp. 13–32). Palgrave Macmillan.

Lewis-Cameron, A., & Brown-Williams, T. (2023). Civil society and the pursuit of sustainable tourism in the small island jurisdiction of Tobago. In L. Briguglio, M. Briguglio, & C. Slatter (Eds.), *Handbook of civil society and social movements in small states* (pp. 13–25). Routledge.

Lindstrom, L. (2018). Surprising times on Tanna, Vanuatu. In J. Connell & H. Lee (Eds.), *Change and Continuity in the Pacific* (pp. 55–68). Routledge.

Loehr, J., & Becken, S. (2025). Institutionalised tourism policy goals: Fit to address climate change. *Journal of Sustainable Tourism, 33*(10), 2075–2093.

MacCarthy, M. (2023). "It will be like a town here, things are really coming up!": Inequality in village-based cruise ship tourism in the Trobriand Islands. *The Contemporary Pacific, 35*(1), 61–93.

McLeod, M. (2024). *Island tourism policy and sustainable development.* Routledge.

McLeod, M., Dodds, R., & Butler, R. (2024). *Island tourism sustainability and resiliency*. Routledge.

Macnaught, T. (1982). Mass tourism and the dilemmas of modernization in Pacific island communities. *Annals of Tourism Research, 9*(3), 359–381.

Mahon, R., Petrie, J., Trotman, A., et al. (2021). Climate services for tourism: Insights from Caribbean small island developing states. *Climate Services, 24*, 100262.

Mahoney, I., & Collins, V. (2020). The capitalist voyeur: Commodification, consumption and the spectacle of the cruise. *Leisure Studies, 39*(2), 280–293.

Minnis, J., Rolle, S., & Bethell-Bennett, I. (2020). The impact of tourism development on small island communities in The Bahamas. The case of Abaco, Bimini and Exuma. In S. Rolle, J. Minnis, & I. Bethell-Bennett (Eds.), *Tourism development, governance and sustainability in the Bahamas* (pp. 31–51). Routledge.

Moghal, Z., & O'Connell, E. (2018). Multiple stressors impacting a small island tourism destination-community: A nested vulnerability assessment of Oistins, Barbados. *Tourism Management Perspectives, 26*, 78–88.
Mohan, P. (2022a). Climate finance to support Caribbean small island developing states efforts in achieving their Nationally Determined Contributions in the energy sector. *Energy Policy, 169*, 113208.
Mohan, P. (2022b). Sustainable tourism and the sustainable development goals in sub-national island jurisdictions: The case of Tobago. *Island Studies Journal, 17*(2), 168–191.
Molu, K., Pratt, S., & Tarte, S. (2024). Sport tourism and local sustainable development in small island territories: The case of Kiribati. In D. Van Rheenen, O. Naria, R. Melo, & C. Sobry (Eds.), *Sport tourism, island territories and sustainable development. A comparative perspective* (pp. 285–295). Springer.
Moore, A. (2019). *Destination anthropocene: Science and tourism in The Bahamas.* University of California Press.
Moorhead, A. (2021). Small steps for sustainable tourism in Fiji. ANUDevPolicyBlog, 13 December.
Mortley, N. (2011). Strategic opportunities from diaspora tourism: The Jamaican perspective. *Canadian Foreign Policy Journal, 17*(2), 171–185.
Movono, A., & Hughes, E. (2022). Tourism partnerships: Localizing the SDG agenda in Fiji. *Journal of Sustainable Tourism, 30*(10), 2318–2332.
Movono, A., & Scheyvens, R. (2021). Tourism in a world of disorder: A return to the Vanua and kinship with nature in Fiji. In Y. Campbell & J. Connell (Eds.), *COVID in the islands: A comparative perspective on the Caribbean and the Pacific* (pp. 265–278). Palgrave Macmillan.
Movono, A., Scheyvens, R., & Auckram, S. (2022). Silver linings around dark clouds: Tourism, Covid-19 and a return to traditional values, villages and the *vanua. Asia Pacific Viewpoint, 63*(2), 164–179.
Mycoo, M. (2014). Sustainable tourism, climate change and sea level rise adaptation policies in Barbados. *Natural Resources Forum, 38*, 47–45.
Mycoo, M. (2022). Caribbean island cities: Urban issues, urbanization processes and opportunities for transformation. In J. González-Pérez, C. Irazábal, & R. Lois-González (Eds.), *The Routledge handbook of urban studies in Latin America and the Caribbean* (pp. 579–602). Routledge.
Naupa, A., Mecartney, S., Pechan, L., & Howlett, N. (2021). An industry in crisis: How Vanuatu's tourism sector is seeking economic recovery. In Y. Campbell & J. Connell (Eds.), *COVID in the Islands: A comparative perspective on the Caribbean and the Pacific* (pp. 231–252). Palgrave Macmillan.

Neef, A. (2021). The contentious role of tourism in disaster response and recovery. *Frontiers in Earth Science, 9*, 771345.

Newell, A., Nuttall, P., & Newell, P. (2025). *Cruise liners – are they good for Aotearoa and the Pacific and what more do we need to know?* Micronesian Center for Sustainable Transport, Working Paper 5.

Orams, M. (2013). Economic activity derived from whale-based tourism in Vava'u, Tonga. *Coastal Management, 41*(6), 481–500.

Pacific Private Sector Development Initiative. (2021). *Looking forward: Evaluating the challenges for Pacific tourism after COVID-19*. PPSDI.

Payne, A., & Sutton, P. (2007). *Repositioning the Caribbean within globalisation*. The Centre for International Governance Innovation, Caribbean Papers No. 1.

Peterson, R., DiPietro, R., & Harrill, R. (2020). In search of inclusive tourism in the Caribbean: Insights from Aruba. *Worldwide Hospitality and Tourism Themes, 12*(3), 225–243.

Pratt, S., Magbalot-Fernandez, A., & Ohe, Y. (2022). Motivations and constraints of developing agritourism under the challenges of climate change: The case of Samoa. *International Journal of Tourism Research, 24*(4), 610–622.

Prentice, C., Kundra, S., Alam, M., et al. (2023). Utopia or dystopia–deterrents to ecotourism development in Fiji. *Tourism Geographies, 25*(2–3), 843–864.

Pugh, J. (2005). Social transformation and participatory planning in St Lucia. *Area, 37*(4), 384–392.

Reddy, M., & Wilkes, K. (2012). *Tourism, Climate Change and Sustainability*. Routledge.

Rhiney, K. (2011). Agri-tourism linkages in Jamaica: Case study of the Negril all-inclusive hotel sub-sector. In R. Momsen (Ed.), *Tourism and agriculture: New geographies of consumption, production and rural restructuring* (pp. 117–138). Routledge.

Rhiney, K. (2012). The Negril tourism industry: Growth, challenges and future prospects. *Caribbean Journal of Earth Science, 43*, 25–34.

Robinson, R., Martins, A., Solnet, D., & Baum, T. (2019). Sustaining precarity: Critically examining tourism and employment. *Journal of Sustainable Tourism, 27*(7), 1008–1025.

Scheyvens, R., & Cheer, J. (2021). Tourism, the SDGs and partnerships. *Journal of Sustainable Tourism, 30*(10), 2271–2281.

Scheyvens, R., & Laeis, G. (2021). Linkages between tourist resorts, local food production and the sustainable development goals. *Tourism Geographies, 23*(4), 787–809.

Scheyvens, R., & Momsen, J. (2007). Tourism in small island states: From vulnerability to strengths. *Journal of Sustainable Tourism, 16*(5), 491–510.

Scheyvens, R., & Russell, M. (2012). Tourism, land tenure and poverty alleviation in Fiji. *Tourism Geographies, 14*, 1–25.

Scott, K. (2022, June 5). Cruising set to launch as new rules come into place. *Sun-Herald,* 24.

Sheller, M. (2021). Reconstructing tourism in the Caribbean: Connecting pandemic recovery, climate resilience and sustainable tourism through mobility justice. *Journal of Sustainable Tourism, 29*(9), 1436–1449.

Spencer, N., Strobl, E., & Campbell, A. (2022). Sea level rise under climate change: Implications for beach tourism in the Caribbean. *Ocean & Coastal Management, 225*, 106207.

Spennemann, D. (2021). Groundwater, graves and golf: Layers of heritage tourism on a Fiji resort island. *Sustainability, 13*, 5863.

Timms, B. (2006). Caribbean agriculture-tourism linkages in a neoliberal world: Problems and prospects for St. Lucia. *International Development Planning Review, 28*(1), 35–56.

Timms, B., & Conway, D. (2012). Slow tourism at the Caribbean's geographical margins. *Tourism Geographies, 14*(3), 396–418.

Tolkach, D. (2023, April 8). Pacific Island Tourism is on the Surge again but its character is changing. *Australian Outlook.*

Tyllianakis, E., Grilli, G., Gibson, D., et al. (2019). Policy options to achieve culturally-aware and environmentally sustainable tourism in Fiji. *Marine Pollution Bulletin, 148*, 107–115.

UNFPA. (2025). *Pulse on progress: Women in Caribbean Tourism Survey Report.* UNFPA.

Uyarra, M., Cote, I., Gill, J., Tinch, R., et al. (2005). Island-specific preferences of tourists for environmental features: Implications of climate change for tourist-dependent states. *Environmental Conservation, 32*(1), 11–19.

Walker, B., Lee, T., & Li, X. (2021). Sustainable development for small island tourism: Developing slow tourism in the Caribbean. *Journal of Travel & Tourism Marketing, 38*(1), 1–15.

Wolf, F., Filho, W., Singh, P., et al. (2021). Influences of climate change on tourism development in small Pacific Island States. *Sustainability, 13*(8), 4223.

Wolf, F., Moncada, S., Surroop, D., et al. (2024). Small island developing states, tourism and climate change. *Journal of Sustainable Tourism, 32*(9), 1965–1983.

Wong, E., Jiang, M., Klint, L., et al. (2013). Evaluation of policy environment for climate change adaptation in tourism. *Tourism and Hospitality Research, 13*(4), 201–225.

World Bank. (2023). *The future of Pacific tourism.* World Bank.

5

People, Governance and State Capacity

This chapter first analyses the growth and distribution of populations in the SIDS, pointing to their increasing urbanisation, where employment may be 'modern' but unemployment and poverty are of concern. That leads on to considering the structures of governance that influence development within the SIDS, the political forces within island states and the role of politicians as instigators of change, related to the tasks of management in small states. One implication of small populations is that national income is also small so that the fixed costs associated with the provision of public goods and services are spread across a very narrow base. These constraints often result in small administrations without the capacity to function efficiently, circumstances that vary considerably, while absent economies of scale emphasise the challenges of providing services effectively and equitably, where infrastructure must be resilient to hazards and where community expectations are higher. The chapter concludes by focusing on social inclusion, and the role of civil society, to be taken further in the subsequent chapter that examines the implications of governance for service delivery and how this relates to equity and the SDGs.

Urban centres grew steadily in the post-war and post-independence years, but rural–urban migration is now primarily a phenomenon of the

J. Connell, Y. Campbell, *Island Challenges*,
https://doi.org/10.1007/978-981-95-8177-1_5

larger SIDS and the atoll states. Most urban residents are in the primate capital cities of the larger SIDS, and a growing proportion are now of a second or higher generation, more so in the Caribbean, with few ties to 'home' areas and limited ability or will to 'return' even where that is possible. The Caribbean differs from the Pacific by greater migration between SIDS, as from the smaller SIDS of the eastern Caribbean to Trinidad, and from the Haiti and Dominican Republic LIDS to the Bahamas and Jamaica. Urbanisation has been a partial consequence of fragile economies, declining interest in agriculture (and the 'slow money' it provides), and the transfer of fertility to urban areas so that many urban residents are not migrants, although sometimes treated as such. Despite towns being established by the nineteenth century in most SIDS, contemporary urbanisation often arrived as a rather unwelcome surprise.

Virtually since independence, urban populations have increased faster than those in rural areas, through a combination of rural-urban migration and urban fertility, posing challenges for land use, planning and equitable access to services (notably housing), as well as the need for effective social, health and employment policies. Economic growth and limited resources have proved too limited to effectively mitigate or manage urban development problems; hence, interest in international migration has consequently grown, and urban centres are simply bypassed, a theme we return to in the final substantive chapter (Chap. 9).

5.1 Population

The average annual population increase in the Pacific SIDS (excluding PNG, which is exceptional in having a growth rate more than double that of most other SIDS) is less than 1% and is barely 0.4% in the Caribbean SIDS (Table 1.1). It is no higher than 1.1% in Antigua and the Bahamas, the fastest-growing Caribbean SIDS, and declining in St. Vincent and the Grenadines. In the Pacific, it is greatest in the larger Melanesian SIDS (with much higher fertility rates), reaching 1.6% in Vanuatu, and lowest in the Polynesian states, where population trends are like those in the Caribbean. Several SIDS populations are declining and that trend is increasing, especially, as in the Caribbean, where this is matched with low fertility, partly a function of those of fertile age being the dominant international migrants (Chap. 9). In the Pacific, the Marshall Islands, FSM, Tuvalu and Tonga are all losing population. Emigration and population decline now occur in all the northern Micronesian states, with persistent flows to the United States since the first signing of their Compacts of Free Association (COFA) with the United States in the 1980s. Thus, in the Marshall Islands, the 2021 census revealed a 26% decline between 2011 and 2021, an exceptional decline that has continued and is shared by the FSM. Accelerated international migration from several SIDS alongside universal rural-urban migration has marked the last half century, with declining populations in all national peripheries and growing high-density urban concentrations, contributing to pressure on coastal resources.

Fertility rates differ between the Caribbean, where they are low (with only Barbados and St. Vincent reaching 1.7), whereas in the Pacific most rates are high with Samoa, Solomon Islands and Vanuatu having more than double that rate. Most Caribbean SIDS are below the fertility rate required to ensure population replacement. The demographic transition from high to low fertility and low mortality rates, with increasing contraceptive prevalence rates, has yet to occur in the Melanesian SIDS, other than Fiji, partly because of limited economic development (Harrington, Harvey, Larkins, & Redman-MacLaren, 2021). From a sample of eight SIDS in the Pacific and Caribbean, only two had experienced growth in contraceptive prevalence rates through the 2010s. Adolescent birth rates have continued to increase in the Pacific, while high rates in the Caribbean were only slowly decreasing. Reducing unmet need for modern contraception and expanding access to maternal health services would benefit both women's health, economic empowerment and education (Kelly, Walsh, Delport, et al., 2023). High teenage fertility rates in the Melanesian SIDS, linked to poor education and unemployment, have resulted in large numbers of young people, a 'youth bulge' that requires education, training and job opportunities. The minimum legal age for marriage is 18 years in Solomon Islands; however, marriage at age 15 is legal with parental consent. As a result, 21% of girls are married before the age of 18 and 6% before 15 (Fitzgerald, Moriarty, Basas, & Shagazatova, 2022); hence, high fertility rates are again the outcome. In most PICs, more than half the population are aged below 24 and urban space seems characterised by 'unattached youths': young urban men wandering around, *kilim taem* (killing time) in Vanuatu, because of lack of work. Growing numbers of single-parent households complicate matters further. Fully a third (35%) of all households in the Caribbean are headed by women, but proportions are as high as 44% in Barbados and 42% in Antigua and Barbuda. Such contexts usually require social welfare programmes.

While single-parent households and the youth bulge developed from and created social problems, centred around unemployment, most SIDS are also characterised by an ageing population. With substantial and growing proportions of the population experiencing non-communicable diseases (NCDs) (Chap. 6), this puts additional pressure on the budgets and human resources of health systems and on social support from kin, yet many of those kin have migrated. SIDS cannot rely on a demographic dividend to finance and strengthen their health systems. Older women are more likely than men to experience ill-health, and gendered inequalities and patriarchal norms affect both men's and women's experiences of ageing. Welfare provisions for older people are weak in most SIDS, especially in rural and remote areas, posing challenges for future equitable development (Anderson & Irava, 2017; Dodd, Shanthosh, Lung, et al., 2021; Plange & Alam, 2024).

The increased relative significance of fertility points to more permanent urban populations. Urban residents may express intentions to return 'home', but few do so: cultural and linguistic competence declines, village kin have died, greater subsistence orientation holds few attractions, and the proximity of wage employment, hospitals, stores and markets is valuable (Petrou, 2020; Petrou & Connell, 2017). Migration has been particularly pervasive in the atoll states, creating very high urban densities (in South Tarawa, Funafuti and Majuro). As urban centres have grown, newer migrants and newer housing areas, often informal, have located in marginal sites that are vulnerable to hazard, whether in the Caribbean (Carby, 2021; Mycoo, 2022), or in the Pacific (well evident in the 2014 Mataniko river floods in Honiara when more than 20 people died). That event became a catastrophe indirectly because of climate change but directly because of rapid run-off from the deforested hinterland and from settlements being on flood-prone riverbanks.

Notions that Pacific and Caribbean islanders are or should be rural people have been difficult to displace. Tourist images of bucolic rural islands are far from reality. Most people in most SIDS live in urban areas. The converse of urbanisation is slow population decline in outer islands—sometimes hastened by environmental change—and mainly in the Pacific. 'Outer islands', a post-war neologism, reflects a new phase in modernity and increasingly significant uneven development, as transport infrastructure fades and services are unevenly distributed. Employment is scarce, and mainly in a declining public sector, skilled workers are reluctant to be posted there and service delivery is limited (Connell, 2013; Chap. 9). Almost every Pacific atoll (around 80 of them in Pacific SIDS), other than the 3 urban centres, has lost population in the twenty-first century. Small islands generally are losing population. Thus, the population of Ono-i-Lau, the most southerly island in Fiji, declined from 767 in 1982 to 583 in 2002 despite household income roughly doubling over the same period (Kuster, Vuki, & Zann, 2005). Where migration opportunities are blocked, for whatever reason, livelihoods may be threatened, as in the densely populated Carteret Islands in PNG (Connell, 2016). Decline and uneven development are more acute and obvious in the archipelagic Pacific states and the Bahamas, where most of the smaller Family Islands are steadily losing population, resulting in loss of local knowledge and management skills amounting to a path dependency that has proved virtually irreversible.

An economic rationale dominates migration, but environmental changes are putting further pressure on SIDS, especially those that are losing population as the margins of settlement contract. Migration has taken increasingly diverse forms as it has gone from temporary to more permanent, unskilled to skilled and organised by households and governments, a function of population growth and rising expectations of what constitute satisfying lives and livelihoods (Chap. 9). Rural–urban migration is now mainly a Melanesian phenomenon and something of an alternative to international migration, where that cannot easily be achieved. Outside Melanesia, urbanisation has slowed, and in several SIDS it is stabilising and declining, partly as commuting grows.

Access to public services—or services of any kind—is more difficult in outer islands (e.g. Currenti, Pearce, Salabogi, Vuli, et al., 2019), and as people move away, the incentive to provide services declines: a vicious circle reducing quality of life and hastening further migration. Where hazards have displaced populations from small islands, governments have frequently discouraged them from returning (Chap. 2) to avoid this problem. Delivering health care, education and other services to people regardless of where they live is a universal challenge, more difficult where populations are ageing, budgets are tight, transportation is poor and management expertise is limited, all in the face of more pressure for public spending in growing urban centres. Costs of service provision increase with remoteness and with sparse populations. All costs are high; typically, on Kadavu island (Fiji), neither particularly remote nor small, store goods cost 15% more than in the capital city (Sofer, 2018), while choice and availability are limited. Urban bias feeds upon itself.

5.2 Urban Land and Livelihoods

With the lone exception of PNG, and to a minor extent Jamaica, all towns and cities in SIDS are coastal and thus at particular risk from environmental hazards. Integrating urban development and management with resilience in the face of hazards has proved difficult (e.g. Carby, 2021; Mycoo, 2018, 2021) but has been relatively successful where local knowledge and experiences have been involved (Kiddle, Zari, Blaschke, et al., 2021; Le Dé, Gaillard, Friesen, & Matautia Smith, 2015; Rey, Le Dé, Leone, & Gilbert, 2017). Invariably, as in Trinidad, the main institutional challenges relate to lack of coordination and communication between relevant actors (Middelbeek, Kolle, & Verrest, 2014). Actions that local people have taken, based on experience and knowledge, may be superior to those devolved from above, especially during crises. Yet in most SIDS traditional knowledge, whether of navigation, biodiversity or hazard responses, is becoming obsolete and sometimes confined to more remote islands and places (Chandra & Gaganis, 2016) as 'school' knowledge, store goods and aid take over. Invocations to include local ecological knowledge, and indeed local people with that knowledge, may be aspirational rather than practical (e.g. Schuermann & Lauer, 2018; Walshe & Nunn, 2012).

Coastal towns and cities are growing at rates beyond national population growth, and in most SIDS more than half the national population are in urban areas, further contributing to coastal squeeze and problems of urban management. Rapid growth has resulted in 'overurbanisation' characterised by high levels of unemployment, formal housing unable to meet the needs of migrants (and others), inadequate urban services, informal housing areas, alongside intensified environmental problems, associated with poor sewerage, water supply and waste management, and occasional flooding. Land shortages are an accelerating urban and coastal phenomenon, alongside land tenure disputes and conflicts, resulting in informal settlements being on unsuitable land (to which services cannot easily be extended), illegal dumping and other harmful practices.

Primate cities are the norm, leaving minimal scope for regional development. Unregulated urban expansion has increased the costs of urbanisation, especially where infrastructure provision only occurs after land is occupied. Constraints to urban management reflect those at national level: shortages of capital, skills and plans, allied with land shortages and tenure constraints, and tensions between urban and national political regimes. Few effective housing policies exist. In this century, several disasters suggest that urban communities are becoming more vulnerable to environmental hazards. In some Pacific SIDS, urbanisation was long opposed by elite urban residents and national governments, informal-sector workers were marginalised and settlements bulldozed, to 'encourage' people to return to rural areas, so that planning and welfare provision firmly excluded the relatively poor and unemployed (Connell, 2011). Especially in Melanesian SIDS, informal settlements developed where people settled on marginal land, generally due to the lack of affordable housing options. Such urban settlements had inadequate access to essential services and infrastructure, from sealed roads to garbage collection.

Urban livelihoods partly reflect the capacities of states. While urban centres are increasingly like cities and towns in other parts of the world (especially where CBDs are visibly dominated by tower blocks), large fresh food markets, a proliferation of street vendors, and the youthful unemployed indicate something rather different. Urban economies are at least visibly dominated by the much maligned informal sector that characteristically dominates in urban spaces, involving small-scale marketing (stalls outside houses), handicraft making, peddling of batteries, noodles and SIM cards (Maebuta & Maebuta, 2009), and small-scale services including cleaning, repairing and security guards. The informal sector could offer fast money schemes, but transience was indicative of uncertainties over 'modern' strategies.

The service sector, notably the bureaucracy, dominates many towns and cities. Although the public sector bureaucracy has scarcely grown in this century, most SIDS have public service employment cores, dominated by education and health, and a private sector, usually led by tourism and retailing. The private sector—financial services, security, construction, restaurants and bars, hairdressing, taxis, real estate—has grown faster while a relatively small proportion of employment is in manufacturing (Chap. 3). Acquiring the skills and capital to progress from dead-end jobs has been difficult. Wages can be barely adequate and unions absent, outside the public sector, to press the case for better circumstances.

In Melanesian towns and cities, and especially in settlements, many urban residents have few sources of income, no legal residential status or security of tenure, limited access to services and live at high densities, a combination which poses social and health risks (Trundle, Barth, & McEvoy, 2019). Settlers make up more than a third of the population of large capital cities such as Port Moresby, Port Vila and Honiara. In such marginal areas, occupied by the relatively poor, urban management is absent, services and water security are scarce, and continued unregulated growth worsens problems (Kiddle, McEvoy, Mitchell, et al., 2017). Rapid urbanisation has created 'standard' problems of service provision. High population densities challenge water supply systems, sanitation and solid waste management, presenting serious environmental and health risks further complicated by climate change and additional hazards. Climate change has emphasised a new dimension of urban coastal management, but diverse and piecemeal strategies have been poorly coordinated at every scale and been ineffective, especially where marginal population groups, usually associated with ethnicity and geography, are most affected (Dedekorkut-Howes, Torabi, & Howes, 2020). Exceptions exist in rare cases where local stakeholders and all community members have been involved over time, nature-based solutions make sense to all, and expertise and external finance are supportive (Ho, Dias Baptista, & McEvoy, 2023; McEvoy, Tara, Vahanvati, et al., 2024). Most SIDS have relied on external funding for urban programmes, an indication of limited financial resources but also the failure to prioritise informal housing for local funding. Yet in so many places, given some degree of tenure, residents have proved entirely capable of developing and managing their own services (Rooney, 2021). Moreover, they have adapted traditional practices of housebuilding, such as using local building materials with flexibility, to improve resilience in the event of hazards (Trundle, 2020). That has been essential since few effective housing policies exist, and with formal housing beyond the reach of many urban residents, people must provide for themselves, yet their own responses and innovations are ignored. Management has tended to operate at an urban scale; indigenous activities and responses are more localised.

Informal settlements are also common in the larger Caribbean cities in Jamaica and Trinidad due to the high cost of housing, creating a demand for informal housing. Indeed, Jamaica is estimated to have about 750 squatter settlements occupied by 600,000 people, as many as 20% of the national population. Some settlements, regarded as 'garrisons', typically lack access to basic services, including security, drainage and sewerage systems, but have their own informal systems that generate a 'local citizenship' (Campbell, 2020; Campbell & Clarke, 2017) and are enmeshed in transnational processes (Bowling 2010; Munroe, 2023). In Trinidad the problem is no less acute, with over 350 settlements catering for around 40% of the national population (Mycoo, 2022). In these spaces there is typically no security of land tenure, poor infrastructure and lack of access to basic amenities. Smaller Caribbean countries, such as Antigua and Barbados, have mainly encountered this on a smaller scale, through other Caribbean nationals moving from Jamaica, Guyana and St. Vincent, with such 'foreigners' rarely garnering positive attention.

Rapid urban growth and the expansion of settlements have been particularly prevalent in Melanesia but without integrated planning, let alone empathy, respect, trust or a sense of community and urban citizenship for the poor. In Port Moresby (PNG) and Port Vila (Vanuatu), particularly on valuable land near the urban core, evictions continue to occur. Opposition to urbanisation has continued, from urban and national authorities and influential leaders in the guise of achieving order and cleanliness, reducing crime and unemployment, freeing land for business development and demonstrating that the state is not weak: neo-liberal and neo-colonial objectives that persist. In 2022, the Jamaican government demolished sections of an illegal settlement, Bernard Lodge: a rare instance of government acting against illegal settlements. Moral panics and achieving 'moral order' take different forms. In Vanuatu, by contrast, after a National Land Summit in 2006, Ralph Regenvanu founded the Land and Justice Party and, as Minister for Lands, introduced stricter legislation to protect indigenous land from alienation by commercial and overseas interests.

Opposition to, and neglect of, settlements remain dominant, if unspoken, anti-urbanisation policy platforms in SIDS. Some settlers are seen erroneously as temporary residents (despite being second-, third- or more-generation urbanites, for whom formal housing is out of reach). The rise of the informal sector and urban poverty has been marked by repressions of the poor and marginalised, rather than by concerted attempts to devise welfare and employment policies that might reduce social problems (Koczberski et al., 2001, Connell, 2003). Urban social safety nets are weak, with informal social protection stretched, as costs of urban living increase and many struggle to gain steady wage employment. Resilience becomes unusually difficult without effective support, alongside the ability of low-income residents to participate in planning, however loosely defined. Urban planning and management, and a sense of community, rarely cover all urban residents, evident in opposition to informal settlements. In the Caribbean informal settlements are seen as 'incubators of crime' associated with the narcotics trade, as governments no longer control parts of cities (Mycoo, 2022), and in the Pacific they are likewise wrongly regarded as centres of crime, lawlessness, unemployment and violence (Hukula, 2017). Meanwhile settlements have continued to reverse the priorities of conventional urban planning, with land occupied despite lack of tenure, dwellings built in the absence of regulation, and services supplied last or in some cases not at all. That disjuncture and lack of knowledge is evident where few of those in informal settlements interact with government officials or even, where these exist, with local councillors; in Honiara that remains true even after settlements have been in place for more than 60 years (Bruce & Marshall, 2020). Such failures of governance inhibit sustainable urban development and management.

Urban land has come under intense pressure, yet some security of tenure is crucial for more resilient housing and livelihoods and often for access to services and employment. Without recognised title to land, occupants cannot obtain loans (let alone mortgages); hence, 'illegal' homes may be only partly built as settlers manage the slow process of income generation and capital formation. In turn, this structural inequality makes providing services to such houses more difficult and in turn makes them doubly susceptible to hazards (McEvoy, Mitchell, & Trundle, 2020; Mycoo, 2022; Williams, Costa, Sutherland, et al., 2019). Lack of resources makes occupants less able to adapt and respond, especially where they are reluctant to move from established sites (Nolet, 2016). Further problems occur where a high percentage of buildings are constructed without adequate planning permission as with over 80% of buildings in Trinidad and Tobago and 25% in Dominica. Upgrading is one approach to illegality, literally building on the owners' own efforts, but usually founders on lack of political will and resolve and weak implementation. Land tenure and boundaries are often ill-defined and contested by various claimants. In Honiara, disputes over land tenure in peri-urban areas were significant factors in the genesis of the 'tensions' that resulted in bitter and costly (in lives and economic growth) internecine violence for several years over the turn of the century. Many Gwale people, ethnically from Guadalcanal (where Honiara was located), resented the influence of settlers from other islands, especially densely populated Malaita, and their occupation of land in and around Honiara (Allen, 2012). Tensions also occur in Caribbean SIDS, including Barbados and Trinidad, with resentment towards migrants from other Caribbean countries.

Frustration over unemployment can become violent, surfacing in 2006 when Tongan youth rioted in the streets of Nuku'alofa. As in Honiara, dissatisfaction with governance, perceptions of corruption and resentment over the role of foreign migrants were catalysts, linked to the presence of disenchanted unemployed youth in urban areas (Ride, 2022).

Frustration, despondency and *kilim taem* could boil over. Protest among the urban poor in the Caribbean has occurred in the context of rising costs (e.g. the 1999 riots over gas prices in Jamaica) or over demands for justice and respect for human rights. Hope is diminished where 'the government is dysfunctional and corruption a problem. Poverty, poor governance, and stagnation. A recipe for resentment' (Wood, 2021). Women shoulder a greater share of the burdens associated with poverty, not only experiencing higher levels of unemployment and wage discrimination but assuming higher levels of responsibility for the social wellbeing of households than men. In Jamaica, households headed by women account for two thirds of single-headed households in poverty (Mullings, 2009). At the very centre of several SIDS critical urban problems exist, stemming from high levels of unemployment, and inadequate access to land, housing and services, spilling over into other forms of inadequate welfare service provision, watched over by disdainful governments. Even so, urban bias prevails and smaller towns, as in Jamaica (Dodman, 2009), fare at least as badly in terms of urban neglect, management and planning.

Achieving harmony between different cultural groups in urban areas can be a long and difficult process. The juxtaposition of tenure problems, deprivation, unemployment and uncertainty can contribute to anomie and crime, real threats to a resilience that demand community collaboration, social stability and trust. Fractious, unemployed youth, seemingly merely hanging around in town (Mitchell, 2004; Vakaoti, 2018), are invariably perceived as problems rather than as community members, let alone potential assets (see below). Even open spaces for passive or active recreation are often lacking, with space only for netball, basketball or volleyball, denying the possibility for many to participate in urban life through sport. Open space could help reduce urban temperatures, and nature-based solutions provide cheap and reliable alternative sources of water and reduce risks.

Most Melanesian cities experience inadequate water, sanitation and waste collection services, as do the urban poor in Caribbean cities (Chap. 6). The three atoll states, especially Kiribati, face particular urban environmental management problems. Their urban centres occupy almost the entirety of islands, with scarce open space, literally putting pressure on underground lenses of fresh water. Access to fresh, clean water is threatened by sewage, pollution and unchecked water extraction (Moglia, Perez, & Burn, 2008). Lagoons and fringing-reef systems which provide invaluable livelihood resources are degraded by the discharge of untreated waste. Flooding is not unusual. Coastal erosion is exacerbated by beach sand mining and seawall construction. Deforestation, the breakdown of traditional subsistence production, and low incomes have contributed to poor nutrition and health-related problems (Storey & Hunter, 2010) to the extent that in the 2010s, the nutritional deficiency disease, beriberi, reappeared in Kiribati after an absence of 50 years (Nilles, Manaia, Ruaia, et al., 2018). Imported foods constitute up to 70% of Kiribati diets (Sievert, Lawrence, Naika, & Baker, 2019). In the humid urban heat islands, higher temperatures have increased the extent of respiratory diseases, and greater sedentarism and diets of imported processed foods have worsened the incidence of NCDs, such as diabetes and obesity, to levels higher than in other parts of the world (Chap. 6). In Tarawa especially, lagoon health has continued to deteriorate in this century, near anthropogenic pollution sources: the main port, a large landfill site, inner lagoon aquaculture ponds, beach defecation and the national hospital (Duvat, Magnan, & Pouget, 2013; Graves, Powell, Stone, Redfern, et al., 2021). In both Tarawa and Majuro high population densities have resulted in land being reclaimed from the lagoons for residences.

Despite multiple problems at the centres of the SIDS, there has been no great interest in urban management (and efforts by the Pacific Forum Secretariat in the 2010s to develop regional policies and learn from success stories and from mistakes across the region, were ignored as, in many places, that would imply unwelcome recognition of the significance of urban growth and mismanagement). While urbanisation remains a decidedly unwelcome 'elephant in the room', as invisible as equally unwanted poverty, the possibility of national and regional scalar directions and solutions has at least temporarily been lost (Connell, 2011; Keen & Connell, 2019; Keen, Sanderson, Osborne, Deo, & Pranaya, 2022). The capacity to implement plans and policies in all arenas is constrained by finance, weak institutional management capacity, human resources, factions, short project cycles and ineffective institutional arrangements, weak government capacity and political will (Hirano, 2008; Jupiter, Wenger, Klein, et al., 2017; Remling & Veiteyaki, 2016). Effective planning and management would involve greater inclusivity, recognition that many people live in informal settlements, that urban services are often inadequate, and that all citizens have the right of urban residence.

5.3 Provinces and Towns??

The scale of development in the Pacific, and the distance between islands, even with small populations, has meant that larger states—Solomon Islands, Vanuatu, Fiji—have created provincial governments, while the FSM is an amalgam, often unwieldy, of four states. So distant are the Phoenix Islands and the Line Islands groups from the main island group in Kiribati that they have their own Ministry of Line and Phoenix Islands Development. The geographically smaller Caribbean SIDS have partial equivalents in Nevis, Barbuda and Tobago, with Nevis having its own premier and government. Shortages of finance, scarce human resources and uncertainty over their role have meant that few provincial governments have functioned effectively. PNG has necessarily made the most substantial attempts to devolve the legal, financial and administration arrangements between central, provincial and local levels of government although their performance is very uneven.

Similar problems have affected attempts to establish urban governments, where conflicts between modern planning and management and customary land rights and values are most severe, evident where cities are growing quickly (Mecartney & Connell, 2017). Building national institutions has been emphasised at the expense of building local government, and urban administrators struggle under outdated laws and regulations, the cross-cutting authority of indigenous land ownership, and occasional opposition from national MPs and government departments (Hassall, 2020). Ensuring that local authorities have the capacity to engage with their communities, deliver essential services and foster social cohesion faces obstacles: limited and shrinking civic engagement, with weak mechanisms for community involvement, preventing local perspectives from being integrated into governance processes and weakening the effectiveness of SDG implementation. Local governments often struggle with limited resources, skills and adaptability, and without coordination between local and central government agencies, impeding their ability to function adequately.

A major challenge concerns the delineation of power, roles, responsibilities, funding and the effectiveness of service delivery. Local government bodies are critically under-resourced. Few town and city councils in the Pacific region will be able to significantly improve their capacities for service delivery or for infrastructural development in the short to medium term (Hassall & Tipu, 2008: 2). All SIDS have struggled with the issue of what powers to devolve to lower levels of government, and the delegation of authority and resources to urban centres, including capital cities, is uneven and usually limited as national governments compete for authority in the same space, as do traditional landowners, who still exercise substantial authority in most towns and cities. Thus, in Port Vila, managing urbanisation is difficult since local landowners have authority over many tracts of land in the capital, the SHEFA province (covering several islands) is responsible for the considerable urban development and service provision beyond outdated formal boundaries, while the urban council is constrained by finance and skilled human resources. In Palau and FSM, new post-independence capital cities were designated to avoid such tenure problems. There is little prospect that many state, provincial and city councils in the Pacific region will be able to significantly improve their capacities for service delivery in the short to medium term, as poorly resourced institutions fail to cope with the demands placed on them to deliver a range of services (Chap. 6). Caribbean SIDS have similarly struggled over many decades to reform local governance and effectively decentralise decision-making (Schoburgh & Martin, 2016). Despite the many governance challenges that require local responses and solutions, local government has largely been unsuccessful, partly because of scarce human and financial resources, with limited implementation of policies necessary for realising meaningful change at local level. Even where local councillors are elected and decentralisation more of a reality, problems of central political control and scarce financial resources persist.

5.4 Unemployment, Poverty and Social Protection

Formal employment, unemployment and poverty are predominantly urban phenomena. Creating economic growth, productive employment and income-generating activity is central to development challenges facing SIDS. This is particularly difficult where population growth is often faster than economic growth, where the number of school leavers is greater than the rate of employment creation (and most seek formal employment) and where rural-urban migration remains significant. Public service employment is widely favoured because of its usually superior wage levels and conditions of employment (prestigious, clean and relatively secure), but in many SIDS public service employment has stabilised or been cut back in recent years, yet exhibits low productivity, unmatched by growth in the private sector.

Unemployment and underemployment are particularly visible in urban areas, and not easily measurable in rural areas, typified by the youth bulge. Urban settlements especially can be seedbeds of discontent and incentives to international migration but amongst those least able to achieve it or gain from it (Connell, 2013). Yet urban populations are highly variable in the Pacific and the Caribbean, spanning the extremes of wealth, from the 'big shots' of Melanesian and Caribbean cities, to informal settlers, scraping a livelihood, and better thought of as a precariat (Gewertz & Errington, 2021). Social extremes are increasingly visible as cities in SIDS become spaces where wealthy people display status symbols and others aspire to such symbols. Urban ethnic divisions, and especially incorporated urban villages, emphasise differences between landowners and newcomers. Many settlements are of complex ethnicity, quite unlike urban villages, that challenge existing modes of urban governance and often face marginalisation and exclusion. Employment practices, aided by churches and schools, have cut across traditional social divisions so that even in ethnically diverse Melanesian cities, where once they followed *wantok* (one language) ties, class and spatial divisions are steadily emerging (Gewertz & Errington, 2021; Gibson, 2021; Jourdan, 2017; Mausio, 2021).

Until quite recently there was a widespread belief that poverty, however defined, did not exist in most island states, notably in the Pacific, because of the existence of urban and rural 'safety nets' where extended families could and would support those who, for whatever reason, experienced problems. However, safety nets are fracturing and it is no longer possible, if it once were, for urban people simply to return and be supported by rural kin. Many Pacific SIDS have been reluctant to concede that poverty existed (with a series of ADB reports early in the century being titled *Hardship in …* [rather than *Poverty in …*]) in order not to offend local politicians. Samoan politicians have bluntly stated that Samoa had no poverty when growing numbers were sleeping rough on the streets of Apia or seeking food in refuse dumps. In mid-2021, the CEO of the Fiji Bureau of Statistics was sacked for publishing a detailed report on the incidence of poverty in Fiji by ethnicity and religion. Denial augured badly for policy formation.

Poverty was undeniable. By the mid-2000s Kiribati had half its population below the poverty line, Fiji had 30% below the poverty line and only Vanuatu of the nine other Pacific SIDS had less than 20% of its population below the poverty line. But in semi-subsistence contexts formal definitions of poverty are inadequate. Samoa, which a decade earlier had rejected any notion of poverty, had 20% below its own poverty line. Grenada, regarded as having the highest proportion of people in extreme poverty in the eastern Caribbean, had 32% in poverty and 5% in extreme poverty in 2002. Rural communities and large families were over-represented amongst the poor (Barrientos, 2010). Dominica, St. Kitts, St. Lucia and St. Vincent all had more than 25% below the poverty line. Only the Bahamas had a rate less than 10% (Bourne, 2009; Connell, 2013). In many places households struggle to achieve adequate livelihoods and access services such as clean water and sanitation. Poverty can be hidden in outer islands and remote areas, with poor access to education, health services and formal employment. While income poverty is more easily measurable, a non-income component of poverty consists of people's capabilities, vulnerabilities and participation: some households are poor by not having enough food, clean water or access to adequate housing or basic education, an evident poverty of opportunity. Overall poverty may be increasing, though formal data are inadequate.

Poverty and the outcome of inequality are most visible in urban settlements. In Honiara as many as a third of the urban population are below the national poverty line and perhaps half of these are in urban settlements so that 'many families are mired in poverty [leading to] family breakdown, street children, sleeping out, sexual exploitation, alcohol and drug abuse, delinquency, violence and lawlessness' (Russell, 2009: 16; Maebuta & Maebuta, 2009). In Suva (Fiji) the pattern was the same whilst in Tonga in 2009, the largest numbers in poverty, many of whom were migrants from outer islands, were on the urban fringes of the capital, Nuku'alofa. Social disruption and domestic violence have followed poverty, and youth are especially vulnerable to risky sexual activity, early pregnancy and a difficult transition from school to work, where work exists, while violence can be the outcome of poverty and unemployment, especially amongst urban youth, as in Port Vila. Begging too occurs in many large urban centres. Such visible evidence of inequality and a growing realisation of relative deprivation have created tensions and frustrations. In some SIDS, poverty and income inequality have resulted in a downward spiral—an urban 'poverty trap' from which there appears scant hope of escape.

Income poverty is usually greater in rural areas, as in Vanuatu, exemplifying the poverty of opportunity evident in many places. In Kiribati, while 28% of those below the poverty line were in the drought-prone southern atolls, 42% were in the capital, South Tarawa (Kwansing, 2025). Generally, 'extreme poverty' (living on less than US$3 a day) is over-represented in urban areas. In Vanuatu, young couples with children, female-headed households and those renting were particularly vulnerable, even though most households had at least one source of income. Their expressed needs, which have changed little over time, centred on core elements of the SDGs: finding a house to rent, access to finance, having land to live on, education, adequate toilet facilities, finding a way to start a business, transportation and access to electricity (Government of Vanuatu, 2002: 22, 63–65). Then and now, poverty was more than a lack of money but, for children, a lack of adequate nutrition, water, shelter, education and health care that limit future opportunities (UNICEF, 2025). Many frustrations, aspirations and pleas echoed and reverberated around the Pacific SIDS, where the urban poor expressed concern about unemployment, governance, declining standards in the delivery of services and lack of economic opportunities (Storey & Connell, 2013). Many basic needs were unmet.

Strategies to reduce poverty and achieve greater equity are rare and not obviously successful, either in promoting inclusion or providing protection. Social protection features in numerous country policies and development agency strategies and in several SDGs. Throughout the SIDS, effective governance and equitable development (especially provision of services to outer regions) remain critical issues (Chap. 6). Traditional protection systems have come under stress through rising urbanisation and food scarcity. Until quite recently most SIDS, and the PICs especially, devoted few resources to social security and welfare. Thus, FSM had no formal social protection programme (at least prior to COVID-19), simply relying on old social safety nets (Patagan, 2022). Caribbean SIDS were generally more advanced, initially focusing on age pensions, but extending into disability and widow pensions, as well as benefits for sickness, maternity and workplace injury, and all have national insurance schemes.

In this century social protection programmes have expanded in the Pacific SIDS, with age and disability benefits as key components (Perch & Roy, 2010). In Fiji the national social protection system provided an effective way to respond to Cyclone Winston in 2016, channelling cash and non-cash humanitarian relief to those in need (Mansur, Doyle, & Ivaschenko, 2017). Recognition of this success and similar responses to COVID-19 led to widespread support for investment in social protection systems. Countries such as the Marshall Islands and Tonga have experimented with new kinds of cash benefits for families with children, providing coverage for a group that previously went without formal government support.

In 2025 the Marshall Islands introduced an ambitious nationwide universal basic income scheme to provide regular cash transfers to all resident Marshallese citizens, including children. Given its status as a remote island state with a small tax base, high aid dependence and acute climate vulnerability, the scheme represents a bold effort to transform external resources, being financed primarily through its Compact of Free Association (COFA), into a direct citizen entitlement. Against a backdrop of rising climate adaptation costs, limited formal employment opportunities and significant outmigration to the United States and elsewhere, the scheme potentially redistributes income 'to reshape the state-citizen relationship in a setting where the tax base is narrow, labour markets are fragmented and many residents of outer atolls have limited access to formal banking services' (Taylor, 2025). By establishing a direct, individualised transfer from the national government to every resident citizen, it reformulates external resources as a shared dividend rather than discretionary aid.

Overall expenditure on social protection in Pacific SIDS has more than doubled since 2013, from 0.9% to 2.3% of Gross National Income, with the biggest increase in Kiribati, which has introduced a large-scale unemployment benefit scheme and expanded the coverage of its Senior Citizens Allowance. Its two largest social assistance programmes are estimated to have more than halved the national poverty rate (Doyle & Knox-Vydmanov, 2024; Kwansing, 2025). Social protection in Solomon Islands and Vanuatu lagged, with national expenditure directed to other facets of development (especially physical infrastructure) with more of the population engaged in 'traditional' semi-subsistence areas (Prasad & Kausimae, 2012). However, all SIDS are now involved, with social protection systems supporting households after crises and in normal times.

What constitutes social protection has broadened substantially to include various government policies that contribute to superior incomes and welfare, that might involve gender equity or, as in Kiribati, where the Copra Price Subsidy is the largest formal national social protection programme in Kiribati (although not conceptualised in this way), aimed at supporting livelihoods in outer islands. All Pacific SIDS have increased investment in social protection schemes, mainly through age pensions and disability support, with assistance from the Australian programme Partnerships for Social Protection (P4SP), contributing to reducing poverty and inequality. The Caribbean SIDS generally have more developed and wide-ranging programmes than in the Pacific, recently influenced by Hurricanes Irma and Maria and, unlike the Pacific, by earlier global financial crises and a more obvious legacy of British systems. Growing evidence of poverty, fragmenting family structures and domestic violence means that more progress is needed on the inclusion of particularly vulnerable groups (see below) and extending programmes, such as early childhood education, to achieve the development dividend of youth. Such strategies, consistent with an agenda of No One Left Behind, require guidance and support, with the provision of social protection raising questions about the role and extent of governance direction and leadership and its effectiveness in a range of areas from obvious interventions in education and health care (Chap. 6) to food security, training to enhance the employability of workers, at home or overseas (Chap. 9), and thus how best to empower the entirety of the population and particular social, geographical and marginalised groups.

5.5 Governance

State structures and political systems are inevitably centred on the primate capital cities (and rare attempts at decentralisation and devolution have been weak and unsuccessful). Size is a key factor in the structure and management of governance, and small size and management issues contribute to urban bias in policy and practice. Constraints to service delivery point to weaknesses in state structures, linked to past histories, which have resulted in limited attainment of the SDGs and before that the MDGs, contributing in several cases to political instability, further hampering development. We start this section by examining the politics of the SIDS and whether smallness complicates and challenges political achievement, stability and progress. The following chapter traces the evolving role of regional organisations since smallness has resulted in some recourse to regional institutions, which have proliferated since independence.

Although democracy can prove elusive, a characteristic of SIDS is the persistence of broadly democratic institutions. In both regions, political systems resemble those of the colonial powers, other than there usually being only a single chamber. A perennial question has been whether such introduced political (and legal) systems are suitable for small states. Other than those in the northern Micronesian SIDS, which took on American models, most newly independent Caribbean and Pacific nations retained broadly British political systems, characterised by a first-past-the-post, winner-takes-all electoral model, three branches of government in the form of the executive, the legislature and the judiciary. Excess power in the hands of the Prime Minister was also an imitation of British politics. This brought calls in the Caribbean for a more hybrid political and constitutional identity (Wheatle & Campbell, 2020), a situation that exists in some Pacific SIDS. In Samoa stability in post-colonial times has been attributed to hybridity (where universal suffrage exists, but only *matai* (chiefs) can stand for parliament, and only 7% of *matai* are women), despite its contradictions coming to a head in a constitutional crisis in 2021 centred on debates over female representation in parliament (Fraenkel & Iati, 2022). In Tonga, kings have regularly sought to impose their will on elected governments (and continued to do so in 2025, as the king dismissed cabinet members and a democratic party system foundered) while nobles, hereditary members of the 'aristocracy', have nine reserved seats—a third of the total—in parliament. In Fiji, the Great Council of Chiefs, which had existed since 1876, was abolished in 2012, on the grounds that it had perpetuated elitism and created divisive ethnic politics, but was revived in 2023. Contradictions and distinctions that challenged hybridity were most apparent in Melanesia where modern economies were relatively recent, local 'traditional' leaders significant, missions powerful (Forsyth & Atkinson, 2024) and such Westminster notions as a 'loyal opposition' largely incomprehensible.

Smallness has undoubtedly complicated the application of Westminster politics. In the Caribbean, the strong two-party political system has often led to a politics of patronage where parties distribute benefits to their supporters in exchange for votes. This is more easily achieved in small societies, still responding to the socio-economic effects of colonialism and where political leaders are relied upon for hand-outs and social relief by the urban and rural poor. With limited institutional capacity and regulation, even the least corrupt SIDS, such as Barbados, experienced problems (Alleyne & Barrow-Giles, 2021). A lack of separation of powers between the executive and legislative arms of government has also fostered corruption, lack of accountability and poor governance in some SIDS. Such corruption is more problematic in the SIDS since socio-economic issues are greater in a context of more limited resources, poor health care and education systems, and higher levels of inequality (Larmour, 2012), manifest in a 'political reality of patronage and corruption that occurs in daily encounters with constituents', as evident in St. Lucia, but largely tolerated because of its near universality (Saati, 2022).

Departures from the British system have occurred. Like Kiribati, Trinidad and Tobago and Fiji, with presidents or multi-party systems, some SIDS have reformed their political systems, becoming more 'modern' hybrid models, sitting somewhere between presidential and parliamentary systems. The most radical reforms were attempted and implemented in Grenada in 1979 when Maurice Bishop sought revolutionary change, moving the country away from western-style democracy to socialism, but the Marxist-Leninist communist regime eventually collapsed (following the killing of Bishop and US invasion in 1983). Smaller SIDS, such as Nauru and Tuvalu, have only a single chamber. Others, such as Vanuatu and Fiji, have a second tier (the Malvatumauri in Vanuatu and the Great Council of Chiefs in Fiji) that is intended to focus on indigenous cultural issues and be advisory. Barbados amended its constitution in 2021 to become a republic, moving away from the monarchy as head of state to a ceremonial (female) president. Debates on constitutional change occur in other SIDS, many of which still retain the monarchy as head of state and the Privy Council as their final appellate court despite the establishment of the Caribbean Court of Justice, in the case of the Caribbean.

5.5.1 Politics, Politicians and Political Instability

With or without leaders, policies and parties, which usually only appear at election times (see below), the democratic institutions in most SIDS have proved persistent and viable. Regular monitored elections exist, with proper voting lists, despite frequent votes of no-confidence (that have been legislated against in some SIDS constitutions). This provides governments with stability and platforms, as parliaments meet only infrequently. The notion of a loyal 'opposition' to the government is a foreign concept in several SIDS, especially since, in traditional Pacific societies, leaders sat together to consult, argue and restore relationships without institutionalised 'government' and 'opposition' to divide loyalties. Despite violence and threats to the rule of law in a few SIDS, such as Jamaica and Trinidad in the twentieth century, most political systems remain stable and democratic governments are elected free from violence. Four coups between 1987 and 2006 brought down governments in Fiji, but it returned to democracy. In Samoa in 2021, national tensions followed the eventually unsuccessful attempt of the culturally elite Prime Minister to resist the national election results and retain power. The distinguished historian, Brij Lal, was expelled from Fiji in 2009, for criticising the expulsion of the Australian and New Zealand High Commissioners, and died in exile. But, with rare and infrequent exceptions (such as Nauru for a period in the 2010s), democratic governments have permitted parliamentary oppositions to operate without hindrance. Exposure to a wider world (and to the Commonwealth Parliamentary Association) has enhanced the propensity to adopt and sustain democratic institutions.

The numbers of politicians can be small (with Tuvalu having just 12—the smallest of all—but a ratio of politicians to voters more generous than in other SIDS), yet the range of issues required to be covered by SIDS governments is no less large or complex than in much larger states. That national (and international) focus may be complicated by the expectations of electors. In SIDS, most voters are influenced by local factors. In Vanuatu in 2024, a third (35%) of voters believed that MPs should focus on community projects, 16% felt that MPs should promote national development, while only 8% felt that governing the country was important. Voters prioritised the ability of candidates to represent them and local community interests, regarding their role in law-making and managing the national budget as 'very low' (Naupa, 2025). While Vanuatu is a complex archipelago where services are unequally distributed, that augured badly for establishing national priorities.

In very small SIDS (that are not archipelagos) there is a sense that 'everybody knows everybody' so that family and kin are central to political dynamics rather than any ideological differences, a situation that has both advantages and disadvantages (Corbett, 2015; Saati, 2022; Veenendaal, 2013). One outcome is that SIDS tend to cluster on the centre-right of the political spectrum, through a process of 'managed intimacy' (Lowenthal, 1987). Thus, when Surangel Whipps was involved in a close contest for the presidency of Palau in 2024 (the other candidate being his brother-in-law), it was expected, somewhat improbably, that he would doorknock each household as many as four times, perhaps because it was recognised that there were no ideological differences between the candidates (Smith, 2024). Almost all ni-Vanuatu politicians had constant direct contact with their constituents. Personality then triumphs over formal structures, and local concerns are emphasised as a political campaign strategy.

Small legislatures may lack the range of expertise and experience required for parliamentary business among its MPs. A small number of parliamentary seats also mean smaller numbers of professional staff, some of whom may be closely related to MPs and to public servants. Throughout the SIDS, most MPs (or their equivalent) are men. Women politicians are rare, especially in the Pacific. Around 22% of ministerial positions in the Anglophone Caribbean are held by women who occupy fewer than 30% of elected positions (UN Women, 2018: 5), limited by global standards, but substantially ahead of the Pacific where fewer than 7% of Pacific politicians are women, compared to 27% globally. Many MPs were previously, and remained, prominent leaders, usually in business. In both gender and social status MPs are atypical of their countries and constituencies.

Some governments, such as that of the Autonomous Region of Bougainville (PNG) and Samoa, have reserved seats for women. A milestone was reached late in 2022, when, for the first time, there was at least one elected woman in every Pacific parliament, but at other times in the early 2020s, there were no women in the parliaments of the FSM, PNG and Tuvalu, and very few in other SIDS. Independent states in the Pacific region have the lowest levels of women's political representation in the world. Nonetheless, both the Marshall Islands and Samoa have acquired female leaders, the latter relatively short-lived (So'o & Jackson, 2025). Numerous obstacles based on culture and socio-economic status in patriarchal societies limited opportunities, and in most contexts (where political parties are few) women lack the financial resources to campaign on an equal par with men and have more restricted social networks. In the Caribbean, potential women political candidates in Barbados, Jamaica and Trinidad were deterred primarily by financial constraints, and their exclusion from influential funding networks predominantly controlled by the 'old boys' club' and so had to rely on their own resources, balancing childcare and other gendered roles with demanding fundraising activities, and navigating systems of political patronage (Campbell & Johnson-Myers, 2023). That was especially so for would-be newcomers (Barrow-Giles, 2015; Johnson-Myers & Campbell, 2024). Nonetheless, Jamaica, Barbados and Trinidad have all had female Prime Ministers. In Vanuatu, and elsewhere in the Pacific, women have similarly been unable to raise and distribute enough finance and gifts to participate effectively. While Samoa, the only PIC with a quota for women, acquired a female Prime Minister in 2021, she faced sexist heckling and a vote of no-confidence from men anxious to re-establish a gendered 'politics as normal' (Palmieri, 2025). Even in Fiji, seen as a positive outlier in terms of women's representation, fewer women were elected in the 2022 elections, evidence that gains in women's representation are never assured.

Several women have been elected to Pacific parliaments, notably in Solomon Islands, as part of a 'widows and wives' phenomenon (and dubbed 'legacy politicians') because of their marriage to deceased former male MPs (Baker, 2022) or, as in Samoa and the Marshall Islands, because of their own chiefly elite status. Thus, FSM gained its first woman parliamentarian in 2022, a medical doctor but more significantly the widow of a member of Congress (Puas & Oliver, 2022). Even where legislatures have mandated a number or proportion of women candidates, progress has been slow.

Absence from positions of power means that issues that may particularly concern women are also marginalised; hence, the absence of women's voices in political decision-making has been frequently raised in regional forums, but exactly what the differences and disadvantages might be is difficult to discern. While women may have a low profile in parliaments throughout the SIDS, women's organisations and NGOs of various kinds play a substantial role, more so precisely because of the absence of women from parliaments. In Chuuk (FSM), where women have lobbied for legal change, superior health and other social services and changed perspectives on gender, the participants were empowered through such activities (Scheyvens, 2003; Smith & Katzman, 2020). In contributing to instigating and sustaining peace in Bougainville after the civil war (Chap. 8), women played a vital role (George, 2016; Spriggs, 2010). That eventually contributed to their having reserved seats in the Bougainville parliament.

The absence of effective political parties has contributed to a rapid turnover of politicians in some SIDS, which has contributed to temptations to corruption, as a motivation for many candidates seeking election is to gain access to state resources, for themselves and their constituencies. That has become, in PNG and Solomon Islands, a routine and standardised allocation of resources to MPs to spend in their own constituencies (officially because they are the best arbiters of what is needed there). In 2015, Tonga joined in with a similar scheme. Possible access to such constituency funds suggested why an average of 30 candidates per constituency stood in the 2017 PNG elections. Such 'pork barrels' rarely contribute to effective development. In Jamaica and St. Lucia, numerous politicians have been accused of nepotism, corruption and lack of accountability in the use of constituency funds, contributing to their rapid turnover when promises are not delivered. In archipelagic SIDS, many politicians only rarely visit their constituencies. In the atoll states, inadequate transport infrastructure makes visiting electorates extremely difficult, evident after the 2023 Tuvalu general election when most new MPs were stranded on their home islands for several weeks before they could return to the capital and elect a Prime Minister. Politicians are often regarded as worthless by their constituents. Bougainville Island (PNG) electors regarded politicians as being adept at *toktok suga* (sweet talk) but incapable of keeping promises or delivering policies so contributing to rapid turnover. That has usually meant a lack of commitment, little learning on the job and no focus on national and international matters. Many electorates are very small, and voting patterns are known: candidates have means of identifying the loyalty or otherwise of family members, kin-groups and others in their areas of interest. Isolated places have little contact with the centre and limited means of following politics or monitoring their representatives' performances. It constitutes a transactional politics, where voters tend to hold clientelistic views about their MPs, who they have elected to parliament to procure essential goods and services rather than merely 'represent' them. But, at worst, clientelism can become cronyism, nepotism and corruption (Bishop, Bouhia, Carter, et al., 2024).

Perceptions of corruption in the SIDS are commonly directed at politicians despite many other sources of corruption, notably in business, because of their insecurity of tenure. That of politicians is often more visible, as in Nauru, where lavish gift-giving is part of political life and people expect money and favours from their politicians, who respond, even flying constituents on international spending sprees (Morris, 2023). However, some blatant corruption has brought prosecutions. In 2015, 14 members of the Vanuatu parliament were found guilty of accepting bribes in a parliamentary vote and imprisoned, albeit released early. In 2022, three Tongan MPs (including the Minister of Finance and a former Prime Minister) were found guilty of attempting to bribe voters with money and gifts of food, prior to the 2021 national elections, ejected from parliament (RNZ, 6 May 2022) and eventually jailed. In the Caribbean too, politicians are widely seen to be venal and corrupt; a legacy which can be traced to a political context in which tolerance of corruption has evolved and is as universal as power. The structure of plantation society still permeates contemporary Caribbean societies, combined with a 'Boss Politics' model that emerged in the United States, arguably more congruent with Anglo-Caribbean political systems than with the 'Westminster' model (De Coteau, 2024).

Corruption has negative consequences, where policy distortions may deprive people of access to resources, but it is almost impossible to police and secure retribution. At a national scale, corruption may mean 'deals being done with foreign powers with seriously negative long-term consequences for national and regional security' as, arguably, the case of relationships between Solomon Islands and China (Chap. 8). In Vanuatu 'numerous' former Ministers of Lands, ensnared in webs of patronage, have corruptly leased land for themselves or business partners. The Nauruan government engaged in various corrupt and improper practices to host offshore refugee processing, with outlaw motorcycle gang members (bikies) providing security services at the refugee detention centre (Kampmark, 2025; McKenzie, 2025). Local 'masters of modernity'—all men—facilitated major land subdivisions and sales in collaboration with various Ministers of Lands and private investors, responding to international desire for 'a piece of paradise' while simultaneously dispossessing their own families and communities of access to customary land (McDonnell, 2017, 2023). In elite Port Vila suburbs property was a tool through which foreigners reasserted capitalism and neo-colonial control over Indigenous landscapes, as individualised rights over land became the norm. Property also created sites for a masculine authority, enabling men to manipulate claims to land and entrench their personal power. No obvious reason exists to assume that corruption will be easily curbed.

Corruption in politics and the public sector has led to declining trust in several SIDS, in turn eroding government legitimacy and leading to challenges to state authority and everyday acts of resistance visibly enacted over social media. Social media are replete with questions on government legitimacy that point to a lack of support for politicians and the political system. SIDS have experienced a decline in voter turnout since the 1990s, with most countries now experiencing the lowest voter turnout in their political history (but much as in 'western' states). Trust may be depleted, but no group of citizens, however large, wields power over national governments.

5.5.2 Parties

In the existence or otherwise of political parties, the Caribbean and Pacific SIDS are poles apart. Both successful in terms of democracy, the former, almost without exception, have the world's most stable two-party systems, while in the latter political parties are weakly institutionalised, prone to transience or simply absent. Since both regions have similar institutions, conventional explanations that highlight social divisions or electoral systems cannot explain this difference. It may be related to colonial history and the related labour movement in the Caribbean that led to the emergence of parties associated with the labour movement, compared to the Pacific where labour movements (outside Fiji) have been non-existent (Bishop, Corbett, & Veenendaal, 2020). Political parties remain an indispensable institutional framework for representation and governance, but the democracies of many Pacific SIDS are undermined by the weakness and ineffectiveness of their local political parties (Rich,

Hambly, & Morgan, 2008), consequent frequent votes of no-confidence in governments and corruption. By contrast, the two-party systems of the Caribbean are unusually rigid but defined by confrontation and 'winner-take-all' electoral systems, as in St. Lucia (Saati, 2022).

Smaller SIDS, like Tuvalu, Kiribati, Nauru and the northern Micronesian states (except the Marshall Islands) have no formal parties. Elsewhere political parties have mostly emerged in support of a charismatic or well-resourced leader and have had little to do with clear objectives or philosophical positions. Most are fluid, populist, policy deficient, and without an ideological origin or administrative base. At the 2024 elections in Solomon Islands, contested by 13 parties, the most successful was the Ownership, Unity and Responsibility (OUR) Party, founded two years earlier, but more seats were won by independents. Such contexts limit stability of governance.

Despite more obvious party distinctions in larger SIDS such as Fiji and Jamaica, parties fit uneasily in systems where 'politics' is undertaken in different ways and where the parties that operate in much of the west with defined distinctions, have limited relevance (and decreasingly so as 'strong' leaders seemingly become more powerful). In the Pacific at least, without ideological differences members shift allegiances and parties; government requires coalitions and being in opposition is unwelcome. As a former rural Vanuatu MP, who had changed parties three times in the previous three governments, stated: 'My voters pressure me to stay in government, even if it means changing parties. Being [on the] government side means I can lobby better and deliver results to my communities' (quoted in Naupa, 2025). There, in Kiribati, and elsewhere, being on the government side also brings additional allowances. Several Pacific SIDS, like Samoa, have needed to develop rules designed to maintain political stability and ensure that MPs adhere to the party alignment under which they were elected, rather than crossing the floor. In Vanuatu, two rulings in 2024 sought to curb political instability by locking MPs into their political parties and forcing independents and single MP parties to affiliate themselves with larger political parties for the life of a parliament.

Kiribati had only two political parties in the 2024 national elections, but most MPs were independents who came from outer islands and did not belong to either party and chose sides (mainly joining the dominant party) only after the election. As one local journalist observed of the election:

> Party affiliation is, especially during election time, more like 'the big elephant in the room'. It's right in front of you, but nobody really mentions it because candidates running for parliament would want to get re-elected or elected first. It's hard to gauge, for example, for one island, which way are they aligning, to whether it's with the current ruling party or with opposition. It's a safe bet for candidates when they run, and most of them are doing that, to say that 'when I get re-elected, my party affiliation will be decided by you. I will come back again with you, once I'm elected. And then I will choose my party, which you want'. A lot of these people running .... I know the affiliations, and values and principles, what they support. I know where they stand. But these can all change when they get elected because ultimately the people decide where they want their candidate or the elected nominee to be within parliament, whether in opposition or within the government. (RNZ, 5 August 2024)

That has not discouraged attempts to build parties and coalitions after elections. In 2024, Vanuatu had 19 political parties; none could form government without a complicated coalition, making for instability. Early in 2025, five political parties signed a memorandum of agreement to form the next government of Vanuatu, with the Prime Minister being the fifth in 4 years. Similar hyper-fragmentation occurs in Solomon Islands, with similar negative impacts, but it ensured that no group dominated the political arena (Veenendaal & Corbett, 2021). In each of these archipelagic states, this is partly a function of many populated islands. Some parties gain no parliamentary representation and dissolve within a year, while others gain parliamentary representation but, having achieved that purpose, are then disbanded.

Occasional mimicking of western contexts occurs. Green parties emerged in some Pacific SIDS in the 2020s but may well be as transient as other ventures to adopt catchy titles (and with a vision of local control of their natural resources and concerns over climate change and deforestation that do not reflect the more immediate issues that concern constituents). Factions appear and disappear and few parties, where they have existed, last for long, other than in the larger Caribbean SIDS, despite extreme political fragmentation and instability contributing to unstable coalitions and clientelist politics where parties simply seek power. 'Most political parties active in Pacific Island states are thus fluid, populist, legally unrestrained, policy-poor, and administratively bare' (Hassall, 2020: 67). They are not then consistent with consensus politics because they highlight division not unity, but nor are they stable or long-lasting, especially in Melanesia. Democracy exists and persists without parties in the Pacific, but similarities exist between SIDS in the two regions, including the absence of ideology, personalisation (with personalities the basis for political contestation, voting behaviour and political attachment) and patron-client politics, which reflect geographic and economic factors (Corbett & Veenendaal, 2016). Parliaments often meet infrequently, sometimes only a few weeks in the year, so that what might be parliamentary business in large states is conducted more informally.

Just as the number of politicians is small, so too are cabinets. In the extreme case of Tuvalu, 9 of the 12 MPs make up the Cabinet. There and elsewhere cabinet members have multiple responsibilities and hold multiple portfolios. In cabinets in the Pacific—whether single-party or multi-party cabinets—the Prime Minister (or equivalent) must often accommodate the aspirations of members who are rivals as much as colleagues (Corbett, Veenendaal, & Connell, 2021). By contrast, in Kiribati in 2024, all members of the government who were members of the President's party were either in cabinet or sitting on select committees, whereas no other MPs were on such committees. Being in a committee provided an additional income to its members (RNZ, 20 January 2025). Ministers, who themselves often lack specialised expertise, can acquire 'super ministries' to lead. The policy areas under their direction can be quite varied. Tuvalu has a Minister for Justice, Communication, and Foreign Affairs. St. Lucia has a Ministry of Commerce, Manufacturing, Business Development, Cooperatives and Consumer Affairs, while much larger Jamaica has a Ministry of Science, Energy, Telecommunications and Transport. Some are unusually grand, such as Barbados' Ministry of the Environment and National Beautification. Others, such as Solomon

Islands' Ministry of Traditional Governance, Peace and Ecclesiastical Affairs, reflect local needs. In some such super-ministries, the tasks of individual divisions may be less onerous and less complex than in much larger countries, but, equally, in other cases—such as fisheries—they may be more demanding. When these are combined into a single ministry, the demands on the minister and on the ministerial staff to acquire expertise are extreme, conceivably limiting their effectiveness, perhaps particularly crucial where that entails external matters (as, in small states, often it does). Moreover, scarce skilled human resources are often depleted by migration.

5.5.3 The Practice of Politics

Despite electoral stability, political turbulence and rapid changes in government have hampered economic growth in several SIDS. Grenada was invaded in 1983 after a Marxist coup. Fiji went through a series of coups. Nauru has banned and exiled politicians for creating dissent while Tonga has gradually transitioned from a parliament largely composed of 'nobles' towards ordinary citizens. In Samoa in 2021 national tensions followed the eventually unsuccessful attempt by the elite Prime Minister to resist the national election results and retain power. The Solomon Islands Prime Minister deferred elections on dubious grounds in 2023. Through this instability all the SIDS have maintained basic democratic systems and have multi-party elections (however much some parties are nominal). Political stability has not always followed, with persistent votes or practices of no-confidence disrupting continuity in the Melanesian states. In Solomon Islands instability resulted in external intervention at the end of the twentieth century, in what were euphemistically known as the 'tensions'. A civil war ensued in PNG after the violent closure of the Panguna mine on Bougainville (Chap. 8). Still, despite various periods of violence and external interventions, SIDS have experienced less political turbulence than in other world regions, and democracies have been restored.

Some Caribbean SIDS, and Fiji, have been troubled by issues of ethnic and political identity and the exclusion of women. Ethnic politics dominates Trinidad where ethnic Indians support Indian political parties while blacks usually support parties dominated by black leaders. 'Safe seats' and strong partisan identities are common features in Jamaica, especially in low-income communities. Patronage has been pervasive in many SIDS, including St. Vincent, St. Lucia and Antigua, with politicians accused of vote-buying to encourage electoral victory and soften the electoral impact of poor governance. Beyond matters of ethnic identity, voters support candidates they think will directly help them or their community. Such clientelist and transactional political systems, which often hinder women's involvement (Baker, 2022), dominate SIDS. Politicians, however, must increasingly manoeuvre in a more dynamic international environment that requires dealing with transnational and complex problems such as organised crime (Chap. 6), making regionalism potentially valuable for state solidity (Chap. 7). Devolution of power upwards has not usually been welcomed. Nor has it been devolved downwards despite the existence of provincial and local governments.

5.6 The Public Sector

SIDS require bureaucracies to implement and advise on development strategies, including those of state enterprises, but they are constrained by small size and by budgets, despite the needs of populations notably in the archipelagic states. Bureaucracy is costly and concerns exist over public sector efficiency and effectiveness. Small bureaucracies are not specialised, cannot easily concentrate on particular issues and face challenges in accessing an adequate range and depth of technical skills (or even the technology) to fulfil the basic functions required for such crucial issues as monitoring public expenditure and maintaining financial accountability (Corbett, 2024). Training and retaining skilled staff remain perpetual challenges for all small states. Their officials lack the interchanges, access to new knowledge and stimulus provided by professional associations and meetings and may 'live in a condition of professional loneliness' (Jacobs, 1975: 141). Key people may often be abroad for substantial time periods and unavailable to manage and discuss national development. Workers may have heavy workloads and limited resources to perform the tasks (Roach, Sabharwal, Abraham, & Charles-Soverall, 2022). Performance management, effective monitoring and evaluation are limited. Moreover, the government role extends into what is the private sector in larger states, broadening the scope and dimensions of executive interests and diffusing core competences while, conversely, public servants may also be involved in the private sector. Gaining sovereign recognition also represents a disproportionate cost for the smallest SIDS since they must represent their interests in international forums despite human resource constraints.

The public sector also includes various authorities, boards and commissions, often related to the supply of services such as energy, water and transport. In larger states such service provision is provided by the private sector (and often competitively), but this is less often true of smaller states, where profitability is unlikely, so that state-owned enterprises (SOEs) are required to provide services. SOEs have generally performed poorly in SIDS. Several governments established and maintained airlines, to enhance national prestige rather than meet market demand. The Government of Samoa established the airline Polynesian Blue as a joint venture with Virgin Blue after the operating costs of its own business, Polynesian Airlines, grew to more than half the government's annual budget (Hassall, 2020). Other airlines, including Air Vanuatu, Royal Tongan Airlines and LIAT, have become bankrupt (and required government bailouts) but that has not discouraged governments continuing to hopefully establish national airlines in search of prestige and soft power (Guthrie, 2013). A combination of low and irregular passenger numbers and freight volumes, high maintenance costs, few appropriate landing rights and metropolitan competition were too much for small airlines.

Relatively few domestic skills and knowledge bases can limit effective policy formation and implementation and require dependence on external advice (that can be inappropriate and/or based on non-national objectives). Shortage of skills, and limited means of training in specialist skills, where the numbers required are small, means that some governance activities must be outsourced at some cost to the SIDS. Solomon Islands has had an agreement for over 20 years with Airservices Australia for the safe management of its upper airspace to deliver 'state-of-the-art' air traffic control services to airlines using it: a necessary regulatory function that ensures that planes continue to use the airspace and so contribute financially to Solomon Islands. Air traffic control in upper airspace requires specialised skills. Officially, the agreement provides for sharing of knowledge between the two countries and provision of national training opportunities (Gina, 2024), but there is little indication that national skills have been created. If not outsourced, then some activities must be forgone. Vanuatu, whose airspace was managed by Fiji, had no official information on whether it was receiving any income from this (*Fiji Sun*, 18 November 2025). Most SIDS have few lawyers—or others with high-level legal skills - hence SIDS can be disadvantaged in international negotiations (or in formulating appropriate legal regimes). During negotiations of PACER Plus—a trade agreement between the PICs and Australia and New Zealand—the PICs were unable to 'assert their position' within a disadvantageous 'asymmetrical power' relationship, despite multiple training modules produced specifically to manage this problem (Llivina, 2018: 49; Crump, 2024). Few skilled workers in some areas have resulted in bureaucracies incorporating expatriates, sometimes with humorous consequences (Ells, 2000), but emphasising differences in resolving national and expatriate perspectives on structures, priorities and processes.

An increasing need for regional and global interaction places strains on small governments and key public servants may not always be in place, posing problems for integrated management. As Epeli Hau'ofa once commented bleakly of the Central Planning Department in Tonga:

> We have six excellent people in the Department but one is reviewing a project on an outer island, one is sick and another on leave, the fourth is at an overseas capacity building workshop, a fifth is in a lengthy in-country review meeting with the World Bank, and the last is too overwhelmed with administrivia and cannot coordinate policy alone. (Pers. Comm. 1998)

Much like politicians, the allure of distant conferences (and per diems) may frequently take them away from the office. An acerbic and established Marshall Islands journalist observed: 'In my experience it is a rare government worker who can avoid the offers of frequent international travel to stay at home and deliver work on the ground where it counts' (Johnson, 2015: 6).

Each international agreement entered into by a state, whether with a global or regional agency or inter-governmental body, requires management and some degree of commitment (that may be to meetings or conferences, or to implementing resolutions and suggestions from distant bodies, such as the UN and the World Bank), putting strain on local and regional relationships, so much so that in the Pacific a CROP (Council of Regional Organisations of the Pacific) was needed to coordinate such agencies. Globalisation has put massive demands on local bureaucracies, reflecting the complexity of governance arrangements and the challenges in responding to them adequately while simultaneously ensuring that they do not disrupt or take too much time away from domestic matters (Corbett & Connell, 2015). Both Samoa and Kiribati took the unusual step in 2024 of requesting that official visitors from international agencies avoid visiting these countries at particular periods of the year since they had inadequate absorptive capacity to cope with them and exercise 'normal' functions at the same time (McNeill & Koro, 2024). Inevitably, global and regional policy networks, as they should, influence domestic policy, but they may also hinder national development strategy implementation.

5.7 Social Inclusion

Although SIDS have long-established policies on social inclusion, practices invariably fall short of plan objectives and significant gender divides exist, let alone policies and practices that better integrate youth and minority groups. PICs have the lowest level of female political representation in the world; there are significantly fewer women in the workforce than men, and about two in every three women and girls experience domestic violence. In relatively developed Fiji, women comprise just 34% of the labour force, where they are concentrated in employment with extremely low average incomes, and women also do 73% of unpaid household work while men have more time for leisure. Unpaid care work limits time available for income-earning activities (Narsey, 2023, ADB,

2023). Social development generally is not attuned to economic development so that capacity development efforts associated with women's economic empowerment programmes are often premised on the problematic idea of the economy being separate from society and social relations (Spann & Maetala, 2024). Yet women are more likely to be obliged to support cultural activities, and entrepreneurial ventures occur in cultural contexts that both enable and impede the empowerment of women (Persson, Zampoukos, & Ljunggren, 2022). Relative exclusion of women from the public sector, and to a lesser extent the private sector, stems from the presence of patriarchal cultures.

Since governments are the largest employer in the SIDS, diversity within the public sector workforce is an indicator of government sensitivity to the issue and an indicator of its willingness to take the lead. Movement towards greater inclusion of women in senior government positions is largely unrealised. Women face higher levels of discrimination compared to men and are significantly underrepresented in leadership positions. Even within the public service women are likely to be concentrated in education and health. Gender stereotypes prevail. Rhetoric far exceeds reality, and women's inclusion remains an issue in both regions. Younger people, too, remain notably absent from political decision-making positions. Men are over-represented in governance (and in prison populations, emphatically so in the Caribbean).

Women are marginalised, if much less so, in educational systems that might have contributed to reducing inequities. In almost every educational context, other than nursing schools, there is a gender bias in favour of males, which increases at higher levels of education in the Pacific SIDS, although the reverse is true in Caribbean SIDS. Still, few women are engaged in Science, Technology, Engineering and Mathematics (STEM) courses. In difficult times, such as during COVID-19 in the Pacific, girls were withdrawn from education rather than boys. Moreover, a consequence of there being more male than female students is that there are usually relatively few female teachers, and this inequity increases in such categories as college lecturers and head teachers, reiterating the problem. Men simply acquire more education and, without inclusion policies, are more employable.

Women are also disadvantaged in the private sector, though they may succeed better there, especially in small businesses, including retailing, marketing and petty commodity production (previously assisted through microfinance), and are a dominant presence in the tourist industry. Women entrepreneurs are common in the Pacific SIDS, especially in family businesses, but are typically concentrated in industries that experience intense competition and generate lower returns such as agriculture, retailing, tourism and handicrafts, and face greater barriers in seeking access to capital (ADB 2023). The presence of women in business has often been supported through training in areas such as bookkeeping and budgeting, business management, marketing skills and obtaining start-up capital (Steven, Banks, & Scheyvens, 2019). However, that can easily ignore diversity and be centred on a developmental logic of 'Cinderella projects', where women in SIDS are targeted for interventions intended to expand their unrecognised economic and entrepreneurial potential: a version of a 'development feminism' that constructs women as both impoverished victims and nascent market-oriented actors (Alexeyeff, 2020). Age, socio-economic status and hereditary rank in some SIDS challenge pigeon-holing of women in narrow contexts. However, Pacific and Caribbean women are as hardworking, creative, resourceful and resilient as anywhere else, evident in their leadership in response to hazards. Women, literally and metaphorically, are more grounded and more able to turn to and benefit from local knowledge when crises come, yet they can too easily be portrayed as vulnerable victims, with limited opportunities for empowerment. When women's leadership is effectively resourced and recognised, it can transform the way societies and communities view women over the longer term, driving progress for gender equality and more inclusive humanitarian responses. However, unequal access to resources and over-representation in informal and precarious employment continues to hamper resilience. Women may 'fall through the net' if social protection is unduly yoked to the public sphere of the state and the formal commodity economy (Jolly, Lee, Lepani, et al., 2015).

In the Pacific especially, gender equality is a key policy concern for international donors and civil society organisations, if less so for governments, but decades of investment in gender equality programmes have seen only modest gains. Women remain underrepresented in leadership and violence against women, especially domestic violence, is seemingly endemic, embedded in the making and remaking of the colonial state, part of a history of labour exploitation, maintained by racial, sexual and gendered stratification that shapes lives and is a key relation of power and difference (DeShong, 2024; Underhill-Sem & Guttenbeil-Likiliki, 2025). In Vanuatu, women who have their own sources of income are about 150% more likely to experience physical and sexual violence than those who do not (ADB 2023), indicating real constraints to equity and effective participation. Beyond patriarchy and education, a further obstacle to greater gender equity is that gender issues are seen by some as a foreign agenda, imposed by donors and distant governments, and so given lip-service in target cultures, rather than perceived as a means of drawing on values such as harmony and sharing resources. Men are often overlooked in efforts to promote women's rights and women's empowerment, with many programmes focused exclusively on women, but powerful men's support for gender equality is critical. As a male development practitioner from Vanuatu reflected: 'We don't need to build capacity, we need to give women space. Men are acting as gatekeepers. We need to open up pathways for women to allow them to reach their potential as leaders' (quoted in Masta & Jackson, 2024). Views about gender and women's leadership have evolved with positive impacts on key livelihood issues. However, forms of discrimination that privilege males exist in all SIDS, however incidentally or subtly. In gender terms, access to and achievement of the SDGs fail on a daily basis: an ongoing human rights problem.

5.8 Youth

Less attention has been attached to youth issues in SIDS despite the youth bulge that has a clear relationship to unemployment and anomie. In the wider Caribbean, and in the Pacific, about a quarter of young people are unemployed, with unemployment among young women more than 30%, compared to 20% for young men. Lack of jobs and lack of qualifications are immediate causes. Youth experience frustration with joblessness, allied to an unwillingness to engage in agriculture that has contributed to rural-urban migration. That may merely result in groups of youth aimlessly walking the streets, but it may also lead to gangs and intermittent crime and violence, where youth lack basic human rights. In the Pacific this tends to be seen as a male problem, though concerns exist over girls drifting into trouble and early parenthood. In some urban centres, notably Suva and Port-au-Prince, joblessness and dropping out of school are marked by increased numbers of 'street children' disengaged from extended households, associated with drug issues and petty crime, who engage in harmful and risky behaviour and are caught in a cycle of abuse and poverty. Such youth, from diverse backgrounds, can too easily be seen as ignorant and lazy, further impacting their sense of self-worth and belonging, occasionally contributing to high youth suicide rates. Interventions to help youth are, however, hampered by lack of understanding of the reasons for problems and by authoritarian structures. Youth are not the direct cause of conflict, but the combination of the lack of socio-economic opportunities for youth, insufficient inclusive engagement, disengagement from school, an educational focus on administrative work, stagnating economies resulting in hardship, conflict between traditional and modern cultures and lack of youth voice in social dialogues can lead to negative or destructive behaviour (Dawkins-Cabrazos, 2024; Noble, Pereira, & Saune, 2011). In Port Vila (Vanuatu), moral panic brought excessive police violence towards youth after an urban riot, on the assumption they were guilty parties, marginalising youth further, turning some to sorcery and furthering the negative image of urban settlements (Mitchell, 2011).

In most SIDS, there are more youth, loosely those aged between 15 and 25, than in any other age group. Yet, in SIDS that can border on social and political gerontocracies, there is limited regard for their needs and desires, particularly from their own perspectives, although some will be the leaders of the future. The youth bulge is more likely to be seen as a security threat rather than as potential for developing their communities. Too often, as in Fiji and Solomon Islands, the 'structural minimisation' of youth creates a 'rational ignorance' where youth are neither trained to be leaders nor choose to engage as active citizens because they have been ignored in the past (Craney, 2022). However, as in Tonga, where youth were previously strongly criticised for their role in burning down several city centre buildings in frustration over unemployment, in a more supportive context they demonstrated their ability to make positive community contributions, especially following the 2022 volcanic eruption (de Waegh, House, Duarte, et al., 2023). Lack of open space in most urban areas denies the possibility of participating in and through sport. Yet it is through sport, whether rugby in the Pacific or athletics and cricket in the Caribbean, that SIDS teams and youthful individuals have staked a place in the world and stimulated national sport diplomacy (Chap. 8). With respect, youth have instigated their own inclusion, often contributing through more sophisticated technical skills and use of social media.

Respect and inclusivity must be combined with employment and development opportunities, integrating youth into all facets of society and economy, broadly a 'holistic livelihoods' approach to development (Craney, 2022) that considers the physical needs of youth while also providing opportunities to envisage and bring change in their own communities. Support for entities such as National Youth Councils and youth-focused NGOs can create sustainable partnerships for long-term social development. In the Caribbean SIDS, CARICOM established a Caribbean Regional Youth Council in 2013, to advocate for and advance the development agenda for youth, and support national councils, as a catalyst towards democracy, social transformation and good governance. Successful interventions typically incorporate elements of participation, self-help and mutual support, and abandon stigmatisation. Providing opportunities for future leaders to develop and use individual and collective capacities is vital.

5.9 Disability

While significant, the extent of disability in the SIDS is largely unknown, since inadequate data prevent assessments of the numbers and categories involved, and the quality of life and socio-economic participation of persons with disabilities, that are crucial to designing effective policies and programmes. Many disabled, particularly those with intellectual disabilities, experience blatant discrimination and stigma. Positive policies to make societies more inclusive and equal and to ensure the full participation of people living with disabilities, have been rare. Few governments have had the expertise, enthusiasm or will to develop effective provisions for disabled groups. Definitions vary over what constitutes a disability, but in the Pacific it has been estimated that as many as 17% of the population live with some form of disability, representing at least 1.7 million people (including PNG and dependent territories), whereas in Caribbean SIDS rates vary from just 2.5–6% (Stephenson, 2023), suggesting that meaningful rates are probably somewhere between the two regional estimates and that better data are needed.

Disability, gender inequality, data deficits and discrimination are closely interlinked, with women experiencing disability differently and being more likely than men to have a disability in their lifetime, due largely to their longer life expectancy. Mobility and sight are key issues, particularly in smaller islands where assistive devices and support are scarce. Women with disabilities are subject to various forms of abuse, stigma, teasing, bullying and harassment, including sexual abuse, and are twice as likely to live in hardship compared to men with disability (Tavola, 2023). That reflected the situation in many SIDS, where the disabled may experience shame within their own societies, alongside physical barriers that reduce mobility, lack of public transport provision, and knowledge and support within the able-bodied community, to enable them to attend school, access health care and obtain jobs. They are much more likely to be affected by hazards. People with disabilities are among the more marginalised groups, overrepresented in poverty, and underrepresented in social, economic and public life.

Relatively few SIDS, just four in the Caribbean (Stephenson, 2023), have enacted anti-discrimination legislation, which prohibits discrimination based on disability in employment, education, and the provision of goods and services. Some distance separates legislation and practice. Medical 'models' of disability, which focus on 'problems', predominate over social and economic models that are linked to inclusion. Low rates of employment predominate in both the Caribbean and Pacific SIDS, relative to those of persons without reported disabilities, partly the outcome of stereotypes and discrimination by organisations and human resource workers, but made more likely where unemployment exists. Difficulties exist in accommodating those with disabilities in some sectors, such as coastal fisheries, but efforts to do so have been non-existent (House, Delise, Malverus, et al., 2024). People with disabilities can be routinely ignored in all phases of disaster management due to stigma, discrimination, hidden disabilities and limited understanding of their needs, as in Solomon Islands (King, Edwards, Watling, & Hair, 2019). Quite simply, throughout the SIDS, few organisations, including governments, have designed their buildings and open spaces with accessibility in mind (Jones & Serieux-Lubin, 2018), and transport is rarely accommodating. In Trinidad, Jamaica and the Bahamas (and no doubt elsewhere), employers were reluctant to make the physical changes needed to accommodate people with disabilities, and legislation that would have enabled this to happen was not enforced (Stephenson, 2023). Too often, as in Tuvalu, children with disabilities are poorly served by the education sector. Children attend school, but most do not receive any specialised attention and are left behind, leaving with inadequate education. Even the best legislation needs to be implemented.

A necessary systematic and multi-pronged approach to addressing the difficulties experienced by people with disabilities, from legal, social, political and economic perspectives, is unlikely to be imminent. Much of the impetus for change and inclusivity has come from external donors. Many mainstream development donors have disability inclusion policies or strategies, while in the Pacific, the regional Pacific Disability Forum, established in 2002, has gained substantial external support through, for example, the Australian government's International Disability Equity and Rights Strategy (IDEARS). It remains central to Forum objectives, if not to effective national policies, that there are six prerequisites for Pacific SIDS: accessibility, assistive devices, support services, non-discrimination, social protection payments and community-based inclusive development, to ensure access and connection to essential services particularly given the geographical spread of most PICs. Some have been at least partially achieved, but equity is a distant dream.

5.10 LGBTQI+

If the disabled experience stigma and marginalisation, and legislation has done little to substantially improve their circumstances, the situation of lesbian, gay, bisexual and transgender (LGBTQI+) people in the SIDS is substantially worse. Despite a long history of non-binary gender and its affirmation, especially in Polynesian SIDS, legislation continues to evolve to oppress and exclude rather than recognise LGBTQI+ rights and inclusion. Prior to colonisation, parts of the Pacific had a fluidity of gender and sexuality that was largely replaced by strict binary and patriarchal European norms as colonial expansionism brought with it a stigmatisation and criminalisation of diverse gender identities, expressions and sexualities, and that remains in place. Changes in societal perceptions of gender, the labelling of varying gender identities and greater campaigning for transgender rights have largely failed to occur in the SIDS. Legislation on equality and non-binary gender recognition is absent.

Throughout the SIDS, LGBTQI+ people experience high levels of community and systemic violence, familial ostracism and societal discrimination (Idris, 2021), and thus have low levels of trust in institutions. This results in their not being able to reach their full potential with reduced access to health care, education and stable employment preventing their being integrated and respected members of their societies: a vulnerability to disasters, for example, that is shaped by social marginalisation, discrimination and stigma, and exacerbated by dominant value systems. Disaster management policies have traditionally been written by and for predominant groups, excluding the needs of minorities, while the SDGs ignore gender and sexual diversity. The language of SDG5 ('Gender equality') is exclusionary and reinforces a male/female binary perspective on gender that excludes the lived experiences of transgender and sexual minorities and acts as a disincentive for states to devise inclusive policies and practices (Haworth, McKinnon, & Eriksen, 2022).

Homosexuality is not illegal in the SIDS but same-sex activities are criminalised, policed and regulated informally in most SIDS, and although laws are rarely enforced, they create an environment that fosters discrimination and violence towards LGBTQI+ people. Those in

Caribbean SIDS fare worst, but violence is common elsewhere and seemingly everywhere (Idris, 2021; Ravulo, Hollier, Waqa, et al., 2024). Gendered and sexual minority issues cannot be publicly discussed in countries like Dominica so that individuals must conceal their identities to maintain community acceptance (Hill, 2025). Only Fiji, Nauru, Vanuatu, FSM, Palau and the Marshall Islands have decriminalised same-sex sexual conduct. In Kiribati, homosexuality is criminalised, discrimination and violence against gay men occur, and there is a lack of sensitisation to diverse sexual orientation, gender identity and expression of this. It retains legislation that provides for a 15-year prison sentence (like that in other SIDS) for same-sex acts if caught and convicted (although there is no evidence of its being used). However, that has not prevented the active existence of BIMBA, an umbrella organisation for the Kiribati LGBTQI+ community. Late in 2024, Vanuatu moved to ban LGBTQI+ advocacy by shutting down V-Pride, which was said to contradict the Vanuatu constitution's preambles, Christian principles and Melanesian values. According to the Vanuatu Christian Council Vice-Chairman, Pastor Thomson Aki, the V-Pride movement in the country posed a threat to its Christian values: 'The Church does not support this lifestyle, as mentioned in the Bible when God destroyed Sodom and Gomorrah for their wickedness' (quoted in RNZ, 5 November 2024). Soon afterwards Vanuatu amended its Marriage Act to ban same-sex marriage. Indeed, same-sex marriage is generally illegal throughout the SIDS, where conservative churches have enormous authority.

Resilience has proved difficult in an oppositional social context and is weak and fragmented at national and regional levels (Ravulo et al., 2024) despite the existence of regional and international bodies that include the Pacific Sexual and Gender Diversity Network (PSGDN) and the International Lesbian, Gay, Bisexual and Transgender Association (ILGA). Civil society movements are still at the nascent stage, encountering a legal tradition inherited from their former colonisers, funding limitations and widespread prejudice. Unlike international aid agencies' advocacy for disabled people, advocacy for LGBTQI+ people has been much less evident and, in this arena particularly, the notion of a human development that leaves no one behind is absent.

5.11 Civil Society

Although governments are aware of social inequalities, a crucial issue has been their ability to respond in a systematic manner. Civil society in Pacific SIDS, while not yet as cohesive as it might be, has at least provided a voice for the poor, the marginalised, and the oppressed. The Pacific Islands Association of Non-Government Organizations (PIANGO) provides an umbrella organisation for 24 'national-liaison units', to undertake research and policy advocacy on social justice issues, notably good governance and development effectiveness. In societies where 'mass media' are weak, there is not so much 'public opinion' as numerous communities with greater or lesser consciousness of their circumstances, depending on their levels of education, contact with other communities, and the quantity and quality of interactions with public servants and politicians. Without the transparency in government that a strong and independent media can help foster, the potential for patronage and corruption increases. The limited ability of government to provide welfare services in some Pacific countries, and the more extensive reach of faith-based organisations, has meant that a significant proportion of social services, including health care, is delivered by NGOs (Hassall, 2020). Human rights NGOs are also very active in Caribbean SIDS, especially around issues of policing, the death penalty (Campbell, 2025), and sexual and reproductive rights (Lazarus, 2020). Many have contributed to shaping the outcomes of various laws, and in some SIDS, including Barbados and Jamaica, are accommodated through joint select committees of parliament where they can present their views and contribute to legislative debates.

More than in any other world region, churches play a significant role in civil (and, of course, religious) society throughout the Pacific SIDS through language and actions, as in Samoa, whose national motto is 'Samoa is founded on God' (*Fa'avae I le Atua Samoa*), and Vanuatu ('*Long God yumi stanap*'), are reflected in active church attendance, reverence for the clergy and financial contributions to church institutions. Both regions are identified with Christianity. Faith-based broadcasters comprise half of all community radio stations in the Pacific, reflecting the deep indigenisation of Christianity and its central role in Pacific cultural identity. Church leaders are key to discussing governance and society's multifaceted issues. PICs have not separated Christianity from political, economic and daily society, and as in other SIDS religion remains essential to understanding development processes (Alofaituli, 2023). Local tensions and divisions, however, are represented in the foundation of new churches, especially in Melanesia. Indeed, 'Christianity and politics, have redefined each other in ways that make the two categories inseparable' (Tomlinson & McDougall, 2012: 2). In most SIDS, eminently so in the Pacific, churches are more cohesive and more powerful than communities.

Churches play a central role in the provision of social services in most SIDS. Even though they do not necessarily distinguish between activities focused on spiritual development and those focused on community development, churches are active in working to reduce poverty, as in Vanuatu and Samoa and in humanitarian assistance (Byron & Montoute, 2023; Cassells, 2019; Clarke, 2013; Thornton, Sakai, & Hassall, 2012). Churches and faith-based organisations play a role within civil society combining advocacy with moral authority, so providing a normative framework for governance, operating between the state and the market. Churches are also key to sustaining or changing underlying social norms around crucial issues such as gender, but most churches have opposed same-sex relationships and abortion rights and several have complicated interpretations of, and responses to, climate change. Several international and national NGOs have partnered with churches to assist them in implementing development projects, but partnership with faith-based organisations around development agendas brought concerns that aid would be used for evangelism (Austin, 2014). Churches are significant providers of health and education services, have contributed to thinking about social policy options, and have played some part in conflict resolution and peace-building when the state showed signs of fragility, especially in Melanesia (Hauck, 2010). However, different denominations may be built on and intensify social divisions, despite practical directions being quite similar. Away from urban centres, they have often had more influence, and certainly more presence, than state institutions and officials. At a local scale, it is less evident that faith-based organisations are more effective at securing development (Lyons, Walters, & Riddell, 2015) and are often as centred in modernist asymmetries of power as other forms of aid delivery (Chap. 8).

5.12 Media

Traditional print media are constrained by financial circumstances with few journalists to cover ongoing situations. Only larger SIDS have newspapers, and national broadcasters are likewise disadvantaged (to some extent the outcome of scarce 'news' with investigative journalism largely absent), yet independent media are necessary to provide reportage and monitor policies and practices. Journalists face significant economic and political pressures, bribes and corruption (and self-censorship) so that state-funded media may ignore stories regarded as too sensitive for governments and that might result in pressures on their existence, through withdrawal of advertising or subsidies (Pacific Freedom Forum, 2024). The growth of an independent media conflicts with governance systems that have sought to control the flow of information for their survival. From the 1980s journalists have been physically assaulted, arrested and jailed, and threatened with deportation. Governments have not welcomed critical coverage by news media that have pointed to irregularities of decision-making and corrupt behaviour in leadership, yet circumstances have generally improved (despite Samoa's banning the country's only newspaper from government access in 2025), perhaps through competition with emerging social media.

As media freedom has tentatively improved in the Pacific region, the reliance on corporate funding has become greater, just when it is invaluable to have an 'insider' perspective on development in the SIDS and the wider regions. As Kalafi Moala, the President of the Pacific Islands News Association, has said:

> Outsiders over the ages have controlled the information content and flow because they owned and designed the information systems—of education, of media, of religion, and in many cases of government. The way things are set up is that information gathering and communication still has a strong

> colonial overhang. Instead of appropriate information being determined by storytellers among their own people, it is being determined by those who tell those stories often in a language foreign to the local culture, in a context borrowed from another culture, and with issues and priorities also borrowed from outside. News, as such, is being written and broadcast primarily for people outside the region. But that's how media money is made! The primary markets are those on the outside and the locals are secondary. (2024)

Media have only intermittently been able to establish a neutral and critical presence in the SIDS. Political decisions may be shrouded in secrecy, denying basic freedom of information and making it difficult to make government processes transparent and accountable to citizens. That may also reflect cultural distinctions. The President of the Journalists Association of Western Samoa (JAWS), Lagi Kerosoma, has pointed out that journalism can be more difficult where particular cultural values exist. Thus, in Samoa, 'the belief that freedom of expression is not part of the Samoan culture makes it harder for journalists to carry out their work' (RNZ, 3 May 2022). Some observers have detected a 'shift to autocracy in several countries across the [Pacific] region. This manifests in several ways, including by imposing restrictions on media, threatening to close down social media, constraining civil society and attacking the independence of the judiciary' (Cain, 2022). The Solomon Islands Broadcasting Corporation, the national broadcaster, was ordered in 2012 (at a time of debate over the signing of a controversial security deal with China), not to broadcast any material critical of the government. Pacific media are characterised by resource constraints, small audiences, limited advertising markets, dependence on the state for revenue, lack of capacity and lack of research. Typically, journalists are young, inexperienced and trained 'on the job'. Only in Fiji, of the Pacific SIDS, is there more than one newspaper. However, in most SIDS, television and radio stations—usually closely linked to governments—are increasingly more important, along with social media. In a few SIDS, such as Tuvalu, the costs of broadcasting limit coverage to the main island only. Information laws in some SIDS, such as Jamaica, provide legislative backing necessary for media reporting and enable greater potential for monitoring government transparency and accountability, invaluable in a deeply divided society.

As legacy media have stagnated, social media have expanded and are of rapidly growing importance in enabling new platforms for policy debate and for individuals and groups to challenge the official narratives that are more likely to dominate newspapers, television and radio. With a growing youth population, high rates of smartphone ownership and weak literacy, islanders have turned to social media to discuss and share political ideas. Digital platforms shape social and political discourse, enabling youth participation and new forms of political assembly and activism, creating distinctive channels for civil society engagement, bypassing conventional print and broadcast media. To some extent that has been propelled by a pervasive distrust of governments and, by association, traditional media, creating something of an ideological divide (Singh, 2017). Caribbean media have increasingly faced pressures and criticism from a politically polarised public, with an active online presence, about their style of reporting and accusations of bias in favour of one political party or another. Some politicians speak directly to voters on Facebook, X and Instagram. Facebook groups have been influential in linking the diaspora with homelands (e.g. Chambers & Chambers, 2018; Howard, 2019). In Solomon Islands, at least, they have created friction with government sources and led to tensions over the validity of sources and the relationships between gossip and news (Hobbis, 2019). Some governments have called for a ban on Facebook. Social media have enabled framings of the struggle for independence in West Papua and of the politics of climate change quite different from those in legacy media that point to different versions of climate justice (Burch, 2020). However, social media are also vehicles for disinformation, with China, Indonesia and Russia all involved in the Pacific with particular geopolitical messages (Ride, Ballard, & Smith, 2025). Although unresolved issues include the use and impact of disinformation campaigns, the role of digital literacy in sustaining democratic norms and the erosion of trust in democratic institutions due to manipulated digital narratives, social media have unrealised potential to contribute to fostering more resilient, inclusive and democratic societies. Integration of AI may speed the expansion of new media and accelerate creativity. While AI can transform government services, increase efficiency and enhance public interaction, it is yet to be much used, but it will set legal and operational hurdles for governments such as controlling the knowledge base to prevent misinformation, addressing procurement procedures and meeting the need for better data. Nonetheless, the legacy media will continue to play a valuable role in governance.

5.13 The Decline of the State?

The SIDS have continued to urbanise, now more quickly in the Pacific, to the exclusion of outer islands, yet urban management policies and practices are too often inadequate for modernising smart cities. Urban planning largely remains in infancy and has faced limited funding, scarce skilled human resources, inadequate data and 'competition' from national governments located in the capital cities. It has consequently tended to serve narrowing economic and sectoral interests and contributed to urban bias. A weak culture of governance is manifest in the widespread absence of even urban land use plans, or lack of effective housing policies, and disengagement from the population. Surprisingly, SDG11 is the only one that mentions housing: 'Make cities and human settlements inclusive, safe, resilient and sustainable' and by 2030 'ensure access for all to adequate, safe, and affordable housing and basic services and upgrade slums'. Few plans exist over how to attain such lofty goals. Finance and effective political will are constraints, and effective inclusive policies absent, despite improved social protection.

Beyond economic and social problems, and consequent environmental problems (Chap. 6), climate change is affecting urban areas, posing health issues through hazards, heat islands and waste. Mobile people and increasing needs (and wants) test governance structures. In an ideal world,

> If managed well, cities can be engines of growth and provide workers and residents with quality job opportunities, better healthcare, improved housing, safety and social development. Cities can help contribute to national growth through increased revenue generation, political stability and post-conflict reconciliation. Opportunities do exist to harness the transforma-

> tive potential of cities to promote development, implement effective disaster response systems and reduce violence. (De Boer, 2016: 114)

That is undeniable but few sustained examples exist where that has been achieved and the towns and cities of the SIDS are far from ideal. Much hangs on the phrase 'If managed well'. That involves more effective social inclusion, beyond even the obvious imperatives of workforce development, cross-departmental integration and citizen trust. In places achieving holistic development even confronts contexts where 'committee members on policy creation bodies may lack engagement, expertise and sometimes interest' (Archer & Gibson, 2023:198). Commitment can be scarce. Governments and governance are both victims and exemplars of urban bias.

Governance is elitist rather than inclusive. A culture has emerged in several SIDS, where there is routine disregard for the wellbeing of ordinary people, out of sight and mind, and a failure to recognise the needs of the poor, realise that poverty exists or that inequality has wider costs in achieving development and that this is shaped and might be remedied through politics. Yet there is occasionally a wilful blindness like that which denied the existence of poverty in Samoa and elsewhere. In the early 2020s the first gated communities in the Pacific were established in Fiji, a new form of segregation, if following precedents in parts of the Caribbean (where crime rates provided some justification). Weak gendered governance has left a place for civil society groups, where women play a key role, and for churches that should be, but are not always, ethical and egalitarian. Marginal people—notably youth and women, scarcely numerically insignificant—are too easily excluded, left outside or 'left-behind'.

A distinct trait of public policy-making and administration is manifest in limited resources and constraints on specialisation, smaller numbers of actors and more informal working procedures, which can occasionally contribute to holistic successes but are more likely to place burdens on fewer staff with sometimes scarce skills (Corbett et al., 2021). Governments are imperfect, not always well-informed and not averse to corruption. Politicians may be in place for too short a time to make a positive impact or support bureaucrats to develop national policies, as their concerns are local. What was written on Kiribati retains general relevance: 'It has been difficult to get broader planning and urban management on the national agenda, as most of the population is preoccupied with the day-to-day survival needs of an over-populated atoll, as well as trying to supplement cash incomes with subsistence activities' (Jones & Lea, 2007: 483): a sharp disjuncture with the significant gap between the large amount of attention given to climate change on the international scene and the everyday concerns of vulnerable communities. Giff Johnson has not been alone in pointing to government 'international grandstanding to excuse a lack of action on domestic issues' (2015). Local governments rarely have the finance or capacity to manage growing cities where overurbanisation is manifest in inadequate housing, employment and environmental management. In the following chapter we examine what has however been possible in a range of contexts of service delivery, despite the shrinking of effective governance.

Bibliography

ADB. (2023). *Invisible Hands: Enabling Pacific Women in the Informal Economy*. ADB.

Alexeyeff, K. (2020). Cinderella of the south seas? Virtuous victims, empowerment and other fables of development feminism. *Women's Studies International Forum, 80*, 102368.

Allen, M. (2012). Land, identity and conflict on Guadalcanal, Solomon Islands. *Australian Geographer, 43*(2), 163–180.

Alleyne, P., & Barrow-Giles, C. (2021). An examination of the institutional governance mechanisms to promote political integrity in Barbados. In L. Briguglio, J. Byron, S. Moncada, & W. Veenendaal (Eds.), *Handbook of governance in small states* (pp. 64–76). Routledge.

Alofaituli, B. (2023). Religion and development in Samoa: Time to draw on the strength of local culture? *Development Policy Review, 41*(Suppl. 2), e12746.

Anderson, I., & Irava, W. (2017). The implications of aging on the health systems of the Pacific Islands: Changes and opportunities. *Health Systems and Reform, 3*(3), 191–202.

Archer, L., & Gibson, J. (2023). Policy dialogue on SDG implementation in an archipelagic nation. In W. Filho, N. Aguilar-Rivera, B. Borsari, et al. (Eds.), *SDGs in the Americas and Caribbean Region* (pp. 1301–1328). Springer Nature.

Austin, L. (2014). Faith-based community radio and development in the South Pacific Islands. *Media International Australia, 150*(1), 114–121.

Baker, K. (2022). 'Copper and solar': The gendered politics of service delivery in Solomon Islands. *Third World Quarterly, 43*(4), 864–878.

Barrientos, A. (2010). Vulnerability and social protection in small island states: The case of Grenada. *Social and Economic Studies, 59*, 3–30.

Barrow-Giles, C. (2015). Beyond the status quo, centring women in the Westminster system in the Commonwealth Caribbean: A preliminary analysis. *Commonwealth & Comparative Politics, 53*(1), 49–70.

Bishop, M., Corbett, J., & Veenendaal, W. (2020). Labor movements and party system development: Why does the Caribbean have stable two-party systems, but the Pacific does not? *World Development, 126*, 104719.

Bishop, M., Bouhia, R., Carter, G., et al. (2024). *Sustaining development in Small Islands: Climate change, geopolitical security and the permissive liberal order*. Cambridge University Press.

Bowling, B. (2010). *Policing the Caribbean: Transnational security cooperation in practice*. Oxford University Press.

Bourne, C. (2009). Economic growth, poverty and income inequality. *Journal of Eastern Caribbean Studies, 34*, 21–48.

Bruce, L., & Marshall, L. (2020). What does neighbourhood level urban resilience look like in Honiara? In D. Sanderson & L. Bruce (Eds.), *Urbanisation at Risk in the Pacific and Asia* (pp. 74–89). Routledge.

Burch, E. (2020). A sea change for climate refugees in the South Pacific. How social media – not journalism – tells their real story. *Environmental Communication, 15*(2), 250–263.

Byron, J., & Montoute, A. (2023). The role of churches in regional cooperation and governance in the Caribbean. In L. Briguglio, M. Briguglio, & C. Slatter (Eds.), *Handbook of civil society and social movements in small states* (pp. 189–203). Routledge.

Cain, T. (2022). *A drop in the ocean? Organised crime dynamics in Pacific Island countries*. Global Initiative.

Campbell, Y. (2020). *Citizenship on the margins: State power, security and precariousness in 21st-century Jamaica*. Springer Nature.

Campbell, Y. (2025). Colonial Continuities in Citizenship and the Role of Civil Society Organisations. In *Routledge Handbook of Caribbean Studies* (pp. 148–159). Routledge.

Campbell, Y., & Clarke, C. (2017). The garrison community in Kingston and its implications for violence, policing, de facto rights, and security in Jamaica. In L. Macdonald & T. Hilgers (Eds.), *Violence in Latin America and the*

Caribbean: Subnational structures, institutions, and clientelistic networks (pp. 93–111). Cambridge University Press.

Campbell, Y., & Johnson-Myers, T. A. (2023). *Gender Inequality and Women's Citizenship: Evidence from the Caribbean*. Routledge.

Carby, B. (2021). Integrating disaster risk reduction in national development planning: Experiences and challenges of Jamaica. *Environmental Hazards, 17*, 219–233.

Cassells, R. (2019). Engaging with Churches to address development-related challenges in Solomon Islands. *Sites, 16*(1), 109–121.

Chambers, A., & Chambers, K. (2018). Engaging the world: Four decades of intensifying change in Tuvalu. In J. Connell & H. Lee (Eds.), *Change and continuity in the Pacific: Revisiting the region* (pp. 17–36). Routledge.

Chandra, A., & Gaganis, P. (2016). Deconstructing vulnerability and adaptation in a coastal river basin ecosystem: A participatory analysis of flood risk in Nadi, Fiji Islands. *Climate and Development, 8*(3), 256–269.

Clarke, M. (2013). Good works and God's work: A case study of Churches and community development in Vanuatu. *Asia Pacific Viewpoint, 54*(3), 340–351.

Connell, J. (2003). Regulation of space in the contemporary postcolonial Pacific city: Port Moresby and Suva. *Asia Pacific Viewpoint, 44*, 243–258.

Connell, J. (2011). Elephants in the Pacific? Pacific urbanisation and its discontents. *Asia Pacific Viewpoint, 52*(2), 121–135.

Connell, J. (2013). *Islands at risk: Environments, economies and contemporary change*. Edward Elgar.

Connell, J. (2016). Last days in the Carteret Islands? Climate change, livelihoods and migration on coral atolls. *Asia Pacific Viewpoint, 57*(1), 3–15.

Corbett, J. (2015). "Everybody knows everybody": Practising politics in the Pacific Islands. *Democratization, 22*(1), 51–72.

Corbett, J. (2024). *Statehood à la carte in the Caribbean and the Pacific*. Oxford University Press.

Corbett, J., & Connell, J. (2015). All the world is a stage: Global governance, human resources and the 'problem' of smallness. *The Pacific Review, 28*(3), 435–459.

Corbett, J., & Veenendaal, W. (2016). Westminster in small states: Comparing the Caribbean and Pacific experience. *Contemporary Politics, 22*(4), 432–449.

Corbett, J., Veenendaal, W., & Connell, J. (2021). The core executive and small states: Is coordination the primary challenge? *Public Administration, 99*(1), 103–117.

Craney, A. (2022). *Youth in Fiji and Solomon Islands: Livelihoods, leadership and civic engagement*. ANU Press.

Crump, L. (2024). *Unity and discord in the Pacific: Negotiating the Pacific Agreement on closer economic relations plus*. Griffith University Asia Institute.

Currenti, R., Pearce, T., Salabogi, T., Vuli, L., et al. (2019). Adaptation to climate change in an interior Pacific Island village: A case study of Nawairuku, Ra, Fiji. *Human Ecology, 47*, 65–80.

Dawkins-Cabrazos, M. (2024). An overview of youth crime and violence in the Caribbean community (CARICOM). In S. Morris-Francis, C. Gibson, & L. Grant (Eds.), *Youth crime and violence in the Caribbean* (pp. 19–32). Lexington.

De Boer, J. (2016). Risk, resilience and the fragile city. In D. Sanderson, J. Kayden, & J. Leis (Eds.), *Urban disaster resilience: New dimensions from international practice in the built environment* (pp. 113–127). Routledge.

De Coteau, D. (2024). *Black fins white sharks: Unmasking the genealogy of Caribbean political corruption*. Palgrave Macmillan.

de Waegh, R., House, J., Duarte, A., et al. (2023). Social inclusion and resilience. In C. Ooi, R. de Waegh, C. Trifan, & Y. Zhang (Eds.), *Islands and resilience. Experiences from the pandemic era* (pp. 17–34). Springer.

Dedekorkut-Howes, A., Torabi, E., & Howes, M. (2020). When the tide gets high: A review of adaptive responses to sea level rise and coastal flooding. *Journal of Environmental Planning and Management, 63*(12), 2102–2143.

DeShong, H. (2024). Gendered violence and gender as a site of violence: Reflections from the Caribbean. *Frontiers: A Journal of Women Studies, 45*(2), 79–101.

Dodman, D. (2009). Globalization, tourism and local living conditions on Jamaica's north coast. *Singapore Journal of Tropical Geography, 30*(2), 204–219.

Dodd, R., Shanthosh, J., Lung, T., et al. (2021). Gender, health and ageing in Fiji: A mixed methods analysis. *International Journal for Equity in Health, 20*, 205.

Doyle, J., & Knox-Vydmanov, C. (2024). Financing social protection in the Pacific and Timor-Leste. *ANU DevPolicyBlog*, 12 August.

Duvat, V., Magnan, A., & Pouget, F. (2013). Exposure of atoll population to coastal erosion and flooding: A South Tarawa assessment, Kiribati. *Sustainability Science, 8*(3), 423–440.

Ells, P. (2000). *Where the hell is Tuvalu?* Virgin.

Fitzgerald, S., Moriarty, A., Basas, R., & Shagazatova, M. (2022). *Women's participation in higher education in Solomon Islands*. ADB Policy Brief.

Forsyth, G., & Atkinson, Q. (2024). A brief history of political instability in Vanuatu. *Anthropological Forum, 34*(3), 253–278.

Fraenkel, J., & Iati, I. (2022). Introduction: What Can We Learn from the 2021 Political Crisis About Long-run Stability and Democracy in Sāmoa?. *Journal of Pacific History, 57*(4), 393–406.

George, N. (2016). Light, heat and shadows: Women's reflections on peace-building in post-conflict Bougainville. *Peacebuilding, 4*(2), 166–179.

Gewertz, D., & Errington, F. (2021). Stingy galitarianism: Precarity and Jealousy at the Sisiak Settlement, Madang, Papua New Guinea. In N. Bainton, D. McDougall, K. Alexeyeff, & J. Cox (Eds.), *Unequal lives: Gender, race and class in the Western Pacific* (pp. 213–235). ANU Press.

Gibson, C. (2021). Theorising tourism in crisis: Writing and relating to place. *Tourist Studies, 21*(1), 84–95.

Gina. (2024, December 17). Solomons renews safe skies partnership with Australia. *In-depth Solomons.*

Government of Vanuatu. (2002). *Vanuatu poverty survey: Analysis report.* Vanuatu Statistics Office.

Graves, C., Powell, A., Stone, M., Redfern, et al. (2021). Marine water quality of a densely populated Pacific atoll (Tarawa, Kiribati): Cumulative pressures and resulting impacts on ecosystem and human health. *Marine Pollution Bulletin, 163,* 111951.

Guthrie, K. (2013). Aviation regionalism in the Pacific. *Journal of Pacific History, 48*(3), 294–308.

Harrington, R., Harvey, N., Larkins, S., & Redman-MacLaren, M. (2021). Family planning in Pacific Island Countries and Territories (PICTs): A scoping review. *PLoS One, 16*(8), e0255080.

Hassall, G. (2020). *Government and public policy in the Pacific Islands.* Emerald.

Hassall, G., & Tipu, F. (2008). Local government in the South Pacific islands. *Commonwealth Journal of Local Governance, 1,* 6–29.

Hauck, V. (2010). The role of churches in creating social capital and improving governance in Papua New Guinea: Lessons for working in fragile situations. *Public Administration and Development, 30*(1), 49–65.

Haworth, B., McKinnon, S., & Eriksen, C. (2022). Advancing disaster geographies: From marginalisation to inclusion of gender and sexual minorities. *Geography Compass, 16*(11), e12664.

Hill, C. (2025). Realists, re-imaginers and reformers: Aid in the post-America multiverse. *ANU DevPolicyBlog,* 11 June.

Hirano, S. (2008). The development of national sustainable development strategies in small island developing states. In J. Strachan & C. Vigilance (Eds.), *Sustainable development in small island developing states* (pp. 4–9). Commonwealth Secretariat.

Ho, S., Dias Baptista, M., & McEvoy, D. (2023). Community profiling to support inclusive urban community-based climate adaptation: Experiences of a survey-based approach in urban informal settlements in Honiara, Solomon Islands. *Climate and Development, 15*(4), 325–339.

Hobbis, G. (2019). New media, new Melanesia? In E. Hirsch & W. Rollason (Eds.), *The Melanesian world* (pp. 546–560). Routledge.

House, J., Delise, A., Malverus, V., et al. (2024). Towards inclusion of persons with disabilities in coastal fisheries in the Pacific: A review of current practice and approaches. *SPC Fisheries Newsletter, 174*, 28–35.

Howard, A. (2019). Diaspora no more? The role of Facebook in the development of a global Rotuman community. *Diaspora, 20*(2), 177–203.

Hukula, F. (2017). Making a home in the city: Ideas of employment and work. *Development Bulletin, 78*, 45–48.

Idris, I. (2021). *LGBT rights and inclusion in small island developing states (SIDS)*. K4D Helpdesk Report 970. Institute of Development Studies.

Jacobs, B. (1975). Administrative problems of small countries. In P. Selwyn (Ed.), *Development Policy in Small Countries* (pp. 134–143). Croom Helm.

Johnson, G. (2015). *Idyllic no More*. Create Space.

Johnson-Myers, T., & Campbell, Y. (2024). The cost of politics: How money acts as a barrier for women in politics in the Caribbean. *Canadian Journal of Latin American and Caribbean Studies, 49*(2), 220–239.

Jolly, M., Lee, H., Lepani, K., et al. (2015). *Falling through the net? Gender and Social Protection in the Pacific*. UNWomen.

Jones, P., & Lea, J. (2007). What has happened to urban reform in the island Pacific? Some lessons from Kiribati and Samoa. *Pacific Affairs, 80*(3), 473–491.

Jones, F., & Serieux-Lubin, L. (2018). *Disability, human rights and public policy in the Caribbean: A situation analysis*. ECLAC.

Jourdan, C. (2017). Haosgel: Parenté, classes sociales et transformations urbaines à Honiara. *Journal de la Société des Océanistes, 144–145*, 131–146.

Jupiter, S., Wenger, A., Klein, C., et al. (2017). Opportunities and constraints for implementing integrated land-sea management on islands. *Environmental Conservation, 44*(3), 254–266.

Kampmark, B. (2025). *Contract bikies and bribed politicians: Australia's Nauru refugee deal* (pp. 1–3). *Independent Australia*.

Keen, M., & Connell, J. (2019). Regionalism and resilience? Meeting urban challenges in Pacific Island states. *Urban Planning and Research, 37*(3), 322–337.

Keen, M., Sanderson, D., Osborne, K., Deo, R., & Pranaya, U. (2022). Area-based approaches and urban recovery in the Pacific: Lessons from Fiji, Solomon Islands and Vanuatu. *Environment and Urbanization, 34*(2), 489–507.

Kelly, S., Walsh, T., Delport, D., et al. (2023). Health and economic benefits of achieving contraceptive and maternal health targets in small island developing states in the Pacific and Caribbean. *BMJ Global Health, 8*, e010018.

Kiddle, L., McEvoy, D., Mitchell, D., et al. (2017). Unpacking the Pacific urban agenda: Resilience challenges and opportunities. *Sustainability, 9*(10), 1878.

Kiddle, L., Zari, M., Blaschke, P., et al. (2021). An Oceania urban design agenda linking ecosystem services, nature-based solutions, traditional ecological knowledge and wellbeing. *Sustainability, 13*(22), 12660.

King, J., Edwards, N., Watling, H., & Hair, S. (2019). Barriers to disability-inclusive disaster management in the Solomon Islands: Perspectives of people with disability. *International Journal of Disaster Risk Reduction, 34*, 459–466.

Koczberski, G., Curry, G., & Connell, J. (2001). Full circle or spiralling out of control? State violence and the control of urbanisation in Papua New Guinea. *Urban Studies, 38*(11), 2017–2036.

Kuster, C., Vuki, V., & Zann, L. (2005). Long-term trends in subsistence fishing patterns and coral reef fisheries yield from a remote Fijian island. *Fisheries Research, 76*(2), 221–228.

Kwansing, R. (2025). Leave no-one behind: Social protection reducing poverty in Kiribati. *ANU DevPolicyBlog*, 5 August.

Larmour, P. (2012). *Interpreting corruption. Culture and politics in the Pacific Islands*. University of Hawai'i Press.

Lazarus, L. (2020). Enacting citizenship, debating sex and sexuality: conservative Christians' participation in legal processes in Jamaica and Belize. *Commonwealth & Comparative Politics, 58*(3), 366–386.

Le Dé, L., Gaillard, J.-C., Friesen, W., & Matautia Smith, F. (2015). Remittances in the face of disasters: A case study of rural Samoa. *Environment, Development and Sustainability, 17*(3), 653–672.

Llivina, C. (2018). Small states and regional dispute settlement mechanisms: The Caribbean and Pacific experiences. In P. Butler, E. Lein, & R. Salim (Eds.), *Integration and international dispute resolution in small states* (pp. 27–59). Springer Nature.

Lowenthal, D. (1987). Social features. In C. Clarke & T. Payne (Eds.), *Politics, security and development in small states* (pp. 26–49). Allen and Unwin.
Lyons, K., Walters, P., & Riddell, E. (2015). The role of faith-based organizations in environmental governance: The case of forestry in Solomon Islands. *Journal of Environmental Policy & Planning, 18*(3), 342–360.
McDonnell, S. (2017). Urban land grabbing by political elites: Exploring the political economy of land and the challenges of regulation. In S. McDonnell et al. (Eds.), *Kastom, property and ideology: Land transformations in Melanesia* (pp. 283–304). ANU Press.
McDonnell, S. (2023). *My land, my life. Dispossession at the frontier of desire.* University of Hawai'i Press.
McEvoy, D., Mitchell, D., & Trundle, A. (2020). Land tenure and urban climate resilience in the South Pacific. *Climate and Development, 12*(1), 1–11.
McEvoy, D., Tara, A., Vahanvati, M., et al. (2024). Localized nature-based solutions for enhanced climate resilience and community wellbeing in urban informal settlements. *Climate and Development, 16*(7), 600–612.
McKenzie, N. (2025, June 19). AFP probes firm's insured luxury. *Sydney Morning Herald.*
McNeill, H., & Koro, M. (2024). Pacific states are setting the terms of diplomatic engagement. *ANU DevPolicyBlog*, 11 September.
Maebuta, H., & Maebuta, J. (2009). Generating livelihoods: A study of urban squatter settlements in Solomon Islands. *Pacific Economic Bulletin, 24*(3), 118–131.
Mansur, A., Doyle, J., & Ivaschenko, O. (2017). *Social protection and humanitarian assistance nexus for disaster response: Lessons learnt from Fiji's Tropical Cyclone Winston*. World Bank.
Masta, M., & Jackson, E. (2024, July 31). Gender equality, the Pacific way. *Interpreter.*
Mausio, A. (2021). Spectre of neoliberal gentrification in Fiji: The price Suva's poor must pay. *Asia Pacific Viewpoint, 62*(1), 2–14.
Mecartney, S., & Connell, J. (2017). Urban Melanesia: The challenges of managing land, modernity and tradition. In S. McDonnell, M. Allen, & C. Filer (Eds.), *Kastom, property and ideology: Land transformations in Melanesia* (pp. 57–84). ANU Press.
Middelbeek, L., Kolle, K., & Verrest, H. (2014). Built to last? Local climate change adaptation and governance in the Caribbean: The case of an informal urban settlement in Trinidad and Tobago. *Urban Climate, 8*, 138–154.

Mitchell, J. (2004). "Killing Time" in a postcolonial town: Young people and settlements in Port Vila, Vanuatu. In V. Lockwood (Ed.), *Globalisation and cultural change in the Pacific Islands* (pp. 338–376). Pearson.

Mitchell, J. (2011). "Operation restore public hope": Youth and the magic of modernity in Vanuatu. *Oceania, 81*(1), 36–30.

Moala, K. (2024, October 23). The Pacific INA - at 50. *ANU DevPolicyBlog.*

Moglia, M., Perez, P., & Burn, S. (2008). Water troubles in a Pacific atoll town. *Water Policy, 10*, 613–637.

Morris, J. (2023). *Asylum and extraction in the Republic of Nauru.* Cornell University Press.

Mullings, B. (2009). Neoliberalization, social reproduction and the limits to labour in Jamaica. *Singapore Journal of Tropical Geography, 30*(2), 174–188.

Munroe, M. (2023). Jamaica's transnational violence: When geography matters the most. *The Round Table, 112*(6), 613–628.

Mycoo, M. (2018). Urban sustainability in Caribbean small island developing states: A conceptual framework for urban planning using a case study of Trinidad. *International Development Planning Review, 40*(2), 143–174.

Mycoo, M. (2021). Vulnerabilities to climate change and enhancing resilience in Caribbean small island developing states: A spatial planning framework. In S. Moncada et al. (Eds.), *Small Island developing states: Vulnerability and resilience under climate change* (pp. 259–283). Springer.

Mycoo, M. (2022). Caribbean island cities: Urban issues, urbanization processes and opportunities for transformation. In J. González-Pérez, C. Irazábal, & R. Lois-González (Eds.), *The Routledge handbook of urban studies in Latin America and the Caribbean* (pp. 579–602). Routledge.

Narsey, W. (2023). *Beyond 33%. The economic empowerment of Fiji women and girls.* Fiji Women's Rights Movement.

Naupa, A. (2025). How can Vanuatu's new parliament rebuild public trust? *ANU DevPolicyBlog*, 28 January.

Nilles, E., Manaia, A., Ruaia, B., et al. (2018). Re-emergence of thiamine deficiency disease in the Pacific islands (2014-15): A case-control study. *PLoS One, 13*(6), e0198590.

Noble, C., Pereira, N., & Saune, N. (2011). *Urban youth in the Pacific: Increasing resilience and reducing risk for involvement in crime and violence.* UNDP Pacific Centre and Pacific Island Forum Secretariat.

Nolet, E. (2016). 'Are you prepared?' Representations and management of floods in Lomainikoro, Rewa (Fiji). *Disasters, 40*(4), 720–739.

Pacific Freedom Forum. (2024). *For us, by us, about us.* Suva.

Palmieri, S. (2025, July 10). Remarks reveal enduring sexist norms in Samoan politics. *East Asia Forum.*

Patagan, R. (2022, December, 13–16). Social protection in the Federated States of Micronesia: Spotlight on subsistence and food security. *Pacific Economic Monitor.*

Perch, L., & Roy, R. (2010). *Social policy in the post-crisis context of small island developing states: A synthesis.* International Policy Centre for Inclusive Growth.

Persson, K., Zampoukos, K., & Ljunggren, I. (2022). No (wo)man is an island – socio-cultural context and women's empowerment in Samoa. *Gender, Place and Culture, 29*(4), 482–501.

Petrou, K. (2020). *If everyone returned, the island would sink: Urbanisation and migration in Vanuatu.* Berghahn.

Petrou, K., & Connell, J. (2017). Rural-urban migrants, translocal communities and the myth of return migration in Vanuatu: The case of Paama. *Journal de la Société des Océanistes, 144–145*, 51–62.

Plange, N., & Alam, M. (2024). Averting the old age crisis in Fiji: Policy, providence and the crisis of livelihood. *Sage Open, 14*(2).

Prasad, B., & Kausimae, P. (2012). *Social policies in Solomon Islands and Vanuatu.* Commonwealth Secretariat and United Nations Research Institute for Social Development.

Puas, G., & Oliver, N. (2022). *Electing the first woman to the Congress of the Federated States of Micronesia.* ANU Department of Pacific Affairs InBrief 2022/1.

Ravulo, J., Hollier, J., Waqa, M., et al. (2024). A queer resilience: Reviving indigenous-Pacific perspectives and practices. *ETropic: Electronic Journal of Studies in the Tropics, 23*(2), 213–234.

Remling, E., & Veiteyaki, J. (2016). Community-based action in Fiji's Gau Island: A model for the Pacific? *International Journal of Climate Change Strategies and Management, 8*(3), 375–398.

Rey, T., Le Dé, L., Leone, F., & Gilbert, D. (2017). An integrative approach to understand vulnerability and resilience post-disaster: The 2015 Cyclone Pam in urban Vanuatu as case study. *Disaster Prevention and Management, 26*(3), 259–275.

Rich, R., Hambly, L., & Morgan, M. (Eds.). (2008). *Political parties in the Pacific Islands.* ANU Press.

Ride, A. (2022). *Riots in the Pacific: Control and change.* Canberra: Australia Pacific Security College Discussion Paper.

Ride, A., Ballard, T., & Smith, G. (2025, March 13). Foreign state-sponsored disinformation in the Pacific Islands. *Australian Outlook*.

Roach, C., Sabharwal, M., Abraham, R., & Charles-Soverall, W. (2022). Does the public service motivation model hold in the Caribbean? *Review of Public Personnel Administration, 42*(3), 514–536.

Rooney, M. (2021). "We want development": Land and water (dis)connections in Port Moresby, Urban Papua New Guinea. *The Contemporary Pacific, 33*(1), 1–30.

Russell, L. (2009). *Stayin' alive. A report on urban livelihoods in Honiara, Solomon Islands*. APHEDA.

Saati, A. (2022). The political consequences of smallness: The case of St Lucia. *Commonwealth and Comparative Politics, 60*(4), 339–356.

Scheyvens, R. (2003). Church women's groups and the empowerment of women in Solomon Islands. *Oceania, 74*(1–2), 24–43.

Schoburgh, E., & Martin, J. (2016). From developmental local government to developmental local governance. In E. Schoburgh, J. Martin, & S. Gatchair (Eds.), *Developmental local governance: A critical discourse in 'alternative development'* (pp. 227–243). Palgrave Macmillan.

Schuermann, S., & Lauer, M. (2018). Disaster recovery in the Western Pacific: Scale vulnerability and traditional exchange practices. *Natural Hazards, 84*(2), 1287–1306.

Sievert, K., Lawrence, M., Naika, A., & Baker, P. (2019). Processed foods and nutrition transition in the Pacific: Regional trends, patterns and food system drivers. *Nutrients, 11*, 1328.

Singh, S. (2017). The media and journalism challenges in Melanesia: Addressing the impacts of external and internal threats in Fiji, Papua New Guinea, the Solomon Islands and Vanuatu. *Pacific Journalism Review, 26*(1), 48–62.

Smith, G. (2024, October 15). The US isn't the only country voting on Nov 5. This small nation is also holding an election – and China is watching. *The Conversation*.

Smith, S., & Katzman, F. (2020). The collective power of women's organisations in Chuuk, FSM. *Global Public Health, 15*(8), 1144–1156.

Sofer, M. (2018). Revisiting the Fijian periphery: A 30-year perspective on Kadavu Island. In J. Connell & H. Lee (Eds.), *Change and continuity in the Pacific: Revisiting the Region* (pp. 166–186). Routledge.

So'o, A., & Jackson, L. (2025). Fiame Naomi Mata'afa's tumultuous tenure has ended – what's next? *ANU DevPolicyBlog*, 30 May.

Spann, M., & Maetala, R. (2024). Economic empowerment programs for women and culturally embedded entrepreneurism: A viewpoint from Solomon Islands. *Development in Practice, 34*(4), 527–531.

Spriggs, R. (2010). Bougainville women's role in conflict resolution in the Bougainville Peace Process. In S. Dinnen (Ed.), *A kind of mending: Restorative justice in the Pacific Islands* (pp. 195–253). ANU Press.

Stephenson, J. (2023). Examining disability and work in small island developing states. In S. Nachmias & V. Caven (Eds.), *Inequality and organizational practice* (pp. 235–264). Palgrave Macmillan.

Steven, H., Banks, G., & Scheyvens, R. (2019). Women leadership in business based on customary land: The concept of Wanbel. *Development Bulletin, 81*, 58–61.

Storey, D., & Connell, J. (2013). Urban challenges. In M. Rapaport (Ed.), *The Pacific Islands: Environment and society* (pp. 310–322). Bess Press.

Storey, D., & Hunter, S. (2010). Kiribati: An environmental perfect storm. *Australian Geographer, 41*, 167–181.

Tavola, H. (2023). *Tuvalu Study on People with Disability*. DFAT/Pacific Women Shaping Pacific Development.

Taylor, M. (2025, December 8). The Marshall Islands' experiment with a universal basic income. *Interpreter*.

Thornton, A., Sakai, M., & Hassall, G. (2012). Givers and governance: The potential of faith-based development in the Asia-Pacific. *Development in Practice, 22*(5–6), 779–791.

Tomlinson, M., & McDougall, D. (2012). *Christian politics in Oceania*. Berghahn.

Trundle, A. (2020). Resilient cities in a sea of islands: Informality and climate change in the South Pacific. *Cities, 97*, 102496.

Trundle, A., Barth, B., & McEvoy, D. (2019). Leveraging endogenous climate resilience: Urban adaptation in Pacific small island developing states. *Environment and Urbanization, 31(1), 53–74.*

Underhill-Sem, Y., & Guttenbeil-Likiliki, O. (2025). Challenging colonialism and patriarchy: Oceanic Pacific feminists on development, women and international law. In J. Dawuni, N. Grossman, J. Ramji-Nogale, & H. Fabri (Eds.), *The Oxford handbook of women and international law* (pp. 413–433). Oxford University Press.

UNICEF. (2025). *Multidimensional child poverty in Fiji*. UNICEF.

UN Women (2018). *Promoting women's economic empowerment: Recognizing and investing in the care economy*. UN.

Vakaoti, P. (2018). *Street-frequenting young people in Fiji*. Palgrave Macmillan.

Veenendaal, W. (2013). Size and institutional legitimacy: The case of St Kitts and Nevis. *Commonwealth & Comparative Politics, 51*(3), 343–361.

Veenendaal, W., & Corbett, J. (2021). Navigating hyper-fragmentation in small states: A comparison of Solomon Islands and Vanuatu. In L. Briguglio, J. Byron, S. Moncada, & W. Veenendaal (Eds.), *Handbook of governance in small states* (pp. 26–37). Routledge.

Walshe, R., & Nunn, P. (2012). Integration of indigenous knowledge and disaster risk reduction: A case study from Baie Martelli, Pentecost Island, Vanuatu. *International Journal of Disaster Risk Science, 3*(4), 185–194.

Wheatle, S., & Campbell, Y. (2020). Constitutional faith and identity in the Caribbean: Tradition, politics and the creolisation of Caribbean constitutional law. *Commonwealth & Comparative Politics, 58*(3), 344–365.

Williams, D., Costa, M., Sutherland, C., et al. (2019). Vulnerability of informal settlements in the context of rapid urbanization and climate change. *Environment and Urbanization, 31*(1), 157–176.

Wood, T. (2021). The cruel ironies of the 2021 Honiara riots. *ANU DevPolicyBlog*, 3 December.

6

Services, Public Management and Uneven Development

This chapter takes up the themes of governance initiated in the previous chapter and their influence on development, and especially access to services. Particular attention is given to health and education, key basic needs and central to the SDGs, and then to questions of access to clean water, energy, and other needs and services. This highlights challenges that reflect issues of management, access and equity, some relating to urban bias, and others to transport. Symptomatic of emerging challenges and responses to evolving structures of development is that the chapter concludes with a brief discussion of crime and the scourge of drugs. It is too simplistic, however, to simply deny and decry service provision and point solely to problems, since most SIDS have welcomed and benefited from developments, such as the internet and solar energy, that have transformed lives and livelihoods. Much depends on service allocation and provision.

For many years it was assumed that the now independent states were too small to achieve independence and that colonial administration was necessary and adequate so that when the tide of independence swept through, in the 1960s in the Caribbean and 1970s in the Pacific, many SIDS were unprepared. Few of their populations had had opportunities to travel, acquire a tertiary education or significant knowledge of

J. Connell, Y. Campbell, *Island Challenges*,
https://doi.org/10.1007/978-981-95-8177-1_6

government systems (and much else), while the infrastructure of an independent state (transport and energy systems, for example) was poorly developed. Prior to independence, there seemed no immediate need to develop a significant educated workforce (McIntyre, 2014). Ministries in Tuvalu were fortunate to have a single college graduate at independence. The belated independence of the COFA states of the North Pacific was accompanied by a colonial perception that many of the people who lived on outer islands were some of the most 'traditional' in the world, invoking a racist attitude to islanders, where even missionisation had been belated. Only in the later years of colonialism were efforts made to put in place systems necessary for independent states. Belatedly educating and training citizens and workers was difficult, and SIDS experienced deficits of skilled workers, infrastructure, and development plans and strategies.

A partial consequence of fragile economies, and not simply a consequence of limited economic growth or smallness, has been weak service delivery. In several SIDS, provision of services to rural or remote areas—especially transport, with the multiple consequences this entails—even worsened rather than improved since independence. As urbanisation gained pace, the tasks and complexity of service provision mounted. Progress to meeting the SDGs has been poor with limited finance and skilled human resources imposing limitations on what is possible, affecting SIDS' ability to provide basic services, whether in health, education, infrastructure, technology or housing, too often exemplifying the inverse care law where services were weakest where most needed. In turn, that pointed to problems of infrastructure provision (such as roads and electricity) and the extent of urban bias and digital divides.

Some SIDS have succeeded in reducing child mortality and achieving universal primary education, but few have made gains in reducing inequality or accomplishing gender equity (Chap. 5). In most SIDS income inequalities have increased in this century, with poverty emerging in both disadvantaged remote areas and urban centres. As we suggest below, inadequate access to services raises basic questions about social justice, equity and citizenship, and the role of government in the developmental process. Demographic pressures from ageing populations, tight fiscal budgets and remoteness alongside pressure to increase spending in growing urban areas have enhanced urban bias and denied equity in service delivery. State capacity has been tested and found wanting. Failures in extending services, the most crucial being health, have demonstrated that some SIDS have been incapable of meeting a basic criterion of statehood: effectively delivering services throughout the nation.

6.1 Human Resources and Service Provision

Underpinning development concerns in the SIDS is a scarcity of capital and human resources, especially skilled workers, the outcome of absolutely limited numbers, limited high school output and emigration (Chap. 9). Schools in some SIDS produce too few graduates, stemming from 'a lack of educational opportunities and, as a result, a dearth of qualified people for adaptation planning and implementation' (Petzold & Magnan, 2019: 150). A serious and obvious consequence is limited capacity to deliver effective services. Where ministers have multiple portfolios, not only is it difficult for one minister to provide policy direction and leadership in these broad and differentiated areas but government organisations lack information necessary for sound decisions. Even in the critical fisheries sector, the majority of SIDS fisheries 'lack formal management and assessment due to limitations in data, management capacity or both [so that] some combination of limited infrastructure, personnel or expertise hampers the ability of fisheries administrations to develop, implement and maintain desired management structures [hence] are at greater risk of performing poorly in terms of sustainability and economic benefits' (Hill et al., 2018: 85). Public sector entities in many small states also struggle with productivity. In Jamaica, Trinidad and Barbados, labour productivity has declined annually since 2020, hampered by bureaucratic inefficiences and institutional challenges (World Bank, 2023a).

The complexity of development and the policy process means that multiple actors are expected to offer assessments and solutions, relying heavily on the most skilled or qualified. It is difficult for public employees to assess the volume and complexity of 'data' and advice from external agencies, consultants, academics and NGOs, some with limited expertise in SIDS. In rare circumstances where external advisers remain, linking external ideas and local contexts becomes possible. As one volunteer in Vanuatu observed, 'the real value of my role came from the embedded nature of the volunteer position. Being situated within the Department—able to pop in to discuss ideas or attend meetings at a moment's notice—was critical. It's something you simply don't get as a FIFO advisor or someone working externally to a department: that deep, day-to-day understanding of how things actually work. It helps you see not only why things are the way they are, but also where the real opportunities lie for change and growth' (*Vanuatu Daily Post*, 23 August 2025). That connection proved invaluable, even more so where human resources are scarce. In Vanuatu only two staff are qualified to manage environmental impact assessment processes; hence, there as elsewhere, human resource shortfalls hinder effective EIA administration and implementation pointing to 'the reality that environment departments/agencies in PICs are unlikely to ever have the ideal complement of human resources for every regulatory stage of EIA' (Bradley & Swaddling, 2018: 358). In several PICs limited human resources have meant that 'weaknesses in areas of project management have constrained efforts of Pacific [states] to mobilize finance at scale … and ensure that finance accessed is invested effectively' (Taloiburi & Homasi, 2022: 37). In Samoa, bureaucrats pointed out that in their one job they wore 'about twelve different hats' so that 'there are simply a limited number of qualified and/or experienced staff in the relevant agencies and resources are also limited. Thus staff working directly on climate adaptation are required to serve in a multitude of other capacities' so that it is hard to manage new approaches and coordination 'when staff are already overworked'. That also tended to result in staff being transient so that 'institutional memory was minimal' (McGinn & Solofa, 2020: 246–7). New international directions and pressures tended to add new tasks without removing previous ones.

To reduce the 'capacity problem' in new areas, like deep-sea mining, the Marshall Islands government insisted that any deep-sea mining companies would be required to support national education and skills training of Marshallese workers (Waiti & Lorrenij, 2018). At the same time, limited scope exists for upgrading skills even as technology changes and as climate change (and global economic shifts) require new responses

(Kelman, 2018) amidst 'competing agendas and limited resources and capacities to focus [on]' (Atteridge et al., 2020: 486). As renewable energy became a possibility in SIDS, a constraint to its successful and sustainable implementation was the concern that 'education and training programs are still scarce in the small island states, and those that exist are localized in the bigger islands' (Lucas et al., 2017: 46). 'Capacity building' and institutional strengthening are ever present. One consequence has been inadequate and ineffective service provision, weak monitoring of projects (such as their environmental impacts) and minimal assessment and supervision of external projects. Civil society and NGOs also lack capacity to monitor these tasks. Simply keeping abreast of changes in relevant fields is not easy. Policy implementation can fail, and it can be difficult to decide who is responsible for failure to achieve objectives where adequate resources are lacking, and tasks are spread across multiple agencies and departments.

6.2 (Un)Healthy Islands

With few exceptions most SIDS have health care systems that are under-financed, under-resourced and lacking in effective primary health care systems and preventative health care. Their ability to provide basic services and deal with health emergencies is limited. Because of fixed costs and the need for technology, financial demands on smaller SIDS are considerable. Thus, 18% of the budgets of Nauru and Tuvalu are spent on health whereas the global average is just less than 10%. Even so, health financing in several SIDS (notably Fiji, Grenada, St. Lucia and St. Vincent) falls below the WHO recommended 6% of GDP, leaving citizens to fund their own health care, often resulting in high out-of-pocket costs. Here, as much as in any other sector, a minimal amount of resources, including people and time, is required, yet many smaller SIDS have limited training facilities and thus inadequate mentoring, while the absence of highly technical care means a need for referrals to health care abroad, costly to countries and households, reflecting the distinctive circumstances of SIDS (Meessen et al., 2024).

Beyond its impact on mortality and morbidity, the greatest impact of COVID-19 was in drawing attention to weaknesses of 'standard' service delivery and thus trust in government. Cultural complexity, urban poverty, weak health care systems, overburdened staff and a double disease burden (and thus co-morbidity) posed wide-ranging challenges. Health sectors struggled with lack of hospital facilities (notably beds and ICU facilities), overworked, undercompensated and sometimes fearful frontline staff, and lack of access to diverse health care services. It became visibly evident that health facilities were far from evenly distributed and typified urban bias, as in Fiji (Wiseman et al., 2017). The pandemic drew urgent attention to the longstanding need for investment in human capital in both health and education.

Ordinarily, SIDS face health financing challenges that affect development. Health care systems have long been underfunded, with shortages of drugs and equipment, a lack of adequately trained health workers: a long existing problem of human and physical resource shortages (Connell, 2010). The COVID crisis crystallised the need for more effective health care systems, with more workers and more robust health infrastructure (which would require substantial financial and management resources) and external aid to finance and support whole of government interventions and reforms extending beyond the confines of the health system itself. However, public expenditure on health including external grants is already greater as a percentage of GDP in the Pacific SIDS than in lower-middle-income countries globally. Tuvalu's expenditure on health was more than six times the global average for lower-middle-income countries. Positively, nine Pacific SIDS have a larger share of public expenditure allocated to health as a percentage of GDP than the global average for upper-middle-income countries. Although 8 of the 11 Pacific SIDS, for which comparable data is available, have seen an increase in the share of expenditure on health, questions remain over the overall effectiveness, efficiency, equity and quality of existing service expenditure. Even where countries have increased their allocations, most budgets do not actually go towards improving service delivery, purchasing equipment or preventive care. Recurrent expenditure in the form of salaries can account for two thirds of budgets, but migration from the health sector suggests this is still inadequate (Chap. 9). Having access to adequate health care is critical to meeting the most basic SDGs and is a significant electoral issue in most SIDS.

Infant mortality remains high in several SIDS, such as Kiribati and the Marshall Islands, partly the outcome of crowded urban households and poor sanitation practices (accounting for the prevalence of diarrhoea as a cause of death) and the absence of early medical intervention (Culpin, 2017). Kiribati has an under-5 mortality rate more than twice the regional average of 27 per 1000 live births, a health care crisis that puts the lives of its youngest and most vulnerable at risk (Nunn et al., 2024). Infant mortality has steadily declined in Caribbean SIDS, with the number of infant deaths more than halving from 1990 to 2021 (World Bank, 2023a, b), whereas non-communicable diseases (NCDs) have increased (Abdulkadri et al., 2021). Cancer, diabetes and obesity are major problems, especially in Nauru and the Polynesian SIDS. In Fiji, diabetes accounts for 70% of all deaths each year, being one of the three leading causes of mortality in the country, alongside cardiovascular diseases and cancer: a limb is amputated every 12 hours (*Fiji Village*, 14 November 2024). Half of all surgical procedures in the Marshall Islands are amputations because of diabetes (Culpin, 2017). NCDs have reached epidemic proportions in Solomon Islands, where two thirds of the population were overweight in 2024 (and half of these were obese), indicative of Melanesians now following Polynesian patterns. In the Marshall Islands, where NCDs were responsible for over 80% of deaths, that proportion was still increasing in 2024 (RNZ, 2 December 2024). The threat of drug-resistant tuberculosis still hangs over several SIDS. People with the least social, economic and historical capital bear the brunt, hence the slogan: 'poverty is the disease, TB the symptom'.

The growing incidence of NCDs has driven up the cost of health care for both citizens and governments. Most of the ten countries with the highest levels of diabetes and obesity are in the Pacific, with women more likely to be obese, and more than 50% of mortality (as much as 83% in Tonga) linked to NCDs (Hawley & McGarvey, 2015). NCDs are the main cause of premature disability and death in Pacific SIDS. Along with obesity and mental health, these chronic diseases are responsible for more than 80% of all deaths in the Pacific, and the situation is little better in the Caribbean, where NCDs are responsible for 75% of all deaths. An ongoing transition from epidemic and infectious diseases to NCDs continues. In several SIDS, including PNG, Fiji and Kiribati, an incomplete epidemiological transition has resulted in a double burden of both infectious and non-communicable diseases (e.g. Ahlgren et al., 2014; Phillips & Narayan, 2017). NCDs are linked to more sedentary lifestyles, alongside poor nutrition (various processed foods, such as two-minute noodles, with high sugar, salt and saturated fat content), smoking and excessive alcohol consumption. Processed foods spread rapidly in islands where previous choices were few and a transition to imported foods was part of the colonisation process (McLennan & Ulijaszek, 2015). Health care systems alone are incapable of coping with NCDs, since it is not only a health problem. The economic costs of NCDs are substantial—both directly, in terms of the costs of care, complex technology and referrals, and indirectly, in terms of lost labour and productivity, substantial enough to have had measurable negative effects on output, as in Fiji, Barbados, Jamaica and Trinidad (Chand et al., 2020; Braithwaite, 2022).

Many SIDS, especially those with small populations, have neither the facilities nor the specialised personnel to treat more complicated NCDs. Most therefore support and finance overseas referrals for specialised care. NCDs are a major driver of overseas medical referrals: a significant and often fast-growing component of government health expenditure that benefits a small fraction of the population. In the Pacific, as much as a third of health budgets have been allocated to overseas referrals for the benefit of about 1% of the population, but many patients refer themselves (in the Pacific, increasingly to India rather than more costly New Zealand). The World Bank has calculated that without action being taken, economic losses due to NCD mortality across 11 PICs would reach between 8.5% and 14.3% of gross domestic product by 2040 (World Bank, 2016). The increasing incidence of injuries and trauma following disasters, road traffic accidents and personal injury further stresses already under-resourced regional health systems and points to the increasingly 'modern' disease pattern. COVID-19 forced many countries to immediately increase spending on health care. While that was encouraging, few governments reformed systems to deal with broader structural and equity issues. Government policies in some SIDS take the form of 'free' health care and the removal of user fees. This has not solved the problem of overcrowded hospitals and clinics, high transport costs, cost of medication and the quality of local health care.

Access to health care workers varies considerably between and within SIDS. The number of nurses in all Pacific SIDS and most Caribbean SIDS, and of doctors in two thirds of Pacific SIDS, falls below WHO recommendations, making it difficult to achieve practical universal health coverage (UHC). While some countries experience training problems, with smaller SIDS such as Tuvalu training all their health personnel overseas, the already longstanding emigration of all categories of health workers became a critical problem after COVID-19, as nurses, dentists and allied health workers especially sought better pay and the opportunity to work in better quality health systems and in other economic sectors (Chap. 9).

The shortage of a nursing workforce is a problem throughout the SIDS, with rare exceptions including the Bahamas, as nurses are numerically the greatest contributors of health care. In Solomon Islands, Vanuatu, Samoa, Marshall Islands, Tonga and Fiji, the 'nurse density' is far below the recommended ratio of 45 nurses per 10,000 population. By mid-2024 Vanuatu's health sector was facing significant gaps, including 400 nurse vacancies and a shortage of 50 doctors (with only 69 in place), exacerbated by the deregistration of the country's nursing college. Solomon Islands, which faced its own problems, had signed an agreement to supply both doctors and nurses to Vanuatu (RNZ, 12 September 2024). Outside the capital, Port Vila, the shortage of beds, technology and human resources was particularly acute. At the same time in Solomon Islands, for the second time in two years, the Operating Theatre at the Solomon Islands National Referral Hospital (NRH) went into emergency mode, due to a shortage of anaesthetic drugs for operations, focusing only on life-threatening cases. The NRH was also trying to obtain urgent supplies from local pharmacies while waiting for the release of overseas supplies stuck in containers on the main wharf. Meanwhile, the only mammogram machine in the country had been unusable for a year because a part had not been sourced overseas (RNZ, 3 September 2024; *Solomon Star*, 19 September 2024). Several SIDS are awkwardly placed in global and regional supply chains, and as the *Solomon Star* observed, Solomon Islands also faced challenges distributing medicines within the country.

Nurse (and medicine) shortages have discouraged accessibility to adequate health care, especially on outer islands. Various factors, typified in Vanuatu, have slowed the growth of an adequate workforce, including a distaste for nursing, rapid nurse turnover, poor working conditions, low wages and lack of administrative support. These contributed to absenteeism (posing problems for those who remained), work overload and eventual resignation (Tamata & Mohammadnezhad, 2022). However, the advent of drones may be particularly beneficial for archipelagic SIDS, such as Vanuatu, for such tasks as the delivery of vaccines and other medical supplies (Enayati et al., 2023). Nonetheless, most SIDS experience unequal nurse distribution and urban bias that in turn became factors in rural-urban migration. Typically, staffing of facilities in regional and remote areas is particularly difficult, since supervision and support are limited, allowances for higher costs and difficulties of remote work are absent, promotion is rare since remote workers are 'out of sight and out of mind', and equipment and medicines are often absent and inadequate (Tamata & Mohammadnezhad, 2022; Awa, 2025). Many experienced Solomon Islands' doctors and nurses choose to work in Honiara rather than in remote hospitals, to avoid social and professional isolation and the lack of opportunity for ongoing professional development. Even in Suva, the capital of Fiji where health care provision might be expected to be relatively good, health workers are short of space, have inadequate and unreliable technology and unreliable supplies, while those involved in outreach have poor access to cars, using public transport to reach patients (Phillips & Narayan, 2017). In Fiji, in mid-2023, St. Giles hospital, the only psychiatric hospital, was operating at 50% capacity due to staff shortage, compromising their services, a situation that had deteriorated a year later with a worsened staff-to-patient ratio and increased patient numbers following the growing impact of substance abuse (RNZ, 10 September 2024; see below). While the health worker-to-population ratio is important, investments in hospitals, equipment and technology and transportation are also needed to ensure greater access and equity.

Regulation of practices that affect health outcomes is valuable, but governments have not always been able to regulate the import, advertising and promotion of unhealthy products (notably alcohol, cigarettes and processed foods), tax them at more prohibitive levels or challenge unfavourable trade agreements. WTO free trade legislation prevented banning inappropriate (but cheap) food imports (Thow et al., 2010), marking 'the double burden of neo-liberalism' and the limited control that SIDS have over their exposure to risk factors and macro-scale processes like trade liberalisation and food marketing by transnational corporations (Glasgow & Schrecker, 2016). The outcomes of tax policies were also less positive than anticipated. In the early to late 2010s, Fiji and Tonga increased taxes on some unhealthy imported foods, and cigarettes and alcohol, and reduced taxes on healthy foods including fruit, feasible only because of a consensus between different ministries and prominent leaders. Policies that discouraged tobacco consumption, as in Fiji and Vanuatu, conflicted with some economic policies and high-level government officials' questionable commitment to protect public health interests (Patay et al., 2023). Smoking was targeted in Tonga since almost half the men smoked, and smoking appeared to be increasing among young women. Taxes generally reduced consumption and increased government revenue. In Tonga at least, the volume of imported soft drinks fell, households switched to purchasing cheaper locally produced soft drinks (a boost to the economy but no benefit to nutrition). The increased tax on imported manufactured cigarettes resulted in many less well-off smokers reducing consumption of manufactured cigarettes, with price, rather than health, the main reason. Consumption shifted to locally manufactured and untaxed hand-rolled tobacco leaves (*Tapaka Tonga*), as harmful to health as manufactured cigarettes. Increased excise tax brought decreased frequency and quantity of alcohol consumption, as many drinkers shifted to kava, but alongside an increased incidence of substance abuse, including marijuana, glue and petrol sniffing, and crystal methamphetamine ('ice'), though the relationship may have been coincidental. While taxes on turkey tails, mutton flaps, and ice cream helped reduce consumption, consumers shifted to substitutes—salted beef and corned beef (which also contained high and unhealthy levels of salt and fat) and tinned fish—and to consuming locally manufactured unhealthy products. Few could afford fresh fish, and other healthy alternatives, on a regular basis due to high prices and lack of availability. Without complementary measures, tax increases may thus merely result in a switch from one form of unhealthy consumption to another (World Bank, 2019;

Sievert et al., 2019; Bell et al., 2021; Reeve et al., 2021). Few other SIDS have tried holistic economic policies but have relied on less evidently successful health promotion, often in the face of strong pressures from the food industry and domestic stores (Reeve et al., 2021). Achieving a shift away from a costly and unhealthy food dependency remains a stubborn challenge.

One of the more ambitious SDGs is universal health coverage (UHC) that calls for all people to have 'access to the full range of quality health services they need, when and where they need them, without financial hardship'. However, access to preventative health care is correlated with wealth (e.g. Lam et al., 2025). Fragmented data on access to health care suggest that SIDS are underperforming on what is needed to deliver UHC: entangled issues of climate change, poverty and food security are worsening, budgets are tight, and the effects of climate change on agriculture and fisheries are likely to drive up the reliance on imported foods of low nutritional quality. Mental health has worsened in several SIDS partly in response to illicit drug consumption and climate change (Charlson et al., 2015), presenting more burdens on staff and exacerbating deficiencies in health care systems. Strengthening health care systems, addressing cross-border health threats, advancing digital health and improving access to treatments and medicines require constant and costly change.

6.3 Educated People

Education is crucial to SIDS for the acquisition of knowledge, skills, attitudes and values necessary for a sustainable future. Education can improve agricultural productivity, enhance the status of women, reduce population growth rates, improve environmental protection and raise standards of living, without necessarily replacing valuable local knowledge. Access to primary education is reasonable in most SIDS, and universal in many, while literacy rates are high, but education quality is often poorer in rural areas and truancy and drop-out rates are higher. However, basic reading skills are weak. Learning poverty, the inability to read and understand a simple, age-appropriate text by age ten, is prevalent in Pacific SIDS. More than half of 10-year-olds in 9 of 11 Pacific SIDS cannot read and understand such texts; in Kiribati, Tuvalu and Tonga more than two thirds could not do so (Cahu & Sondergaard, 2023). Where English is the language of learning in secondary schools, that is a significant disadvantage. Family background is the most important influence on learning outcomes, but at school the most critical factor is the quality of the teachers, yet education systems are struggling to attract and retain enough teachers; hence, class sizes are too large to be effective. Teachers themselves in the Pacific SIDS, required to teach primarily in English, may have limited English competency, a situation that is not easy to remedy. Physical facilities are disadvantageous; in Solomon Islands less than 30% of schools have electricity. Recruitment and retention of qualified teachers remain problems for many SIDS, especially in STEM subjects, with a growing number of untrained and pre-trained teachers in several Caribbean SIDS (UNESCO, 2024; Wright, 2024). Even in better-performing SIDS, transformation of the education system is required to keep pace with workforce demands and the need to develop skills in technology and science (Miller, 2014). Policy has sometimes perceived teacher training as a panacea, but in both regions several SIDS have been affected by the migration of teachers, leaving gaps in the ability to provide quality education. As with health services, the COVID-19 pandemic exposed disparities and technological barriers in education, and, in its aftermath, migration of teachers, notably from Fiji, Barbados and Jamaica, emerged as a policy concern (Crossley et al., 2011; Iredale et al., 2015; see Chap. 9). Education can be undervalued and is less frequently the beneficiary of external aid. Despite progress in access, equity and gender parity in basic education, and increased provision of early childhood education (ECE), post-primary and tertiary education remain problematic in several SIDS.

Remote schools experience problems in recruiting and retaining experienced teachers, maintaining communication with central agencies, including infrequent visits from school authorities to ensure standards, and delays in receiving supplies and salaries (Coxon & Munce, 2008). In Fiji, a third of primary schools, mostly rural, have fewer than 100 pupils, increasing the difficulty and cost of providing good-quality education. Standards are weakest in more remote schools where facilities are poor, getting and supporting teachers more difficult, students' access to schools sometimes itself being difficult (e.g. Volpe, 2021), while the value of education is less obvious and drop-out rates greater, especially where there are fees and costs in accessing education. In some villages no primary schools exist and parents send their children to boarding facilities (Bacchus, 2008), exemplifying a vicious cycle of inadequate opportunity. High drop-out rates even from primary schools reflect inadequate incomes (though, when education has become free, as in Vanuatu, attendance rates quickly increase). In St. Lucia over a third of children leave school without any qualification, there are not enough places in secondary schools and tertiary education is a luxury for a small percentage so that education is inadequate to equip school leavers for local wage employment (Henry-Lee et al., 2010). In Vanuatu, most children in secondary education come from medium and high-income households, a situation broadly similar even for primary school children, which also enabled the better-off to lobby more effectively for improved access to education. Rural areas are relatively disadvantaged.

The value of 'modern' skills and the loss of local skills in a technological world indicate the relevance of appropriate education, crucial yet often ignored as a development issue. As the Vice-Chancellor of the emerging University of Solomon Islands, Transform Aqorau, has said, in disappointment over national budget cuts to education:

> Education is not just a policy issue or a line item in a budget; it is the bedrock of national growth, social stability, and the empowerment of our people. The importance of education in building a resilient and prosperous nation cannot be overstated. As global challenges become more complex, and as we seek to position ourselves in an increasingly interconnected world, the Solomon Islands must prioritise the intellectual growth of its people. Education equips individuals with the knowledge, skills, and values needed to address societal challenges such as poverty, inequality, and environmental sustainability. It prepares citizens for active participation in governance and fosters innovation, entrepreneurship, and economic growth. (quoted in *In-depth Solomons*, 23 October 2024)

Education is an investment in the future for SIDS, many of which have struggled with deep-seated inequalities in education delivery, too little emphasis on ECE and limited access to good quality education in adequate facilities. ECE remains rare, despite its benefits for both wellbeing and learning, for affirming cultural identities, and for strong engagement and connection with families and communities.

In Fiji, as elsewhere, students from poor households suffered most during COVID-19, sometimes simply going without education for months at a time. As education switched to remote learning, children from low-income families, students with disabilities, females and young learners were further disadvantaged due to limited time and access to the necessary technology (Jumo, 2022; Vakalalabure & Saurara, 2022). Students found Zoom classes difficult, and at all levels performance suffered. Intermittent school opening and closing disrupted courses and progress in already difficult times. When schools reopened, in Fiji at least (but certainly elsewhere), girls were initially less likely to return as households were less able to afford the school fees and girls were more expendable (and could engage in 'domestic duties'). The outcome was uneven access to education, inequitable opportunities and the reduced possibility of upward social mobility.

SIDS have mainly retained a loosely neo-colonial system of education perpetuating a system of traditional and non-traditional schools, well-resourced in the case of the former, but which contributed to a huge 'information divide' during COVID-19 (Vakalalabure & Saurara, 2022). Long evident in Nauru, the 'imported education system also evolved stratified systems of education and social structure that led to cultural dissonance between the educated elite and their local communities' (Johansson-Fua, 2022: 7; Burnett & Suluma, 2024). Evidence of real progress in securing positive and more equitable outcomes since independence is rare. Caribbean education systems are partial relics of the past. They emerged under colonialism, and, in its formal aftermath, too little has been done on restructuring for independent nations. Churches retain a large role in the development of the school systems in the post-colonial period alongside the state. Caribbean SIDS, despite investment in education, have left the largely unequal educational system in place. The Patterson Report, commissioned by the Jamaican Government in 2021, found that 'a third of students at the end of primary school could not read, 56 per cent could not write, and 57 per cent could not identify information in a simple sentence' (Jamaica Education Transformation Committee, 2021: 25). In Pacific SIDS, remnants of the colonial past remain present in language systems and in curricula despite countries needing their own educational strategies. In Tonga, even on the main island, education was almost unchanged in terms of facilities and content. Government-run primary schools with decrepit classrooms were poorly resourced, and although education was officially compulsory to the age of 14, children were dropping out earlier, and pass rates were falling. Few vocational skills were taught in schools and leavers faced severe problems in obtaining employment or gaining access to further education (Lee, 2018). Boredom, high opportunity costs, irrelevance and perceived lack of achievement have all played a part.

In Tonga, as elsewhere, the goal of many children and their parents was to gain coveted but scarce public service positions. However, the long-held notion of education as an investment only occasionally worked, despite this being the main source of 'investment' by migrant workers (Chap. 9). That contributed to a pushback on vocational education in schools (Lee, 2018). Educated and semi-educated unemployed youth in many SIDS, for whom jobs were simply unavailable, were frustrated by the promise of education leading to discontent and dissidence. Some, especially those already suffering from marginalisation and feeling opressed by systems of inequality, turn to gangs, violent crime, 'scamming' (Bourne, 2025; Lewis, 2020), and the 'rule of dons' (Jaffe, 2024). Schools fostered and facilitated variants of modernity but at some cost. Widespread frustration over inhospitable and inaccessible job markets has occasionally resulted in reversion to the past: on Manus Island (PNG), as articulated by one village resident: 'It's very hard to get a job. We must teach the ways of the ancestors in school. We have reached the end of our school knowledge' (quoted in Demerath, 1999: 162–163). More frequently, as on Pomponio Island (PNG), education was the road to and symbol of development, however infrequently aspirations might be met (Pomponio, 1993). Emphasis on universal education and considerable efforts to ensure this have raised concerns about the appropriateness of education, more so in the Pacific where indigenous languages are widespread, but curricula follow western epistemologies, values and content. Such curricula train students for urban employment, and therefore migration, often distant from the realities and needs of rural existence, and undervalue and demean indigenous knowledge (Overton & Murray, 2018;

Hicks et al., 2021). Lowly valued (and remunerated) jobs dominate many SIDS, while youth are uninterested in work in 'old-fashioned' technology-free agriculture and fisheries. The quite reasonable recognition by many that access to employment is extremely difficult at the end of high school prompts many youth, especially boys, to discontinue early and seek the financial independence that would enable them to participate in youth social activities, sometimes through migration (Chap. 9), where traditional knowledge provides no part.

Teacher-student ratios are 23 and 19 for the Pacific and Caribbean SIDS, respectively, with higher ratios in Jamaica and the Bahamas, but higher ratios have not necessarily translated into better education systems or outcomes. Ultimately, primary and tertiary education systems in both regions are too often marked by student disengagement and indiscipline, syllabuses detached from economic needs, weak leadership and teacher quality, the absence of standards and accountability, and archaic regulatory frameworks. Uncertainty and ambivalence reflect the cost of education, both to parents and the state, and concerns over the shortage of employment to which no curriculum change can fully respond.

Several SIDS have sought to depart from ancient curricula and technology despite financial constraints. Some of that followed recognition of the extent of problems during COVID-19. The Kiribati Ministry of Education sought to ensure that home-based learning could be accessed by all through the development of a Kiribati Learning Passport, a digital learning platform accessible both online and offline, but lack of smartphones in most of the country has limited its effectiveness. Some SIDS have developed programmes designed to assist students from poor households. The Programme of Advancement Through Health and Education (PATHE) in Jamaica is a conditional cash transfer programme funded by the Government and the World Bank and aimed at delivering cash grants to the most needy and vulnerable students.

The challenge of after-school vocational education has been addressed through Technical Vocational Education and Training (TVET) in the Pacific and Human Employment and Resource Training (HEART) in the Caribbean. It is potentially highly valuable since SIDS face shortages of workers in such activities as plumbing, painting, mechanical and electrical engineering. Australia has invested in training such skilled workers through the Australia Pacific Training Coalition (APTC), which has campuses in most Pacific SIDS and is open to students in all ten SIDS south of the equator. Initially, graduates were more likely to remain in the Pacific, filling obvious employment vacancies, but courses were later altered to enable graduates to acquire skills that would enable them to be employed in Australia, with the entire programme divided into domestic and international strands, and thus of less direct value to the SIDS (see Chap. 9).

At university level, obvious difficulties were attached to the development of higher education, not least because few islanders had tertiary opportunities, many had little knowledge of what it entails, and decisions were needed on whether courses should reflect employment needs to a greater extent than in the global north. University education only arrived in SIDS in 1948, in the case of the University of the West Indies (UWI) and much later, 1968, in the Pacific, with the University of the South Pacific (USP) cautiously emerging and expanding to support the region, although the Micronesian states favoured the Universities of Guam and Hawai'i rather than the new southern competitor. USP was centralised in Fiji, despite two smaller campuses in Vanuatu and Samoa (the former for the law school, the latter for agriculture) intended to counter the view that USP was for Fijians only. UWI was designed as a multi-campus university—with campuses in Jamaica, Barbados, Trinidad and eventually Antigua. Despite expansion, neither the universities nor a proliferation of nursing schools, theological and teachers' colleges provided enough of the skilled graduates that the SIDS needed; hence, scholarships to overseas institutions (for specialised course) were costly and more vocation-oriented universities appeared. Nonetheless, a progressive decentralisation of university education from (and even within) Fiji, borne out of national pride and cultural concerns (and assisted by corruption at the core), meant that USP lost its central dominance and smaller SIDS, notably Vanuatu, Solomon Islands and Samoa, aspired to and created their own universities, each with a limited and more 'vocational' curricula. In the Caribbean, new universities also competed with UWI (with at least six universities in Jamaica). Growth of distance education providers in the Caribbean has sought to 'cash in on opportunities' and exploit the demand for tertiary education further marginalising UWI (Jules, 2008). A period of neoliberalisation of higher education has resulted in financial barriers in tuition fees, and the dispiriting impact of low-value jobs experienced after graduation.

Limitations in access to tertiary education, especially for girls and especially in Pacific SIDS, remain. Overall, enrolment rates for women in tertiary education stood at 25% in the Pacific region in 2020 and at 18% in Solomon Islands in 2019. Educational attainment of at least a bachelor's degree or equivalent among women aged 25 and over ranges from just 2.7% in Kiribati to 10.9% in Samoa. In Solomon Islands, just 4.9% of females aged 12 and over had completed some post-secondary education in 2019 (10.1% in urban areas and 2.8% in rural areas) compared with 10.7% of men. Men also predominate in TVET courses. Barriers to female participation include cost, inadequate infrastructure, distance, increased risk of violence and harassment, high levels of adolescent pregnancy and traditional gender norms that prioritise household investments in men and boys and discourage women and girls from pursuing further education (Fitzgerald et al., 2024). At the same time, gender inequality needs to be addressed by including labour market discrimination and social norms that justify the exclusion and exploitation of women and girls (Chap. 5). Efforts have been made to achieve more equitable access to education, even physically. A new 2024 USP campus in the Solomon Islands capital, Honiara, featured safe spaces, improved lighting and bus shelters, designed for women's safety. Simultaneously, job counselling and employment support improved women's job prospects, despite low female participation rates in formal employment. By contrast, girls are more prominent than boys in tertiary education in the Caribbean. Over 65% of UWI students are female. Only engineering and sports attract a higher female enrolment. Explanations for these differences range from gender socialisation and 'male marginalisation' to classroom experience and insufficient male teachers.

Overall education outcomes in many SIDS are faltering. Across much of the region, participation in formal education is declining, along with graduation rates and academic performance. Domestic and international funding for education has waned as economic shocks and competing priorities have shifted spending away from the education sector. Combined with the region's growing youth bulge, this poses a risk to development and social stability. Both participation and performance in education are declining. Average primary school attendance and performance in most Pacific SIDS have fallen since the early 2010s. Teacher training and support is a problem, with the average training level of primary and secondary school teachers also declining, alongside equipment and resource constraints, partly the outcome of aid donors switching their attention away from education, and partly the result of SIDS spending less on education. In 2022, the per capita expenditure of all but two Pacific SIDS (FSM and Marshall Islands) was less than a decade earlier (Duke & Ahsan, 2024). That decline coincided with the youth bulge, less educated and skilled than previously, and of regional concern.

While education cannot guarantee subsequent success for individuals or nations, effective investment in education is critical for developing the human capital that can end poverty. Uneven education structures reflect the difficult task of delivering services and deciding what services to deliver and where, from ECE to tertiary and vocational education, and perhaps with some hybrid content of local values and 'modern' objectives with curricula and structures of dissemination reflecting socio-economic contexts (Johansson-Fua, 2022). That is a barely possible task, in the SIDS or the global north, even where education is highly valued for individual and household livelihoods and national development.

6.4 Energy

Providing, delivering and obtaining services, including energy, water, sewerage and transport, have been ever-present issues in SIDS. Both an urban bias in distribution and conspicuously uneven access within urban areas are contemporaneous. However, environmental degradation is also an urban phenomenon, and even in the small cities and towns of the SIDS, coasts and waterways are polluted and air pollution is of growing concern (Shah, 2021; Isley & Taylor, 2018). Most urban and national markets are too small for competition; consequently, private or public monopolies exist but are often poorly regulated by the state and do not operate in the public interest, sometimes leaving many citizens literally in the dark, and without regular, reliable and effective services. Because most SOEs operate as monopolies, they have little incentive to reduce operating costs or prices. Wastage in the water sector is a problem in SIDS where maintenance is poor. These problems are normalised in some SIDS, where the media frequently cover people protesting and demanding better services through their political representatives. Digital divides have largely disappeared, but the physical infrastructure of access roads, piped water and electricity, the basic needs that enable human security and development, can be unnecessarily fragmented.

Four decades ago, just three years after independence, a wide-ranging study of energy futures in Vanuatu summarised the key issues:

> During the next eight years, Vanuatu faces the prospect of reducing petroleum imports by 20 to 25 percent ... through the development of wood fuel for power generation ... The opportunity also exists to substitute a small proportion of diesel and motor spirit used in road transportation directly and indirectly with coconut oil and ethanol from cassava and sugar cane. (SPEC, 1982: 1)

It was also anticipated that this would provide a welcome boost to the copra industry, but economic and technical changes ensured that such directions disappeared into a void, energy practices went in the opposition direction (and the copra industry collapsed). Moreover, at least into the 2010s the key concern in most SIDS was that prices would increase as peak oil became the main problem (Niles & Lloyd, 2014). That is indicative of how much has subsequently changed—and one small indication of how the best predictions can fail completely as the tides of change, here both economics and technology, shift direction for whatever reason, and that it is not necessarily in SIDS' best interests to accept all recommendations from external experts. It now seems almost quaint in its anticipation of a different kind of sustainability.

Energy is vital for economic and social activities; hence, SDG7 focuses on 'affordable, reliable, sustainable and modern energy for all' by 2030 with the main targets being rural areas. Establishing interconnected networks is particularly difficult in archipelagic states. In most Caribbean SIDS access is universal, achieved through rural electrification programmes, whereas in the Pacific less than 45% of the population in Solomon Islands and Kiribati have electricity. There, most households enjoy only a limited number of hours of illumination, derived from kerosene or diesel, neither cheap nor with positive environmental effects. Development of the energy sector has not been a gender-neutral process and has evolved in the context of the global political economy, shaped by limited government regulation and increased privatisation. Women are more likely to bear the burden of sourcing, collecting and using energy for household activities, alongside other tasks, and be affected by energy poverty in many ways, including time burdens, health and physical burdens (Walters, 2025), circumstances most prevalent in semi-subsistence economies (Hills et al., 2018).

Few SIDS have indigenous energy resources, other than biofuels and wind and solar power; hence, most rely on the costly import of fossil fuels over considerable distances, exacerbated in archipelagic states, to generate power for homes, businesses and transport. Only Trinidad and Tobago is not a net oil importer. High transport costs mean that energy costs in SIDS are among the highest in the world. Overall, oil prices in islands are 3–4 times higher than those in mainlands and almost all SIDS generate more than 80% of their electricity using imported fuels (Kuang et al., 2016; Shirley & Kammen, 2013). SIDS economies are thus vulnerable to oil price fluctuations, and supply disruptions are a disproportionate economic burden. Fuel for electricity generation often exceeds 10% of GDP in SIDS, and occasionally as much as 30%, yet power shortages sometimes leave communities subject to prolonged blackouts (Raghoo et al., 2018). Management issues, often in SOEs, lower reliability and raise the costs, and therefore affordability and utility, of power supplies. High costs are partly the result of monopoly power companies, regulated by the state, that maintain an advantage in setting prices.

Energy imports are costly hence any movement towards solar, wind or hydro-electric energy would potentially reduce economic and environmental costs and support sustainable development (Dornan, 2015; Genave et al., 2020). Wind power offers considerable potential but is mainly unrealised because of limited policy development and awareness of viable options, lack of technical and maintenance skills, fluctuations in appropriate technology and high capital costs (Haraksingh, 2023). The high initial costs of transition to renewable energy systems are barriers to new energy technologies despite ambitious targets, with small-sized projects leading to high transaction costs and inability to deploy capital at scale (Meschede et al., 2022; Pueyo & Haas, 2025). Yet archipelagic states like Vanuatu and Solomon Islands, both with some of the lowest electricity access rates in the Pacific, have some of the highest electricity costs in the world because of their reliance on diesel oil, distributed across large archipelagos, limited demand, high transport costs and no competition in the markets: barriers to development, education and health care that require off-grid solutions. Other Pacific SIDS display variants of this problem.

Pacific SIDS aim to achieve an energy transition to universal renewable energy by 2030, supported by investment in decentralised, small-scale sources of electricity generation, notably off-grid solar, possibly linked through intra-island microgrids. Connecting isolated sources to a national grid would enhance the reliability and coverage of electricity supply, and a national grid would also standardise energy regulations, utility fees and maintenance operations. Centralised power supply generated by conventional methods is uneconomic and unmanageable for rural areas, particularly islands with small populations; hence, the decentralised approach based on power produced with renewable energy resources available locally is increasingly recognised as more appropriate. That recognition has only slowly been translated into practice, partly hampered by the lack of an energy sector at the national scale. Successful small-scale projects have not been scaled up (Michalena & Hills, 2018) and, until the mid-2020s, targets were vague and progress limited (Praene et al., 2018).

The development of modern technology, like solar photovoltaic (SPV) cells, is invaluable where conventional distribution systems based on fossil fuels were not feasible. Since the 2010s, small-scale solar energy projects have expanded fast in Pacific islands, from small 'solar lanterns', then modular home systems to much larger solar farms that could illuminate schools and hospitals. Subsequently, grid-connected solar photovoltaic systems have become a successful component of electricity networks in urban and peri-urban areas in many Caribbean and Pacific SIDS, with the renewable energy transition best advanced through policies targeting the adoption of small-scale, distributed photovoltaics (Kersey et al., 2021). Thus, for example, between 2009 and 2024, Pentecost Island

(Vanuatu) made a transition from being 75% dependent on kerosene and diesel to being 97% dependent on solar energy, a rapid transition which probably reduced costs to households, let different rooms to be illuminated, and fridges, freezers and radios to become viable, enabling fish to be stored before sales, village stores to stock perishable items (including ice-cream), kava bars to stay open longer, school homework and weaving to be undertaken at night, and occasional disco sessions to disturb the peace. However, solar systems experienced a high level of breakage, especially in storms, spare parts were rarely available and villagers lacked the skills to undertake repairs. Nonetheless, aid projects were training villagers in maintenance and aid donors cherished the notion that supporting programmes had gone from 'development aid' to providing a valuable 'economic commodity' that addressed 'energy poverty' and provided the 'moral good' of 'green energy' (Paul Munro, pers. comm. 2024). Elsewhere batteries wear out, components break and no trained technicians are available to repair them. Communities and vulnerable groups such as female-headed households, located in more remote rural areas continue to experience 'distributional energy injustice' in having less access to energy services necessary to fulfil basic capabilities, such as being healthy, educated and socially connected. This is evident in Jamaica (Walters, 2025). Elsewhere in Melanesian SIDS, unlike Pentecost, typical home systems cost between US$500–700, beyond the reach of most rural households. Prices have fallen and solar energy offers a valuable option, especially for small islands, even for some, like Bougainville with extensive cloud cover. Consequently, despite inevitable initial technical problems, by the 2020s small-scale solar energy was expanding fast through the smaller islands and several SIDS were coming much closer to 100% electrification. The social and economic impact of solar power is much higher in small islands than mainlands.

At a 'mainland' scale, St. Lucia established the first large-scale solar farm in the eastern Caribbean, with multiple private and public stakeholders, linked into eight solar microgrids, powering three schools, a police station, a polyclinic, a hospital, a water treatment facility and the Ministry of Infrastructure building. Such microgrids power critical facilities that can continue to operate during widespread outages from a hurricane or other disaster, without requiring generators. However, grid extension has often been held back by a mix of financial, regulatory, technical and community-level challenges.

Despite the growing significance of solar energy, other options include hydropower, which demands both significant regular rainfall to keep dams filled and mountainous topography, conditions absent in many SIDS. Fiji, a rare SIDS large enough for dam and reservoir construction, developed hydropower in 1983, and it provides more than half the country's electricity (Weir, 2018). Dominica generates about 40% of its energy from hydropower, but even quite mountainous SIDS, like Samoa, are restricted by their topography or rainfall seasonality. Solomon Islands has extended its limited hydropower to supply part of Honiara. In rural areas of islands with low electricity demand and scattered population, small hydropower plants, some 'homemade', supply rural consumers.

St. Lucia, St. Kitts, Dominica and Vanuatu are in various phases of developing geothermal energy, an effective renewable energy resource that is not intermittent. Fiji, Vanuatu and Solomon Islands, located at the junction of the Pacific plate and the Australian-Indian plate, have great potential for geothermal energy. However, their development of geothermal energy has not passed the planning stage with no installed capacity. Various schemes have been suggested for sites close to volcanoes, such as Tanna (Vanuatu), but none have come to fruition, mainly because of the high initial costs of establishment, the lack of technical expertise (Santagata et al., 2024), the absence of working models and inertia. Although St. Lucia began to exploit geothermal energy as early as the 1980s, electricity generation is yet to occur. Dominica was on track to complete its first geothermal power plant in 2025, providing power to around 23,000 homes (about 90% of the population), and supplying electricity to the neighbouring islands of Guadeloupe and Martinique, marking a significant step in the nation's renewable energy development, intended to reduce 96% dependence on imported oil, to become 'the first climate resilient country in the world' (Channel 24, 25 May 2022; OECS, 2024) while also gaining export income.

Ocean energy, in the form of waves, tides and thermal gradients, exists around all islands. Compared to wind and solar energy, ocean energy is less volatile and more predictable, yet blue ocean energy is still in development despite its being accessible, constantly available and renewable. It was attempted in Nauru as early as 1981 when the country was wealthy, with the first Ocean Thermal Energy Conversion (OTEC) pilot plant established by the Japanese Electric Power Company, but it failed because of the impact of extreme weather events (Rashid et al., 2024). With further technical development, ocean energy nonetheless suggests a viable option (where wind or solar installations are more difficult and land is scarce). Without significant tidal ranges, SIDS otherwise have little potential for tidal energy and the technology of harnessing wave power remains expensive and exploratory.

A range of possibilities exists—wind power may enter the scene—but none are necessarily easy to establish. Options have firmly swung towards solar energy. Even then initial costs can be high and Caribbean energy systems have been characterised by policy-making being too easily influenced by fossil-fuel-oriented companies, slowing any transition to environmentally oriented systems (Liu et al., 2022). Expansion of solar energy is a welcome development, minimally enabling diversification of energy sources from a costly dependence on oil and towards renewables with greater resilience to hazards. Rooftop solar is spreading fast but mainly at household level, rather than the greater scale required to power schools and hospitals. Combining a physical product with its relationship to greater sustainability, it has proved an attractive area for aid delivery (Dornan & Shah, 2016) and for many small-scale private companies. Fossil fuels remain part of the energy mix, especially for transport, and remain a costly import for most SIDS. However, solar energy especially has provided greater energy security at relatively low cost: an irony of widespread commitment to renewables where SIDS contribute less than 1% of global greenhouse emissions. The solar journey continues.

6.5 Water

Most SIDS are primarily dependent on rainwater for freshwater supplies, few have significant rivers or lakes, and several—especially the atoll states—are vulnerable to drought. Growing populations, greater affluence and new demands (including those of tourists) have placed increased pressure on existing water supplies, while inadequate and decaying infrastructure has contributed to urban supply problems. Many rivers, lagoons and especially urban watercourses are polluted. Sewage and other discharges into lagoons have contributed to eutrophication, pollution and the loss of habitat in more urbanised sites. Such problems have tended to increase with accelerated urbanisation and vegetation removal and have rarely been effectively remedied. Predicted future climate change suggests that access to fresh water will become more difficult—with greater seasonality of rainfall, occasional drought, fast run-off in storms and a reduction in the size of lenses on atolls. Human pressures on water sources are likely to be even more crucial.

Water is more than a sectoral issue: it is a unifying force, serving and connecting agendas relating to climate change, conservation and economic development and, where women are often the dominant users of water, is a key element of equity and inclusion (Willetts, 2025). Even on high islands in the Caribbean droughts and sea-level rise have contributed to saline intrusion of underground aquifers and surface water sources (Mycoo, 2018). Water has consequently become one of the critical resources of this century and access to clean water for domestic use (and water for agriculture and other activities) is already difficult in many SIDS, particularly in drought-prone islands in the atoll states. As droughts increase, as is likely, lesser rainfall and rivers drying up will mean that allocating water may become necessary. Entire SIDS may become water-deficit, as has already occurred with Nauru and Tuvalu, importing internationally when local sources have dried up (Gheuens et al., 2019). Globally, a strong relationship exists between water insecurity, food insecurity and even increased obesity (Young et al., 2022; Levasseur et al., 2025).

Demand for water has increased substantially where tourism has grown, since tourists have high per capita demands, problematic on islands with low rainfall and contributing to the Caribbean being one of the most water-stressed regions in the world. Severe water challenges are driven by declining surface water quality, escalating water demand and poor water management practices, and droughts as in Barbados in 2009–10 and 2014, which placed pressure on the health sector, with poor water storage leading to gastroenteritis and the proliferation of the *Aedes aegypti* mosquito, the vector responsible for the transmission of dengue and chikungunya (Lester & Rhiney, 2018; Mycoo & Roopnarine, 2024). In the 2020s, every Caribbean island experienced some sort of water scarcity (Chap. 2) alongside increasing demand and changes in annual rainfall patterns, urbanisation (increasing pressure on public water systems), unregulated groundwater extraction, tourist consumption and lack of appropriate management practices (resulting in leaks and lost revenue). Hybrid rainwater harvesting, better governance and investment could reduce water losses from theft, leaks and over-indulgent showerheads. Educating tourists might help reduce waste at hotels and resorts (Nibbs, 2024). Rainwater harvesting is an important alternative water source for individual households and small communities, and water reuse from tourist infrastructure has potential for SIDS to achieve water sustainability while protecting coastal waters from sewage pollution (Crisman & Winters, 2023). Several SIDS have had to invest in expensive desalination plants (where running costs are compounded by high fuel costs). Antigua invested in desalination as early as 1970. Generally costs are too high for 'normal' water production and energy generation; hence, desalination has been mainly used (and subsidised) in various outer islands, as in Tuvalu (where it has gone from being reserved for emergency use to becoming a primary source of fresh water) and the Marshall Islands, where droughts and water shortages are not unusual. Almost all households have experienced drought (and rationing of rainwater for consumption during droughts), with desalination units crucial during drought emergencies, although they provide maintenance challenges (MacDonald et al., 2020). Governance and management are critical.

Economic (and moral) regimes are rarely in place to encourage effective water conservation. Capacity to pay for access to water and sanitation services is limited, especially in expanding urban settlements, where there has also been inadequate government investment. Water management is made difficult in most SIDS by a familiar litany of issues: fragmented governance in the water sector, inadequate water resources legislation, failures of policy and planning (and implementation), insufficient human resources (with technical and management skills), and deficient community education, awareness and participation. Colonial infrastructure has often decayed into obsolescence with many systems running at nearly full capacity and suffering from high levels of non-revenue water (supplied but not billed to customers) alongside water losses. Reluctance and inability to pay, challenges to collecting revenue and constraints of land tenure have reduced the success of some supply schemes. Investment is needed to expand and upgrade services, particularly in underserved rural areas, and to curb the level of illegal connections that contribute to water loss in urban areas.

Indicative of urban atoll environments, water supply in urban Tarawa is particularly problematic, being reliant on access to shallow groundwater, with some rainwater tanks and desalination. Urbanisation and population growth have led to over-extraction, conflicts, severe pollution and high levels of waterborne disease (White et al., 1999). Due to small scale, limited human resources, system complexity, isolation and scarce financial resources, water management capacity is inadequate; water can be polluted while sustainability is uncertain and landowners are resistant to some forms of restructuring (Moglia et al., 2008). Overall, Kiribati experienced all too typical problems of infrastructure deteriorating, with few people having the ability to maintain and repair it. Solar desalination was placed in inappropriate places without discussions over who would have access to the water so that necessary behavioural changes did not happen. By 2024 only 27% of Kiribati households had access to safe drinking water (Pareti, 2024). Water-related problems in Tarawa include endemic waterborne diseases, contributing to one of the highest infant mortality rates in the world, over-extraction of freshwater from groundwater sources, especially at times of drought, social disruption and conflict due to displacement of traditional landowners from water reserves, and inability to finance water services by charging customers for water. Similar but less extreme situations prevail in other urbanised atolls: Ebeye and Majuro (Marshall Islands) and Funafuti (Tuvalu), all atolls with high population densities.

Water governance is a problem for many SIDS, with unreliable water supply systems, old infrastructure and poor maintenance, leakages and high levels of non-revenue water. Achieving water security and its sustainability has therefore been difficult because of the need to balance economic, social and environmental issues (Belmar et al., 2016; Mycoo, 2018). Many SIDS communities have multiple local water sources that have evolved in response to the specific water sustainability challenges faced by remote communities and seasonal constraints. Thus in the Marshall Islands rationing and sharing of private rainwater occurs in the dry season (Elliott et al., 2017). Recycling and more efficient rainwater harvesting are widely needed. Water providers, usually SOEs, have limited finances to invest without expanding their client base and reducing high levels of illegal (unpaid-for) access. Other than in Fiji, fewer than 65% of settlements in Melanesian cities have access to reticulated water (Schrecongost & Wong, 2015). In Honiara, in the mid-2010s, 70% of households experienced water shortages, only 25% had a 24-hour supply, and water was often unsafe to drink after rain (Pendeverana, 2017). In the Marshall Islands in 2015 as few as 3% of households had water piped into their homes; 92% used public taps, standpipes and other sources (Culpin, 2017). While urban populations in Pacific SIDS generally have adequate access to improved water and sanitation infrastructure, thousands of people in informal settlements routinely lack safe water and sanitation.

Access to clean, piped water is least adequate in the Melanesian archipelago, with some of the lowest global rates of access to safely managed or basic drinking water and sanitation. In Vanuatu, rural communities do not necessarily choose the safest water source for drinking, when choosing between groundwater and rainwater (Foster & Willetts, 2018). In Solomon Islands, despite efforts by the government and aid organisations, the provision of sustainable and appropriate water, sanitation and hygiene (WASH) services remains poor. As in Vanuatu, rural-urban, provincial and central-periphery inequalities exist in access to drinking water with central provinces using piped water and remote provinces using rain and surface water (Anthonj et al., 2020). In 2017, only 60% of the rural Solomon Islands population had access to a source of water that met the SDG6 definition of basic service level, in contrast to 91% of the urban population. As with other services, this disparity between rural and urban populations is due to the remoteness of rural communities, the high cost and challenges of transport between islands, limited access to resources and markets, and vulnerability to extreme weather events that can permanently damage WASH infrastructure (Love et al., 2022). At a local scale, water management is inherently political and affected by broader community dynamics, social relations and inequalities that can be perpetuated in access to water. Lack of access impoverishes livelihoods and adds health care expenses, with a higher incidence of waterborne illness or diabetes, lost productivity and other expenditure, including expenditure on bottled water. Severe weather events, which are likely to intensify, further disrupt water and sanitation services.

Informal settlements in Melanesian SIDS experience overcrowded and unreliable standpipes, unhygienic pit latrines and open defecation (Schrecongost & Wong, 2015). Recent efforts to improve access to water supply in the urban village of Hanuabada (Port Moresby, PNG) where no households had individual access to water, and 80% of households used the sea for sanitation, proved largely unsuccessful. Projects failed because of unclear mandates to the providers, too many layers of decision-making, a complex and diverse community (because of considerable in-marriage) without accepted leadership and thus agreements on community matters, and lack of government involvement and accountability. What might have been a straightforward project to provide more adequate water supply and sanitation to perhaps the most famous, and most visible, urban village in the Pacific, was unsuccessful. In microcosm it typified problems elsewhere but in a site where success should have been possible.

Water utilities have sought to improve services to settlements, including provision of community standpipes with special tariff rates that allow users to share a single meter, and placement of meters at the edge of settlements to sidestep land tenure constraints (Schrecongost et al., 2015). Similar participatory and community-based urban water service delivery suited to informal communities has been piloted in Honiara and Suva (Gooden, 2017; Phillips & Keen, 2016). Such initiatives demonstrate the need to accord with the limited resources of 'clients', convey clear community messages on why water charges are necessary and work with communities and non-state actors to build trust between service providers and urban residents. Urban water supply must be considered across sectors, social groups and scales, indicating why achieving positive development is difficult.

Progress against the more ambitious targets of the SDGs, notably to ensure drinking water is safe, and available when needed, and that sanitation includes not just a toilet but a system to safely manage the waste has been slow. Few villages in Melanesian SIDS have access to safe water in households. Most rely on so-called self-supplied water services, often a burden on women and children, with poor water quality and without regulatory oversight. Pacific SIDS have made the least progress towards global drinking water and sanitation targets (MacDonald et al., 2017) and are vulnerable to water shortages in rural and urban areas and to water-related climate change threats. Caribbean SIDS have superior management systems but are even more vulnerable to water stress. Storms may create waves that wash over atolls, contaminating freshwater lenses and agricultural plots for many months. A cyclone affecting the large high island of Vanua Levu (Fiji) resulted in three quarters of the island's water sources, of various kinds, still being contaminated a month later, because of soil, bacteria and other debris washing into rivers and streams and entering groundwater supplies (Mosley et al., 2004). Greater severity of storms will increase the incidence of such problems. Projected impacts of climate change are likely to exacerbate problems of scarcity and equitable access, especially on atolls, if rainfall declines (Wale et al., 2021). More effective rainfall harvesting, conservation of water systems, recycling wastewater, effective distribution and management have become crucial to water security.

6.6 Sanitation and Sewerage

Wastewater management in SIDS is no easier than water management. Most wastewater in SIDS is uncontrolled or untreated. Only a third of people in SIDS are connected to wastewater treatment systems and 45% to wastewater collection systems. The outcome is a vicious cycle of poor health, low economic productivity and reduced water quality. Sewage treatment plants are often inadequate. Indeed, it is estimated that 85% of wastewater discharged into the Caribbean is completely untreated, which contributed to the loss of about 80% of lining coral in the Caribbean over the previous 20 years (Adeoti et al., 2020). In Pohnpei (FSM), raw sewage is discharged directly into the ocean leading to biodiversity loss, acceleration of climate change effects through emissions from waste, reduced tourism revenue and health impacts for local people exposed to the pollution (UNEP, 2021; Rouse, 2013). State capacity and successful governance of waste management, whether of sewerage or solid wastes, is lacking in both regions. Thus, in Solomon Islands, where poverty is significant, open defecation is a problem since about half of the country's population do not have formal toilets or sanitation, and many still use streams, beaches, bushes and mangroves as toilets (*Solomon Business Magazine*, 13 September 2024). In Kiribati, about a quarter of the population are in the same situation. Otherwise, SIDS only experience extreme and localised problems.

Centralised wastewater systems, where they exist, need a high customer density and are vulnerable to sewer infiltration, a problem exacerbated by high coastal water tables and sea-level rise (Fleming et al., 2019). High energy and chemical costs on islands reduce the feasibility and sustainability of conventional activated sludge treatment processes. Consequently, onsite wastewater management, or very little management, prevails. Onsite systems are usually based on septic tanks, which involve a holding tank and a drainfield, but often function poorly because of small property sizes, porous soil and high water tables. Such limitations, alongside underfunding of infrastructure, especially in rural and low-income areas, have left many Pacific SIDS with cesspools (a subsurface porous holding tank of excrements) the norm for onsite wastewater management. Since water resources are limited, especially in the Caribbean, wastewater treatment is inadequate or absent. Nature-based solutions, especially constructed wetlands, are effective treatment options for all SIDS and have added value for recreation, conservation and product development, but low rainfall restricts such possibilities. More sustainable wastewater management would enable the development of local economies via a healthier workforce, and improved agency for women who are disproportionately affected by inadequate WASH services.

WASH problems are most evident in informal urban settlements and mainly in Pacific SIDS. In the two capitals of Fiji and Vanuatu (Suva and Port Vila), in nine informal settlements, although piped water was available, most households (from 56% to 100% of the households in each settlement) relied on unsafe sanitation to manage human waste.

Many used poorly built dry pits—for example, a pedestal above a hole in the ground—or cesspits, without any plumbing or running water to flush. All were onsite (in the house or adjacent to it) so that treated or untreated waste remained in the settlement (Rousso et al., 2024). In such circumstances a third of households lose access to functioning toilets during heavy rain, cyclones or floods. Even when toilets are available and working well, there is often no safe and sustainable way to manage the waste that builds up in pits and tanks, seeps into the ground and can pollute water sources and create environmental and public health risks. Although probably less than 5% of all urban households in Pacific SIDS experience such basic unimproved or unsafe sanitation, they are concentrated in a few urban settlements. Inadequate sanitation is mainly a rural problem where, in the Pacific, circumstances are similar to those in urban settlements, with few households having access to safely managed sanitation systems and poor maintenance of systems (Nasim et al., 2023).

Many countries have small administrations dealing with the varying complexities of main and outer island issues, without economies of scale, and with distinct variations in access in urban areas. Few trained water resources specialists must cope with strategic planning and economic and environmental management, and a lack of water resources expertise and baseline knowledge is a barrier to informed decision-making. In parallel with the management of water systems, communities need comprehensive approaches involving reliable waste removal, treatment and disposal, which are resilient to disasters. That requires increased local expertise, supporting local service providers, ensuring systems are well maintained and fostering community ownership of these systems to guarantee long-term sustainability, but this is complicated by insecure land ownership and limited access to adequate infrastructure. Solving the 'sanitation crisis', centred particularly in the Pacific SIDS, involves protecting health, restoring dignity, supporting livelihoods and building resilience at the level of households and communities, which requires public sector commitment and finance (Rousso et al., 2024). Management and finance are ever-present concerns.

6.7 Solid Waste

Waste disposal can be unusually difficult in island states, and especially the atoll states, where the tide of plastics cannot easily be reversed, land is scarce, and achieving a circular economy is unusually difficult. Buried toxic waste can leak into lagoons and into the lenses that are crucial to water supplies when rainfall is scarce. While the Pacific SIDS especially have low rates of goods consumption, they generate waste that is not biodegradable by traditional methods (so that composting is rare).

Most SIDS are too small (in area and population) to enable financially viable recycling plants; hence, waste is often dumped in official, or sometimes unofficial, landfill dumping sites, which are not necessarily well maintained. Distance and the lack of economies of scale effectively prevent recycling and a circular economy: visually evident for plastics from packaging and single-use containers. Non-biodegradable washing machines, cars and fridges line the roadsides in Funafuti, Tarawa and Pohnpei (FSM), where landfill sites are scarce. Vehicle and white goods ownership has increased greatly in all SIDS in recent years, resulting in a rapidly growing number of so-called end-of-life vehicles whose demise is hastened by rust: aesthetic disasters (Hall & McDonald, 2020), resulting in frustration over a wider 'waste colonialism' where plastics, toxic waste and second-hand goods are dumped in SIDS (Manglou et al., 2022). In 2025 Tonga banned the import of vehicles more than ten years old to reduce this problem. Old appliances and vehicles cannot be recycled and their inadequate disposal causes environmental problems from lead batteries, refrigerant gases, mercury and oil.

Poorly managed waste poses environmental problems for rivers and coastal seas, leaching and seeping into groundwater and rivers, and unsightly aesthetics and visual pollution (conceivably a brake on tourism) while microplastics are consumed by marine species, later entering human digestive systems (below). Because of many islands' small size, limited disposal capacity, prohibitive export costs, low priority attached to waste management and the coastal squeeze, over 80% of mismanaged waste in SIDS ends up in oceans. Multiple land-based activities (mismanagement of household and commercial waste, industries, harbours, unprotected landfills, sewage overflows and littering) are sources of marine litter. Lack of recycling capacity and low awareness of pollution effects contribute to beach litter, with urban beaches near river mouths being particularly polluted with single-use and other plastics and polystyrene, being detrimental to the environment, to marine species and ultimately to health (Binetti et al., 2020; Ferreira et al., 2020; Verlis & Wilson, 2020). While pollution has contributed to the loss of some coastal habitats, notably mangroves, Pacific seagrass habitats have remained relatively unscathed in Pacific SIDS (Brodie et al., 2020). Coastal lagoons close to urban areas, as in Port Vila and Suva, experience several forms of pollution from heavy metals to floating plastics (Pratap et al., 2020; Paris et al., 2022) as urban activities, from sewage disposal systems, piggeries and dredge sites, have resulted in water quality being worst in urban areas where it might otherwise have been a valuable resource (Dehm et al., 2025; Tonga et al., 2025).

SIDS inhabitants generate, on average, 2.3 kg per day per person of 'municipal solid waste', less formally garbage, which is three times the world average, with tourism playing a disproportionate role in increases during seasonal peaks (UNEP, 2021). Just as tourists use more water than local people, they also generate substantially more waste (Mohee et al., 2015). Caribbean SIDS, with higher levels of living, produce proportionately more solid waste. As much as 85% of marine litter in the eastern Caribbean SIDS is from land-based sources (Graham, 2023). Most municipal solid waste in SIDS consists of organic waste (44%) followed by recyclables: paper, plastics, glass and metals (43%) (Mohee et al., 2015). Much of the organic waste, notably food waste, is more manageable, but the proportion of solid waste is growing. Solid waste management is, however, characteristically inadequate and, even in urban areas, only extends to a proportion of the population. Waste disposal via burning (contributing to air pollution), landfilling, illegal dumping (including in rivers and on coasts) and backyard burning are often favoured at the expense of sustainable waste treatment technologies such as composting, anaerobic digestion and recycling, so threatening the livelihoods of people who depend on the sea. Mismanaged solid waste is a major issue, not just as a significant source of marine plastic debris but also because it causes localised flooding due to blocked drains during storms, with water contamination and disease, further increasing vulnerabilities and reducing the quality of life in urban areas. Improving local solid waste management can positively impact urban populations and contribute to resilient communities.

Inadequate waste management is a function of alternative government priorities and political support for action; lack of finance; no long-term planning or business planning; poor landfill siting and physical limitations; poor design, planning and management; lack of skilled personnel; lack of awareness of the problems caused by poor waste management; physical limitations to the establishment of landfill sites; poor handling of clinical waste; insufficient recycling and reuse, including limited reuse of organic wastes, sludge and effluent (Morrison & Munro, 1999; Matagi et al., 2024). Multiple minor schemes have improved local practices. Strategies implemented in Pacific SIDS include container deposit programmes, pre-paid waste collection bag systems, waste and tipping fees. Some open dumpsites have been rehabilitated into semi-aerobic landfills, and garbage collection services have been expanded and made more efficient. Recycling initiatives in several countries, including Palau and Kiribati, have resulted in a substantial volume of materials taken offshore, relieving pressure on limited landfill space. However, the absence of proper waste collection systems discourages any formal waste management flow, resulting in illegal dumping and backyard burning of wastes. While in some SIDS, the infrastructure is well developed, in other SIDS homes may be inaccessible, few garbage trucks operate, or funds have not been provided for proper trucks and garbage bins. Failure to adequately address these problems has contributed to environmental and health problems, more serious where toxic wastes have become more dominant.

6.8 Toxic Waste

Electronic waste such as televisions, computers, phones and solar panels contains toxic chemicals that can contaminate land and water. Batteries are particularly corrosive yet commonplace. As the fastest-growing global waste stream, electronic waste accounts for approximately 54 million tonnes of e-waste, with only about 17% formally collected and recycled.

It has proliferated in recent years, with various metals and rare earth elements accumulating in marine trophic webs, exacerbating ecological concerns, much as with plastics. Medical waste poses further disposal problems (much accentuated during COVID-19), some ingested by sea turtles in the western Pacific (Fukuoka et al., 2022). Four Caribbean SIDS—Barbados, Grenada, Jamaica, Trinidad and Tobago (and Aruba)—over a 60-year period from 1965 to 2025, collectively produced twice the global average of e-waste per capita per year, with 13 kg per capita compared to the global 6.1 kg in 2016. E-waste generation on these islands was projected to increase substantially (Mohammadi et al., 2021a; Wosnick et al., 2024). Disposal of toxic waste—from batteries and electronic waste to medical waste—poses special problems to avoid the risks of contamination and, again, especially so in atoll states where adequate sites are scarce and leaching into water resources is hard to avoid. Fish around Ebeye have long been contaminated by arsenic and other toxic waste used at the US base on Kwajalein. The volume of batteries alone is considerable and standard lead batteries are highly toxic. Integrated strategies are needed to enhance the basic management and regulation of e-waste as its volume increases, to enable environmental resilience and sustainable development, but it poses significant recycling and management difficulties, because of toxicity, few alternative uses and in the context of limited waste infrastructure on individual islands, and in each SIDS, so that regional linkages would help to achieve economies of scale (Mohammadi et al., 2021b).

A grandly entitled project, launched by UNEP in mid-2022, the *Implementing Sustainable Low and Non-chemical Development in Small Island Developing States (ISLANDS), aimed to* prevent the release of over 23,000 tonnes of toxic chemicals and more than 185,000 tonnes of marine litter by 2027. ISLANDS was intended to help SIDS control the import of hazardous substances, soundly dispose of harmful chemicals and waste, and establish circular production systems, dealing with toxic chemicals and waste challenges in a holistic way, by involving companies from the tourism, recycling and shipping sectors as partners. Whether such grand plans can influence human behaviour and local governance is yet to be demonstrated. Basic structural factors such as inadequate waste collection services and recycling infrastructure contribute to low recycling rates and may be remedied, but factors such as negligence, personal responsibility, lack of civic pride and lack of awareness may be less amenable to change.

6.9 Plastics

Single-use plastics are a huge problem as plastic has gone from glamorous hero to villain in half a century, an essential part of the world economy that has grown, and continues to grow, exponentially since 1950. So much has been dumped, with such negative consequences, for waterways and oceans, for health and livelihoods, that it is a global problem. For the SIDS, as with climate change, the key problems originate elsewhere. The World Bank has estimated that Caribbean coasts are contaminated with an average of 2014 objects per kilometre, a fifth of which are plastic bottles: one of the highest levels of plastic pollution of any marine ecosystem (Aranda et al., 2022). Of the top 30 global polluters per capita, 10 are now from the Caribbean region, including 7 SIDS: Trinidad and Tobago, Antigua and Barbuda, St. Kitts and Nevis, Barbados, St. Lucia, the Bahamas and Grenada (*Island Innovation*, 11 March 2022). The Pacific SIDS, less affluent and more distant from the countries and cities of the north, contribute less and are less affected.

Plastic is light, can easily float and drift, affects all countries (including unoccupied islands) and breaks down into microplastics that are ingested by various species, including humans, and which disperse much further. Two thirds of fish consumed in Fiji contain microplastics, an emerging health concern (*Fiji Sun*, 25 October 2025). Plastic mars beautiful environments and endangers marine life and ecosystems. Fish, seabirds and marine mammals ingest small plastic objects; microplastics affect fish and seabird mortality and have become hazardous to human health (Corbin, 2023). Even smaller nanoplastics, often from textiles and tyres, are present in air, water and food. Some wildlife species become entangled in abandoned fishing gear (ghost nets) or plastic packaging. Plastic waste threatens coral reefs by depriving them of oxygen and light and damaging their surfaces. It degrades very slowly, leaving a significant carbon footprint and aggravating climate change. More obviously than with other waste issues, disposal requires international solutions as its impacts have become a major global conservation concern with negative implications for ecosystems, maritime industries, tourism, marine life, and human health and indirectly through the loss of biodiversity and the costs of remedying pollution (McIlgorm et al., 2022; Lachmann et al., 2017; Napper & Thompson, 2020; MacLeod et al., 2021; Beaumont et al., 2019).

Sources of plastics in SIDS vary. Most come from beyond the SIDS; even in a large SIDS like Trinidad and Tobago, 48% of plastics in landfills comes from packaging of imported products rather than intentional domestic use (Millette et al., 2019), constituting a typical yet basic challenge to developing a circular economy. Plastic usage is extremely varied, although the packaging industry accounts for 47%. Recycling and the circular economy are seen as solutions to unlocking the plastic problem, yet neither is of real significance in the SIDS. The majority of waste—mostly plastics—on the atoll of Kayangel (Palau) came from elsewhere, mainly drift based but also brought in by returning non-resident islanders (Owens et al., 2011). In Jamaica, however, most plastic waste was local with only small quantities coming from elsewhere (Garces-Ordonez et al., 2025). Drifted plastic predominates on ocean-facing beaches, as in Eleuthera (Bahamas) where it also constituted 93% of all marine debris (Ambrose et al., 2019). Drifting ensures that few beaches are spared.

Plastic has attracted particular interest and concern because of its visibility, the rapid increase in volume, its presence on beaches (and therefore a deterrent to tourism, though tourists are particularly prone to generate waste) and its health impacts. SIDS are vulnerable to the impacts of marine debris and are susceptible to receiving streams of ocean-based plastic debris, disproportionate to their own consumption levels. Limited waste management infrastructure creates a multifaceted pollution problem, especially since much marine litter emanates from production and consumption elsewhere.

A more complex issue involves reducing plastic production. The amount of plastic in circulation has continued to increase without technologies to deal with waste management or any legally binding global governance arrangement that would effectively limit and control plastic pollution (Williams & Rangel-Buitrago, 2022). In this international context, waste governance is complex, especially when convenience and limited awareness clash with attempted environmental management. As with other environmental issues, hierarchies, budget constraints, the absence of environmentally sound waste collection and treatment systems and inadequate environmental knowledge limit perceived agency, genuine commitment and effective waste governance at every scale.

The enormous significance and mobility of plastic waste means that national solutions cannot suffice. The importance of the plastic waste issue has been recognised by the ratification of a United Nations resolution to create a legally binding international instrument to end plastic pollution, a so-called Plastics Treaty. In 2022, UN member states adopted a historic resolution to develop a legally binding agreement on plastic pollution by the end of 2024. To surmount the patchwork of international regulations, the new treaty developed a comprehensive approach to address the full life cycle of plastics from product design stage to disposal (Kim, 2024). Much like the Paris Agreement on climate change, it sought to reduce plastic production, eliminate unnecessary single-use plastics, promote sustainable alternatives, and ensure a just transition for communities and workers affected by plastic waste and industry shifts. Some progress on less contentious but less critical issues was made (Dauvergne et al., 2025), but the treaty failed to progress in 2025. Although more than 170 countries supported it, several petrochemical-producing nations, including Saudi Arabia, Russia, China, Iran and India, opposed constraints on production, favouring waste management strategies. Ratifying even a watered-down treaty, where words such as 'chemicals', 'harmful' and 'phase-out' had been excised, was postponed with no future action planned. Tuvalu, speaking for 14 Pacific island states, said the lack of a treaty 'fails to protect our people, culture and ecosystem from the existential threat of plastic pollution' (*Island Times,* 15 August 2025). Since so much of plastic waste drifts into SIDS from elsewhere, without a treaty or convention, even best practice in the SIDS would be inadequate to prevent pollution (Farrelly et al., 2021). The UN had established a consensus on the necessity of a transition to a circular economy, with production and consumption systems based on the reuse and regeneration of materials and products, but without agreement on how this transition might be financed. Failure to achieve a wider consensus on the minimisation of plastics pointed to the paralysis of consensus-based decision-making on environmental issues where economic issues held sway. Meanwhile, climate change has made plastic pollution worse. Heat and extreme weather accelerate the breakdown of plastic into tiny particles, while storms and floods stir up old garbage from landfills, and as microplastics mix with air, water and soil, toxic elements such as pesticides are carried with it. At the same time, the manufacture of more plastic generates billions of tonnes of greenhouse gas emissions—creating a vicious cycle (Kelly et al., 2025) and emphasising the need for a treaty.

6.10 Recycling: Towards a Circular Economy?

Recycling waste of any kind is limited or absent in most SIDS. Global recycling has failed to keep pace with a culture focused on economic growth, consumerism and consumption, where even 'fresh' products are packaged. Various factors hinder successful practice: low levels of awareness, the form of collection undertaken, lack of any government subsidies, the low total volume of particular recyclables, lack of interest by the public and by offshore production companies. High costs are involved in collecting and transporting waste to recycling centres, outweighing any income that could be derived from sale and reuse (Mohee et al., 2015). However, practices are slowly changing with greater sensitivity, more volume and therefore greater necessity. While the current focus in developed countries is mainly on three elements (reducing, reusing and recycling), with less attention to recovering materials, each of these is more difficult in SIDS.

Scale is critical. Thus, New Providence, the main island in the Bahamas, with a population of about 250,000 and a very large tourist industry (that generates considerable waste), with more or less current technologies, is regarded as large enough for recycling to be economically viable. That conjuncture (including its location close to Florida) is absent elsewhere in the SIDS, where waste volumes are smaller and archipelagic problems complicate waste management, collection and disposal logistics. Zero waste, where no solid waste enters a landfill or dump, and all materials are composted, incinerated (for energy), repurposed or sorted for return or export markets, is desirable, but unattainable, on most islands. By contrast, on many, including Dominica, Grenada and St. Lucia, household waste remains a substantial contributor to marine litter from illegal dumping and poorly managed waste dumps (Graham, 2023).

The ability of small islands to meet sustainability goals is exacerbated by costs of transport in taking garbage from the islands or otherwise disposing of it. At small scales, recycling can be prohibitive and complicated by labour costs and limited available technologies. For the island of Exuma (Bahamas), attempts to recycle the substantial accumulation of solid waste, associated with one large resort (that had led to an increase in pests such as rats and flies, along with fires associated with unburied solid waste), faced multiple barriers, typical of other places, which were underpinned by economics and included the training of staff (many of whom were temporary migrant workers, while there was rapid turnover in management), to change old habits and embrace new technologies. Without regulatory directives, few incentives existed for resorts to implement recycling, seemingly paradoxically made more difficult by insufficient volumes on smaller islands. Consequently, recycling could seem to be a futile effort. Recycling bins and transfer stations are designed to sort individual commodities, but this reprocessing is very expensive for small islands; hence, the disappointing and implausible conclusion was that 'resorts such as SEB [Sandals Emerald Bay] as well as island residents need to embrace the consumer limitations of island life, and reduce the solid waste accumulation' (Sealey & Smith, 2014: 36). At national scale, dependence on tourism has resulted in a high level of imports and amplified the tasks of developing a circular economy in any SIDS, let alone an archipelagic nation (Del Campo et al., 2023).

Meeting the economic costs of assembling and transporting waste has been no different in Pacific SIDS, even with voluntary private sector involvement. The maritime transport company, Swire Shipping, committed finance to help develop end-of-life vehicle export and recycling businesses, and provide free shipping of recyclables from Pacific countries to regional centres (notably Fiji and Palau) and onwards to Pacific rim countries such as New Zealand and Australia. While shipping usually had spare capacity travelling from the Pacific SIDS, such a proposal was unlikely to be sustainable, even with 'free' assistance from the private sector (Bennett et al., 2024). Further complications relate to the need for entrepreneurial ability to take advantage of small volumes, the diversity of waste (even within a single sector such as plastics) and lack of economies of scale (Millette et al., 2019). Some waste materials do not lend themselves easily to recycling (or effective disposal) even were they without other 'standard' complications, as in the difficulties that Grenada has experienced with plastics, tyres and motor oil (Elgie et al., 2021). While the concept of the 'circular economy' suggests a useful strategy for SIDS to improve solid waste management, remoteness, dependency on imports and limited waste absorption capacity make small islands poorly placed to participate, so that initiatives must be global (Barrowclough & Eugui, 2021; Del Campo et al., 2023). Theoretical solutions and models for recycling, applied to electricity generation from waste in Tonga (Yousefzadeh et al., 2023), are often complex and constrained by management problems.

Greater affluence and more consumption mean more imports which are often more comprehensively packaged, hence greater challenges to recycling. However, several SIDS, from Samoa to Jamaica, have banned single-use plastic bags. Solomon Islands officially banned single-use plastics in 2023, but little happened and the Mataniko urban river was still occasionally blocked with water and soft drink bottles (and household waste) increasing flood hazards. Inadequate waste disposal infrastructure, limited public awareness and a continuous influx of waste from urban areas and informal markets accentuated problems (*In-Depth Solomons,* 30 December 2024).

Greenwashing, limited corporate social responsibility and ignored invocations for public-private collaboration and 'community level behavioural change' (Shah et al., 2019) have met little response. 'Returning' to a more fresh, unpackaged food-oriented lifestyle is neither easy nor necessarily welcome. For one SIDS characterised by considerable illegal dumping, one recommendation involved 'improving data collection, implementation of the polluter pays principle, banning substitutable, problematic materials, and developing waste management plans for problematic materials' (Elgie et al., 2021): a measure of the task rather than of solutions likely to be easily or quickly implemented. Meeting SDG12 (responsible consumption and production) has infrequently become effective policy formation or practice. Waste is, usually, not out of sight but out of mind, reducing the possibility of its receiving priority in government planning with support through appropriate legislative and institutional frameworks. Rarely in the SIDS (or in developed countries), is environmental management a priority, and a shift to a circular economy must emanate from changing practices in the global north.

6.11 Transport

Unlike most other facets of service provision, whether health care, education or WASH, and without railways or national shipping lines, transport is unusual in being primarily provided by the private sector (involving an unusually high component of taxis). The exception that proves the rule is aviation, with many SIDS seeking, and having sought, national airlines, few of which have survived. Long distances between ports, low trade volumes, imbalanced exports versus imports and the widely varying standard of ports and their operations result in relatively high shipping service costs within the region while provision for containers has marginalised smaller ports and excluded international shipping services. Sustainable, reliable and adequate transportation systems are rare in SIDS, whether on land, sea or in the air. Public transportation, where it exists, is affected by poor infrastructure, congestion and rapid urbanisation.

Towns and cities grew in colonial times partly to meet the needs of a small elite with few considerations for possible post-independence aspirations to vehicle ownership. Consequently, few SIDS have developed strategies for sustainable transport, partly a function of inadequate urban planning, so that transport has made less progress towards sustainability than other sectors. Investments in transport have hardly kept pace, with most SIDS in the Caribbean investing less than 3% of GDP on transport. Cars, traffic congestion, unpleasant and harmful diesel fumes all plague even small cities, with transport dependence on fossil fuel. Congestion has meant that aid projects have been entitled 'Capacity Improvement of Urban Transport Management in Honiara' implying a new variant on urban bias. Moreover, imports of expensive cars, buses, motorbikes (and fuel) contribute significantly to the overall trade imbalance. Through a combination of factors including consumerism, lack of proper roads, reckless driving and high road fatalities in some SIDS, people tend to bypass smaller cars, choosing SUVs instead. Bicycles are largely invisible. Those most affected by limited transport availability are urban residents without private transport and rural residents seeking urban access.

The flexible and highly visible informal transport sector is evident in the popularity of 'paratransit transportation', where people share taxis and minibuses that may operate without required licences or proper regulation. These are supplemented by privately owned fleets of buses, which usually operate only in the larger towns and cities, except for Jamaica and Fiji where bus services and pick-up trucks (often collectively owned by Fijian villages) serve rural areas. Where road networks are poorly developed, shared transport is particularly important. In the Pacific SIDS at least, the proliferation of vehicles of all kinds has much to do with investment by returned overseas workers. Regulating fares for private taxis and single-fare schemes for public vehicles are strategies used by Caribbean SIDS but with limited success in easing transport problems and enabling more effective mobility. Problems of sustainability, and reliance on fossil fuels to power public and private transportation, remain. Despite congestion and other costs, the flexibility of urban transportation, symbolised by Vanuatu's minibuses (that run particular routes at fixed fares but deviate according to the needs and requests of passengers), provides an example of cheap transport accommodating the needs of the public, where distances are short.

SIDS are highly dependent on imported diesel, which drives up fuel costs, reduces energy security, and contributes to pollution and greenhouse gas emissions. Despite making up a small percentage of vehicles on the road, diesel trucks and buses account for more than a third of global fuel consumption. In SIDS this is compounded by distance and limited access to fuel markets. Islands are well suited for zero-emission vehicles: driving distances are short and potential exists for renewable energy generation, which is an ideal ecosystem for electric vehicles to thrive, enabling reduced operating costs while cutting emissions. Pilot projects are underway in Antigua and Barbuda, Grenada and St. Lucia, and Barbados has introduced electric buses (Shah et al., 2022). Hindrances to such a transition include high vehicle costs, limited experience of maintenance, lack of charging stations and, almost certainly, a lack of any national priority. Indeed, the increasing size of SUV vehicles ('mobesity') currently undermines potential gains from electrification due to their physical impact, greater energy consumption and associated emissions.

Where transport is costly and roads are poor, distance disadvantages development, especially for small farmers and market women, seeking occasional urban access. In Nawairuku village (Fiji), some 25 kms from a small town, 'restricted access limits villagers' ability to partake in some livelihood activities such as personally selling goods at market or purchasing grocery items from the store, attending workshops and training, engaging in wage employment while retaining family presence and accessing basic health care services' (Currenti et al., 2019: 71). To that list could be added other rural disadvantages: access to electricity, piped water, educational and sporting facilities. Nawairuku is on a road and on the main island of Fiji, but villages and their roads are never priorities when government assistance is required. Fractured communication is worse on smaller islands and a source of frustration and insecurity (Hobbis, 2019a, b), especially in contexts, widespread in Pacific SIDS, where 'finding the road' is a metaphor for development. Over time, as incomes and facilities have become more important, such cumulative disadvantages have discouraged agricultural development and contributed to the slow depopulation of many such inland villages.

6.11.1 Marine Scenes and Carbon Trails

In the Pacific at least, networks of sea transport are thinner now than in the ages of sailing and steam ships, and air transport is costly and mainly dependent on external carriers (often at inconvenient times) emphasising the colonial construction of transport connectivity. Port facilities vary widely, with some lacking modern cargo-handling infrastructure, especially on outer islands (where containers may not easily be unloaded). Several factors make shipping services relatively expensive, especially in the Pacific, including long distances between ports, imbalanced and low trade volumes, which lead to costly container repositioning (or their abandonment). Shipping lanes mostly bypass the islands, resulting in high transport costs that raise the cost of imports, reduce income from exports, limit the growth of a local private sector and deter moving towards a circular economy. What was written several decades ago remains true: 'the microstate is in a double bind: squeezed from the outside by corporate monoliths, and squeezed from the inside by inadequate capital and demanding physical environments' (Kissling, 1984: 2). Regional shipping services have been somewhat more successful, with the early post-independence success of the Pacific Forum Line (PFL), founded in 1978, despite the lack of complementarity between countries and the dominance of metropolitan countries that poses problems, replicated in air transport. The PFL, now jointly owned by Samoa and the Neptune Pacific Line, could not overcome some national interests and serves just four of the larger SIDS (and some dependent territories).

Throughout the Pacific, inter-island and coastal shipping services are usually run by governments or small, independent shipping companies. Most routes are uneconomic and irregular, and are subsidised by governments, occasionally at cost to reliability and safety. Financing shipping investment, either for governments or private operators, is problematic. Consequently, 'for many PICs at a domestic level, shipping services are restricted to an aging and inefficient fleet' (Bola, 2016: 263). At different scales, transport costs affect access to health, education and markets, and influence social and cultural practices, with high costs of living on outer islands where no alternative fuels or sources of transport exist (Bola, 2016). Meanwhile most outer islands, like Ontong Java (Solomon Islands), have experienced a massive increase in their carbon footprints, as eagerness to adopt the new 'energy slaves' of fibreglass canoes, outboard engines and petrol, reduced self-sufficiency while also damaging marine ecosystems (Bayliss-Smith & Christensen, 2024). Similarly, Namdrik atoll (Marshall Islands), served by ship, three or four times a year, and by occasional planes, has a slowly declining population, typical of Marshallese outer islands, that has made outmigration and climate change the two key island problems (Jarillo & Barnett, 2021). Such islands are particularly disadvantaged, despite sea transport being essential at all levels, from fishing and local transport needs to inter-island shipping. Currently, all maritime transport services are fossil-fuel powered and are becoming increasingly unviable as fuel costs increase. Such increases have considerable effect socially and culturally, inhibiting social and economic development (Bola, 2016), and invariably more significant in archipelagic contexts.

Beyond connectivity, global and regional shipping services pose a second problem, that of high carbon emissions, despite pressure from SIDS in both regions urging the International Maritime Organisation (IMO) to adopt a global emissions levy to drive maritime decarbonisation. The shipping industry is responsible for roughly 3% of global emissions. Compared with efforts to reduce fossil fuel use in electricity production little was done to secure similar reductions in fuel use in shipping (Goundar et al., 2017). Eventually, in mid-2024, the Tuvalu Minister for Transport, Energy, Communication and Innovation, Simon Kofe, speaking at a global Marine Environment Protection Committee (MEPC) meeting, called for bold, decisive action: 'The choices made here will determine the future of our people, our cultures, and our very survival. A universal carbon levy is critical—not for revenue, but to put a real cost on pollution and drive the energy transition we so desperately need!' (quoted in PINA, 1 October 2024). Kofe emphasised that a universal carbon levy, followed by equitable revenue distribution, was essential to ensure vulnerable nations were not disadvantaged in the shift towards a sustainable future. Vanuatu's Special Envoy for Climate Change and Environment, Ralph Regenvanu, added: 'We are not waiting for this climate emergency to arrive. It is here, now'. A regional programme, the Pacific Maritime Decarbonisation Program, covering Fiji, Tonga and Vanuatu, funded by the Republic of Korea and implemented by the Global Green Growth Institute, was launched in mid-2025 to improve the regional context (*Fiji Times*, 20 August 2025). The Marshall Islands (the world's third largest shipping nation) has played a crucial role in negotiating the Paris Agreement and emissions reductions at the IMO, through small, critical and focused interventions, linked to equity. Despite eventual disappointment, the Marshall Islands' success in achieving some change could be explained by four factors: vision, focused capacity development, broad regional support and the commitment of key actors to create and seize opportunities in appropriate forums to realise that vision (Corbett et al., 2020; de Beukelaer et al., 2026). Such approaches, with implications for climate governance, highlight the capacity of small states, while individually inhibited by size, to use informal networks and regional organisations to advance their interests and influence international affairs (Corbett et al., 2021; Cumberbatch et al., 2024). Concerted and directed collaborations have become core elements of regional organisations (Chap. 7). Nonetheless, some 15 IMO states, including China, the United States and Australia, remain opposed to a levy, preferring weaker mechanisms, such as greater emphasis on fuel standards, despite the IMO's target of net-zero emissions by 2050. Meanwhile shipping fleets in the Pacific at least remain some of the oldest and worst carbon emitters in the world, expensively plying the longest transport routes.

6.11.2 Aviation

While the costs of land transport and shipping are largely borne by individuals and companies, aviation has been a widespread economic disaster for many SIDS, insistent on having a national carrier, consistent with a national identity and a support for tourism. Limited connectivity, high fares and poorly timed flights have discouraged tourism in both regions since metropolitan airlines see their principal advantages elsewhere and do not prioritise smaller island markets. Governments have awkwardly filled that void.

SIDS national airlines have now largely disappeared, other than with small planes on limited domestic routes; national pride and national interests were not enough to sustain their services and almost all collapsed because of low and irregular volumes of passengers and freight, high maintenance costs, few landing rights and metropolitan competition. That was compounded by lack of local pilots and ground crew and the high costs of overseas expertise. Aviation is further challenged by fuel costs, airport facilities and regulatory compliance costs to the extent that carriers must fly an aircraft at least 16 hours a day with enough passengers to cover the operating costs, a virtual impossibility in both regions (Ganivatu, 2025). LIAT, owned by Antigua and Barbuda, Barbados, Dominica and St. Vincent, went into administration during the COVID pandemic but survived. Air Vanuatu went bankrupt in 2024 to be eventually rescued by the government and reduced to flying domestic routes: a substantial drain on the national economy and providing much less employment. Once national flag carriers like Polynesian Airlines and Royal Tongan Airlines have disappeared or been reduced to local routes. Yet such small, subsidised airlines can play a valuable role in medical evacuations and search and rescue missions. Intra-regional travel is limited, despite multiple regional initiatives. CARICOM governments established a multilateral Air Services Agreement to facilitate the development of intra-regional travel, but costs are usually considerable. Neither region benefits from the absence of low-cost airlines that reduce passenger costs in Asia, Europe and elsewhere.

6.12 The Internet and Information Technology

At national scale, internet connectivity is limited, especially in Pacific SIDS and the outer islands, since cables are costly to install. That became obvious and critical after the Tongan volcanic eruption of January 2022, when the lone international cable was severed and the entire country cut off. The Vanuatu earthquake of late 2024, which damaged the single cable station, resulted in lost internet connection for ten days. Lost national connectivity occurred precisely when it would have been most valuable. Neither governments nor telecommunication companies had then diversified communications (such as by using more satellite-based systems). SIDS have, however, enjoyed greater international connectivity with the liberalisation of the telecommunications sector from the 1980s, which reduced costs to consumers and increased accessibility, but major gains came in the 2010s. While smartphones may seem universal, coverage is uneven, especially in the archipelagic states of Melanesia and Micronesia (Hobbis, 2019a, b). Prior to 2017, social media uptake was slow, but the completion of submarine cable projects and telecommunication infrastructure investment by regional and international organisations drove a surge in subsequent uptake. Cheap smartphone penetration alongside more affordable data plans meant most SIDS experienced a rapid uptake in internet connectivity. About 80% of the population of

Fiji, Nauru, Tuvalu, Samoa and Marshall Islands are online, well above global averages. However, internet services generally remain slow and often unreliable in Pacific SIDS while also being prohibitively expensive for many. Smartphones have altered the dynamics of communication and access to the internet, but wi-fi connectivity and even the availability of charging facilities are often inadequate in rural locations.

Improved connectivity could stimulate economies, notably tourism, e-commerce (but also crime), while welfare would also benefit if telemedicine could extend access to health care in remote areas (as long anticipated). The most beneficial role of smartphones has been social, anticipating and responding to hazards, and linking families in the SIDS with relatives overseas, thereby keeping in contact, learning about job opportunities and responding to needs. Seasonal workers in Australia and New Zealand depend on their phones to resolve employment and welfare issues and keep in touch with their families at home (Petrou & Connell, 2020). Social media can also contribute to social problems and intergenerational differences. Thus, in Fiji, social media provide emotional support to village youth and their aspirations to modernity, but their increased access to Facebook causes family conflict, with social media transforming village dynamics (Volpe, 2021). It has also resulted in moral uncertainties, restructured kinship relations and enabled access to films (some of dubious quality) and useful information on government policies and market prices (Hobbis, 2020). Social media can be domestically disruptive, enabling secret liaisons and illicit affairs, but it can also enable temporary escape from gender-based violence. Mistrust of partners and surveillance of their communications have become common. Yet the connectivity that telephony provides has been valuable in negotiating and benefiting from modernity, for example, including online financial transactions such as the receipt and expenditure of remittances.

The rapid spread of IT and smartphones has made some contribution to improving social equity through moving away from an early extreme 'digital divide', but mobile services can be patchy, and costly, while digital literacy excludes older people. Even in Fiji, 60% of remote islands lack reliable internet access, a difference compounded by only 35% of rural Fijians possessing basic digital skills compared to 68% of urban residents (Pakaang & Gurung, 2025). Nonetheless, Fiji is in advance of other

Pacific SIDS, especially the archipelagic states of PNG, Solomon Islands and Vanuatu (Tjondronegoro et al., 2025). However, the imminent use of Low Earth Orbit (LEO) satellite technologies is likely to reduce the spatial gap significantly, although, here too, in its use of satellite technology for diverse objectives Fiji is ahead (Sanmarti, 2025).

Although computers are not readily available in the poorest households or the poorest schools, a digital generation has emerged. The Pacific Regional Infrastructure Facility noted in 2015 that 90% of households across Fiji, Samoa, Solomon Islands, Tonga and Vanuatu had a mobile phone, while even in PNG, the most physically challenging of nations for market penetration, 44% of the population had mobile phone subscriptions. Most PICs effectively have more than 100% coverage, although coverage extends to less than 50% of households in the Marshall Islands, FSM and Kiribati, reflecting the disadvantages of communication in large archipelagos. However, internet and mobile phone costs in the SIDS remain higher than in continents, but, in the Caribbean especially, coverage and relative costs have improved and equalised access to government and commercial services irrespective of geographic location or economic and social status.

Telephone services have been provided by the private sector following, in the Pacific at least, government failure to provide basic services (including effective communication systems). Small domestic markets have reduced competition, but multinational companies such as Digicel moved in with alacrity, with elaborate advertising campaigns, slogans emphasising equity of access, with school classrooms and educational initiatives consolidating notions of shared 'infrastructural citizenship'. Consequently, a large proportion of the population have acquired mobile phone access (which is usually loaded with Facebook as part of prepaid data plans, popular among those with lower incomes). By 2024 in the Pacific, except in Vanuatu, FSM and Solomon Islands, more than half of all households used Facebook (Stambolie, 2024). Landlines, rarely ever widely available, became an anachronism, with subscription rates declining, as more people use internet services such as email, WhatsApp, Skype or Zoom for personal links. Governments communicated more through Facebook and X. Like people the world over, islanders sought out TikTok and circulated their own photos.

Since the 1990s, SIDS have experimented with e-government and digitisation but with uneven outcomes. As part of their thrust towards public sector reform, Caribbean and Pacific SIDS have attempted to use information technology to improve service delivery. In theory, new technological advances have enabled the provision of at least adequate services in new ways to outer islands, but, in practice, technology has yet to prove invaluable, typified by the long-anticipated but long-term failure of telemedicine to benefit outer islands, frustrated by management and without the skills to implement and maintain such schemes. As in the Caribbean, SIDS are dependent on unreliable and costly external providers, locked into contracts and without economies of scale, so face problems of operating and maintaining outdated software and technology (Bailey et al., 2017). National digital divides have nonetheless declined.

Cybersecurity has become a big challenge with government and the private sector finding it difficult to prevent such abuses and SIDS poorly prepared for cyberattacks. With few resources, even fewer skilled information security personnel, and low levels of internet awareness among populations, SIDS are vulnerable to phishing attempts, malware and ransomware attacks. High-profile cases in the Caribbean are outcomes of government agencies being incapable of guaranteeing the privacy of information stored on their platforms, whether in Trinidad and Tobago and Jamaica in 2021, or Antigua in 2020. Banks in various Caribbean SIDS, including Jamaica and Trinidad, have also suffered from cyberattacks. The prominent Marshall Islands Marine Resources Authority was the target of a partially successful malware attack in 2023 that forced it to shut down its computer system. A cyberattack on Vanuatu in 2022 closed down government departments for 11 days. A month later, frustrated officials were still using private Gmail accounts, personal laptops, pens and paper and typewriters to manage government business. Early in 2025, Samoa's Computer Emergency Response Team (SamCERT) issued a warning about APT40, a Chinese state-backed hacking group operating in the region, the first time that a PIC had formally attributed a cyberattack to a China-linked group. In mid-2025, the Tonga health system was hacked and encrypted, with the hackers demanding millions of dollars, data being handled manually and residents asked to bring essential paper records to the central hospital (RNZ, 20 June 2025). Indicative of access to technical skills in small SIDS, the Australian government sent teams to assist in re-establishing government IT systems in both Tonga and Vanuatu, with other PICs with even weaker cybersecurity seeking their own precautionary assistance. Such problems have weakened trust in governments and caused financial losses. The private sector and private individuals have also been hacked. Less frequently, physical problems disrupt telecommunications on a large scale. In mid-2024, nearly a third of Tonga's population had been without internet access for two weeks. Again, there were technical and infrastructure issues, when repairs to damaged cables in two of the three island groups, Ha'apai and Vava'u, were hindered when a special repair vessel from Singapore broke down in Fiji and had been stuck there for over a week due to a mechanical fault and fishing net entangled in its propeller (*Talanoa 'O Tonga*, 2 August 2024). Old technological problems were combined with new ones. Extending modern technology has nonetheless been rapid and invaluable.

6.13 Keeping the Peace: A Rule of Law

All SIDS have judicial systems. Shortages of legally qualified citizens have resulted in the appointment of expatriate judges to some national courts, and reliance on foreign judges in most appeal courts. Only the older and larger states—Fiji, Jamaica, Trinidad and Tobago and Barbados—make little use of foreign judges (Robinson, 2023). Foreign judging is pervasive in the SIDS, because the final courts of appeal for all countries are extra-territorial, such as the centuries-old Privy Council, and regionally, the Eastern Caribbean Supreme Court that serves six independent states and three overseas territories. The presence of foreign judges affects the ways in which courts adjudicate and decide cases, and therefore achieve independence and impartiality (Dziedzic, 2023). Despite reliance on expatriate judges, a source of political friction in Kiribati (in a small state scenario where a high court judge was the husband of the leader of the opposition), and resulted in the expulsion of three New Zealand judges, legal systems have usually proved effective in support of stable governance.

Expatriate judges are inevitably unfamiliar with local custom, and this has occasionally proved problematic in cases concerning issues such as land and marine tenure. More generally, however, reference to customary law exists in constitutions but is largely untested in court cases, and with limited and rare exceptions (Techera, 2015; Datta & Chaffin, 2022), efforts to include customary processes into law have stalled. That is so at national level, and, as in FSM, tempers the representative qualities of the judiciary and its connections with the state and the community (Puas, 2023). More generally, merging traditional practices into modern government systems has given rise to ambiguities created by attempts to blend Western and traditional processes, as in the ad hoc blending of Samoa's *fa'amatai* (chiefly) and parliamentary democratic systems (Suaalii-Sauni, 2014); hence, political systems have tended to follow Western systems. Justice is often slow although most SIDS have reputable justice systems in terms of the independence of judges and the quality of legal practice. Yet many people face problems of access, limited legal aid and a system of law seen by many as exhibiting colonial origins as in the common law traditions across many SIDS (e.g. Robinson, 2008). Where anonymity for judges is almost impossible, scrutiny can be intense, so that questions of bias and prejudice may appear more frequent in small states.

Apart from Trinidad, Jamaica and Fiji, SIDS do not have armies, although smaller states such as Tonga have token ceremonial forces. That may have contributed to their being generally more peaceful than many much larger states (Goujon & Wagner, 2023). Beyond cost savings, a further advantage is that only in Fiji have there been military coups. In Fiji too, the military have played a substantial role in UN peacekeeping missions in the Middle East and Africa, and in Bougainville and Solomon Islands, rather than at home, and have avoided conventional warfare. The military absence makes the role of police forces unusually important.

Caribbean SIDS have encountered frequent challenges in ensuring peace and security for their citizens, whereas crime rates in Pacific SIDS are lower than in most parts of the world. Crime and violence appear disproportionately greater in larger SIDS where uneven development is more entrenched. Typically, youth have been universally blamed for high levels of crime but have themselves been disadvantaged by socio-economic structures that present few opportunities (Chap. 5) and they too are victims of crime (Dawkins-Cabrazos, 2024). The long experience of colonialism and uneven development has negative consequences. Levels of violence are high in several Caribbean SIDS with homicide rates well above the global average of 6%, much higher than in the Pacific. Jamaica had a rate of 49 per 100,000 in 2022, one of the highest in the world, with St. Lucia, St. Vincent and Trinidad and Tobago not far behind. By contrast, in the Pacific, rates were usually below the global average, with only Palau (11) in double figures. High rates of violence are often associated with gangs and interpersonal conflicts, less frequent in the Pacific. In Trinidad and Jamaica gang violence accounts for 70% of homicides, mainly in inner city urban garrisons associated with lack of access to citizenship rights, power struggles and marginalisation. Armed gangs emerged, involved in various levels of criminal activity, notably drug trading, as urbanisation and unemployment increased, even becoming politicised as instruments of various individuals and groups. Some gangs created no-go areas where police access only occurs during states of emergency, which, in most cases, represent desperate and unsuccessful attempts at reducing violence. Some police themselves are involved (Mullings, 2009; Rampersad, 2024). States of emergency have been touted as a solution to the gang and rule of law problem, but often generate complex policy outcomes, placing further stress on the rule of law. Intense violence, failed policing strategies, ineffective governance and the presence of gangs make police militarisation appealing (Campbell & Harriott, 2024). In the Melanesian SIDS, gangs have been a parallel problem, at least in the largest towns, but this may be stabilising. Several factors have resulted in urban settlements such as Burns Creek (Honiara, Solomon Islands) becoming more peaceful and less crime-ridden. Residents had built their own primary school and numerous churches; inter-ethnic marriages had occurred; 'middle class' residents including teachers, police and nurses, unable to afford formal urban housing, had moved in; more of the people themselves monitored 'troublemakers' to bring 'shame' on offenders (Evans, 2019). Over time once unruly settlements became more established and more law-abiding suburbs, probably assisted by some reduction in the pace of urbanisation.

More serious acts of violence in the SIDS are committed using firearms, often stolen and, in the Caribbean, trafficked illegally from the United States. Regional bodies such as CARICOM, and individual SIDS have struggled to control the flow of US-manufactured weapons through the ports. Guns have become widespread in some parts of PNG, including Port Moresby. Trade in arms, often in exchange for marijuana, has seen the emergence of local warlords, with the ability to determine election outcomes. Human trafficking exists in the Caribbean, with Jamaica considered to have a flourishing sex trafficking market, the outcome of poverty and where society was perceived to be tolerant of these acts, so that it has been identified, controversially, as a country serving as an initiation point, transit and destination site in trafficking. However, absent data and definitions prevent analysis (Kempadoo, 2016; Harrison et al., 2023; Rhoden-Trader, 2024). Pacific SIDS have largely avoided this problem, despite anecdotal reports of children recruited and exploited in their own countries, trafficked into domestic servitude and commercial sexual exploitation, most apparent in Fiji and Palau, but formal data are again absent (UNODC, 2023).

Gender-based violence is a problem in many SIDS, but too often ignored in largely patriarchal societies, despite its higher incidence in Pacific SIDS than in most parts of the world. In PNG, 43% of women aged 15–19 years have experienced physical and/or sexual violence, and 14% of adolescent girls (15–19 years) have experienced sexual violence and coercion. Violence against children is also high, three quarters of which occurs in the home or school environment. More than 60% of women in Fiji, Solomon Islands, Vanuatu and Kiribati report facing physical or sexual violence, or both, in their lifetimes, by an intimate partner, alongside emotional, verbal and psychological abuse that keeps victims subjugated (UNFPA, 2024; SCF, 2025). In Caribbean SIDS, 44% of women in Trinidad, 29% in Grenada and 28% in Jamaica have also experienced physical or sexual violence from an intimate partner in their lifetime. Too many women do not complain and suffer low self-worth. The combined disasters of COVID-19 and tropical cyclones Harold and Yasa in 2020 led to a rise in domestic violence in Fiji, exacerbating the crisis. After the 2024 earthquake in Vanuatu when 14 died and many houses were destroyed, mainly in or around the capital city, Port Vila, women experienced higher levels of domestic violence, and violence against women had generally increased. In two weeks, the police family protection unit had responded to nearly 50 assault cases, mostly against women, including attempted rape, and indecent assault especially on young women, and especially women in sub-standard accommodation (RNZ, 8 January 2025). While crises intensify violence, social changes, including greater mobility of women and female employment, increase the likelihood of domestic violence, as 'traditional' family structures change, as in Chuuk (FSM) (Smith, 2020; Rauchholz, 2023). As in other parts of the world, there is little indication that the extent of domestic violence in SIDS is declining.

Corruption is both a political and economic phenomenon (the latter particularly evident in money laundering). While politicians have been tried and convicted, as in Nauru, Vanuatu and elsewhere (Chap. 5), money laundering especially is poorly policed, the capacity of local systems to detect corruption is weak and cases rarely reach court (Nguyen & Hopkins, 2022). In Pacific SIDS, crimes of theft, stealing property and damage to property are high on any numerical list, but low on the list of what both police and the public are concerned about. Most such crimes are opportunistic and a product of high rates of youth unemployment, increasing inequality and rapid urbanisation (Watson et al., 2023: 47). Here there is a bias towards youth.

Crime and violence, and especially domestic violence, have scarcely been deterred by ineffective police services, especially in relatively impoverished inner-city areas where, as in Jamaica, a form of self-regulation has become the norm. Police services are often ineffective. Somewhat cynically, it has been pointed out that the Royal Solomon Islands Police Force must surely be one of the most frequently and expensively trained in the region, even the world (Peake & Smith, 2022), but was quite incapable of dealing with urban riots in 2006 and 2021. Police forces have their limitations. Overly violent behaviour is not unusual, as in Vanuatu (Mitchell, 2011). In Fiji too, the police force is renowned for violence: 'Between 2015 and 2040, 400 criminal charges were laid against Fijian police or military officers. Sixteen were for rape, two for murder, nine for manslaughter and five for abduction. There were also 100 charges of assault' so that public order became a pretext for 'setting aside due process, the rule of law and fundamental human rights' (O'Sullivan, 2022). An Amnesty International report on Fiji, *Beating Justice* (2016), recommended independent oversight of the police and military, the military's withdrawal from policing duties and a general presumption that the military is not above the law. There was no indication that this had any impact, causing concern to some that at any time circumstances could worsen, and reversion to a coup culture might follow: 'The broader point is that Fiji is a conditional democracy. The democratic norms of free speech, a free press, freedom of association, freedom of religion, conscience and belief, and the public operation of criminal courts may at any time be overridden' (O'Sullivan, 2022). Fiji's political culture and institutions are not sufficiently strong or focused on human rights to allow or be interested in independent scrutiny of the police.

Despite measures to reform police forces, high quality, effective policing is rare, as in Jamaica, where a history of human rights abuse and a colonial orientation towards violent suppression and control have made them prime candidates for citizens' complaints and high levels of distrust. Security is a major state responsibility and where SIDS fail to deliver it, the social contract between citizens and government is breached. Similar problems of inadequate training afflict policing in the Pacific, despite training programmes repeatedly being part of aid packages from different donor countries. That prompted a brand-new regional intervention in 2024, the Pacific Policing Initiative (PPI), formally directed at the rapidly emerging drug problem that had both national and international elements (see Chap. 7).

In Fiji and elsewhere, prison service workers may also resort to violence and come under even less scrutiny. Even with some degree of independent scrutiny, as in Jamaica where there is an Independent Commission of Investigations (INDECOM), fatal shootings by the police and violations of the rule of law have continued. Excess use of force, scarce resources and lack of community trust have been a feature of policing in most SIDS, as limited accountability, vigilante policing and public desperation for solutions have coalesced: all characteristics that prevent effective functioning and proper governance.

Retention of skilled police is an ongoing challenge for police forces. While most police organisations face human resource capacity issues, in SIDS this is exacerbated by decreased prioritisation of resources for police forces, increasing demands for better-qualified officers, a mismatch between the supply of—and demand for—female officers, and low incentives relative to job requirements (Watson et al., 2023). Moreover police officers experience a complex interplay of factors shaping work-life dynamics, including shift work schedules, high-pressure environments, community expectations and the emotional toll of policing in small, tight-knit societies, all of which require informal support networks, cultural empathy and adaptive coping mechanisms to mediate the impact of crime on wellbeing, but which results in a substantial turnover of workers (Watson et al., 2025). In many Pacific SIDS using community leaders is regarded as the most effective, and using police the least effective, way of resolving disputes, especially in rural areas where a police presence is rare. The police and military have also been seen as part of the problem, not the solution: ineffective in many urban areas and overstretched. Police forces have routinely been regarded as short of resources, increasingly dysfunctional, beset with factions and not to be trusted (Dinnen, 2017). This brought the rapid growth of private security firms, with a PNG workforce alone estimated at 30,000: more than three times the number of police officers. While the most pressing issues in SIDS are domestic and centred in growing urban areas, recent emphasis has been on the external threat posed by transnational crime which requires a more technologically advanced and subtle form of policing.

In Pacific SIDS especially, and in outer islands, most ordinary citizens rely for their security on informal and hybrid arrangements in their communities, which might include local *committees,* church organisations and community leaders, extended families, and, perhaps, the local Village Court magistrate and some government employees associated with the particular community, rather than the police. Minor conflicts and everyday security remain local and informally managed in communities, with little resort to police. Security provision has always been plural and multiscalar, but, as in Solomon Islands, often physically absent at local level (Dinnen & Allen, 2018). In many SIDS, especially in urban areas, citizens have organised their own security (just as they have organised housing, access to water and jobs) and government services (and security) are at best a last and unlikely resort, partly through experiencing a range of resource constraints common to public service functions. Constraints on policing starkly emphasise basic questions about social justice, equity and citizenship, especially for women, tasks of local management and the wider role of government.

6.14 Drugs

Drug trafficking represents an established crime problem in Caribbean SIDS and a rapidly growing one in several Pacific SIDS, affecting both the SIDS themselves and the drug flow between the Americas, Asia and Australia. The United States has long prosecuted a 'war on drugs' in the Caribbean, seeking to curtail the problem (for the United States), but with limited gains, and with the SIDS attracting financial aid from the United States to support maritime policing and control drug flows into America.

Drugs have long flowed through the Caribbean to the United States and are seen as a threat to national regional security by the United States (Chap. 8). The UN estimates that 'the illegal drugs market in the Caribbean is worth 3.3 billion U.S. dollars-3.1 percent of the registered Gross Domestic Product in the region–and between 20 to 65 percent of the total economy depending on the country and estimation methods' (UNDP, UNODC, and OHCHR, 2024: 24). Cocaine represents 85% of the drug market while marijuana's share has declined. Although the United States is the main market for drug consumption and drug transhipment, mainly from Jamaica and the Bahamas, transhipment to Europe has increased in this century. That in turn increased corruption and the recruitment of 'unattached youths' into gangs.

Alongside drug trafficking is gun trafficking. Caribbean SIDS suffer from gun violence, evident in high homicide rates. Guns used in violent crimes are trafficked mainly from the United States, but the United States has focused on drug control rather than gun control. However, cocaine frequently transits through Trinidad peacefully whereas guns are increasingly trafficked into what is a 'weapons sink' (Baird et al., 2023). The Caribbean SIDS have effectively called for a shift from the war on drugs, declaring in 2023 a 'war on guns', and supporting Mexico in taking legal action against American gun manufacturers (Campbell et al., 2026). The Prime Minister of the Bahamas, Phillip Davis, has stated that:

> We do not manufacture guns in the Caribbean. Every gun used to commit a crime in the Caribbean is smuggled into our countries. In The Bahamas, 98.6% of all recovered illegal firearms can be traced directly to the United

> States. … In Jamaica, it amounts to 67% of all recovered firearms and here in Trinidad and Tobago it amounts to 52%. We have asked the US government and US-based gun manufacturers to cooperate with CARICOM member states when it comes to identifying weapons purchased in the US, as a part of a wider effort to hold weapons dealers and traffickers accountable for the many lives lost to gun violence each year. … We intend to challenge the laws that previously protected gun manufacturers from lawsuits. (2023)

Small arms are also a problem in the Pacific but almost exclusively in PNG (and then in the highlands). PNG maintains the largest defence and police force in the region but lacks the resources to prevent imports of arms and ammunition, or deter and prosecute violence. Much as in the Caribbean, illegal guns are at the centre of crime and conflict in the Pacific.

In the Pacific SIDS, drug flows are also from America (and Mexico). Criminal threats undermine maritime security, and PICs positioning along global trade routes increases their vulnerability to maritime crime threats (Lindley & Lothian, 2024). Pacific SIDS have experienced a significant spill-over drug problem into domestic markets for consumption and production with greater involvement from local actors (Sousa-Santos, 2019, 2022). In Tonga and Fiji, that has contributed to a shift from soft drugs, marijuana (present for at least half a century), to harder ones, such as ice and cocaine, flowing along the 'Crystal Road' through the two SIDS from Asia and the Americas to Australia, a particularly valuable market, and New Zealand, increasingly made possible by growing indigenous criminal networks. In turn, that has required more complex and intricate policing of the nexus between transnational and local crime, where plural and hybrid policing approaches, some associated with community policing, have been necessary to combat drug-related crime (Sousa-Santos & Howes, 2022).

While still significant, international flows of some drugs have declined in the Caribbean, notably in the Bahamas, as cocaine prices in European markets accelerated beyond those in the United States. Drug flows through the SIDS consequently declined whereas those through European-affiliated territories, notably Guyane and Curaçao, increased.

Tonga and Fiji were simultaneously emerging as strategic Pacific waypoints. In the Pacific drugs have become part of an expanding organised criminal activity that has moved into the region from elsewhere, including from Australia, Mexico, Colombia, Malaysia and New Zealand. Growing internet penetration has made it easier to link small islands within SIDS into global supply chains of illicit goods, including the introduction of new narcotics for which health services are unprepared. Pacific SIDS were poorly organised to withstand growing criminal threats, some linked to corruption (Comolli, 2024). Some Chinese criminal organisations also moved into the region, alongside legitimate Chinese investors, introducing a new complexity to regional geopolitics (Chap. 8). Other organisations that have entered the islands include Australia's Rebels and Comancheros and New Zealand's Head Hunters, all outlaw motorcycle gangs, and the Mexican Sinaloa drug cartel, introducing and extending a new scale of criminal activity.

Deportations for criminal activity have become a contentious issue regionally, blamed for the spread of criminality acquired in the main deporting states: the United States, New Zealand and Australia. By 2021, Tonga had seen over 1000 people returned after criminal activity overseas, significant enough to justify deportation, the majority in the previous decade, often without community links, languages or cultural knowledge, who often returned to crime. That accelerated dramatically and extended to other SIDS, with further negative impacts, in Donald Trump's second presidency (Chap. 9). Various arrivals have exploited favourable tax regimes and limited monitoring and enforcement capabilities, corrupted political connections and contributed to instability. Beyond obvious but immeasurable corruption, with public servants arrested in Tonga and police in Fiji, guns and ammunition have also been imported to Tonga, health status has worsened, and related activities, including human trafficking for prostitution and scamming, have increased. New Zealand, while offering support to Pacific nations to combat transnational organised crime, deported 400 criminals to Pacific SIDS between 2013 and 2018 (when the narcotics trade was in its relative infancy) and conceded that its deportation policies were a significant contributor to the rise of organised crime in the Pacific (RNZ, 21 August 2025). So substantial have these interrelated problems become that some observers see transnational crime, alongside climate change, as the top two regional development problems and priorities, with the Pacific in danger of becoming a 'semi-narco region' (Sousa-Santos, 2019).

Poverty has been blamed for rapidly increasing drug use in Fiji, especially among i-Taukei (Indigenous) Fijian 'street kids' and school students, with health, especially mental health, much affected, alongside 'a lot of unexplained deaths attached to it' with poverty and the stigma of drug use being deterrents to seeking care (Staunton, 2024). In mid-2024 the capacity of the lone psychiatric hospital in Fiji was overflowing. The practice of 'bluetoothing', attempting to get high by withdrawing blood from a host and injecting it into a second person, resulted in record numbers of new cases of HIV and AIDS-related deaths in 2024. By 2021, Tonga had a methamphetamine epidemic that was disrupting families and overrunning the national hospital and jail: convicted individuals were being given suspended prison sentences because of inadequate incarceration facilities. Drug use was the main precipitating factor for mental illnesses and for the frequent relapse of people with chronic psychiatric disorders such as schizophrenia. In both countries prison systems provide minimal prospects or rehabilitation. Corrections officers lack the training and skills for rehabilitation of offenders with criminal histories, mental health challenges or continued engagement in transnational and organised crime activities. Funding for prison services is limited. Overcrowding, understaffing, under-resourcing and recidivism are typical (Watson, 2024).

Increasing drug flows and lack of rehabilitation have contributed to pressures for greater regional policing, with external involvement, that came to some degree of fruition in 2024, with threats from transnational criminals a problem thrust on the region from outside and thus requiring external intervention. Earlier problems of illegal financial flows and money-laundering had brought similar recommendations of more adequate staffing, resourcing and training of Financial Intelligence Units in Pacific SIDS, greater collaboration and commitment between key players in the financial system of each country, and better cross-border cooperation including regional initiatives (Nguyen & Hopkins, 2022). Invariably, where recommendations sought superior staffing, limited human resources meant that they fell on stony ground. Public services almost always needed strengthening, whether in the Caribbean or the Pacific, but, just as frequently, training capacity and finance were lacking.

6.15 A Thousand Cuts?

The rise of a Pacific drug trade and its shift in direction in the Caribbean indicate, like many other issues and interventions, that the SIDS are constantly changing whether through their own initiatives or through the tribulations of the global world, creating new challenges that could not be resolved at national level where state capacity, as in resolving drug issues, is found wanting, and the genesis of problems lies elsewhere. Capacity building and institutional strengthening are recurrent themes in addressing SDG16 (e.g. Ackman et al., 2018). This chapter and its predecessor have pointed to the fragility of government service provision, critically in education or health, but readily evident in most facets of service delivery across the panoply of needs, from water and sewerage to roads and prisons. Uneven development has partly been shaped by archipelagic geographies and by a politics of inattention and amnesia, and by failures of management where human resources are limited (and even declining). Public administration is costly and inefficient, and service provision in remote areas has often worsened rather than improved (Utz, 2021). As urbanisation increased so too did urban bias in service delivery, which has meant that rural areas and outer islands are disadvantaged.

Urban-resident politicians can be corrupt and self-interested (hence the 'slush funds', or constituency funds, of PNG and Solomon Islands) and put little pressure on bureaucrats to develop national policies (Chap. 5). One outcome of urban bias and limited and uneven service provision has been outmigration and an effectively vicious circle: a continued slow depopulation, and the withdrawal, rather than extension of services—death by a thousand cuts. By contrast, the environmental degradation from multiple varieties of plastic has been described by Sarah Dunlop as 'death by a thousand conveniences' (*The Sun-Herald*, 17 August 2025). Technology has yet to overcome distance. Technological innovation in transport has benefited larger states and meant that almost every outer island is less accessible now than at any time in the last 50 years, so local export economies have collapsed, while access to services has worsened, suggesting a bleak future (see Chap. 9).

Although the digital divide is less divisive with the ubiquity of mobile phones, services are invariably superior in the 'better' urban suburbs and poorest in the settlements and garrisons, and the outer islands, where hazards exacerbate marginality. While the rapid contemporary spread of solar energy has brought welcome electricity in many places, it has also been accompanied by a new phrase—'distributional energy injustice'—as distant places never quite catch up and women bear the brunt of uneven extension. Clean water provision too can be erratic, as infrastructure is degraded, and climate change poses a potential threat to urban sustainability. Islanders expect superior infrastructure—early childhood education, diabetes and broken bridges should not have to wait—and that everyday systems be more congenial, resilient, sustainable and equitable. As in policing, and in so many other areas, where challenges come from outside the nation (whether climate change, the tide of plastics, ice or epidemics), imported problems can come without imported solutions. As in each of these contexts, possibilities exist that governance and service provision might be more successfully resolved at a regional scale. We consider this possibility in the next chapter, where Caribbean and Pacific regionalism have provided benefits over half a century, without being wholly successful, but enabling new ways of thinking about cooperation, development and sustainability.

Bibliography

Abdulkadri, A., Cunningham-Myrie, S., Mkrtchyan, I., et al. (2021). *Addressing the adverse impacts of non-communicable diseases on the sustainable development of Caribbean countries*. ECLAC Studies No. 1000.

Ackman, M., Hammond, D., Cooney, C., & Liu, L. (2018). *SDG16 in the Pacific. Strengthening and legitimising institutions to achieve sustainable development*. Institute for Economics and Peace.

Adeoti, T., Fantini, C., Morgan, G., et al. (2020). *Infrastructure for small island developing states*. UNOPS.

Ahlgren, I., Yamada, S., & Wong, A. (2014). Rising oceans, climate change, food aid, and human rights in the Marshall Islands. *Health and Human Rights Journal, 16*(1), 69–80.

Ambrose, K., Box, C., Boxall, J., et al. (2019). Spatial trends and drivers of marine debris accumulation on shorelines in South Eleuthera, The Bahamas using citizen science. *Marine Pollution Bulletin, 142*, 145–154.

Anthonj, C., Tracy, J., Fleming, L., et al. (2020). Geographical inequalities in drinking water in the Solomon Islands. *Science of the Total Environment, 712*, 135241.

Aranda, A., Oxenford, H., Medina, J., et al. (2022). Widespread microplastic pollution across the Caribbean Sea confirmed using queen conch. *Marine Pollution Bulletin, 178*, 113582.

Atteridge, A., Verkuijil, C., & Dzebo, A. (2020). Nationally determined contributions (NDCs) as instruments for promoting national development agendas? An analysis of small island developing states (SIDS). *Climate Policy, 20*(4), 485–498.

Awa, A. (2025, June). Tackling systemic inequities in the Solomon Islands Healthcare System: The race-health-governance nexus. *Visions and Voices*.

Bacchus, M. (2008). The education challenges facing small nation states in the increasingly competitive global economy of the twenty-first century. *Comparative Education, 44*, 127–145.

Bailey, A., Minto-Coy, I., & Thakur, D. (2017). IT governance in E-government Implementations in the Caribbean: Key characteristics and mechanisms. In L. Rusu & G. Viscusi (Eds.), *Information technology governance in public organizations: Theory and practice* (pp. 201–227). Springer.

Baird, A., Bishop, M., & Kerrigan, D. (2023). Differentiating the local impact of global drugs and weapons trafficking: How do gangs mediate 'residual violence' to sustain Trinidad's homicide boom? *Political Geography, 106*, 102966.

Barrowclough, D., & Eugui, D. (2021). *Plastic production and trade in small states and SIDS: The shift towards a circular economy*. Commonwealth Secretariat.

Bayliss-Smith, T., & Christensen, A. (2024). From copra sequestration to outboard engine trawling: Changes in the carbon footprint of Ontong Java Atoll, Solomon Islands. *Journal of Pacific History, 59*(4), 468–490.

Beaumont, N., Aanesen, M., Austen, M., et al. (2019). Global ecological, social and economic impacts of marine plastic. *Marine Pollution Bulletin, 142*, 189–195.

Bell, C., Latu, C., Na'ati, E., Snowdon, W., Moodie, M., & Waqa, G. (2021). Barriers and facilitators to the introduction of import duties designed to prevent noncommunicable disease in Tonga: A case study. *Globalization and Health, 17*, 136.

Belmar, Y., McNamara, K., & Morrison, T. (2016). Water security in small island developing states: The limited utility of evolving governance paradigms. *WIREs Water, 3*(2), 181–193.

Bennett, S., Talouli, A., & Dixit, V. (2024). Towards a circular logistics partnership: Regional trade, waste, and inter-sectoral cooperation. In A. Dansie, H. Alleway, & B. Böer (Eds.), *The water, energy, and food security nexus in Asia and the Pacific: The Pacific* (pp. 197–199). Springer.

Binetti, U., Silburn, B., Russell, J., et al. (2020). First marine litter survey on beaches in Solomon Islands and Vanuatu, South Pacific. *Marine Pollution Bulletin, 161*, 111827.

Bola, A. (2016). Potential for sustainable sea transport: A case study of the Southern Lomaiviti, Fiji islands. *Marine Policy, 75*, 260–270.

Bourne, P. A. (2025). Violence in Western Jamaica: A Secondary Data Analysis. *International Journal of Insights & Transformations in Law, Crime & Justice, 9*(2), 18–44.

Bradley, M., & Swaddling, A. (2018). Addressing environmental impact assessment challenges in Pacific island countries for effective management of deep sea minerals activities. *Marine Policy, 95*, 356–362.

Braithwaite, S. (2022). Estimating the economic costs of noncommunicable diseases in CARICOM. In C. Cannonier & M. Burke (Eds.), *Contemporary issues within Caribbean economies* (pp. 65–82). Palgrave Macmillan.

Brodie, G., Holland, E., N'Yeurt, A., et al. (2020). Seagrasses and seagrass habitats in Pacific small island developing states: Potential loss of benefits via human disturbance and climate change. *Marine Pollution Bulletin, 160*, 111573.

Burnett, G., & Suluma, W. (2024). Formal education in Nauru: A Pacific development assistance story. *Asia Pacific Viewpoint, 65*(2), 276–287.

Cahu, P., & Sondergaard, L. (2023). *Estimating interim learning poverty for Pacific Island countries*. World Bank.

Campbell, Y., & Harriott, A. (2024). The resort to emergency policing to control gang violence in Jamaica: Making the exception the rule. *Journal of Latin American Studies, 56*(1), 115–136.

Campbell, Y., Harriott, A., Grey, F. A., & Blake, D. (2026). From the 'war on drugs' to the 'war on guns': South–South cooperation between Mexico and the Caribbean. *European Journal of International Security, 11*(1), 96–114.

Chand, S., Sing, B., & Kumar, S. (2020). The economic burden of non-communicable disease mortality in the South Pacific: Evidence from Fiji. *Plos One, 15*(7), e0236068.

Charlson, F., Diminic, S., & Whiteford, H. (2015). The rising tide of mental disorders in the Pacific region. *Asia and the Pacific Policy Studies, 2*(2), 280–292.

Comolli, V. (2024, March 14). Geopolitics, influence and crime in the Pacific islands. *The Strategist.*

Connell, J. (2010). *Migration and the globalisation of health care. The health worker exodus?* Edward Elgar.

Corbett, J., Ruwet, M., Xu, Y., & Weller, P. (2020). Climate governance, policy entrepreneurs and small states: Explaining policy change at the International Maritime Organisation. *Environmental Politics, 29*(5), 825–844.

Corbett, J., Yi-Chong, X., & Weller, P. (2021). *International Organizations and Small States: Participation, Legitimacy and Vulnerability.* Bristol University Press.

Corbin, C. (2023). The role of waste management in underpinning the blue economy. In P. Clegg, R. Mahon, P. McConney, & H. Oxenford (Eds.), *The Caribbean blue economy* (pp. 195–209). Routledge.

Coxon, A., & Munce, K. (2008). The global education agenda and the delivery of aid to Pacific education. *Comparative Education, 44*, 147–165.

Crisman, T., & Winters, Z. (2023). Caribbean small island developing states must incorporate water quality and quantity in adaptive management of the water-energy-food nexus. *Frontiers in Environmental Science, 11*, 1212552.

Crossley, M., Bray, M., & Packer, S. (2011). *Education in small states: Policies and priorities.* Commonwealth Secretariat.

Cumberbatch, J., Hinds, C., McConney, P., Speede, R., et al. (2024). Adaptive capacity, governance and small island developing states: A case study of sargassum management in the Eastern Caribbean. *Coastal Management, 52*(6), 340–358.

Culpin, A. (2017). Child health and nutrition in the Marshall Islands. *Development Bulletin, 78*, 83–86.

Currenti, R., Pearce, T., Salabogi, T., Vuli, L., et al. (2019). Adaptation to climate change in an interior Pacific Island village: A case study of Nawairuku, Ra, Fiji. *Human Ecology, 47*, 65–80.

Datta, A., & Chaffin, B. (2022). Evolving adaptive governance: Challenging assumptions through an examination of fisheries law in Solomon Islands. *Ecology and Society, 27*(2), 30.

Dauvergne, P., Allan, J., Beaudoin, S., et al. (2025). Competing axes of power in the global plastics treaty: Understanding the politics of progress and setbacks in negotiating a high-ambition agreement. *Marine Policy, 181*, 106820.

Davis-Mattis, L. (2023). The future of deep-seabed minerals and marine genetic resources in blue economies. In P. Clegg, R. Mahon, P. McConney, & H. Oxenford (Eds.), *The Caribbean blue economy* (pp. 182–194). Routledge.

Dawkins-Cabrazos, M. (2024). An overview of youth crime and violence in the Caribbean community (CARICOM). In S. Morris-Francis, C. Gibson, & L. Grant (Eds.), *Youth crime and violence in the Caribbean* (pp. 19–32). Lexington.

De Beukelaer, C., Taukäve, J., Sahib, M., et al. (2026). Sailing Through Time: Building Pacific Maritime Resilience. *Global Policy*, 1–12.

Dehm, J., Le Gendre, R., Lal, M., et al. (2025). Water quality within the greater Suva urban marine environment through spatial analysis of nutrients and water properties. *Marine Population Bulletin, 213*, 117601.

Del Campo, F., Singh, S., Fishman, T., et al. (2023). Can a small island nation build resilience? The significance of resource-use patterns and socio-metabolic risks in The Bahamas. *Journal of Industrial Ecology, 27*, 491–507.

Demerath, P. (1999). The cultural production of educational utility in Pere Village. *PapuaNew Guinea. Comparative Education Review, 43*(2), 162–192.

Dinnen, S. (2017). The inexorable rise of private security in Papua New Guinea. *Development Bulletin, 78*, 63–66.

Dinnen, S., & Allen, M. (2018). Reflections on hybridity as an analytical lens on state formation: The case of Solomon Islands. In J. Wallis et al. (Eds.), *Hybridity on the ground in peacebuilding and development: Critical conversations* (pp. 129–144). ANU Press.

Dornan, M. (2015). Renewable energy development in small island developing states of the Pacific. *Resources, 4*, 490–506.

Dornan, M., & Shah, K. (2016). Energy policy, aid, and the development of renewable energy resources in small island developing states. *Energy Policy, 98*, 759–767.

Duke, R. & Ahsan, N (2024, November 18). Pacific education: Backsliding outcomes and a financing freefall. *Interpreter*.

Dziedzic, A. (2023). Judicial mobilities. Travelling judges in the Pacific. In A. Dziedzic & S. Young (Eds.), *The Cambridge handbook of foreign judges in domestic courts* (pp. 300–322). Cambridge University Press.

Elgie, A., Singh, S., & Telesford, J. (2021). You can't manage what you can't measure: The potential for circularity in Grenada's waste management system. *Resources, Conservation and Recycling, 164*, 105170.

Elliott, M., MacDonald, M., Chan, T., et al. (2017). Multiple household water sources and their use in remote communities with evidence from Pacific Island Countries. *Water Resources Research, 53*(11), 9106–9117.

Enayati, S., Campbell, J., & Li, H. (2023). Vaccine distribution with drones for less developed countries: A case study in Vanuatu. *Vaccine:X, 14*, 100312.

Evans, D. (2019). Bad places turned good: Security in Melanesia's settlement communities. *ANU DevPolicyBlog*, 4 February.

Farrelly, T., Borrelle, S., & Fuller, S. (2021). The strengths and weaknesses of Pacific Islands plastic pollution policy frameworks. *Sustainability, 13*, 1252.

Ferreira, M., Thompson, J., Paris, A., et al. (2020). Presence of microplastics in water, sediments and fish species in an urban coastal environment of Fiji, a Pacific small island developing state. *Marine Pollution Bulletin, 153*, 110991.

Fleming, L., Anthonj, C., Thakkar, M., Tikoisuva, W. et al. (2019). Urban and rural sanitation in the Solomon Islands: How resilient are these to extreme weather events?. *Science of the Total Environment, 683*, 331–340.

Fitzgerald, S., Moriarty, A., Basas, R., & Shagazatova, M. (2024). *Making Space: Women's participation in higher education in Solomon Islands*. ADB Policy Brief.

Foster, T., & Willetts, J. (2018). Multiple water source use in rural Vanuatu: Are households choosing the safest option for drinking? *International Journal of Environmental Health Research, 28*(6), 579–589.

Fukuoka, T., Sakane, F., Kinoshita, C., Sato, K., et al. (2022). Covid-19-derived plastic debris contaminating marine ecosystem: Alert from a sea turtle. *Marine Pollution Bulletin, 175*, 113389.

Ganivatu, E. (2025). Fuel cost threat. Economic challenge for island aviation. *Islands Business, 51*(5), 14.

Garces-Ordonez, O., Spence, R., Canals, M., & Thiel, M. (2025). Macrolitter contamination in beach, dune, and mangrove ecosystems on a Caribbean island: A comparative analysis. *Marine Pollution Bulletin, 213*, 117616.

Genave, A., Blancard, S., & Garabedian, S. (2020). An assessment of energy vulnerability in small island developing states. *Ecological Economics, 171*, 106595.

Gheuens, J., Nagabhatla, N., & Perera, E. D. P. (2019). Disaster-risk, water security challenges and strategies in small island developing states (SIDS). *Water, 11*(4), 637.

Glasgow, S., & Schrecker, T. (2016). The double burden of neoliberalism? Noncommunicable disease policies and the global political economy of risk. *Health & Place, 39*, 204–211.

Goujon, M., & Wagner, L. (2023). Fragility of small island developing states. In D. Carment & Y. Samy (Eds.), *Handbook of fragile states* (pp. 299–315). Edward Elgar.

Goundar, A., Newell, A., Nuttall, P., et al. (2017). King Canute muses in the South Seas: Why aren't Pacific Islands transitioning to low carbon sea futures? *Marine Policy, 81*, 80–90.

Graham, R. (2023). The marine litter issue in the Windward Islands- a pathway to responses using the DPSIR framework. *Frontiers in Environmental Science, 11*, 1150722.

Hall, I., & McDonald, A. (2020). The Abandoned Cars of Pohnpei. Reflections on a small island economy and environment. *Shima, 14*(2), 121–130.

Haraksingh, I. (2023). Renewable energy: An emerging blue economy sector. In P. Clegg, R. Mahon, P. McConney, & H. Oxenford (Eds.), *The Caribbean blue economy* (pp. 156–168). Routledge.

Harrison, A., James, C., & Pemberton, A. (2023). Human trafficking of adolescents: A Caribbean perspective. *Current Pediatrics Reports, 11*, 116–112.

Hawley, N., & McGarvey, S. (2015). Obesity and diabetes in Pacific Islanders: The current burden and the need for urgent action. *Current Diabetes Reports, 15*(6), 29.

Henry-Lee, A., Watson, P., Osei, P., & Trezelle, S. (2010). Social risk management in a small state: The case of St Lucia. *Social and Economic Studies, 59*, 61–89.

Hicks, R., McDougall, D., & Oakeshott, D. (2021). The promise of education: Schooling journeys in the Southwestern Pacific. *The Contemporary Pacific, 33*(2), 301–336.

Hill, N., Peatman, T., Wakefield, C., et al. (2018). Improving guidelines for implementing harvest strategies in capacity-limited fisheries – Lessons from Tonga's deepwater line fishery. *Marine Policy, 98*, 85–91.

Hills, J., Michalena, E., & Chalvatzis, K. (2018). Innovative technology in the Pacific: Building resilience for vulnerable communities. *Technological Forecasting and Social Change, 129*, 16–26.

Hobbis, G. (2019a). New media, new Melanesia? In E. Hirsch & W. Rollason (Eds.), *The Melanesian world* (pp. 546–560). Routledge.

Hobbis, S. (2019b). A road to development? Rural perspectives on infrastructure maintenance in Solomon Islands. *Development in Practice, 29*(6), 748–759.

Hobbis, G. (2020). *The digitizing family. An ethnography of Melanesian smartphones*. Palgrave Macmillan.

Iredale, R., Voigt-Graf, C., & Khoo, S. (2015). Trends in international and internal teacher mobility in three Pacific Island Countries. *International Migration, 53*(1), 97–114.

Isley, C., & Taylor, M. (2018). Air quality management in the Pacific Islands: A review of past performance and implications for future directions. *Environmental Science & Policy, 84*, 26–33.

Jaffe, R. (2024). *The rule of dons. Criminal leaders and political authority in Urban Jamaica*. Duke University Press.

Jamaica Education Transformation Committee. (2021). *The Reform of Education in Jamaica, 2021*. Kingston.

Jarillo, S., & Barnett, J. (2021). Contingent communality and community-based adaptation to climate change: Lessons from a Pacific rural atoll. *Journal of Rural Studies, 87*, 137–145.

Johansson-Fua, S. (2022). Wansolwara: Sustainable development, education, and regional collaboration in Oceania. *Comparative Education Review, 66*(3), 516–534.

Jules, D. (2008). Rethinking education for the Caribbean: A radical approach. *Comparative Education, 44*(2), 203–214.

Jumo, M. (2022). Time to reprioritise girls' education. *ANU DevPolicy Blog*, 3 May.

Kelly, F., Wright, S., Woodward, G., & Fussell, J. (2025). Plastic pollution under the influence of climate change: Implications for the abundance, distribution, and hazards in terrestrial and aquatic ecosystems. *Frontiers in Science, 3*, 1636665.

Kelman, I. (2018). Islandness within climate change narratives of small island developing states (SIDS). *Island Studies Journal, 13*(1), 149–166.

Kempadoo, K. (2016). The war on humans: Anti-trafficking in the Caribbean. *Social and Economic Studies, 65*(4), 5–32.

Kersey, K., Blechinger, P., & Shirley, R. (2021). A panel data analysis of policy effectiveness for renewable energy expansion on Caribbean islands. *Energy Policy, 155*, 112340.

Kim, K. (2024, June 25). Navigating the intersection of climate change and ocean contamination: Addressing plastic waste issues for the future of the Pacific. *Visions and Voices*.

Kissling, C. (Ed.). (1984). *Transport and Communications for Pacific Microstates: Issues in Organisation and Management*. USP.

Kuang, Y., Zhang, B., Li, C., & Cao, Y. (2016). A review of renewable energy utilization in islands. *Renewable and Sustainable Energy Reviews, 59*, 504–513.

Lachmann, F., Almroth, B., Baumann, H., et al. (2017). *Marine plastic litter on small island developing states (SIDS): Impacts and measures*. Swedish Institute for the Marine Environment, University of Gothenburg.

Lam, C., Islam, M., Devi, R., et al. (2025). Examining wealth-related inequality in childhood vaccination in Fiji using the UNICEF Multiple Indicator Cluster Survey 2021. *International Health, 17*(5).

Lee, H. (2018). Becoming Tongan today. In J. Connell & H. Lee (Eds.), *Change and continuity in the Pacific: Revisiting the region* (pp. 118–135). Routledge.

Lester, S., & Rhiney, K. (2018). Going beyond basic access to improved water sources: Towards deriving a water accessibility index. *Habitat International, 73*, 129–140.

Levasseur, P. Rouly, S & Seetahul, S (2025) Water insecurity and purchases of soft and sweet beverages: A case study in the obesogenic context of the Pacific Islands, *Journal of the Asia Pacific Economy*, 30.

Lewis, J. S. (2020). *Scammer's yard: The crime of black repair in Jamaica*. University of Minnesota Press.

Lindley, J., & Lothian, S. (2024). Exposure to transnational maritime crime in the Pacific Islands Region. *Ocean and Society, 1*.

Liu, X., Peerally, J., De Fuentes, C., et al. (2022). Who makes or breaks energy policymaking in the Caribbean small island jurisdictions? A study of stakeholders' perceptions. *Sustainability, 14*, 1902.

Love, M., Beal, C., Gonzalez, D., & Hagabore, J. (2022). A livelihood analysis of resettlement outcomes: Lessons for climate-induced relocations. *Development Policy Review, 40*(4), e12597.

Lucas, H., Fifita, S., Talab, I., et al. (2017). Critical challenges and capacity building needs for renewable energy deployment in Pacific small island developing states (Pacific SIDS). *Renewable Energy, 107*, 42–52.

MacDonald, M., Chan, T., Elliott, M., Kearton, L., et al. (2017). Temporal and thematic trends in water, sanitation and hygiene (WASH) research in Pacific Island countries: A systematic review. *Journal of Water Sanitation and Hygiene for Development, 7*(3), 352–368.

MacDonald, M., Elliott, M., Langidrik, D., et al. (2020). Mitigating drought impacts in remote island atolls with traditional water usage behaviors and modern technology. *Science of the Total Environment, 741*, 140230.

McGinn, A., & Solofa, A. (2020). Multi-level governance of climate change adaptation: A case study of country-wide adaptation projects in Samoa. In W. Filho (Ed.), *Managing climate change adaptation in the Pacific region* (pp. 151–170). Springer.

McIlgorm, A., Raubenheimer, K., McIlgorm, D., & Nichols, R. (2022). The cost of marine litter damage to the global marine economy: Insights from the Asia-Pacific into prevention and the cost of inaction. *Marine Pollution Bulletin, 174*, 113167.

McIntyre, W. (2014). *Winding up the British Empire in the Pacific Islands*. Oxford University Press.

McLennan, A., & Ulijaszek, S. (2015). Obesity emergence in the Pacific islands: Why understanding colonial history and social change is important. *Public Health Nutrition, 18*(8), 1499–1505.

MacLeod, M., Arp, H., Tekman, M., & Jahnke, A. (2021). The global threat from plastic pollution. *Science, 373*, 61–65.

Manglou, M., Rocher, L., & Bahers, J.-B. (2022). Waste colonialism and metabolic flows in island territories. *Journal of Political Ecology, 29*(1), 1–19.

Matagi, L., Richards, D., Yabar, H., et al. (2024). Enhancing solid waste management in Fiji: A comprehensive approach with LCA, GIS, and waste treatment strategies. *Resources, Conservation & Recycling Advances, 23*, 200228.

Meessen, B., Ancia, A., Gill, D., et al. (2024). When one size does not fit all: Aid and health system strengthening for small island developing states. *Health Policy and Planning, 39*(Suppl.1), i4–i8.

Meschede, H., Bertheau, P., Khalili, S., & Breyer, C. (2022). A review of 100% renewable energy scenarios on islands. *WIREs Energy and Environment, 11*, e450.

Michalena, E., & Hills, J. (2018). Paths of renewable energy development in small island developing states of the South Pacific. *Renewable and Sustainable Energy Reviews, 82*(1), 343–352.

Miller, P. (2014). Education for all in the Caribbean: Promise, paradox and possibility. *Research in Comparative and International Education, 9*(1), 1–3.

Millette, S., Williams, W., & Hull, C. (2019). Materials flow analysis in support of circular economy development: Plastics in Trinidad and Tobago. *Resources, Conservation & Recycling, 150*, 104436.

Mitchell, J. (2011). "Operation restore public hope": Youth and the magic of modernity in Vanuatu. *Oceania, 81*(1), 36–30.

Moglia, M., Perez, P., & Burn, S. (2008). Water troubles in a Pacific atoll town. *Water Policy, 10*, 613–637.

Mohammadi, E., Singh, S., & Habib, K. (2021a). Electronic waste in the Caribbean: An impending environmental disaster or an opportunity for a circular economy?, *Resources. Conservation and Recycling, 164*, 105106.

Mohammadi, E., Singh, S., & Habib, K. (2021b). How big is circular economy potential on Caribbean islands considering e-waste? *Journal of Cleaner Production, 317*, 128457.

Mohee, R., Mauthoor, S., Bundhoo, Z. M., Somaroo, G., et al. (2015). Current status of solid waste management in small island developing states: A review. *Waste Management, 43*, 539–549.

Morrison, R., & Munro, J. (1999). Waste management in the small island developing states of the South Pacific: An overview. *Australian Journal of Environmental Management, 6*(4), 232–246.

Mosley, L., Sharp, D., & Singh, S. (2004). Effect of a tropical cyclone on the drinking water quality of a remote Pacific island. *Disasters, 38*, 405–417.

Mullings, B. (2009). Neoliberalization, social reproduction and the limits to labour in Jamaica. *Singapore Journal of Tropical Geography, 30*(2), 174–188.

Mycoo, M. (2018). Achieving SDG 6: Water resources sustainability in Caribbean small island developing states through improved water governance. *Natural Resources Forum, 42*(1), 54–68.

Mycoo, M., & Roopnarine, R. (2024). Water resource sustainability: Challenges, opportunities and research gaps in the English-speaking Caribbean small island developing states. *PLOS Water, 3*(1), e0000222.

Napper, I., & Thompson, R. (2020). Plastic debris in the marine environment: History and future challenges. *Global Challenges, 4*(6), e1900081.

Nasim, N., Anthony, S., Daurewa, T., et al. (2023). Understanding on-site sanitation in rural Fiji: Where definitions of sanitation back-ends differ. *Environmental Science: Water Research & Technology, 9*(7), 1913–1931.

Nguyen, C., & Hopkins, W. (2022). *Corruption and money laundering in the Pacific: Intertwined challenges and interlinked responses.* Transparency International New Zealand.

Nibbs, F. (2024, May 13). Thirsty in paradise: Water crises are a growing problem across the Caribbean islands. *The Conversation.*

Niles, K., & Lloyd, B. (2014). Using power sector reform as an opportunity to increase the uptake of renewable energy in the power sector: Responding to peak oil and climate change in Caribbean and Pacific small island developing States, between 1970-2010. *Natural Resources Forum, 38*, 14–26.

Nunn, R., Itinteang, T., Ioteba, N., & Tan, W. (2024). *Child mortality in Kiribati: A wake-up call for urgent action.* ANU Devpolicy Blog, 1 November.

OECS. (2024). *Power below the surface. Harnessing geothermal energy in the OECS.* OECS Commission.

O'Sullivan, D. (2022, June 30). Police violence in Fiji: An endemic problem with no simple end in sight. *Australian Institute of International Affairs.*

Overton, J., & Murray, W. (2018). Migration, education and marginality: Networks and strategies in the Pacific Islands. In S. Pelc & M. Koderman (Eds.), *Nature, tourism and ethnicity as drivers of (de)marginalization. Perspectives on geographical marginality* (pp. 215–233). Springer.

Owens, E., Zhang, Q., & Mihelcic, J. (2011). Material flow analysis applied to household solid waste and marine litter on a small island developing state. *Journal of Environmental Engineering, 137*, 937–944.

Pakaang, A., & Gurung, O. (2025). *Bridging Fiji's digital divide: A holistic approach to digital and financial inclusion*. Griffth Asia Institute.

Pareti, K. (2024). Addressing Kiribati's 'alarming' water situation. *Islands Business, 50*(8), 48–50.

Paris, A., Kwaoga, A., & Hewavitharane, C. (2022). An assessment of floating marine debris within the breakwaters of the University of the South Pacific, Marine Studies Campus at Laucala Bay. *Marine Pollution Bulletin, 174*, 113290.

Patay, D., Schram, A., & Friel, S. (2023). The challenges in protecting public health interests in multisectoral governance in the context of small island developing states: The case of tobacco control in Fiji and Vanuatu. *Globalization and Health, 19*(1), 31.

Peake, G., & Smith, G. (2022, January 27). Solomon Islands: Cops bearing gifts. *Interpreter*.

Pendeverana, L. (2017). Rethinking Honiara's water supply: Voices from the settlements. *Development Bulletin, 78*, 59–61.

Petrou, K., & Connell, J. (2020). Overcoming precarity? Social media, agency and Ni-Vanuatu seasonal workers in Australia. *Journal of Australian Political Economy, 84*, 116–146.

Petzold, J., & Magnan, A. (2019). Climate change: Thinking small islands beyond small island developing states (SIDS). *Climatic Change, 152*, 145–165.

Phillips, T., & Keen, M. (2016). *Sharing the City: Urban growth and governance in Suva, Fiji*. ANU State, Society and Governance in Melanesia Discussion Paper 2016/6.

Phillips, T., & Narayan, A. (2017). Meeting the health care challenges posed by rapid urbanisation in the Pacific: The view from Fiji. *Development Bulletin, 78*, 79–82.

Pomponio, A. (1993). Education is development on a ten-acre island. In V. Lockwood et al. (Eds.), *Contemporary Pacific societies* (pp. 197–215). Prentice Hall.

Praene, J., Payet, M., & Bénard-Sora, F. (2018). Sustainable transition in small island developing states: Assessing the current situation. *Utilities Policy, 54*, 86–91.

Pratap, A., Mani, F., & Prasad, S. (2020). Heavy metals contamination and risk assessment in sediments of Laucala Bay, Suva, Fiji. *Marine Pollution Bulletin, 156*, 111238.

Puas, G. (2023). Shaping the legal landscape. Foreign judges in the federated states of Micronesia. In A. Dziedzic & S. Young (Eds.), *The Cambridge handbook of foreign judges in domestic courts* (pp. 447–462). Cambridge University Press.

Pueyo, A., & Haas, J. (2025). Energy transitions in the Pacific, one island at a time: The case of Efate island, Vanuatu. *Energy Policy, 206*, 114751.

Raghoo, P., Surroop, D., Wolf, F., et al. (2018). Dimensions of energy security in small island developing state. *Utilities Policy, 53*, 94–101.

Rampersad, I. (2024). Contemporary dynamics of drug and gang activism in Trinidad and Tobago. In S. Morris-Francis, C. Gibson, & L. Grant (Eds.), *Youth crime and violence in the Caribbean* (pp. 231–250). Lexington.

Rashid, A., Nakib, T., Shahriar, T., et al. (2024). Energy and economic analysis of an ocean thermal energy conversion plant for Bangladesh: A case study. *Ocean Engineering, 293*, 116625.

Rauchholz, M. (2023). Family and gender violence in Chuuk, Micronesia. In A. Kelly-Hanku, P. Aggleton, & A. Malcolm (Eds.), *Sex and gender in the Pacific. Contemporary perspectives on sexuality, gender and health* (pp. 181–191). Routledge.

Reeve, E., Thow, A.-M., Huse, O., et al. (2021). Policy-makers' perspectives on implementation of cross-sectoral nutrition policies, Western Pacific Region. *Bulletin of the WHO, 99*, 865–873.

Rhoden-Trader, J. (2024). Human trafficking of children and youth in the Caribbean. In S. Morris-Francis, C. Gibson, & L. Grant (Eds.), *Youth crime and violence in the Caribbean* (pp. 189–208). Lexington.

Robinson, T. (2008). Gender, nation and the common law constitution. *Oxford Journal of Legal Studies, 28*(4), 735–762.

Robinson, T. (2023). Foreign judging and securing judicial independence in the Anglo Caribbean. In A. Dziedzic & S. Young (Eds.), *The Cambridge handbook of foreign judges in domestic courts* (pp. 324–344). Cambridge University Press.

Rouse, J. (2013). Sustainability of wastewater treatment and excess sludge handling practices in the Federated States of Micronesia. *Sustainability, 5*, 4183–4194.

Rousso, B., Sanderson, R., Love, M., et al. (2024). A multi-country survey on sanitation systems in underserved urban settlements in the Melanesian Pacific region. *Clean Water, 7*(93).

Sanmarti, M. (2025, December 4). Astropolitics of Fiji: An introduction. *Australian Outlook.*

Santagata, E., Regenauer-Lieb, K., & Corkish, R. (2024). Geothermal energy for PICTs. In A. Dansie, H. Alleway, & B. Boer (Eds.), *The water, energy and food security Nexus in Asia and the Pacific: The Pacific* (pp. 416–444). Springer.

SCF. (2025). *Regional child protection situational analysis – Pacific*. SCF.

Schrecongost, A., & Wong, K. (2015). *Unsettled: Water and sanitation in urban settlements communities of the Pacific*. Pacific Regional Infrastructure Facility.

Schrecongost, A., Wong, K., Dutton, P., & Blackett, I. (2015). Delivering water and sanitation to Melanesian informal settlements. *Water: Journal of the Australian Water Association, 42*(3), 40–43.

Sealey, K., & Smith, J. (2014). Recycling for small island tourism developments: Food waste composting at Sandals Emerald Bay, Exuma, Bahamas. *Resources, Conservation and Recycling, 92*, 25–37.

Shah, K. (2021). Air pollution in small island developing states: A growing but avoidable challenge requiring multilevel interventions. *Ramphal Institute Policy Brief, 2*(3), 1–7.

Shah, K., Niles, K., Ali, S., Surroop, D., & Jaggeshar, D. (2019). Plastics waste metabolism in a petro-island state: Towards solving a "wicked problem" in Trinidad and Tobago. *Sustainability, 11*, 6580.

Shah, K., Awojobi, M., & Soomauroo, Z. (2022). Electric vehicle adoption in small island economies: Review from a technology transition perspective. *Wiley Interdisciplinary Reviews: Energy and Environment, 11*(4), e432.

Shirley, R., & Kammen, D. (2013). Renewable energy sector development in the Caribbean: Current trends and lessons from history. *Energy Policy, 57*, 244–252.

Sievert, K., Lawrence, M., Naika, A., & Baker, P. (2019). Processed foods and nutrition transition in the Pacific: Regional trends, patterns and food system drivers. *Nutrients, 11*, 1328.

Smith, S. (2020). Structural vulnerability and marital sexual violence in Chuuk, Micronesia. In M. Torres & K. Yllö (Eds.), *Sexual violence in intimacy* (pp. 44–58). Routledge.

Sousa-Santos, J. (2019, June 26). The Pacific is in danger of becoming a semi-narco region. *The Guardian*.

Sousa-Santos, J. (2022). *Drug trafficking in the Pacific Islands: Impact of transnational crime*. Lowy Institute.

Sousa-Santos, J., & Howes, L. (2022). Policing illicit drugs in the Pacific: The role of culture and community on the frontline. *Journal of Contemporary Criminal Justice, 38*(3), 364–379.

SPEC. (1982). *Pacific energy programme mission report: Vanuatu 1982*. Pacific Islands Development Program, East West Center.

Stambolie, E. (2024). Facebook use in the Pacific: All over the place. *ANU DevPolicyBlog*, 19 June.

Staunton, M. (2024, August 1). 'Government is really working' to fix poverty, the main driver of Fiji's drug crisis - minister. *Radio New Zealand*.

Suaalii-Sauni, T. (2014). Legal pluralism and politics in Samoa: The Faamatai, Monotaga and the Samoa Electoral Act 1963. In P. Butler & C. Morris (Eds.), *Small states in a legal world* (pp. 165–189). Springer.

Taloiburi, E., & Homasi, L.-A. (2022, August). Financing resilience to mitigate the Pacific's climate vulnerability. *Pacific Economic Monitor*, 36–39.

Tamata, A., & Mohammadnezhad, M. (2022). 'Most nurses in the rural or remote health clinics run away from their job due to no remote allowances and poor working equipment and environment'. Health professional leaders' perception of shortages in the nursing workforce in underserved areas in Vanuatu. *Rural and Remote Health, 22*(3), 7229.

Techera, E. (2015). Enhancing legal frameworks for biodiversity conservation in the Pacific. *Pacific Conservation Biology, 21*(1), 87–96.

Thow, A.-M., Swinburn, B., Colagiuri, S., et al. (2010). Trade and food policy: Case studies from three Pacific Island countries. *Food Policy, 35*, 556–564.

Tjondronegoro, D., Marshall, A., Hunter, S., & Yuwono, E. (2025). *Bridging digital financial literacy gaps in the Pacific: A data-driven policy roadmap for inclusive growth*. Griffith University.

Tonga, B., Aguiar, D., Wiegner, T., et al. (2025). Baseline conditions and land-based drivers of nearshore water quality in Pohnpei, Federated States of Micronesia. *Marine Pollution Bulletin, 215*, 117833.

UNEP. (2021). *Small island developing states: Waste management outlook*. United Nations Environment Programme.

United Nations Office on Drugs and Crime (UNODC). (2023). *Regional report on the existing capacities to measure trafficking in persons in the Pacific Islands*. UNODC.

UNDP, UNODC, and OHCHR. (2024). *Global progress report on sustainable development goal 16 indicators. At the crossroads: Breakdown or breakthrough for peace, justice and strong institutions*. UNDP.

UNESCO & International Task Force on Teachers for Education 2030. (2024). *Global report on teachers: Addressing teacher shortages and transforming the profession*. UNESCO.

UNFPA. (2024). *Violence against women, regional snapshot, 2023*. UNFPA.

Utz, R. (Ed.). (2021). *Archipelagic economies. Spatial economic development in the Pacific*. World Bank.

Vakalalabure, E., & Saurara, L. (2022). Fijian education and COVID-19. *Islands Business, 48*(1), 26–27.

Verlis, K., & Wilson, S. (2020). Paradise trashed: Sources and solutions to marine litter in a small island developing state. *Waste Management, 103*, 128–136.

Volpe, C. (2021). 'High tide by boat, low tide we walk': The everyday digital lives of girls in remote villages of Vanua Levu, Fiji. *Children's Geographies, 19*(6), 766–779.

Waiti, D., & Lorrenij, R. (2018). Sustainable management of deep sea mineral activities: A case study of the development of national regulatory frameworks for the Republic of the Marshall Islands. *Marine Policy, 95*, 388–393.

Wale, C., Nagabhatla, N., & Perera, D. (2021). Institutional and policy analysis: Water security and disaster management in small island developing states. In J. Roberts, S. Nath, S. Paul, & Y. Madhoo (Eds.), *Shaping the future of small Islands: Roadmap for sustainable development* (pp. 205–217). Palgrave Macmillan.

Walters, S. (2025). *Energy poverty among women in Jamaica's low-income communities*. Palgrave Macmillan.

Watson, A. (2024, October 29). Pacific Islands need stronger corrections systems to address transnational and organised crime. *Interpreter*.

Watson, D., Howes, L., Dinnen, S., et al. (2023). *Policing in the Pacific Islands*. Palgrave Macmillan.

Watson, D., Pino, N., Amin, S., & Scott, J. (2025). Navigating policing and work-life balance in the South Pacific. *Police Practice and Research, 26*(3), 276–293.

Weir, T. (2018). Renewable energy in the Pacific Islands: Its role and status. *Renewable and Sustainable Energy Reviews, 94*, 762–771.

White, I., Falkland, T., & Scott, D. (1999). *Droughts in small coral islands: Case study, South Tarawa, Kiribati*. UNESCO.

Willetts, J. (2025). Water at the centre of climate residence: At COP30 and beyond. *ANU DevPolicyBlog*, 25 November.

Williams, A., & Rangel-Buitrago, N. (2022). The past, present, and future of plastic pollution. *Marine Pollution Bulletin, 176*, 113429.

Wiseman, V., Lagard, M., Batura, N., et al. (2017). Measuring inequalities in the distribution of the Fiji health workforce. *International Journal for Equity in Health, 16*, 115.

World Bank. (2016). *Pacific possible: Health and non-communicable diseases.* World Bank.

World Bank. (2019). *Using taxation to address noncommunicable diseases: Lessons from Tonga*. World Bank.

World Bank. (2023a). *Improving outcomes of Pacific labor mobility for women, families and communities*. World Bank.

World Bank. (2023b). *The future of Pacific tourism*. World Bank.

Wosnick, N., Curtis, D., & Hauser-Davis, R. (2024). Managing technology-critical elements from electronic waste in small developing Island States: A burden or an opportunity? *Frontiers in Marine Science, 11*, 1459794.

Wright, N. (2024). Assessing the main barriers to student success in the Caribbean, *Focus. ECLAC in the Caribbean*, 8–11.

Young, S., Bethancourt, H., Frongillo, E., et al. (2022). Concurrence of water and food insecurities, 25 low-and middle-income countries. *Bulletin of the World Health Organization, 101*(2), 90–101.

Yousefzadeh, M., Lenzen, M., & Tariq, M. (2023). Cooling and power from waste and agriculture residue as a sustainable strategy for Small Islands – A case study of Tonga. *Sustainability, 15*, 537.

7

A Future for Regionalism?

In both the Caribbean and the Pacific regional aspirations and linkages have gone in and out of favour in the roughly half century since independence. In the Caribbean, where cultural and language differences are slight, regional and sub-regional ties have been more substantial, certainly at a personal level. Regionalism is complicated by varied considerations of what constitute regions, for what objectives, the presence of various dependent territories and LIDS, with several metropolitan powers having different views of what regionalism might entail. It emerged from both disappointments, several discussed in the previous chapter, but rather more from the belief that regional unity and solidarity were invaluable, for sharing visions, policies, practices and identities across diverse issues, from combatting health problems to collectively pressing larger countries to amend practices that disadvantaged them in areas such as trade and climate issues. Consequently, regionalism has long been viewed as one partial solution to the problem of small size, with small states being more powerful economically and politically, when formally associated with the work of regional bodies such as the Pacific Island Forum (PIF) and CARICOM. Regionalism also represents a cultural vision of ties within an interconnected 'sea of islands' (Hau'ofa, 1993), with regional

J. Connell, Y. Campbell, *Island Challenges*,
https://doi.org/10.1007/978-981-95-8177-1_7

empowerment potentially subverting the discursive hegemony of colonial and subsequent development projects that belittled the islands and their people. In similar vein, therefore, SIDS have been more recently seen less as island microstates and more as 'big ocean states'.

Around the time of independence, as regionalism was being promoted and practised, several states faced intermittent secessionist challenges, whether Nevis from St. Kitts, Abaco from the Bahamas or Barbuda from Antigua, or, later and more violently, Bougainville from PNG. These divergencies invariably came from dual concerns centred on identity and perceived lack of support from a distant central government, perspectives that persist (Chap. 8). SIDS have different interests at various scales that inevitably challenge some notions of regionalism or that made competition—for example for tourist markets—appear more logical than cooperation, much as in established regional bodies like the EU. Universally, nations and regional bodies pull in different dimensions and directions while nations can fragment.

In both regions, regionalism has at its core two key formal institutions: the Pacific Islands Forum (PIF) and CARICOM. Alongside these, dozens of other bodies coordinate everything from aviation and customs to public health, environmental management and fisheries, as the foundation of regional governance and management. In addition to these 'government' regulatory institutions is a diverse cultural regionalism, including religious and sporting bodies, more closely linked to how islanders themselves interact and operate, and the sense of affinity this may engender. Sport alone is important in shaping ties and rivalries at different scales, and its significance has contributed to sports diplomacy (Chap. 8).

In oceanic areas, where the same seas envelop the islands, regional relations might reasonably be expected and have been implicit in the rhetoric surrounding past histories and a contemporary blue economy. Practice often diverged from potential, but regionalism has had successes in such divergent areas as tackling the drug trade and managing tuna resources. Cohesive policies and practices have been perhaps most effective for environmental and maritime matters but less so for trade and economic integration. Some activities, such as organising submarine cables, demand regional cooperation while others are imposed on regions by aid donors and international agencies (such as the World Bank, ADB and Economic

Commission for Latin America and the Caribbean [ECLAC]) which establish general regional programmes that minimise or ignore diversity and geography but relate to their own institutional economies of scale.

7.1 Whither Regionalism?

Since independence, and to some extent beforehand, Pacific and Caribbean SIDS have been heavily involved in regional politics, through institutions like the South Pacific Commission (SPC) (later the Pacific Community), the Pacific Islands Forum, the West Indies Federation and CARICOM. Prior to independence the colonial powers envisaged the value of regionalism, with the Pacific particularly seen as a cluster of small and fragmented microstates, disadvantaged in their efforts to achieve economic development. Regionalism was seen as a vehicle to achieve economies of scale, perhaps political integration and rather later environmental management, boosted by and benefiting from shared values and goals.

However, regional integration embodies tensions in the scalar dimensions of regionalism. Small states might benefit from pooling resources, yet the enormity of distances, a weak transport infrastructure and cultural distinctions made the Pacific especially a poor case for integration. Regionalism has been patchy, ad-hoc and amorphous, rising and falling around particular themes, issues and events, rather than based on a strong sense of shared interests or identity, weak even at national level in Melanesia and much of Micronesia. Fragmented and archipelagic SIDS, in Melanesia especially, are characterised by many distinct languages, cultures and forms of social and spatial organisation.

Despite widespread consensus that regional solutions are desirable, both economically and politically, the regional project has frequently floundered, as local interests become salient. Contemporary exceptions to this general trend are environmental issues, where Caribbean and Pacific nations appear to be speaking with one voice, a powerful practical and moral position directed beyond the region. Other partial exceptions, such as the PNA agreement over tuna fisheries (Chap. 3), concern only some PICs while other sub-regional ventures, such as the Melanesian Spearhead (formed in 1988 as a regional trade group but more overtly political), may even annoy some SIDS. Tensions and divisions have always existed, manifest with the politically dominant PIF temporarily splitting into two in 2021, separating Micronesia from the rest, until political divisions were healed a year later. Similar divisions surfaced in CARICOM's decision to vote against Venezuela despite a PetroCaribe agreement between CARICOM and Venezuela, where several SIDS gained concessionary pricing for Venezuelan oil (MacDonald, 2022; Jules et al., 2023). Divisive strategies have been instigated from outside: neo-colonial pressures that venture into divide and rule, as different SIDS later aligned with China or Taiwan.

7.2 Constituting Regions

Regions are constituted for multiple reasons. Like hazards, they are not necessarily natural. In the Pacific, that might include 'rims' such as New Zealand, large island states such as the Philippines and Indonesia (and especially West Papua), and small, remote (even by Pacific standards) islands under colonial powers, such as Pitcairn, Clipperton, Norfolk Island or Rapa Nui (Easter Island). However, ironically, the most familiar region was almost always that 'constructed' by colonial powers for the South Pacific Commission, so excluding West Papua, ignoring Rapa Nui and 'placing' Hawai'i in the United States rather than in the region. The Caribbean has a more obvious geographical structure but often includes the non-island states of Guyana and Belize (that we have excluded here), based on their former colonial status. Likewise, while geography is suggestive of a basic region, politics can disentangle islands, like Cuba or New Caledonia, from regional practices: an indication of the politicisation of regionalism, although many other factors point to socio-economic divisions.

The Pacific region (Fig. 1) contains 22 distinct regional entities. While Australia, France, New Zealand and the United States are also members of significant regional bodies such as SPC, only Australia and New Zealand are members of the PIF and the South Pacific Regional Environmental Programme (SPREP). Distinctively, the earliest regional organisations (and currently PIF and SPC) included all PICs not merely those that were independent states: the tiny territories of Pitcairn, Tokelau and Wallis and Futuna held equal stakes with PNG and Fiji (although rarely choosing to participate). None include West Papua, Rapa Nui or the United States (with the partial exception of the Pacific Islands Development Program, based at the University of Hawai'i). Broad cultural differences, impossible to pin down geographically, differentiate Melanesian, Micronesian and Polynesian regions, which eventually resulted in some formal sub-regional groupings, notably the radical Melanesian Spearhead. Evolving from colonial origins, old divisions have been widely used both by institutions within the region and external bodies with quite different compositions and objectives, from FIFA to UN agencies. It was not until 1997 that the South Pacific Commission (with other organisations) changed its name, turning 'South' into 'Secretariat' to better include PICs north of the Equator. That the change came late was a measure of both widespread external images of the region and quite significant differences between the former American territories in Micronesia and other parts of the Pacific. Borders are inevitably arbitrary and evolving.

While the Caribbean region (Fig. 2) might appear geographically obvious, mainland Central and South American countries are often linked in through different colonial and post-colonial perspectives. That has created an amalgamation of diverse cultures and languages shaped by colonial conquest: Dutch, French, British, Spanish and American islands. The Caribbean is also divided into the Greater (the north) and Lesser Antilles (the east and south), where culture and colonialism see several 'mainland' states, like Guyana, Guyane, Suriname and Belize, being perceived as part of the Caribbean and without strong ties to linguistically distinct mainlands. By contrast, the LIDS of Cuba, Haiti and Dominican Republic have ignored many regional organisations, in contrast to PNG, which is a prominent participant in Pacific regionalism. Both regions share one characteristic: UN organisations include them in a wider region, notably Economic and Social Commission for Asia and the Pacific (ESCAP) for the Pacific and ECLAC for the Caribbean, where both island regions are often peripheral to continental interests and supra-regional policies and practices of limited value—further complexities of regionalism.

7.3 Towards a Blue Pacific? A Short History

Regionalism in the Pacific has gone through six phases, each underpinned by two main discourses: a political and economic perspective which suggests that Pacific island states are small, isolated and vulnerable to globalisation and a contrasting cultural perspective where marginalisation is ideational, and Pacific entities have been variously conceptualised as interlinked parts of a 'large ocean' rather than separated 'small island' states. The six, sometimes subtly distinct, phases of regionalism were initiated in the colonial era when the first regional institutions, notably the South Pacific Commission (SPC), were established in 1947 primarily to facilitate the administration of the Pacific colonies by metropolitan governments.

Secondly, in the post-independence era after the 1970s, island states, assisted and supported by metropolitan powers, established a range of institutions, partly for more effective management, perceiving virtues in 'learning from each other' and cooperating in a new era, but also through the goodwill and later political self-interest and security of metropolitan states. Regionalism was partly imbued with the spirit of the 'Pacific Way', an idealistic assumption that certain regional values, centred on extended kinship, community and consensus, might shape a distinctive regional identity and approach. The Pacific Way emanated from Fiji, whilst PNG went to independence amidst what was loosely called the Melanesian Way, both sometimes empty clichés, romantic allusions and occasionally glimpses of an alternative development (James et al., 2012). However, island states were diverse and resources scarce, so regional cooperation left 'very little room for altruistic commitment' (Neemia, 1986: 32); hence, regional ties were weak, partly because of distance, space and size (several times that of the EU), a limited economy, limited connectivity and lack of economic complementarity.

Creation of the Pacific Islands Forum (PIF) by the few newly independent island states in 1971 partially addressed the barrier to political debate in the 'technical' SPC. Regional identity was encouraged by former colonial powers conscious of the small size and limited modernity of the new nations and of the involvement of 'new' Pacific nations—China,

Japan and the Soviet Union—that 'forced those with established interests in the area- such as the United States, France and Australia [and New Zealand] – to devise policies to maintain their predominant influence' (Fry, 1980: 455). Such objectives scarcely changed thereafter (Chap. 8).

Although less frequently invoking radical visions, regionalism was closely linked to the emergence of a more radical and political Pacific, with the 1970s struggle against nuclear testing, and towards an independent Pacific, with strong opposition to the French presence (Chesneaux & Maclellan, 1998). In the same decade, the Forum Fisheries Agency (FFA) was established, emphasising the significance of EEZs, which gave Pacific countries the rights to marine resources, in areas very much larger than their land masses, and a measure of regional economic significance.

Regional organisations proliferated (alongside membership of international organisations, notably the United Nations and its agencies) as growing economic, political and environmental complexity required increased specialisation. An explosion of regional organisations sometimes distracted governments and senior bureaucrats from national matters (Corbett & Connell, 2015) and extended beyond obvious arenas of economics and governance into religion, sport and culture. In the 1980s, SPREP emerged from a single office within the SPC (Carew-Reid, 1989) to become one of the largest regional organisations, taking a lead on climate change issues (primarily analysis and mitigation) with regional and global significance, but also addressing such themes as retaining biodiversity, national park development, waste management, primarily national issues with considerable variations between (and within) states, but with potential for sharing approaches.

With diversification and proliferation, regional institutions competed for mandates and funding, especially over climate change issues, and the Council of Regional Organisations (CROP) was established in 1971 to harmonise and consolidate regional programmes and organisations. As regional organisations emerged with increasingly specialised mandates, these did not necessarily apply to all Pacific states.. Divisions also attended who might be members of regional bodies and specifically involved the several remaining dependent territories and relevant colonial powers. Both France and the United States were often less than enthusiastic about their colonies' participation in several regional bodies.

Despite the proliferation of regional organisations, within a decade of independence it was evident that regionalism was ineffective. Early on there was 'conflict among the island states themselves over the nature, pace and control of regional ventures', no more than 'a token economic integration' and emerging Melanesian and Polynesian divisions, centred on more forceful Melanesian opposition to metropolitan involvement in the region, partly accounted for by their greater resource base (Neemia, 1986; Fry, 1980). The option of creating a free trade area and a shared currency had been quickly passed over and shared imports (such as oil and pharmaceuticals), and airlines had failed to materialise as 'selfishness', 'parochialism' and prestige dominated regional institutions (Neemia, 1986; Guthrie, 2013). That some services might be more effectively provided at regional level constituted a rationale for regionalism, including the shared provision of air and shipping services, training programmes (and a shared university) and shared knowledge. However, regional service delivery, or pooling, advocated as a means of addressing various challenges, never met the optimistic expectations of advocates, including development partners, that were relied upon for donor funding (Dornan & Cain, 2014).

Geography posed problems. The centrality of Fiji, larger than all other Pacific states except PNG, was resented, as was the perceived domination of Fiji-based regional institutions, like the University of the South Pacific (USP). The sheer size of the region limited cooperation and communication across great distances, especially as transport corridors were oriented towards former colonial powers rather than Pacific neighbours. The legacy of geography and colonial history was particularly evident in the separation of the 'northern' Pacific, most of Micronesia, which was oriented towards the United States with few connections across the equator. Differences were accentuated by culture and the lack of complementarity between countries which reduced trade initiatives. The relative size of countries also proved divisive and indicative of distinct interests and mandates. In 1978, Niue's government called for recognition of the distinct problems of the smallest PICs, a realistic recognition that national interests were diverse, and that larger PICs could take advantage of smaller ones.

National issues remained dominant, not merely because of the significance of national sovereignty but because development issues varied enormously between states according to physical geography, natural resource endowments (with Melanesian states being best placed), colonial legacies (of physical and human resources), and health, educational and linguistic status. That provoked some reaction towards the nature of regionalism, as externally imposed; hence, a third overlapping phase of regionalism involved a greater focus on culture within the region.

The third phase of regionalism represented something of a 'traditional' backlash, through an emerging perspective focused on the way colonialism was argued to have disrupted traditional ties, imposed borders and metropolitan linkages, ignored local values and forced economic strategies that had minimal relationship to existing local livelihoods and opportunities (Hau'ofa, 1998; Teiawa, 2018). Epeli Hau'ofa provided a perspective both cultural and economic:

> A common identity that would help us act together for the advancement of our collective interests, including the protection of the ocean for the general good, is necessary for the quality of our survival in the so-called Pacific century when important developments in the global economy will be concentrated in huge regions that encircle us. As individual tiny countries created by colonial powers and acting alone, we could indeed 'fall off the map' or disappear into the black hole of a gigantic pan-Pacific doughnut. Acting together, as a region, for the interest of the region as a whole, and above those of our individual countries, we would enhance our chances of survival in the century that is already dawning. (1998: 33)

The notion that regionalism could reduce marginalisation (Hau'ofa, 1998) remained a persistent theme, but it had hitherto been addressed in a colonial and neo-colonial way at odds with the Pacific world. Hau'ofa thus sought a 'genuinely independent regionalism' distinct from imposed colonial regionalism and 'anchored in our common inheritance' (2008: 41). It was both traditional and aspirational and subsumed intra-regional contradictions. Indeed, the rise of sub-regional groupings (the Melanesian Spearhead Group, the Polynesian Leaders' Summit and others) and the persistence of national and local interests (that Hau'ofa asked islanders to set aside) undermined the sense of shared strength and unity. Unity was often most visible to analysts beyond the region (Connell & Corbett, 2016). While some islanders professed to see the world as 'a web of ocean pathways, connecting places' (Steinberg, 2001), such pathways were usually hierarchical, localised and not necessarily harmonious.

Cultural perspectives could not induce a greater sense of regional identity, where economic relations were divisive rather than unifying, but as physical security began to pose problems, a new fourth phase of regionalism emerged. In contrast to interest in cultural similarities and linkages, the region had been disturbed by violent civil war in Bougainville from the 1980s, followed a decade later by the 'tensions' in Solomon Islands, partly hinging on cultural divisions and distinctions. Both crises had resulted in the intervention of regional forces—some from Australia and New Zealand, some from other Pacific SIDS (Chap. 8). That prompted Australian concern over a notional 'arc of instability' to its north and over the stability of the wider region.

Response to these concerns (emphasised in the context of a military coup in Fiji in 2000) was evident in the Biketawa Declaration (2000) supported by PIF and intended to constitute a framework for coordinating response to regional crises, an acknowledgement that problems existed. It committed Forum members to eight core values, including good governance, liberty of the individual, democratic processes, indigenous rights and cultural values, while recognising the importance and urgency of equitable economic, social and cultural development to satisfy basic needs and aspirations. Despite references to culture, it was largely focused on 'modern' concerns over physical security and invoked to endorse the PIF's regional and highly political Regional Assistance Mission to Solomon Island (RAMSI) mission in Solomon Islands. In due course the Forum was involved in resolving a 2011 Fiji-Tonga territorial dispute, the Nauru government's 2014 crackdown on opposition, and other politically unethical or violent issues. Nearly two decades later in 2018, the Biketawa Declaration was expanded into the Boe Declaration, where a broadly based security became paramount (Chap. 8). While climate change issues had become a PIC priority, underpinning this fourth phase were continued regional and metropolitan concerns over security issues, with suggestions of unstable, failed states and coups within the 'arc of instability', resulting in more direct external intervention and a metropolitan desire for greater control and involvement (Ayson, 2007; Connell & Corbett, 2016). Opposition to external intervention was complicated by the outcome of coups in Fiji and the temporary expulsion of Fiji from the PIF in 2009, after the Fijian military regime had reneged on its promise to return the country to democracy.

From the end of the twentieth century, various external agencies such as the World Bank, UNDP, ADB, AusAID/DFAT (Australia's aid agency), extended their presence in the region and developed Pacific plans and strategies that again centred on the virtues of a broad region, albeit sometimes excluding the northern Pacific or dependent territories. Many agencies did not cross the French-English language divide, which was partly a divide between SIDS and dependent territories. Most deliberately supported efforts at regionalism, evident in the ADB's report, jointly with the Commonwealth Secretariat, *Towards a New Pacific Regionalism* (ADB, 2005), which discussed service provision and market integration, while recognising that varied approaches and sub-regional groupings were appropriate for different objectives. Other external actors, from the Australian Labor Party to ANU's Pacific 2020 project, sponsored by AusAID, developed their own variants of plans. None ever developed sub-regional (as opposed to regional or national) plans. Such plans, which sought some degree of island state participation and ownership—many set out in the PIF's own 2005 *Pacific Plan*—returned to a technocratic emphasis on the importance of economic growth and good governance, alongside a stronger emphasis on regional security. That was wholly different from Hau'ofa's orientation and therefore criticised for being distant from islander values and concerns (Huffer, 2006). The *Pacific Plan*, like most national development plans, broadly reiterated a top-down perspective where politics was central.

Even with all-encompassing regional concerns over climate change, its significance varied, with atoll states particularly involved. Countries only occasionally shared similar interests, actively competing against each other in several ways, readily apparent in economic tensions over trade, transport and tourism (Connell & Corbett, 2016), which occasionally verged on conflict, as in the 'kava-biscuit war' between Vanuatu and Fiji (Connell, 2007). Two main trading groups emerged with movement towards a WTO free trade regime: PICTA and PACER were perceived as a means of engaging with the world economy. Pacific states were given little choice but to adapt to the WTO's 'new global trading order', once described as nothing less than 'the recolonisation' of the Pacific by 'free' trade (Kelsey, 2004). Trade both challenged regional unity, through countries having different trade possibilities and goals, and unified it through resentment at the domination of the former colonial powers in trade negotiations and through 'possible trade and aid reprisals as well as of intimidation in the negotiating room' (Jones et al., 2010: xiv).

Economic issues thus emphasised an emerging fifth phase where fragmentation and division overcame common interests. That accompanied a rise in sub-regionalism, more evidently cultural and political than economic, although PNA was held out as a technical model (Cain, 2015). Nonetheless, the value of and scope for regionalism was still avowedly economic. As the PIF's *Pacific Plan Review* of 2013 argued, 'Pacific economies, with small populations and tax bases on which to draw, need to do all they can to leverage voice, influence and competitiveness, and to overcome their inherent geographical and demographic disadvantages' (PIFS, 2013: 15). Constant slippage from such ideals and some notion of distinctiveness meant that, as late as 2012, the President of Kiribati was saying: 'I believe the Pacific is now entering a new phase – a new paradigm shift where the Pacific needs to chart its own course' (Anote Tong, quoted in Fry, 2015). But collective themes were weakening, partly under pressure from international organisations, with their own agendas, and divisive metropolitan nations, with external military interventions creating a quite different Pacific. Frustration over the leadership of PIF brought the departure of the Micronesian states (a brief MicXit) eventually resolved with the appointment of a Micronesian PIF leader (Maclellan, 2022).

Quite quickly, and unsurprisingly, that evolved into a more positive sixth phase that once again triumphed Pacific values and interests. This contemporary phase sought a Pacific development that again tried to take more account of cultural dimensions and distinctiveness while still focusing on economic matters that were at the core of Pacific Islander livelihoods. The *Framework for Pacific Regionalism* (PIFS, 2014) once again sought to better reflect Pacific interests in a more complex geopolitical arena, which also led to the PIF Secretary-General, Dame Meg Taylor, encouraging it to be more attentive to broader stakeholders, particularly civil society.

Regionalism was increasingly influenced by geopolitics. By 2017 it was being argued by the PIFS that geopolitical trends 'place the Pacific at the centre of contemporary global geopolitics' and that the core of Pacific regionalism was the shared Pacific Ocean (PIFS, 2017). That was further emphasised by the PIF endorsing the Blue Pacific narrative in the same year (Kabutaulaka, 2021). Moreover, French Polynesia and New Caledonia had joined the Forum, and other territories were moving towards membership. China had begun its maritime Belt and Road Initiative, although PICs were less concerned than the metropolitan states over possible security risks from 'non-traditional regional partners' (Chap. 8). Broader elements of security were becoming involved at regional level, from violence against women to urban poverty, food security and NCDs. That aligned more closely with civil society perspectives, but elites in some states were more committed to securing defence forces.

At each iteration of development strategy, scales, objectives and directions subtly changed, but with a greater focus on the value and centrality of the shared Pacific Ocean. PIF was concerned that the identity of the Pacific had returned to 'an emphasis on smallness and vulnerability' to garner finance where a focus on domestic resources would be the means of generating development and reasserting identity. That gradually focused on a Blue Pacific narrative more oriented to a strategic repositioning rather than primarily regional cooperation, and to an economic focus, that culminated in the *2050 Strategy for the Blue Pacific Continent* (PIFS, 2022). This offered a blueprint for the future, to emphasise that it could not be left to chance (Tekiteki, 2024), and whose cover depicted islanders in traditional canoes sailing towards a bright star, a graphic indication that it also sought to ensure that the diversity and heritage of island cultural values and traditional knowledge were to be honoured and protected (Naupa, 2023). The *2050 Strategy* stressed broad values and began from concern over climate change but largely repeated truisms and established objectives that set no new directions. It elaborated what had become the region's long-term approach to regional cooperation, to 'maintain environmental integrity and drive sustainable economic prosperity and development for all Pacific peoples'. The notion of a Blue Pacific sought to build on the potential of shared ownership and management of the Pacific Ocean, combining geography, identity and resources within a cultural 'sea of islands' perspective. By the mid-2020s it was increasingly difficult to maintain a semblance of unity, especially in political matters; the notion of 'friends to all' had not overcome preferences for China or Taiwan, with repercussions for aid delivery (Chap. 8). The slogan of the 2025 PIF Conference 'Iumi Tugeda [All Of Us Together]: Act Now for an Integrated Blue Pacific Continent' was something of a plea for PICs to bury differences and forcefully act together on climate change, ocean protection and the rejection of militarisation. The success of the PNA across so much of the region might have been a viable, hard-won and practical model, but few obvious parallel success stories followed. Politics was obscuring economics. Regionalism always accommodated both rhetoric and reality.

7.4 Caribbean Regionalism: A Core in CARICOM

The Caribbean went through similar phases to those in the Pacific, driven by similar factors, and from the same perspective that small size needed regionalism and even federation. Early attempts at regionalism dated back to British colonialism in the seventeenth and eighteenth centuries when the UK sought to form political unions linking the Windward and Leeward Islands with their small populations (Bardouille, 2018). Problems with federation were just as early and reflected in parochial sentiments:

> The Islands were too jealous of one another, too unwilling to allow Imperial control, and too fond of financial control by their own treasurers to consent to surrender any powers save in time of war into the hands of a body representing the whole group… Insular prejudice and general dislike of the idea of closer union by the white landowners who dominated island poli-

> tics was so predominate that no strong federal government could be established. (Hughes, 1958: 88)

The immediate pre-independence period could be seen as the second important phase of regionalism, and this time, on a wider scale and mainly dominated by politics. The 1958 West Indies Federation was an externally initiated process, with support from important Caribbean leaders, aiming to prepare the British colonies for independence through a political union, with the UK devolving its economic and political responsibilities, as its economic interests in the region faded. Caribbean governments were more open to the idea of federation but as a precursor to self-government, full independence and control over their own affairs. The 1938–39 Royal Commission that investigated the 1930s riots that spread across the region in response to harsh socio-economic and labour conditions, made the question of self-government even more urgent. The Royal Commission concluded that the idea of federation was popular, but the time was not yet right owing to the distances between the countries, parochialism and national pride. Federation eventually came through the 1956 agreement signed by the Windward and Leeward Islands, Jamaica, Trinidad and Tobago, and Barbados. It lasted only six years, as the largest states—Jamaica followed by Trinidad—withdrew and the late 1950s became a period of struggle between nationalism and regionalism, characterised by fears that the larger countries would be burdened by smaller ones with small populations and limited resources.

The third phase of Caribbean regionalism was less acrimonious, with consensus that intra-regional trade was important and could be advanced through a free trade area. Viability became synonymous with strong economic relations at regional scale. Priority was given to economic rather than political issues, and to the establishment of the Caribbean Free Trade Association (CARIFTA) in 1968, negotiated between 11 Caribbean governments. Although trade within the sub-region was relatively small, it was anticipated that intensification of intra-regional trade could stimulate mutual development. All the independent states, and Montserrat, joined CARIFTA, which resulted in increased trade, and even product diversification, with smaller countries benefiting from larger markets, and larger countries, such as Trinidad and Jamaica, exporting a wider variety of products to new markets. The main challenges that emerged from the creation of a free trade area in the Caribbean surrounded the removal of tariffs and the introduction of national fiscal incentives in relation to foreign capital (Hope, 1974). Some countries were better able to attract foreign capital than others, and smaller countries found themselves offering incentives far too generous for any benefits that might follow. No common foreign investment policy could be established, reflecting previous failed attempts at integration.

The fourth phase of integration in the Caribbean quickly followed and broadly involved the transition from CARIFTA to CARICOM in 1973, by far the most consequential regional body in the Caribbean. CARICOM consisted of both member states (comprising 20 countries) and 5 associate members. Although cultural diversity within CARICOM was less than in the Pacific region, it nonetheless included diverse ethnic and language groups despite culture having less influence on its operation. Most people spoke English, along with their country's creole, but with Dutch and French locally important. Its objectives followed from those of CARIFTA, directed at strengthening, coordination and regulation of the economic and trade regulations, but with a gesture towards harmonious and equitable development and the need to provide special opportunities for the Caribbean's less developed countries. It went beyond CARIFTA in seeking to provide and develop common services and activities in social, cultural and technological development but, despite the 1983 invasion of Grenada, gave little attention to political security.

In due course, the CARICOM Single Market and Economy (CSME), launched in 2006, pushed the regionalism agenda in a neoliberal direction (Payne & Sutton, 2007) while it adopted a top-down approach to decision-making to the near exclusion of civil society and private sector engagement, as in the Pacific. CARICOM's interests were also often compromised as it attempted to coordinate many countries, players with competing interests, circumstances eventually exacerbated by variations in support of the One China Policy (Gonzalez-Vicente & Montoute, 2021) and in relations with Venezuela and the United States. Politics was beginning to exercise a stranglehold on economic development.

A fifth stage involved recognition that specialist agencies were required to cover technical issues. In the Caribbean, as in the Pacific, these centred on fisheries and later on environmental themes. CARICOM mandates expanded into areas not initially contemplated, resulting in newer institutions and agencies in such areas as health, disaster management, agriculture, climate and renewable energy. Much as SOPAC and SPREP emerged in the Pacific, so CARICOM spawned a considerable number of environmental agencies, addressing issues that were regional or occurring simultaneously in different SIDS. CARICOM devolved agencies in fisheries (notably the Caribbean Regional Fisheries Mechanism, particularly concerned with IUU fishing, and the pursuit of 'blue justice'), health (the Caribbean Public Health Agency) and the Caribbean Meteorological Institute. The Caribbean Disaster Emergency Management Agency (CDEMA) deals with disaster management while the Caribbean Community Climate Change Centre (CCCCC) coordinates the Caribbean region's response to climate change. Some had success, as environmental governance was universally regarded favourably, but were nonetheless influenced by political divisions that did not always enable policy coherence (Scobie, 2016).

Over time domestic security and policing gained increased significance with CARICOM leaders perceiving increasing rates of violence and crime, including the drug trade and trafficking of illegal firearms as security problems. Small size was a central concern, and emphasis was placed on the need to pool resources and implement regional security strategies, including sharing judges, pooling forensic capabilities and implementing intelligence systems aimed at controlling the influx of illegal weapons from the United States into the region. Acting alone, Caribbean countries lacked capacity to address transnational problems or national problems of high rates of violence (Campbell, 2018). A CARICOM Implementing Agency for Security was initiated, but pooling security efforts was hardly successful, despite its utility for the smaller states (a problem which later recurred with cybersecurity and AI). Much like the SPC and the PIFS in the Pacific, the strength of CARICOM remained its ability to fill gaps in technical, financial and human capacity, until these became particularly complex, and to provide a valuable regional forum for strengthening regional policy development. CARICOM addressed both the need to address vulnerabilities and to adapt to a rapidly changing and increasingly volatile political environment. Still, regionalism was no panacea, and it continued to suffer from nationalist sentiments, limited financial investment to support the growing numbers of regional institutions as well as external intervention.

Failure to implement a single economy and the lack of shared foreign policy objectives captured basic limitations of regionalism. Although Caribbean SIDS had agreed in 2001 to create a CARICOM Single Market and Economy (CSME) by 2006, with the free movement of goods, services, people, capital and technology, it never eventuated, despite being regarded as crucial after the erosion and ultimate elimination of preferential market access for the region's major agricultural exports. Its importance was reinforced by CARICOM countries' obligations under WTO rulings, to fully liberalise their domestic markets in goods and services. Arguably, as the former Jamaican Prime Minister, Bruce Golding, suggested, that was due to 'a lack of competitiveness, diversification of goods produced and the absence of leadership with the will to implement a new market structure ... [so that] the region has about 10 different fiscal regimes, monetary policies and differing debt dynamics' (*The Gleaner*, 26 April 2024). This restructuring of the global economy with its emphasis on trade liberalisation forced CARICOM states to critically evaluate their economic prospects. Removing barriers to trade highlighted the SIDS' weak competitive position in terms of ability to defend home markets from an influx of extra-regional products and compete successfully in regional and other markets.

7.5 Global Regionalism?

Despite many shared contexts and objectives, in the post-independence era little consideration had been given to Caribbean and Pacific states (and those of the Indian Ocean) developing shared objectives or even evaluating shared contexts and practices. Beyond global United Nations entities and the Commonwealth Secretariat, which perceived parallels, produced occasional reports and comparative studies, and commissioned others (e.g. Dommen & Hein, 1985; Fairbairn, T. & Worrell, D 1996), little wider regional cooperation and no integration occurred until the advent of

AOSIS and the SIDS. AOSIS drew the regions somewhat more closely together, invaluable for later UN and COP negotiations, but the regions and individual states made no direct gains from the new international linkages.

Constraints to regional development existed and stymied various proposals. When Caribbean states produced and launched 'Actioning Blue: A Caribbean 30x30 Vision' at the United Nations Ocean Conference in mid-2025, it was apparent that the initiative had actually come from OECS, rather than all Caribbean SIDS or CARICOM member states, despite its being a call to stimulate a resurgence of regional cooperation and multilateral action (and in so doing protect at least 30% of the region's 'shared marine and coastal areas'). The proposal itself stressed the need for greater technical capacity, more effective governance and better funding to implement the plan and 'harness the power of data, science and indigenous knowledge' (James & Clarke, 2025). Emphasis was placed on the need for technical and financial support for initiatives, themes that were inherent in all that had gone before.

Regionalism was further complicated by the existence of such global organisations as the UN and its many agencies, and their regional counterparts, or the Commonwealth Secretariat, and increasingly, NGOs. The interests of global organisations did not necessarily align with those of the SIDS. Conversely, the SIDS and AOSIS have supra-national objectives that necessarily focused on common issues, especially environmental ones, that may detract, distort and deflect from local livelihood issues. Merely providing the human resources to staff these many organisations, participate in their meetings and other activities, and accommodate their visits, activities and projects, could be enormously demanding on small states and even conflict with efforts at national development.

Within and beyond notions of the blue economy, as plastics floated into the regions, and development shifted towards the high seas, oceans became of greater significance in regional discourse. Regional organisations produced numerous plans and strategies such as the *Framework for a Pacific Oceanscape* (PIFS), a somewhat general motherhood statement initiated by Kiribati, and intended to be a catalyst for the Pacific Islands Regional Ocean Policy (PIROP), which incorporated straightforward themes adorned by quotations from Epeli Hau'ofa. It eventually led to an

Office of the Pacific Ocean Commissioner, based within the PIFS. Sensible though such strategies could be, too many were phrased in such general terms that it was far from obvious who might adopt which strategies, where and over what time period, and where technical and financial support might come from. Remarkably, despite the first Pacific environmental management organisation, SPREP, being regional, no environmental issue coincided with the SPREP region, being either global, sub-regional, national or local, or encompassed every scale. Rhetoric could not easily be turned into reality.

7.6 A Kind of Blue: Resource Regionalism and the PNA

The signing of the Parties to the Nauru Agreement (PNA) in 1982, under the aegis of the Forum Fisheries Agency (FFA), regulating tuna fishing and resource management in the Pacific marked the most important regional agreement in the Pacific region, 'at the centre of a geopolitical tussle between the richest and most powerful states in the world, and some of the smallest and least developed countries' (Aqorau, 2019: 128). Political and economic ambitions overrode any lingering regional cultural differences to enable the one abiding and incontrovertible economic success story: a new regional order and scale of development, with a mobile and valuable resource and equally mobile fishers (Chap. 3). In the Pacific SIDS at least little is more crucial than the economic role of fisheries—and above all the tuna fishery—even for those SIDS not involved in the PNA scheme. Revenue is so substantial in these 'tuna dependent' economies that only in the largest states are there significant alternative revenue streams. The PNA SIDS consequently sought to lock in the system for the foreseeable future to guard against climate change and other possible external influences.

Despite the lack of parallel successes, tuna fisheries management offers a potential model for other regional agreements on the conservation and exploitation of resources and for navigating the complex relationships between social, economic, political and biological systems, emphasising the virtues and values of collective action. As a specialist agency, it evolved from SPC and the FFA, integrating the biological understanding of migratory tuna species with the technical components of deep-sea fishing (including international satellite vessel tracking) into an economic and political achievement. Problems existed in the twentieth century when SIDS were powerless to stop Japanese driftnet fishing and Japan and the United States were unwilling to pay access fees. The United States had also taken a hardline stand on UNCLOS (whose start in December 1982 imposed a global obligation on fishing states to cooperate in the management and conservation of fish stocks) and rejected the jurisdiction of island states over highly migratory tuna (eventually resulting in the arrest of the US purse-seiner Jeanette Diana by the Solomon Islands in 1984 that provoked a retaliatory trade embargo). SIDS were not without teeth.

PNA transformed what had hitherto been a fragmented open-access fishery, without limits to volumes caught or days fished, and with many illegal operators. This resulted from frustration with the United States and Japan, which tended to play the countries off against each other, and supported by Tonga and Samoa within the FFA, despite their both having limited tuna resources (Aqorau, 2019). The United States, with its powerful tuna fishing lobby, tried to persuade FSM, Palau and the Marshall Islands (then yet to sign their Compacts of Free Association) not to join. Others claimed it would engender a new political division in the region. The strongest opposition came from DWFNs, notably Japan and the United States, who argued that tuna were migratory so belonged to no particular country and that they had historic rights to access these stocks. The United States ignored UNCLOS rulings and even reimbursed American tuna boat owners for any fines they incurred and deducted these from development aid (Johnson, 2021). Gradually acquiring data on fisheries that made negotiations and management more effective, which the FFA had been without, gave PNA and the SIDS a political edge and a valuable resource worth fighting for, avoiding any 'tragedy of the commons'. After PNA had eventually overcome the metropolitan nations, Transform Aqorau, the first CEO of the PNA office in the Marshall Islands, starkly observed 'it's a nice change to see Australia's and New Zealand's role largely reduced to that of spectator' (2019: 10). It proved the first Pacific instance of a small group of SIDS establishing a breakthrough in governance and management not just without any help from metropolitan powers but in opposition to their interests, a triumph for the success of rights owners with skills, organisation and enthusiasm, beyond those of outsiders.

7.7 Scalar Interventions

As the number of regional bodies increased and served more specific functions, it became evident that only some SIDS were concerned with most issues. Fragmentation, specialisation and diverse functions indicated that no single organisation—especially the largest—could possibly accommodate all regional pressures, issues, needs, challenges and diversity, evident in the achievements of the PNA and which had some resonance in environmental domains. One size could never fit all, and universal approaches were rarely feasible.

Environmental issues showed how multiple scales and approaches both undercut the rationale for some regional organisations while also pointing to the value of regional (and global) umbrella organisations. While an ocean might seem to be a region, neither the Pacific nor the Caribbean is an ecological region; hence, environmental management is almost always local (sub-national) or national rather than regional. Even then the PICs have come together in opposition to nuclear waste dumping and (mostly) to potential seabed mining. Nor are oceans political, cultural or economic regions. However, when environmental issues are global and transboundary, most obviously climate change, then regionalism plays a valid and valuable role, whether in delegations to COP or a different kind of advocacy from Climate Change Warriors, in the face of a wicked global problem that cannot be resolved by small political units, and that has given all SIDS a shared (moral and practical) purpose. Yet environmental degradation is also a local phenomenon, and mitigation and remediation may also be local, while marine protected areas are valuable at many scales.

The origins and impacts of climate change obviously transcend national boundaries, accounting for the rise of global environmentalism and politics, from the 1992 Global Earth Summit in Rio onwards, eventually to

UNEP and the IPCC, two of a range of organisations with different if related objectives. The SIDS quickly united in opposition to climate change. Although none are without threatened coastal zones and rising sea levels, coastal communities and cyclones are not evenly distributed while physical responses occurred at both national and local levels—taking a variety of forms (Chap. 2). While all SIDS were affected by climate change and the need for mitigation, atoll states were particularly expressive and regional divisions again followed. In 2015 the Polynesian Leaders Group issued an urgent plea for action, prior to the Paris COP, while a Smaller Island States (SIS) group began within and through the PIF in 2006. SIS membership initially comprised Cook Islands, Kiribati, Marshall Islands, Nauru, Niue, Palau and Tuvalu (so including three of the four atoll states in the Pacific, the other being Tokelau, a New Zealand territory and not a member of PIF). The establishment of SIS reflected the frustration of small states that their objectives were not being met by PIFS and that a new grouping was necessary to focus on their own particular objectives: climate finance, transport, fisheries and equity (Connell & Corbett, 2016). A *SIS Regional Strategy 2016–2020* emerged but little followed.

Environmental concerns do not match political regions and have quite different kinds of linkage at different scales (Gruby & Campbell, 2013; Gruby, 2017). Thus, a Coral Triangle Initiative began in 2009 as a multilateral partnership of six Asia-Pacific countries was designed to address threats to the coastal and marine resources of what has been called 'one of the most biologically diverse and ecologically rich regions on earth' and 'the global epicentre of marine biodiversity' (Green et al., 2011: 174). The six states (Indonesia, Philippines, Malaysia, Timor Leste, PNG and Solomon Islands) thus crossed the 'standard' Pacific border (and that of ASEAN) while also excluding parts of Indonesia and most of Malaysia. In 2023, ASEAN and the PIF signed a Memorandum of Understanding committing the two regional organisations to increased collaboration on shared priorities: cooperation on maritime issues, regional connectivity, sustainable development and economic partnerships. More specifically, PIF and ASEAN sought to collaborate on implementing the 'blue economy', including promoting sustainable use of oceans, disaster risk reduction, humanitarian assistance, and developing digital and green economies across the Indo-Pacific. This loose framework guided both the PIF's *2050 Strategy* and ASEAN's *Blue Economy Framework* (Ballagh, 2024). However, linkages were more rhetorical than real as 'regional cooperation on infrastructure development and disaster management required aligning diverse priorities and mobilizing large amounts of financial and technical resources' (Ballagh, 2024). The economic gains from seeking to align policies and practices were considerable where countries were vastly different. Political gestures were more evident than structures for development.

Likewise all Caribbean SIDS are members of the Organisation of American States, which includes all the North American and Latin American states, in a loose political grouping, and in the Economic Commission for Latin America and the Caribbean (ECLAC), a regional branch of the UN, which plays a role in wide-ranging economic matters, and also in security and policing. Typically, in the Caribbean, the LIDS and the continental states of Guyana, Belize and usually Suriname were included in regional organisations, occasionally alongside such small dependent territories as Saba and Bonaire, thus extending the range of contexts and interests.

Hazards do not adhere to national boundaries and vary in form and severity, and no two hazards are identical. Cyclones are regional and volcanic eruptions are local, but, according to severity, response occurs at several scales, often simultaneously, whether as health care or resettlement. It is not always clear who might take the lead. While the Caribbean Disaster Emergency Management Agency (CDEMA) often takes the lead in the Caribbean, the extent of damage can warrant immediate intervention from other national and international agencies. A structured division of responsibility was well summarised:

> The sovereignty of the State must first and foremost be respected by all external partners. State sovereignty also requires that States do not [cannot] abdicate their responsibility to either prepare for and/or mount a response to the effects of a natural or man-made hazard event. Their preparation for and response to an event should, when possible, be first executed using all available means at their disposal. When local means are or expected to be exhausted, regional and international support resources may be requested.

> Accepting external assistance does not, and should not imply handing over or relegating control of the coordination of the national response and responsibility to regional or international entities. (CDEMA, 2016)

In PNG, a critical problem in delaying response to the needs of temporarily relocated refugees from the volcanic eruption of Manam Island was uncertainty over who was responsible and therefore at what scale responsibility should be accepted and responses initiated (Connell & Lutkehaus, 2017). No national or regional position exists on this or other hazard responses, hence the common intervention and role of NGOs in response. Invariably, some functions are best undertaken at sub-national scale because of local knowledge and experience.

7.8 A Politics of Regionalism

Regional organisations that existed in one region had no necessary parallel in the other. The Caribbean Development Bank was unique, and the Caribbean came much closer to having a regional currency than the Pacific. The Pacific had several regional fisheries organisations while religious organisations were less evident in the Caribbean. Many organisations did not neatly conform to regional borders. There could be no blueprint or model for what was necessary or what would work. Small newly independent island countries sorted this out themselves. Politically dependent territories were often excluded from regional bodies (or had observer status). That only changed belatedly with dependent territories recognising the virtues of linking into the experience of the SIDS (e.g. in hazard management and response) and the SIDS seeing strength in numbers. Early on, by 1991, CARICOM included dependent territories, whether Dutch, French or British, as associate members, in a context where their decolonisation seemed unlikely. In this century, PIF has sought to strengthen regionalism and oppose colonialism, by allowing dependent Pacific territories to go from observer status to full membership. In 2016, the admission of New Caledonia and French P`olynesia as full members marked a complete departure from the Forum's founding principles as a political grouping and a riposte to France when deep concern existed over the French colonial presence (Chap. 8). The 2024 PIF Leaders Meeting accorded both American Samoa and Guam Associate Membership after the United States stated that it would support this change provided it did not require them 'to take on or exercise any rights or responsibilities inconsistent' with their status as a US territory. Conversely, since they had played little part even in the 'non-political' SPC, a concern of the PIF was over their general commitment to Pacific regional affairs rather than being a proxy US geopolitical influence within it.

Less than a decade after CARICOM was established, cracks began to appear in the political structure, most obviously when the Organisation of Eastern Caribbean States (OECS) came into being in 1981 when seven eastern Caribbean states (Antigua, Dominica, Grenada, St. Kitts and Nevis, St. Lucia, St. Vincent and Montserrat, which was then seemingly moving towards independence) signed a treaty agreeing to cooperate and promote unity and solidarity amongst them. The large French territories of Martinique and Guadeloupe were not included. Three decades later, the OECS removed tariffs and other trade barriers between them and created an economic union, facilitating the free movement of goods, people and capital, becoming the single example of a more integrated economic region in either the Pacific or the Caribbean.

7.9 A Defensive Regionalism and Intervention

In the Pacific especially, questions of security began to trouble both the SIDS and metropolitan countries by the end of the century. The insurrection and civil war in Bougainville, which began in 1989, resulted in direct Australian military assistance to PNG, and as Bougainville was settling into an uncertain peace, a Regional Assistance Mission to Solomon Islands (RAMSI) was despatched to combat violence there during the euphemistically described 'tensions'. Four coups in Fiji between 1987 and 2006 added to the uncertainties, with problems in several states suggesting an 'arc of instability' north-east of Australia.

Regional disarray was further exemplified by Australian intervention in Bougainville, after the violent and divisive secessionist struggle and civil war. The Bougainville Revolutionary Army (BRA) centred around the mine site proclaimed Bougainville independent from PNG in 1989, resulting in fighting between the BRA and the PNG Defence Force, with the latter receiving logistical support from Australia. In early 1990 PNG withdrew and imposed a blockade on Bougainville, enforced with helicopters and patrol boats supplied to PNG by the Australian government as part of a defence cooperation programme. The blockade remained in effect until the ceasefire in 1994 (although fighting continued in some parts of Bougainville until 1997). Bougainville eventually became an autonomous province of PNG but continued to press claims for independence (Chap. 8), and many Bougainvilleans remain critical of Australia's role as part of the first significant external intervention in an independent PIC.

Unresolved ethnic violence in Solomon Islands in the late 1990s eventually brought the intervention of RAMSI, sometimes known as Operation Helpem Fren. It was an initiative of the PIF, authorised under the Biketawa Declaration of 2000, which encouraged regional responses to regional crises. It sought to restore peace and stability in a country divided by violent ethnic tensions, which had mainly pitted Malaitan islanders against the people of the main island of Guadalcanal and therefore partly centred on the capital, Honiara, where two rival militia groups terrorised the town. That was complicated by urban riots that burned down much of Chinatown. Although all the PICs participated, with some military and police forces, Australia led and funded the mission. Political leaders in Solomon Islands were divided over their support for the role of the special forces in ending violence; some criticised RAMSI's intrusion into Solomon Islands governance. A subsequent history of opposition to Australia followed on the part of Prime Ministers Solomon Mamaloni and Manasseh Sogavare (Chap. 8). In Solomon Islands, popular support for RAMSI was consistently high throughout its life span, even as support from political elites waxed and waned.

Outside Solomon Islands, RAMSI was generally seen as successful in achieving its short-term security objectives, but less effective in reaching more ambitious state-building goals following on from 'state failure'. The decade-long involvement was unable to influence a more inclusive political settlement. RAMSI's technocratic and militarised approach, while necessary, limited its ability to effect long-term developmental change and tackle what were (and are) entrenched, complex and above all political issues around governance in Solomon Islands, related to 'the political economy of patronage and power' (Allen & Dinnen, 2016; Kabutaulaka, 2005; Wesley, 2023). RAMSI was most effective at strengthening law and justice, had some success in stabilising the economy, but grand ideas about capacity building and improved bureaucratic functionality largely failed, leading to growing scepticism over interventionism. As the first-ever Regional Mission, it was enthusiastically endorsed by PIF.

The violence in Bougainville and Solomon Islands, the rise of transnational crime and the dysfunctional role of the Solomon Islands police force meant that regionalism took on security overtones, at least partly directed by metropolitan partners. Since few SIDS had military forces, police forces received more 'military' training, often directed from Australia, and Australian police officers were embedded within PIC police forces to encourage and assist in capacity building. PIC SIDS had few objections to additional support for local police forces, and despite the role of the *regional* mission in Solomon Islands, it was recognised that this had been led by metropolitan countries (Australia and New Zealand), as in Bougainville, and the region itself could not have provided security and pacification at the necessary scale and with adequate intensity.

7.10 Police and Policing

Substantial Australian assistance has been directed at strengthening state policing structures in the Melanesian SIDS, from PNG to Fiji. The increased scale and intensity of this assistance was exemplified in RAMSI and in an Enhanced Cooperation Program with PNG, both of which contained significant state policing components. A 2017 decision to re-arm Solomon Islands police following the withdrawal of RAMSI raised concerns about the possibility of renewed gun violence. Arguments in favour of arming the police in Solomon Islands focused on the need to maintain order without the need for foreign assistance, but reforming national police services was not matched by local demands for such reform. Solomon Islands has a relatively small police force, after PNG with the second-lowest ratio of police to people in the Pacific region. Moreover, a focus at national level distracted from the need to engage creatively with other elements of policing and justice provision if it was to extend the effectiveness of the local police in rural and peri-urban areas and achieve real and lasting improvements to security. This brought unsustainable levels of dependence on external assistance (Dinnen & McLeod, 2009; Dinnen & Allen, 2012). Despite the inevitable limitations of police reform, involvement in Solomon Islands was the start of a continued and fluctuating engagement between Australia and the Pacific SIDS, which became increasingly bound up with geopolitics, on a regional scale that was not activated by the PICs. Police increasingly filled a hybrid role that merged law enforcement with national security but was made difficult because of corruption and political instability, and with too few officers to cope with rising crime. As in other government sectors, retention and rapid turnover proved problems, and smaller PICs were without domestic training facilities.

Increased Australian focus on policing had initially intensified in the 2010s, following China's offer of support for the Solomon Islands force. China's Pacific policing assistance was diverse, providing anti-riot equipment for Fiji before the 2014 elections, a police training centre for Samoa, and vehicles for Asia-Pacific Economic Cooperation meetings in PNG. While Fiji had hosted Chinese police on short-term secondments, in-country assistance was novel, and followed China appointing its first Pacific police liaison officer to Fiji and its first defence attachés to Fiji and PNG in 2021 (Peake & Smith, 2022). China also sent police personnel and equipment to Vanuatu and offered to do the same in Solomon Islands in 2023 (including drones for riot control and a forensic laboratory, completed a year later). Australia had previously sent in troops in 2021 after major riots in Honiara and subsequently contributed vehicles and semi-automatic weapons. China countered with water cannon trucks, and more police cars and motorcycles, an indicator of a more direct militarisation, a bidding war replicated in other SIDS, and a more complex geopolitics, as distant pressures on the region intensified (Chap. 8). Police training now came with a regional and a geopolitical edge, premised on the rise of international drug trafficking and local police being overwhelmed but even more obviously in response to Chinese overtures. Meanwhile Australia had funded key defence infrastructure, including military bases in PNG and Fiji, and delivered military equipment, such as patrol boats and bushmaster vehicles, while liaison officers were embedded in forces across the region.

A new scale of policing was required to combat drug trafficking, which was somewhat thrust upon the PICs, when late in 2024 the PIF launched its Regional Transnational Organised Crime Strategy, to combat importation, transshipment, domestic trafficking and use of illicit drugs, all forms of transnational criminal activity in the region. That was combined with an Australian and New Zealand-directed Pacific Policing Initiative to provide new training centres in the region. While drug trafficking had become a major problem (Chap. 6), PIC leaders had reservations about the Australian initiative (which involved four new regional policing centres providing specialist training, and a Pacific policing support group able to deploy trained officers to countries in response to crises, and a sophisticated training centre in Brisbane). The Vanuatu Prime Minister, Charlot Salwai, queried whether the policing initiative was 'framed to fit our purposes and not developed to suit the geostrategic interests and geostrategic denial security postures of our big partners' (quoted in O'Keefe, 2024). The threats from transnational criminals were framed as a problem that was thrust upon the region from outside, but which had become a major 'inside' problem, despite few countries being involved in trafficking. Pacific SIDS proved better able to monitor migratory fish than migratory drugs.

What began as pressures for and interest in 'good governance' evolved into a form of deterritorialisation, partly dislodging linkages between territory and political authority and involving transnational forms of regulation and surveillance through military and police linkages and regional institutions: regulatory regionalism in the form of 'capacity building' (Hameiri, 2009a). Not only did this represent a transition from the role of aid in economic and social development to external assistance intended to prevent state fragility and failure, and ensure regional security, but in Solomon Islands it represented an early version of many subsequent 'political' incursions. Policing was directed and superimposed rather than spontaneous or emerging from the region. Such emerging versions of assistance emphasised regionalism by external intervention, so evidently less to do with economic issues than politics, and the outcome of divisions and external perceptions of weakness in the regional political architecture.

7.11 Regionalism Thwarted: Seas of Islands or Oceans Divided?

Enthusiasm for regionalism came and went, sometimes inspired from within, sometimes directed from beyond, and never wholly one or the other: too often a marriage of convenience or rhetorical device rather than a partnership with positive and tangible outcomes. Yet regionalism, finding and operationalising a 'Pacific voice' or a 'Caribbean voice' and transcending smallness, has been essential—yet difficult. Smaller SIDS have been overwhelmed by the regionalism of multiple organisations, as much as benefiting from it. However, CARICOM and the PIF provided and developed commitment to regional values and directions, and regional bodies played valuable roles, often largely unnoticed and unheralded, since they were distant from most SIDS, but supported national development. (The SPC, for example, assisted every PIC with undertaking censuses and income and expenditure surveys; the FFA monitored fish populations and marketing strategies; and SPREP provided PICs with technical support for developing waste management strategies). Conversely, many regional agencies, often with 'narrow' specialisms, such as religious bodies, perhaps closer to 'ordinary people' than many entities (e.g. Byron & Montoute, 2023), provided valuable advice and direction to small states. Collaboration has worked in agriculture and health (less so in the wake of COVID-19) and to some extent in regulating food systems (Thow et al., 2022) but not in developing a Pacific urban agenda, despite countries having similar issues and objectives (Keen & Connell, 2019), or in formulating regional responses to disasters (Thompson, 2015) and education policies (Johansson-Fua, 2022) or in achieving useful skilled labour mobility within either region. Incorporating regional universities into practical research and policy making has been disappointing, once again because of scarce human resources, although NGOs such as the Caribbean Natural Resources Institute (CANARI) have proved valuable. The PIF played an essential role in coordinating regional police and military operations, notably during RAMSI (but was unable to resolve some other regional disputes). Regional sporting and cultural events (with the South Pacific Games emerging from SPC and, to a much lesser extent, cricket in the Caribbean) contributed to a sense of unity and identity, but regional groupings, such as the Melanesian Spearhead, and Micronesian frustration over Pacific leadership spoke to a different sense and scale of unity and regionalism.

Perhaps the greatest achievement of SIDS has been the success of small states in global-scale climate change negotiations, where the SIDS have necessarily worked together through their sharing of one abiding, overriding, crucial and existential concern. The greatest (yet incomplete) success of regionalism has been beyond the regions, while at national level there is substantial fragmentation in the application of climate policy in individual SIDS (Scobie, 2016). A pervasive belief in, and practice of, regionalism remains, as it must, in regions of small states confronted by climate change. Unity is strength and, while dissimilar, SIDS are quite different from the massive metropolitan colonial and neo-colonial powers on the Pacific and Caribbean rims. It is primarily in adversity and in the context of opposition to climate change, where this unity is most effective.

The PNA proved the striking example of successful regionalism—more accurately sub-regionalism—tied to a valuable resource, where commitment provided economic benefits. However, its substantial economic achievements, the one great success story of regionalism, have never been used or perceived as a model. Indeed, Dame Meg Taylor, then Secretary-General of PIF, concluded that it was 'the result of a fortuitous combination of the right set of people' (2019: xiv). It was technical, disadvantaged no one in the region, yet took almost two decades to bring to fruition: perhaps the exception that proves the rule since no other comparable examples of similar resource-based regionalism exist. Plans emerged but with limited policy coherence (Scobie, 2016, 2019) and without technical and financial support, while political objectives could be short-lived.

While Hau'ofa pointed to a sea of islands tied together by an extensive oceanic network, this was localised and never without significant schisms and differences. Broad regions exist only in a political rather than in an economic, cultural or environmental context. Dissident voices and objectives always existed: most significant scales of management and practice, if not policy formation, are local, where lives are lived, and agency exists, rather than either national or regional. Regionalism implies some sacrifice of national autonomy, rarely welcomed, counteracting always inflated bureaucratic optimism (and consultancy recommendations) on what might be feasible. Both regions are most clearly seen from afar. Within the islands unity and shared interests seemed less obvious, evident in trade disputes, competition for tourist markets and the 'tensions' in Solomon Islands. The Coral Triangle initiative with ASEAN states, debates within the Melanesian Spearhead Group over whether a West Papuan organisation might be a member, indicated that no 'natural' boundaries or sub-divisions existed. The inclusion of entities from Central and South America in CARICOM emphasised the point. Conversely, in the Pacific, Australia, much the largest member of the PIF, could be seen as an inappropriate member of any regional organisation. For most purposes, SIDS have essentially acquiesced to neo-colonial versions of 'the Pacific' and the West Indies, established centuries ago and accepted by most people most of the time.

Grand regional visions, in a vast ocean, with multiple islands and nations, were destined to disappoint, just as in such blocs as the EU (exemplified, eventually, in Brexit), ASEAN, ECLAC and elsewhere. The Pacific had no great interest in a more 'political' regionalism (although a common currency, open borders and a joint parliament were all mooted at some point) despite support for the Melanesian Spearhead. A regional airline was anathema, after token efforts at unity, as smaller states feared Fiji domination. Size mattered. Neither region acquired a regional assembly, central bank or single currency, as the EU had done, although both were established in the eastern Caribbean SIDS. In the Pacific, proposed joint import of oil or pharmaceuticals failed to occur, efforts to share skilled human resources never took off and USP fractured. CARICOM and the Caribbean SIDS failed to determine what a blue economy might mean to the region and how that related to regional ocean management

(Hassanali, 2020), or developed and implemented regional policy coherence on climate change governance (Scobie, 2016, 2019), or collectively responded to the hazard of sargassum seaweed influxes (van der Plank et al., 2022). Yet, the significance and value of working together, above all on climate change and high-sea fisheries, have counteracted disadvantages and divisions, and the morality of mitigating and minimising climate change has been inescapable. Here the discursive construction of smallness as vulnerable, once so pervasive, has been both revitalised and counteracted.

Regional bodies move slowly, especially in response to emergent transboundary issues, despite that being their raison d'etre. Power differences and uncertain responsibilities can be difficult to resolve. While integrated management is invaluable, 'harder' economic issues have proved more difficult. Regional trade is limited and has foundered on production of the same goods, despite, in the Caribbean, CARIFTA (the antecedent of CARICOM) being established in 1965. Quite simply, few economic benefits to integration exist (Warner & Anatol, 2014). Neither CARICOM nor the PIF have achieved a common external trade policy, and in the Pacific at least, it has proved a divisive issue. The trade agreement, PACER Plus, between PIF members and Australia and New Zealand, was signed in 2020, but the two largest PICs, Fiji and PNG, declined to join, concerned that tariff removal on imports would cause a significant loss of national revenue and that key provisions on aid and labour mobility were lacking. The greater integration in the eastern Caribbean states suggests that sub-regionalism is possible and that it works best where states are of similar size and share similar concerns. That too may be a function of geography, where the nine members of OECS are closer together than just half of the archipelagic Solomon Islands.

In both the Pacific and the Caribbean, the value and role of a 'big' region were most evident in the years immediately after independence as small states sought to benefit from cooperation primarily in economic matters and with metropolitan support. Despite strong incentives and pressures to integrate, as in the Caribbean, the actual process of integration was 'often haunted by historical ghosts; compromised by conditions of vulnerability, poverty and insecurity; undermined by the lack of political will and popular support and stymied by the absence of common institutions' (Grenade, 2011: 4). Over time smaller and specialised organisations emerged—especially in key arenas such as fisheries and the environment that were unusually valuable. However, much of the agenda of Caribbean and Pacific regionalism has been driven in a top-down manner, divorced from culture. Thus, 'despite strong pan-Caribbean sentiment it has proved hard to convince citizens of member states of the merit of more intensive regionalism' (Warner & Anatol, 2014: 193), a conclusion even more valid for the more diverse Pacific. CARICOM, for example, has been seen as a neo-colonial structure, founded on western modernist principles, unresponsive to the challenges of decolonisation, and effectively reduced to developing regional trade deals (Jules, 2024). Similar criticisms have been directed at most Pacific regional bodies, emphasised after regional quasi-military missions.

In this century, external interventions intensified in increasingly politicised regions, still somewhat aloof from civil society, and with only tangential relationships to SDGs. The Pacific and the Caribbean were never free of external interventions of one kind or another, most at least intendedly benign variants on aid (from other nations, development banks and NGOs) through to the soft interventions of multiple Pacific Plans (many unsolicited) or the 'hard' intervention of RAMSI (Hameiri, 2009b). Few such interventions sought or enabled indigenous ownership or took more than token note of local cultures and sensitivities. Demonstrably the changing role of policing as it served security and international interests pointed to a continued 'hard' intervention as securitisation and militarisation followed.

Regionalism faded as countries increasingly pursued individual goals, as they inevitably did, competition as much as collaboration, accentuated where human resources proved a critical constraint to effective governance and management. Political leaders met infrequently, few SIDS bureaucrats gave regional programmes preference over national ones or saw the regions as related to their needs and identities; hence, 'while it is important to reinforce the power of the region as one political bloc, it comes off as self-serving. Pacific islanders generally do not think in these terms. Our communal values are much stronger' (Tekiteki, 2024). Few are driving regionalism. Political will and trust lie elsewhere. Regional organisations are distant, often anonymous, populated by outsiders, with minimal link to local identity. Rather than a growing regional unity that might parallel accelerated globalisation, the converse of a 'progressive balkanisation of the islands' was often more evident (Grynberg, 2010). Regionalism can easily be no more than a rhetorical mantra—a vision of outsiders (and PIF and CARICOM) but much shaped from beyond and responding to issues that came from beyond.

Over time achieving a positive regionalism became more difficult, as participants and interests diversified. As the Samoan Prime Minister, Fiame Naomi Mata'afa, observed, 'More than ever, there is increased interest and jostling for attention in our blue Pacific region thus creating a very crowded and complex geopolitical landscape for all of us and our regional architecture' (RNZ, 9 June 2023). That was more apparent as the Pacific especially became of increased geopolitical significance, and other countries sought greater involvement, increasingly through 'harder' manifestations than through soft power. The consequences and merits of these new and varied engagements, most centred on security, evident in the changing role of policing, some thrust upon the SIDS, and some welcomed, and the SIDS' strategies and responses, are discussed in the next chapter.

Bibliography

Allen, M., & Dinnen, S. (2016). Beyond life support? Reflections on Solomon Islands after the Regional Assistance Mission. *Asia & the Pacific Policy Studies, 3*(1), 6–15.

Aqorau, T. (2019). *Fishing for success: Lessons in Pacific regionalism*. ANU Department of Pacific Affairs.

Asian Development Bank. (2005). *Towards a new Pacific Regionalism*. ADB.

Ayson, R. (2007). The 'arc of instability' and Australia's strategic policy. *Australian Journal of International Affairs, 61*(2), 215–231.

Ballagh, A. (2024). *ASEAN and the Pacific Islands: Pathways for Collaboration on climate and sustainable development*. Visions and Voices.

Bardouille, N. (2018). Caribbean regionalism*s* in a comparative-historical perspective: The making of four regional systems. *Canadian Journal of Latin American and Caribbean Studies, 43*(2), 171–211.

Byron, J., & Montoute, A. (2023). The role of churches in regional cooperation and governance in the Caribbean. In L. Briguglio, M. Briguglio, & C. Slatter (Eds.), *Handbook of civil society and social movements in small states* (pp. 189–203). Routledge.
Cain, T. (2015). Rebuild or reform; regional or sub-regional architecture in the Pacific island region. *Journal de la Société des Océanistes, 140*(1), 49–58.
Campbell, Y. (2018). Security cooperation in Latin America and the Caribbean: Threats, institutions and challenges. In E. Fiddian-Qasmiyeh, & P. Daley (Eds.), *Routledge handbook of south-south relations* (pp. 309–313). Routledge.
Carew-Reid, J. (1989). *Environment, aid and regionalism in the South Pacific.* National Centre for Development Studies, ANU.
Caribbean Disaster Emergency Management Authority. (2016). *The regional response mechanism.* CDEMA.
Chesneaux, J., & Maclellan, N. (1998). *After Moruroa: France in the South Pacific.* Ocean Press.
Connell, J. (2007). Towards free trade in the Pacific? The genesis of the 'Kava-Biscuit War' between Fiji and Vanuatu. *Geographical Research, 45*, 1–12.
Connell, J., & Corbett, J. (2016). Deterritorialisation: Reconceptualising development in the Pacific Islands. *Global Society, 30*(4), 583–604.
Connell, J., & Lutkehaus, N. (2017). Escaping Zaria's fire? The volcano resettlement problem of Manam Island, Papua New Guinea. *Asia Pacific Viewpoint, 58*(1), 14–26.
Corbett, J., & Connell, J. (2015). All the world is a stage: Global governance, human resources and the 'problem' of smallness. *The Pacific Review, 28*(3), 435–459.
Dinnen, S., & Allen, M. (2012). Paradoxes of postcolonial police-building: Solomon Islands. *Policing and Society, 23*(2), 222–242.
Dinnen, S., & McLeod, A. (2009). Policing Melanesia – international expectations and local realities. *Policing and Society, 19*(4), 333–353.
Dommen, E., & Hein, P. (1985). *States, microstates and islands.* Croom Helm.
Dornan, M., & Cain, T. (2014). Regional service delivery among Pacific Island Countries: A case study. *Asia and the Pacific Policy Studies, 1*(3), 541–560.
Fairbairn, T., & Worrell, D. (1996). *South Pacific and Caribbean Island Economies: A Comparative Study.* Foundation for Development Cooperation.
Fry, G. (1980). Regionalism and international politics of the South Pacific. *Pacific Affairs, 54*(3), 455–484.

Fry, G. (2015). *Recapturing the Spirit of 1971: Towards a new regional policy settlement in the Pacific*. ANU SSGM Discussion Paper 2015/3.
Gonzalez-Vicente, R., & Montoute, A. (2021). A Caribbean perspective on China–Caribbean relations: Global IR, dependency and the postcolonial condition. *Third World Quarterly, 42*(2), 219–238.
Green, S., White, A., Christie, P., et al. (2011). Emerging marine protected area networks in the coral triangle: Lessons and way forward. *Conservation and Society, 9*(3), 173–188.
Grenade, W. (2011). *Regionalism and sub-regionalism in the Caribbean: Challenges and prospects – Any insights from Europe?* University of Miami Jean Monnet/ Robert Schuman Paper Series 11/4.
Gruby, R. (2017). Macropolitics of Micronesia: Toward a critical theory of regional environmental governance. *Global Environmental Politics, 17*(4), 9–27.
Gruby, R., & Campbell, L. (2013). Scalar politics and the region: Strategies for transcending Pacific Island smallness on a global environmental governance stage. *Environment and Planning A, 45*(9), 2046–2063.
Grynberg, R. (2010). The trade and health debate. *Islands Business, 36*(9), 34–35.
Guthrie, K. (2013). Aviation regionalism in the Pacific. *Journal of Pacific History, 48*(3), 294–308.
Hameiri, S. (2009a). Governing disorder: The Australian Federal Police and Australia's new regional frontier. *Pacific Review, 22*(5), 549–557.
Hameiri, S. (2009b). The region within: RAMSI, the Pacific Plan and new modes of governance in the Southwest Pacific. *Australian Journal of International Affairs, 63*, 348–360.
Hassanali, K. (2020). CARICOM and the blue economy- Multiple understandings and their implications for global engagement. *Marine Policy, 120*, 104137.
Hau'ofa, E. (1993). Our sea of islands. In E. Waddell, V. Naidu, & E. Hau'ofa (Eds.), *A new Oceania: Rediscovering our sea of islands* (pp. 2–16). USP.
Hau'ofa, E. (1998). The ocean in us. *The Contemporary Pacific, 10*(2), 392–410.
Huffer, E. (2006). Regionalism and cultural identity: Putting the Pacific back into the plan. In S. Firth (Ed.), *Globalisation and governance in the Pacific Islands* (pp. 25–48). ANU Press.
Hughes, C. (1958). Experiments towards closer union in the British West Indies. *The Journal of Negro History, 43*(2), 85–104.
Hope, K. R. (1974). CARIFTA and Caribbean trade: An overview. *Caribbean Studies, 14*(1), 169–179.

James, K., & Clarke, J. (2025, June 11). Securing our blue future: Caribbean leadership for global marine protection. *IISD*.

James, P., Nadarajah, Y., Haive, K., & Stead, V. (2012). *Sustainable communities, sustainable development: Other paths for Papua New Guinea*. University of Hawai'i Press.

Johansson-Fua, S. (2022). Wansolwara: Sustainable development, education, and regional collaboration in Oceania. *Comparative Education Review, 66*(3), 516–534.

Johnson, G. (2021). *Our ocean's promise: From aspirations to inspirations: The Marshall Islands fishing story*. Create Space.

Jones, E., Deere-Birkbeck, C., & Woods, N. (2010). *Manoeuvring at the margins: Constraints faced by small states in international trade negotiations*. Commonwealth Secretariat.

Jules, T. (2024). Introduction: CARICOM @ 50- is it time to decolonise southern regionalism? *The Round Table, 113*(1), 1–14.

Jules, T., Arnold, R., & Smith, A. (2023). Caribbean exceptionalism and the rise of Sino-CARICOM relations in the post-bureaucratic era. *The Round Table, 112*(1), 57–72.

Kabutaulaka, T. (2005). Australian foreign policy and the RAMSI intervention in Solomon Islands. *The Contemporary Pacific*, 283–308.

Kabutaulaka, T. (2021). Mapping the blue Pacific in a changing regional order. In G. Smith & T. Wesley-Smith (Eds.), *The China alternative. Changing regional order in the Pacific Islands* (pp. 41–70). ANU Press.

Keen, M., & Connell, J. (2019). Regionalism and resilience? Meeting urban challenges in Pacific Island states. *Urban Planning and Research, 37*(3), 322–337.

Kelsey, J. (2004). World trade and small nations in the South Pacific region. *Kansas Journal of Law and Public Policy, 14*, 247–306.

MacDonald, S. (2022). *The New Cold War, China, and the Caribbean*. Palgrave Macmillan.

Maclellan, N. (2022). The region in review: International issues and events 2021. *The Contemporary Pacific, 34*(2), 422–446.

Naupa, A. (2023, August 31). Cultural security in the Pacific: Why it matters for regional security. *Interpreter*.

Neemia, U. (1986). *Cooperation and conflict: Costs, benefits and national interests in Pacific regional cooperation*. Institute of Pacific Studies, USP.

O'Keefe, M. (2024, August 30). 'Hot mic' moment aside, the Albanese government's Pacific policing deal is a masterstroke of diplomacy. *The Conversation*.

Pacific Islands Forum Secretariat. (2013). *The Pacific plan review*. PIFS.

Pacific Islands Forum Secretariat. (2014). *Framework for Pacific regionalism*. PIFS.

Pacific Islands Forum Secretariat. (2017). *State of Pacific regionalism*. PIFS.

Pacific Islands Forum Secretariat. (2022). *2050 strategy for the blue Pacific continent*. PIFS.

Payne, A., & Sutton, P. (2007). *Repositioning the Caribbean within globalisation*. The Centre for International Governance Innovation, Caribbean Papers No. 1.

Peake, G., & Smith, G. (2022, January 27). Solomon Islands: Cops bearing gifts. *Interpreter*.

Scobie, M. (2016). Policy coherence in climate governance in Caribbean small island developing states. *Environmental Science & Policy, 58*, 16–28.

Scobie, M. (2019). *Global environmental governance and small states*. Edward Elgar.

Steinberg, P. (2001). *The social construction of the ocean*. Cambridge University Press.

Taylor, M. (2019). Foreword. In T. Aqorau (Ed.), *Fishing for success: Lessons in Pacific regionalism* (pp. xiii–xiv). Department of Pacific Affairs, ANU.

Tekiteki, S. (2024). Reflections on Pacific regionalism. *ANU DevPolicyBlog*, 17 October.

Teiawa, K. (2018). Our rising sea of islands: Pan-Pacific regionalism in the age of climate change. *Pacific Studies, 41*(1), 26–54.

Thompson, D. (2015). Disaster logistics in small island developing states: Caribbean perspectives. *Disaster Prevention and Management, 24*(2), 166–184.

Thow, A.-M., Ravuvu, A., Iese, V., et al. (2022). Regional governance for food system transformations: Learning from the Pacific Island region. *Sustainability, 14*, 12700.

van der Plank, S. (2022). Polycentric governance, coordination and capacity: The case of Sargassum Influxes in the Caribbean. *Coastal Management, 50*(4), 285–305.

Warner, R., & Anatol, M. (2014). Caribbean integration – Lessons for the Pacific? *Asia and the Pacific Policy Studies, 2*(1), 183–196.

Wesley, M. (2023). *Helpem Fren: Australia and the regional assistance mission to Solomon Islands 2003-2017*. Melbourne University Press.

8

Geopolitical Manoeuvres: Whose Islands?

Geopolitical tensions, resulting from the exercise of military or economic power in pursuit of, or to sustain, regional and global dominance, or from secessionist movements and great power rivalries, have wide-ranging effects in SIDS. While external interventions in favour of regionalism were usually benign, or at least well-intentioned, some even virtuous, over time both the SIDS and metropolitan powers became involved in a regional and global geopolitics that was sometimes less well-intentioned. Intervention was scarcely new and seen in both regions as one more phase of neo-colonialism. Metropolitan powers had long viewed the islands as places to control, to support economic and later military interests. In the Caribbean, the islands were all too close to North America, and in the Pacific, less close to America but dangerously close (at least geographically) to an Asia where first the Soviet Union and then China might challenge and destabilise metropolitan interests. The United States had viewed the northern Pacific region as a buffer zone and occasional source of resources, through its nineteenth-century Monroe Doctrine, where it would control the western hemisphere. In this century, as mounting acrimony, internecine violence and conflict over policing in the Pacific region indicated (Chap. 7), what might once have seemed inconsequential

J. Connell, Y. Campbell, *Island Challenges*,
https://doi.org/10.1007/978-981-95-8177-1_8

fluctuations in diplomacy became caught up in a flurry of tense geopolitical exchanges, some of which lacked immediate significance or consequences for the lives and livelihoods of island citizens.

This chapter traces the evolving geopolitics of development in the SIDS, first recording how threats to national integrity have been domestic, with ongoing struggles for secession and independence that suggest a degree of instability and an intermittent politics of identity. However, the core of the chapter and the greatest source of instability has external origins, exemplified in the twenty-first-century rise of China's maritime Belt and Road Initiative (BRI), creating a new political environment that challenged the hegemony of the western powers. SIDS have been increasingly affected by geopolitical relations and the uncertainties of global politics and institutions. How they have responded to these and will respond to them is crucial to future development.

8.1 Evolving States? Struggles for Secession, Independence and Identity

Fragmentation and aspirations to secession, the converse of regionalism, have accompanied rumblings of dissent in several islands (Corbett, 2023). Fractures in island peripheries suggested social, political and development problems. Several SIDS, especially the archipelagos, have experienced secession movements, with some islands claiming a distinct identity, seeking a more exclusive use of island resources or simply frustrated at their marginalisation by the centre. Bougainville has made the strongest claim. Many in West Papua have made similar claims, but success currently seems implausible. Much geopolitical dissent occurred around times of uncertainty and instability, often around independence. Abaco sought to leave the Bahamas early on, but there was no interest after independence. Western Province sought to secede from Solomon Islands before independence (Premdas et al., 1984), the Polynesian island of Anuta similarly considered separation from the remainder of the Solomon Islands (Feinberg, 1986), and Rotuma considered leaving Fiji.

Anuta and Rotuma were small, isolated islands of different ethnicity from the rest of the country, but separation would have been problematic.

While most island states have seemingly obvious, well-defined borders, several inherited colonial borders with only a weak relationship to culture or geography (the entirely artificial borders of PNG with its three different neighbours being an extreme example). That was most evident in West Papua, where a colonial border drew a straight line through a complex island. Indonesia retained the western half, after a corrupt referendum, leaving it with the world's largest gold mine, where indigenous Papuan people continue to fight for political independence, with repercussions for violent civil unrest, refugee flows into PNG and the politics of various regional organisations (Chap. 7). Another colonial border split Samoa in two. However, the greatest and most persistent dissent over colonial borders came from Bougainville.

Conflict over identity, geography and ownership and sharing of resources dominates Bougainville's recent history, factors shared in several SIDS contexts. Bougainville claimed its own independence as the Republic of the North Solomons a few weeks before that of PNG in 1975. It was ignored. The quest for independence was revived in 1989, simultaneously with the closure of the Panguna copper mine, the core of the island economy, followed by a civil war and more than a thousand deaths, the most violent contemporary struggle in the Pacific since the Second World War. In 2001 a peace agreement was signed with PNG, bringing hostilities to an end (Boege, 2022). Eventually, an internationally recognised but non-binding referendum in 2019 resulted in a 97.7% vote in favour of independence, but PNG argued that independence could only be granted by the PNG parliament. Six years later, the PNG government and the government of the Autonomous Region of Bougainville remained engaged in dialogue to determine Bougainville's future political status, with Bougainville demanding independence in 2027. Neither PNG (because of the loss of access to Bougainville's resources and the potential for similar demands from other island provinces),nor Australia, through its concern for the arrival of another small state in the midst of the Melanesian 'arc', so close to Australia, was receptive. China is interested in the mineral resources and has contributed development assistance. The future of mining in Bougainville remains sensitive, but it has potential for economic development, despite disputes over compensation and ownership (Connell, 2021b; Regan et al., 2022; Nobetau, 2025). More than any other aspirant to independence, Bougainville, with a population of about 380,000, has potential for independent development.

Other occasional claimants to independence minimally faced geographical challenges. Malaita is not poor by Solomon Islands standards but is densely populated and has relatively fewer services and opportunities than the main island of Guadalcanal. The province has a history of association with Maasinu Rule, an indigenous movement that opposed British colonial rule prior to independence. As the largest migrant group in Honiara, Malaitans were aware that other parts of the country, and some other ethnic groups, were better off. Malaitans in Guadalcanal were a focus of the violent 'tensions' that broke out in 1999, leading to the RAMSI intervention. Two decades later, in 2020, after Solomon Islands switched diplomatic recognition from Taiwan to China, a delegation to Malaita, led by Prime Minister Manasseh Sogavare, was met with great hostility and pro-Taiwan protests broke out throughout Malaita, directed at China's animosity towards Christians and its undemocratic political system. Malaita had long enjoyed close ties with Taiwan, and Malaita for Democracy, a movement supportive of Taiwan and with links to the provincial premier, Daniel Suidani, called for anti-China protests and asked the Solomon Islands government for independence. With a population of around 165,000, but in islands embedded within a larger nation, the claim to independence was much more a rhetorical stance of the Premier than a serious demand. More significantly, as was reported at the time, 'the ongoing Malaitan agitation exposed how debate over China's influence polarised a fragile Solomon Islands, exacerbated historical tensions, and granted political activists an opportunity to leverage fear over China for domestic political aspirations' (Cavanough, 2020). SIDS could become entangled in global geopolitics and power plays.

Chuuk's (formerly Truk) interest in secession from the FSM had similarities with that in Malaita, becoming in the 2020s a movement to return it to its status prior to the formation of the FSM, which was evidently a neo-colonial construct. Ironically, while isolated Yap state had previously contemplated separation, the strongest attempt came from the most populous of the four states and in the centre of the country. Chuukese supporters of independence, with their own distinctive language and culture, believed that their economic development was hampered by decisions being made elsewhere by other Micronesians without stakes in Chuuk (Corbett, 2023). Independence supporters called for a referendum, sought a separate Compact of Free Association (see below) with the United States and suggested that were they not to achieve secession, they would explore the idea of a similar relationship with China, which had constructed roads and government buildings there. An independent Chuuk state would control a large EEZ in a strategic location, and Chuuk has a deep lagoon and potential port. The United States warned Chuuk against independence, and no referendum occurred. After COVID-19 the idea failed to regain local resonance until erratic American foreign policy saw a limited revival.

The unlikely secession of Chuuk has been paralleled by minor secession movements elsewhere in FSM, in Pohnpei but especially Yap (where unusual geography means that it is only linked by air to the rest of FSM via Guam), which may have been stimulated by tensions within Micronesia over recognition of China or Taiwan. As Guam's former delegate to the US House of Representatives suggested, 'there is a secessionist movement; there is a Yap independence movement, there is the Pohnpei independence movement and it is rumoured that some of it has been fed by Chinese agents, but those movements have been around for a while' (RNZ, 15 March 2023). As in Malaita, there was no real evidence of any Chinese pressure, but invoking Chinese involvement could prove a useful strategy.

Secessionist impulses have likewise waxed and more recently waned in the Caribbean. Prior to Anguilla seceding from St. Kitts in 1969 and remaining a British colony with overwhelming local support, Barbuda, with a population of around 1200, and a weak economy, had sought to secede from the much larger island of Antigua. A delegation from Barbuda, demanding separate associated statehood, was rejected by the UK in 1967, strongly supported by Antigua. Independence was intermittently referred to in subsequent years, but the Barbuda People's Movement—which dominates the island's politics—has led opposition to deleterious economic ventures in the island, including sand mining and a casino, while autonomy has provided extra resources from the centre, which Antigua claimed to be more than appropriate or proportionate (Corbett, 2023).

Like Anguilla and Barbuda, Nevis also protested against being forced into a unitary state with St. Kitts on the grounds that it would entrench inequalities between the two islands. While Nevis failed to gain independence, it gained considerable autonomy and its own administration and economic benefits. As its economy grew, support for secession declined but revived in the 1990s, leading to a 1998 referendum that fell just short of the two-thirds success rate that would have made Nevis a state. Those who voted against independence were swayed by concerns over economic viability, and opposition from CARICOM and the United States. Economic development in the twenty-first century and greater integration of the two island economies dampened further pressures for independence (Corbett & Byron-Reid, 2024).

Although the UK countenanced Anguillan secession prior to St. Kitts independence and had agreed to Tuvalu separating from Kiribati, most Caribbean SIDS aligned themselves with the UK position against secession for Nevis and Barbuda, fearing that if Barbuda and Nevis were to go their own ways, then other peripheral islands—Bequia, Tobago or Carriacou perhaps—might seek to follow. It might be costly, but no country was willing to lose territory. Intermittent rhetorical calls for secession have come from other islands, such as New Ireland (PNG), but none have been more than rhetoric frustrated by limited service provision, and none have any real possibility of success. Greater autonomy has often been achieved and most islands have disavowed continued pressure for independence. As geopolitical entities, discounting Bougainville, present SIDS are stable.

8.2 Independence Movements

Much like secession movements, accompanying the late independence in the 1970s of many SIDS, other dependent territories also sought independence. Some like Montserrat (and Bermuda) came relatively close; others like Anguilla and the Turks and Caicos, Guam and American

Samoa barely countenanced the possibility (Aldrich & Connell, 1998). In this century most have drawn further away from the prospect, regularly claiming greater autonomy but not an outcome that would result in the loss of migration or other rights (Clegg, 2012; Connell & Aldrich, 2021). Many nonetheless harboured small independence movements as in the French Caribbean territories of Martinique and Guadeloupe, Guam (and West Papua), with no realistic chance of success but valuable if infrequently persuasive in raising issues around inequality, identity and culture.

Beyond Bougainville, only one significant ongoing independence struggle remains, that in New Caledonia. France is the last European colonial power maintaining a territorial presence in the South Pacific, other than the UK's problematic Pitcairn Island, with a population of fewer than 60 people (administered in absentia from New Zealand). France has a more established colonial presence in the Caribbean (and the Indian Ocean). In the Pacific, New Caledonia, French Polynesia and Wallis and Futuna comprise a land area of 23,000 square kilometres and an EEZ of more than 7.6 million square kilometres. New Caledonia and French Polynesia are on the UN List of Non-Self-Governing Territories, and both have active indigenous independence movements. In New Caledonia, named Kanaky by the independence movement, violent anticolonial struggles in the 1980s by the indigenous Kanak people eventually brought a truce between the Kanak Socialist National Liberation Front (FLNKS), loyalist parties and the French state. The French government agreed to transfer most powers to New Caledonia except sovereign powers (defence, foreign policy, police, courts and currency). Many loyalists welcomed the transfer of authority from Paris to Noumea but fiercely opposed separating from the French Republic. Concern that independence from France would inevitably result in New Caledonia being strategically dominated by China was a popular theme in the rhetoric of 'loyalists' who sought to remain French, and who used the 'China card' to try to influence Australia's position on New Caledonia in their favour. French government officials also played up the threat of China to the French Pacific, but evidence of China's alleged designs on New Caledonia itself was barely circumstantial.

New Caledonia is the only Pacific settler colony and its demography has so far produced a majority population opposed to independence. Roughly 41% of the population are Kanaks, 24% are European, with 18% who see themselves as Caledonian (usually the dependents of earlier settlers). Some 8% are Wallisians and Futunans, from two small Polynesian French islands, who have mostly opposed independence, fearful of any enforced return to their islands. Since the 2010s, Europeans have been leaving New Caledonia, and the overall population has fallen slightly, but Kanaks are unlikely to become a majority in the foreseeable future, and not all Kanaks support independence. In two referendums in 2018 and 2020, the proportions supporting independence grew, reaching 46% in 2020 (Connell, 2021a). Disputes followed a third referendum, held late in 2021 during COVID-19, which Kanaks boycotted, being engaged in mourning requirements and fearing further contagion, so that 'loyalists' inevitably carried the day with 96% of the vote (Kowasch et al., 2022).

France accepted the results of the third referendum, but Kanaks objected. That came to a head in mid-2024 when France sought to change the electoral rolls, to include more recent migrants to the territory, which would have resulted in greater opposition to independence. Violence erupted, barricades cut off suburbs, and supermarkets, schools and factories were set on fire and destroyed. Air links to New Caledonia closed, the valuable tourist economy collapsed, several people were killed, and some pro-independence leaders were deported and jailed in France. Electoral matters occurred in a context of uneven development, and deep frustration among Kanak youth unable to find decent employment in their own land, alongside a realisation that after more than 30 years of slow movement to 'rebalancing' society and economy towards what might have been a more equal state, too little had changed for too many. Coincidentally, the main island's nickel economy, faced with competition from Indonesia and falling prices, was experiencing a major downturn and rising unemployment.

Kanaks took their case to the United Nations Decolonisation Committee. External players became involved. The PIF argued for independence, queried the validity of the third referendum and undertook a mission to New Caledonia. In mid-2025, after intense discussions between all parties in France, a Bougival Accord was agreed, for a year of consideration and discussion in New Caledonia. France suggested that New Caledonia move towards a 'sovereignty in partnership with France', a status like that of the northern Micronesian states (see below). New Caledonia might then be a state within France (a hitherto unknown status) with control over international relations, subject to ratification by France (but not control over defence, currency and justice), and would be able to negotiate a new name and flag. France would secure a stable supply of nickel and keep Chinese influence at bay. It was a consensus, even an exercise in semantics, where each side failed to gain what it wanted, and within weeks it was rejected by the FLNKS as inadequate.

With France, and its military presence, welcomed by metropolitan states in the Pacific region as one potential bulwark in containing an expansive China, French Polynesia too was caught up in global geopolitics, despite being more geographically marginal to great power interests, and despite five distinct archipelagos and 75 populated islands. Both French Polynesia and New Caledonia have vast EEZs with fish and probable seabed mineral resources. Named 'Tahiti Nui' by the independence movement, and reinstated on the UN List of Non-Self-Governing Territories in 2012, French Polynesia has a smaller proportion of its population in favour of independence. After the end of nuclear tests on Mururoa in 1974, the interests of France became less obviously either strategic or economic. France's Pacific presence has been welcomed by Australia and other 'western' powers as it strengthens their own presence and interests in the region. But, in the Pacific, there was considerable support for decolonisation in the French territories, particularly from the Melanesian Spearhead Group.

In a culturally and politically diverse region, uncertainties abound where geopolitics is significant. The third French dependency, Wallis and Futuna, is unlikely to seek a change in status. However, in 2023, the head of government (*ulu*) of Tokelau, a New Zealand territory of 1500 people on 3 atolls, expressed a wish for change and identified 2025, the anniversary of 100 years of New Zealand rule, as the date for a referendum on independence (RNZ, 13 March 2023). The remote possibility of island states disappearing as sea-level rises, the most likely being Tuvalu, raises questions about their legal status where a state requires a permanent population, a defined territory, a government and the capacity to have relationships with other states. Absent islands would seem to rule that out (Yamamoto & Esteban, 2014). Tuvalu's perspective on the hitherto definitive Montevideo Convention ruling is that 'if Tuvalu were to have no physical territory due to climate change and sea-level rise, Tuvalu would be missing one of the conditions under the Montevideo Convention that would make it recognised as a state. This is why Tuvalu's statehood would be challenged if it lost its physical territory' (Simon Kofe, Minister for Justice and Foreign Affairs, quoted in *PacNews*, 25 April 2022). However, the ICJ advisory opinion on the obligations of states under international law with reference to climate change also argued that once a state had been established 'the disappearance of one of its constituent elements would not necessarily entail the loss of statehood' and it would retain its borders (Pouponneau & Northrop, 2025). Long before any physical disappearance, the population would have left and a virtual Tuvalu would exist in the metaverse (Rothe et al., 2024).

In some places international borders have yet to be finalised, with, for example, Fiji, the Solomon Islands and Vanuatu submitting in mid-2025 three extended continental shelf claims to the United Nations, covering 453,400 sq kms, to strengthen Pacific sovereignty and sustainable resource management (RNZ, 14 July 2015). However, no borders are subject to disagreements, except in the case of the Matthew and Hunter Islands, tiny uninhabited rocky volcanic islands, claimed by both Vanuatu and France (Song et al., 2023). While it is unlikely that political instability would follow any resurgence of claims, and France has seemed likely to accept Vanuatu's claim, the islands are one more site in the uncertain geopolitics of the Pacific.

In some respects, Caribbean and Pacific nationhood is still emerging, reflecting the continued legacy and presence of colonialism, at the ends of empires. Although both regions experienced belated and incomplete decolonisation, circumstances differ. Tensions over the borders of nation states in some Pacific SIDS have international significance and pose one distinctive strand of insecurity. While a few places may seek a different political future, more independent from colonial powers, others are strengthening ties with metropolises, through memoranda of understanding and even treaties. That was exemplified by the treaty between

Australia and Tuvalu in 2023, which created new and unusual political ties, although a subsequent Nakamal Agreement between Australia and Vanuatu of 2025 was initially rejected (see below). New Caledonia and Bougainville are distinctive in attempting to break ties but lack support from metropolitan powers concerned over any possibility of small, unstable states and further unwanted external interventions.

8.3 New Senses and Strategies of Place

In the last quarter of the twentieth century, just as the United States, Australia and New Zealand saw the Pacific as 'their domains', so the United States had equally assumed inherent mastery of the Caribbean. In political terms that was sealed by its invasion of Grenada in 1983, ending Maurice Bishop's Marxist-Leninist New Jewel Movement, which had been in power for four years and rather later by the demise of the Soviet Union and thus a perhaps less dangerous Cuba. During the Cold War era, the United States put pressure on the Caribbean SIDS to withdraw any support for Cuba. Grenada tragically demonstrated how SIDS could be caught in a geopolitical world that pitted liberal democracy against state-led development, and political and civil rights against social and economic rights (Puri, 2014; Lewis, 1999). It provided a violent example of how small states are affected by geopolitics determined elsewhere, how they are positioned by prevailing global forces, but could fall prey to the ambitions of global powers (and to pressure, and forced austerity measures, from such institutions as the IMF), which offered little scope for flexibility (Handa & King, 1997). Such extremes of political violence never recurred. Ironically, many Caribbean SIDS were aligned with a non-aligned movement, the Group of 77 formed in 1964, which refused to side with any of the great powers. SIDS were not, however, powerless; Jamaica influenced US policy in such areas as development aid and drug certification during the 1990s (Bernal, 2015), and Antigua succeeded in disputing internet gambling regulation (Cooper, 2023).

A second phase of external intervention concerned the drug trade in and through the region, perceived as a threat to Caribbean political stability and so to regional security, sovereignty and democracy (Griffith &

Munroe, 1995; Griffith, 2010). This provided an explanation for a US war on drugs and was a primary focus for US cooperation, hegemonic exercise of power and a pragmatic foreign policy in the region. Heavy reliance on the United States to control imports, ineffective state management of crime, when combined with gangs who embedded themselves in marginalised communities, had made it difficult for regional bodies to cope with the problem. Countries that did not cooperate with the United States, were denied foreign aid. For decades its war on drugs was criticised for over-riding the rights and sovereignty of other countries (Corva, 2008) while gun trafficking in the Caribbean, that had contributed to high homicide rates, was relatively ignored despite most guns emanating from the United States. Drug flows and their control came decades later in the Pacific, contributing to the crisis in policing (Chap. 7).

At the end of the twentieth century physical interventions in both regions remained limited. In the Caribbean, the invasion of Grenada was old history, the country was stable, and in terms of security the United States was much involved in countering drug trafficking. The Cold War was over, and Russia had lost interest in both regions but remained a reluctant supporter of Cuba. Warfare in Bougainville and the RAMSI intervention in Solomon Islands had heralded a quarter of a century of metropolitan nervousness and watchful vigilance. RAMSI provided a critical juncture in Australia's relationship with the Pacific, from positioning itself as an external observer to being a regional leader and, eventually, a presumed part of the 'family' (Baker, 2023): positions not necessarily welcomed by the PICs.

By the end of the century, amidst greater economic and political complexity, as island economies struggled, Nauru was deemed by some the first Pacific failed state (Connell, 2006), while a Melanesian 'arc of instability', a much-critiqued notion, was seen by some as extending from West Papua to Fiji. Nauru was later viewed as a suitable candidate to host one of Australia's Regional Processing Centres, a 'Pacific Solution' for unwanted refugees (Morris, 2023). In this uncertain context, the United States had lost interest in the region, focusing only on the 'tip of the spear' of Guam and Micronesia, and the UK had effectively departed. New Zealand had limited interest in Melanesia, focusing mainly on Polynesia, but taking an approach like that of Australia.

In the twenty-first century, China emerged as a partner in trade and aid in hitherto largely Anglophone and Francophone regions. Chinese interest contributed to the rethinking of the Pacific especially, but also the Caribbean, despite its distance from China and the particularly watchful nearby US presence. Bordering on the Pacific and with expansionist objectives, China proved a powerful presence that had threatened near neighbours. With broad economic and strategic objectives, it had concluded a war with Vietnam, disputed Vietnam, Japan and the Philippines over sovereignty of the East and South China Seas (where its 'nine-dash line' map was minimally regarded as a 'tongue of passive cartographic aggression'), and the ownership and control of rights to fishing that challenged UNCLOS norms, which China ignored (Valencia, 2007; Raine, 2017). It increasingly threatened the existence of Taiwan as a separate state, while it expanded as an aid donor and diplomatic presence in both regions.

The notion of the seas being military regions was not new. Both regions were places of weapons testing, bombing practice, high-security jails and even chemical weapons dumping, from Johnston, Kwajalein and the Northern Marianas (Davis et al., 2020), to Vieques and Guantanamo, with the worst practices reserved for dependent territories. Countries with different philosophies had briefly flexed their muscles before, primarily in the Pacific, notably the Soviet Union, with a border and an interest in fisheries. Scale, significance and players changed in the twenty-first century. As the interests of seasoned participants in the Pacific region stabilised or declined, those of China increased, initially centred on terrestrial mining but soon to expand. By the 2010s, the government and state-owned companies were playing a role in finance, trade, investment, aid, cultural diplomacy (education scholarships) and infrastructure provision, all perceived in the United States and Australia as unaccustomed and dangerous challenges in their historic 'backyards'.

As Chinese interests moved southwards and its naval capability expanded, this provoked 'middle power', especially Australian, rivalry in a more contested ocean. India, Indonesia, Korea and Japan took greater interest and Chile looked westwards. The UK reversed its dwindling interest in the region, with a new diplomatic presence, but unlike Australia its role was cultural and economic rather than political, as it was in the Caribbean, being the main destination for Caribbean exports (Cullum, 2021). Distance enabled detachment. In contrast, by 2012 as security issues appeared critical, the Pacific was witnessing the new context of Foreign Ministers of several countries (notably China's Wang followed by Australia's Wong) anxiously traipsing through the region's capitals delivering (or promising) largesse and wisdom, winning favour and promoting their own solidarity with the needs and aspirations of island states. That stream persisted in subsequent years. In this evolving geopolitics, Australia had some soft power advantages whereas China's role and vision were always more opaque.

Chinese interest in the Caribbean was simultaneously increasing, even suggesting a slide into 'a new Cold War' (MacDonald, 2022). When Grenada established diplomatic relations with China in 2005, it was assumed that it was motivated by a bid for Chinese aid after Hurricane Ivan (and assistance duly arrived) since, like other SIDS, Grenada had no interest in ideology or which was the 'real China' (Yang, 2011). Chinese expansionism was underpinned by external assumptions that SIDS were vulnerable, financially weak and could be tempted by generous loans and infrastructure, but intervention was also linked to control, notably in the Pacific with its fish and mineral resources. Chinese interests were consequently seen in the Anglophone world as threatening 'traditional interests'. Big power manoeuvring in the regions was never primarily in the interests of the island states or initiated by them. However, by the 2010s, although geopolitical shifts and manipulations were beginning to occur at a frenetic pace, it had become complicated by the need to accommodate the wishes and interests of increasingly assertive SIDS.

The Pacific especially became immersed in a 'polycrisis' that partly reflected the travails of island states rebounding from the economic and social exigencies of COVID-19, the shifting geopolitics following Chinese involvement in the region, seen, somewhat fancifully, as the new 'Great Game' (Kowasch & Lindenmann, 2014; Sora et al., 2024), or even, as in the Caribbean, 'a new cold war' (Wesley-Smith, 2021). Rather later, that was allied to the possibility of Bougainville securing independence and the violence that engulfed and divided New Caledonia. Change was less complex in the Caribbean, where challenges to US 'control' were limited, quests for independence now absent and no obvious uncontrolled resources. A constant process of securing and shoring up support and aid delivery created a growing complexity of international involvements.

8.4 Enter the Dragon? Chinese Investment

An expanding Chinese presence in both regions has brought strategic complications and contestations, evident in the competition over policing (Chap. 7), more demonstrable in the Pacific, more subtle in the Caribbean. China's role in both regions has been broadly similar to that in Africa, Asia and Latin America, but in the Pacific alone (and to a lesser extent in the Indian Ocean) it created unusual tensions.

8.4.1 The Caribbean

In the late twentieth century a significant Chinese indirect trading presence brought the collapse of attempts to develop a manufacturing industry in the Caribbean, effectively resulting in the region returning to where it had a comparative advantage: natural resources and tourism (Gonzalez-Vicente, 2021). At the start of the century, China deliberately began to grant concessional loans in the region, a pattern of 'dollar diplomacy' designed to shift the SIDS that recognised Taiwan, including Grenada and St. Lucia, to recognition of China. Bilateral economic cooperation between Jamaica and China rapidly developed. Following the launch of the 1999 Chinese 'going out' policy, it became China's biggest trading partner in the Anglophone Caribbean. By 2017 only the United States exported more to Jamaica. China had become the second-largest trading partner in the region, weakening intra-CARICOM trade, and moving CARICOM towards an earlier pattern of raw material exports. China supported institutional initiatives (such as the China Caribbean Business Council) and bilateral relationships, especially with resource-rich Jamaica (Bernal, 2010). Investment, especially in mining, was motivated by the same market and resource seeking as elsewhere in the Caribbean and the Pacific. That included acquiring Jamaica's largest bauxite mining refinery from a Russian conglomerate and, more surprisingly, an indebted sugar estate (Minto, 2019). Investments in the oil sector of Trinidad, the second-largest Caribbean SIDS, followed. China was able to invest in the declining 'sunset industries' of sugar and bauxite since companies had access to cheap credit and used Chinese workers who were (and are) paid lower wages and worked longer hours than local workers. An increasing focus on infrastructural projects since 2013 brought stadiums, hotels and resorts (in Barbados and the Bahamas), highways, a cricket stadium in Grenada, convention centres and government buildings (as in Antigua), a pattern identical to that in the Pacific. In both regions some projects were criticised for shoddy construction work and environmental damage, and resentment grew against the use of Chinese labour. It was rarely clear why China chose such a range of projects to invest in, fund and construct, or how they related to Caribbean development priorities (Jules et al., 2023), suggesting the primary goal was to have a presence in the region.

Barbados became the first Caribbean SIDS to join China's Belt and Road Initiative (BRI), a modern maritime re-imagination of the Silk Road that was effectively a global infrastructure project to boost China's influence and international connectivity. Barbados and Jamaica gained valuable infrastructure for tourism development, and Chinese companies flooded the Jamaican market, focusing on roads and energy projects, including China Harbour's investment in the North-South Highway in exchange for a 99-year lease on valuable adjacent land. Many companies were involved in island-wide infrastructure projects, facilitating the migration of a new wave of Chinese migrants, resulting in Jamaican media critiques of 'economic colonialism'. China's capacity to develop new projects in indebted economies proved valuable however and resulted in more traditional 'partners', such as the IMF, the World Bank and the EU, losing their 'undisputed capacity for influence and coercion' (Gonzalez-Vicente, 2021: 171). However, it also helped entrench long-established structural and post-colonial impediments to development.

BRI initiatives in the Caribbean as elsewhere primarily focused on infrastructure projects, often criticised as a form of 'debt-trap diplomacy' characterised by unmanageable loans, where the debt becomes a bargaining chip for political concessions. However, little evidence exists in either region of debt traps, while some debts and loans have been written off (Baud et al., 2022). For heavily indebted countries like Jamaica, the key benefit of the BRI was access to funds through alternative means other than 'traditional' lenders, who used access to loans to impose conditions on the national economies. While some Jamaicans welcomed BRI opportunities for much-needed infrastructure development and economic advancement, others remained sceptical, fearing that these initiatives would compromise national interests, undercut national contractors, affect relationships with the United States and be a threat to tourism.

Unlike other SIDS in the Caribbean or the Pacific, Barbados accepted military equipment and engaged in military cooperation with China. Jamaica received quasi-military assistance to 'build capacity' and acquire superior uniforms, and Trinidad gained police motorbikes and a coastal patrol vehicle, much as in Solomon Islands (Chap. 7). Most Caribbean SIDS gained some infrastructure development: in Trinidad this centred on oil and gas infrastructure, while the Bahamas gained a deepwater port. Several Caribbean SIDS welcomed Chinese loans for development that offered some prospect of reducing reliance on tourism. After Antigua and Barbuda had signed into the BRI in 2018, so gaining several critical infrastructure projects, including ports and water systems, some Antiguans expressed concern over China's growing influence on the island, suggesting that it had 'traded in its sovereignty'; others feared that such projects created major environmental damage potentially threatening a tourism-based economy. Environmental problems had emerged elsewhere, with several SIDS cancelling projects that had, or seemed likely to have, negative environmental impacts, but with China responding by financing clean energy projects, albeit alongside fossil fuel investments (Shah & Chiorazzi, 2024). China also provided loans for housing construction and renewable energy including the building of solar farms in Barbados.

Caribbean concerns were matched in greater American response and vigilance, with ports seen as potentially used for resupply and military ship operations in times of crisis or conflict, or for simply gathering information on ships and goods movements through the region, enabling the possibility of crippling supply chains by withholding or seizing key shipments. That was likened to Africa: as one US Army general remarked in

2024: 'The PRC's negative influence in this region could soon resemble the predatory and self-serving influence it now has in Africa' (quoted in *Dialogo America*, 20 July 2024). Further concerns mounted as Huawei took an interest in the region, resulting in the US ambassador to Jamaica stating on Twitter, and on the national radio, that Huawei was not to build any 5G technology in Jamaica. Concerns over debts and the nature of development existed but little concern was otherwise attached to broader geopolitical issues, including the recognition of China.

With the growing dependence on China for infrastructure and loans, China itself faced problems with 'overstretch' as economic, military and political resources were extended to gain global influence. The BRI enabled China to establish strategic partnerships, gain control over key resources, project power globally and squeeze out Taiwan (MacDonald, 2022). While infrastructural benefits were welcome, the more visible presence of Chinese activity 'triggered an anti-China discourse' that contributed to a 'relative decline in Chinese soft power' (Gonzalez-Vicente, 2021: 185). Much of that discourse related to the secrecy of Chinese investment and projects, inappropriate priorities and the lack of even unskilled employment for local people on large projects. Many new buildings had visual impact but no obvious development function, whether economic or social, including under-utilised sports and conference facilities, and several had design flaws. Such patterns and problems were almost identical to those in the Pacific.

8.4.2 The Pacific

Although Chinese traders were a familiar presence in the late-nineteenth-century Pacific and small Chinatowns were not far behind, Chinese involvement in the Pacific took a different turn in the late twentieth century with trade and investment, notably in mining, and aid all playing a role. In the early twentieth century Chinatowns were growing with a new wave of Chinese, buying out small stores in Tonga, Fiji and elsewhere and participating in a range of economic arenas, which some viewed with alarm, and others saw as offering new opportunities (Crocombe, 2007; D'Arcy, 2014; Cheung & Day, 2025). China recognised that some entrepreneurs were harming their national interests and took unusual steps, with Chinese police bypassing Pacific legal systems in Fiji in 2017 and in Vanuatu in 2019, 'to round up Chinese nationals, label them as criminals, and bundle them onto planes' (Peake & Smith, 2022). Similar problems followed in Palau (below), both unhelpful to soft power. Chinese interest in the Pacific was part of its global expansionism, with clear that parallels in the Caribbean, including the construction of roads, stadiums and hospitals (Zhang, 2018). As American interests in the Caribbean and most of the Pacific had waned, such developments initially took place without active strategic rivalries but were part of an offensive to decrease the number of Taiwan's allies in the region.

China provided considerable aid to Fiji after the 2006 coup, the first time that Chinese support was greater than that of 'established donors', the United States, Australia and New Zealand. Between 2006 and 2013, China provided almost half of all development assistance to Fiji and close to 30% of total overseas development assistance to Samoa and Tonga (Dornan & Pryke, 2017). China had joined the big donors, becoming, after 2011, the second largest aid donor in the Pacific after Australia. While much of that came in the form of loans and grants, not dissimilar to that of other donors (see below), China did not disclose its practices but claimed that they came within a framework of south-south cooperation 'so our origin, our policy approach and our practice are very different from those of the traditional donor countries' (http://news.xinhuanet.com). That difference was further centred on arguments that China was not a colonial power (unlike Australia and the United States) and that there were shared benefits through the BRI (Xie, 2025). Although, as China became wealthy, that argument proved difficult to sustain (Sou & Day, 2025). PNG became the first PIC to join the BRI, and several others followed. China astutely introduced a major new multilateral bank, the Asia Infrastructure Investment Bank, while Australia countered with its own Australian Infrastructure Financing Facility for the Pacific. As in the Caribbean, aid centred on loans to fund infrastructure projects, built by Chinese companies (Bernal, 2015), the latter practice differentiating Chinese aid from that of other donors, contributing to recurrent suggestions of debt-trap diplomacy.

China's trade in the South Pacific increased sevenfold over the first decade of the twentieth century, mainly in PNG where it was interested in the various mineral resources. Its investment in the Ramu Nickel mine became its single largest investment in the Pacific, and it later acquired Gold Ridge, the only mine in Solomon Islands. It had long sought mining opportunities, a legitimate interest where its own resources were depleted, and by the end of the twentieth century was involved in undertaking exploration in the northern Pacific for deep-sea mining opportunities (Zhang, 2018). Tuna-processing attracted Chinese investment, and Chinese aid was becoming visible through various infrastructure projects: exact parallels with the Caribbean. It extended into other contexts in 2024 when Vanuatu signed an agreement with China to enhance bilateral ties in aviation, tourism and trade. China's representative in Vanuatu stated: 'Over the past 42 years, the relationship between the two [countries] resulted in something stronger, and we are appreciative of how Vanuatu has stood and supported the One-China Policy' (*Daily Post*, 4 December 2024). The agreement allowed Chinese planes to make direct flights between China and Vanuatu.

It was early evident that Chinese aid had strategic dimensions. Between 2006 and 2011 China disbursed about US$850 million in bilateral aid to the eight Caribbean island states that recognised the PRC. Its Pacific interests were similarly driven by competition with Taiwan for diplomatic recognition. At the start of the 2010s, Taiwan was recognised by six SIDS: Kiribati, Marshall Islands, Nauru, Palau, Solomon Islands and Tuvalu. Taiwan offered more low-key blandishments, notably agricultural projects, but crude 'chequebook diplomacy' died down with both promising implausibly that they would no longer seek to persuade states to switch recognition. In 2021 China announced plans to upgrade an airstrip on the remote atoll of Canton (Kiribati), about 3000 kms southwest of Hawaii, claimed rather improbably by Kiribati to be a non-military project designed to improve transport links and bolster tourism, by turning it into a 'high-end niche tourism destination'. In the same year Kiribati switched its recognition from Taiwan to China, part of continued 'chequebook diplomacy'. By then the promised truce was over, with Taiwan 'losing' Kiribati, Nauru and Solomon Islands.

Diplomatic trends (as several nations, not only in the Pacific, broke longstanding ties with Taiwan to recognise China), a growing physical presence (evident in diverse shipping movements) or economic shifts with increased trade, aid (soft and hard loans, infrastructure provision and port construction) and investment (especially in the mining sector as in PNG and Solomon Islands, where two mines are majority Chinese-owned) were all part of the maritime BRI, prompting Pacific SIDS to 'look north' with greater frequency (Szadziewski, 2020; Mausio, 2023) and other metropolitan nations to look north and west with some anxiety. By 2011, the US Secretary of State, Hillary Clinton, had declared to the Senate's Foreign Relations Committee that China and the United States were vying for supremacy in the Pacific: 'Let's put aside the moral, humanitarian, do-good side of what we believe in and let's just talk straight, realpolitik. We are in a competition with China'. The same sentiments were matched in Australia. By then, as Chinese infrastructure offers increased, rumours existed of Chinese 'hard power' involved in establishing ports, military bases and rocket launching pads in the region. The arrival of COVID-19 in 2020 brought visible rivalry through 'mask' and then 'vaccine diplomacy'; a geopolitics of health care intensified, as both China and Australia sought to use the crisis to boost their presence in the region by delivering medical supplies, personal protective equipment and technical assistance. China had both expanded and diversified its regional interests with a more competitive, politically targeted model of aid engagement, centred on some key SIDS, notably Palau, Kiribati and Solomon Islands, to secure diplomatic support in strategic locations.

8.4.3 Palau

As one of the SIDS closest to China and partly dependent on the United States through a Compact of Free Association (COFA) (see below), Palau was put under some pressure by China, centred around tourism (that accounted for half of its GDP), to shift its allegiance from Taiwan. China was accused of making 'empty investments' in real estate, bribing local politicians and sending Chinese organised crime to Palau, a variant on 'dollar diplomacy' to the extent that it might pose a 'challenge to the sovereignty of this US-aligned country' (Caterson, 2024). In 2019–2020 Palau had deported more than a hundred Chinese citizens working in illegal online gambling operations that did little to boost China's interests and image. Such disruptions and diversions proved unsuccessful.

China's main objective was to deter tourism from Taiwan and minimise tourism income (as Taiwan was the largest market) and redirect tourism to the China market. In 2017, the Chinese government ordered tour operators to stop selling package tours to Palau. In mid-2024, Palau's President, Surangel Whipps, claimed that China was further 'weaponising' tourism to get the nation to sever its ties with Taiwan, promising that 'the sky is the limit' if they ditched Taiwan, while simultaneously promoting the notion that Palau was unsafe. Palau was unable to attend the Pacific Asia Travel Association's annual summit, hosted in Macao, because it did not have diplomatic relations with China, and its representatives were unable to receive visas. There was some suggestion that Palau was unexcited about Chinese tourism since its visitors caused environmental damage, spent most of their money in Chinese-owned businesses, caused inflation and upward pressure on prices, and locked up land in China-linked real estate investments (Bock & Parilla, 2024). Taiwan increased the number of flights to Palau while Palau sought to diversify tourism sources, signing a deal with Japan for direct flights from Tokyo in 2025. Three days later, the Australian airline Qantas agreed to take over direct flights from Brisbane from Air Nauru and increase their frequency. Whipps stated that 'we need economic development but at the same time we have values, and we have partnerships and the relationship we have with Taiwan we treasure' (RNZ, 15 August 2024). Soon afterwards Palau received US$80 million in COFA grants from the United States, which gave the US exclusive rights to Palau's waters, airspace and land (RNZ, 7 October 2024). Palau remained one of the three PICs to recognise Taiwan.

8.4.4 Solomon Islands

The changing geopolitics of the Pacific was reflected in the diversity of Chinese interests and interventions in the SIDS closest to Australia, where the former colonial power, a distant UK, had minimal continued interest. The recent history of Solomon Islands had involved Australia leading the RAMSI intervention (the culmination of decades of friction between the islands of Malaita and Guadalcanal), corruption and environmental degradation in the forestry industry, and mismanagement (Allen & Dinnen, 2017; Cavanough, 2023). An uneasy political relationship was accentuated, with the Prime Minister, Manasseh Sogavare, who had four terms as PM between 2000 and 2024, being highly critical of RAMSI, which he accused of being dominated by Australia and undermining Solomon Islands' sovereignty. Meanwhile neighbouring Bougainvillean politicians, some of whom had fled to Solomon Islands during the blockade, also resented Australian intervention in the island's recent history (Regan, 2022).

China had long been involved in the economy of Solomon Islands, especially with its state-owned industries' engagements with the logging industry (Kabutaulaka, 2006). In 2017, Huawei Marine was endorsed to lay an internet cable connecting to Sydney, to replace satellites with fast internet connections. Australia had security and strategic concerns about the Chinese telco plugging into Australia's internet so that it quickly offered to pay for the cable, costing at least A$86 million, through the overseas aid programme. Since the cost was excessive for few national benefits, the strategic rationale was apparent. The original Chinese plan, arranged by Sogavare, was replaced by the Australian version in a brief period when Sogavare was not Prime Minister. Huawei had previously been banned on the advice of Australia's security agency, ASIO, from being involved in its own National Broadband Network (SMH, 25 January 2018). A year later Sogavare returned to power and Chinese involvement again increased after the diplomatic switch from Taiwan to China in 2019. That caused some disappointment since Taiwan had constructed the main national hospital and precipitated the rise in tensions between Malaita and Guadalcanal (see above). China gave financial support to MPs who had supported the switch and was also investigating the potential of a port site on Kolombangara island. When China provided Solomon Islands with a new stadium for the 2023 Pacific Games, the independent Malaita MP, Peter Kenilorea, expressed concern that it would come at a cost to the nation's health care and education systems. The visually seductive stadium was rarely used. By contrast, a substantial

Comprehensive Medical Centre donated by China was opened early in 2025 (with both specialised NCD facilities and Chinese traditional medicine) alongside the dominant National Referral Hospital, a gift from Taiwan in a previous era. Huawei had reappeared in 2022 with a contract to build 161 transmission towers despite these too having little obvious utility at considerable cost. That also involved more infrastructure construction and related concern, as elsewhere, that employment was dominated by Chinese (reducing local incomes and skill transfer). Aid from Australia and other countries had tended to focus on education, training and 'capacity building'.

Outside Solomon Islands the switch from recognising Taiwan to China prompted international condemnation, especially in Australia, where its long-assumed advantage in its Melanesian 'backyard' seemed threatened. Perceived Australian inattention and strategic blunders have been linked to China's undue economic influence and its interest in extractive industries (Wood, 2021). International rivalries between China and Australia were matched at local level: China could be a valuable and benevolent donor, and the nation's main trading partner, but others sought the old post-colonial order and resented Chinese involvement in local business. Riots and arson in 2021 targeted businesses in Honiara's Chinatown, much of which was burned down. Despite the riots being largely centred on domestic issues, notably unemployment, an underlying current was a 'New Cold War' where the Premier of the Malaita Province, Daniel Suidani, opposed centralised decision-making and Solomon Islands switching allegiance from Taiwan, to make a show of preventing Chinese investment in Malaita (Ride & Zhang, 2024). After the 2021 riots, the Solomon Islands government stressed that it needed to strengthen the Royal Solomon Islands Police Force capability and capacity to respond to future unrest (Chap. 7). A contingent of Australian police and military arrived following further riots in Honiara late in 2022, and an Australian Federal Police commander was sworn in as Deputy Commissioner of the Police Force. That was followed by the Solomon Islands government accepting China's offer of six police advisers and emergency riot equipment, a suggestion of militarisation that generated consternation in Australia. The Chinese announcement and consequent concerns suggested that police assistance programmes were not so much about the technical matters of capability building, that they were promoted to be, but were 'presence missions', more to do with geopolitical 'shadow boxing' and currying favour with local elites (Peake & Smith, 2022). Meanwhile Sogavare had concluded a security pact with China in 2022 that potentially would involve logistical support for Chinese naval ships but promised Australia that no military base would be involved or needed: rhetorically querying 'who are our enemies?'.

The security pact and the offer of police training brought wider concerns, although Sogavare stated that it promoted trade and humanitarian cooperation, not the militarisation of the Pacific. On the same day, a US delegation arrived in Honiara and indicated that were China to establish any kind of 'military installation', the United States would have significant concerns. New Zealand, whose Pacific policy direction tended to be subordinate to both Australia and the United States (Steff, 2024), argued that it would undermine stability and 'fundamentally alter the strategic balance of the region', a position shared by the United States and Australia (Brady, 2022). A few days later, a Chinese Foreign Ministry spokesperson, Wang Wenbin, said: 'Island countries in the South Pacific are independent and sovereign states, not a backyard of the US or Australia. Their attempt to revive the Monroe Doctrine in the South Pacific region will get no support and lead to nowhere'. After the 2023 Solomon Islands elections, Sogavare was again ousted and replaced by Jeremiah Manele, who offered a somewhat more conciliatory role to Australia, despite making a high-profile visit to China in mid-2024, alongside the Vanuatu Prime Minister, Charlot Salwai, during which both vowed to 'deepen the comprehensive strategic partnership' with China (*Vanuatu Daily Post* 2 August 2024). Great, and lesser-power competition was rarely far away. However, fluidity and flexibility, manipulations and machinations, and varied regional and personal approaches were inexplicable without understanding the complex convolutions and particular personal style of Melanesian leadership and politics (Köllner, 2025).

Over just a few years Solomon Islands had become increasingly linked to China in terms of its economy, aid delivery (especially for infrastructure), government scholarships, inter-province cooperation and security, with each of these elements being viewed with suspicion and discouraged by Australia and other 'western' (sometimes referred to as 'legacy') countries. That was accentuated as Vanuatu took up similar positions and Kiribati and Nauru switched diplomatic recognition. As several SIDS looked towards China, domestic opposition sometimes intensified, while aid delivery moved towards security, disadvantaging education, health, water and energy. SDGs were of no concern. A more transactional political system was in place. Solomon Islands was a more extreme version of trends in other SIDS—because of the scale of Chinese economic involvement, the strong hint of militarisation and its closeness to Australia. Such activities provoked new strategies of external engagement, especially from Australia, where countries that most people had barely heard of loomed in importance, and 'doorsteps', 'backyards', 'neighbours' and eventually 'families' became common currency. The Pacific was reinventing itself as SIDS could simultaneously disparage external intervention and play off rival offers and contenders. Solomon Islands had never hitherto been so important on the international stage.

8.5 Diplomatic Shifts and Metropolitan Responses

In the Pacific especially, China's expansive arrival, competitiveness and military potential raised regional diplomatic stakes, stimulating greater Australian interest (the 'step-up'), New Zealand interest (the 'reset'), renewed American interest through a 'Pacific Pledge' (but particularly in Guam and the Micronesian Compact states) and even the return of the UK to the region (the 'Pacific Uplift') after a lengthy hiatus and in anticipation of Brexit. In 2019, Indonesia launched its own 'Pacific Elevation', claiming a Pacific islands identity and touting its trade with the region, downplaying violence in West Papua. A little later Japan began to emphasise its 'Pacific Bond'. Albeit slogans, they represented a greater and more considered regional involvement and a recognition that 'neighbours' might have been ignored and taken for granted directly and indirectly. In the Caribbean, where Chinese involvement was more limited and without the same economic, military and strategic elements, concern was centred almost exclusively in the recipient countries and on US-Caribbean relations.

8.5.1 Australia

Australia, contemplating at the turn of the century an 'arc of instability' in its northern Melanesian neighbours and with a preferred future in Asia, was more likely to see Pacific SIDS as problematic rather than partners or simply to forget they existed, an amnesia particularly reserved for its lone former colony, PNG (Dorney, 2016). Chinese interests discomfited a country hitherto certain of its primacy in the region. However, Australian aid to the region had declined and sometimes was confounded with labour migration, as in the supposed benevolence of a seasonal migration policy that was primarily designed to assist Australian farmers (Petrou & Connell, 2023a). Unwanted competition for hegemony was evident, as in 2006 when the Minister of Defence, Brendan Nelson, gave a parliamentary speech on the arc:

> We cannot afford to have failing states in our region. The so-called 'arc of instability', which basically goes from East Timor through to the south-west Pacific states, means that not only does Australia have a responsibility in preventing and indeed assisting with humanitarian and disaster relief, but also that we cannot allow any of these countries to become havens for transnational crime, nor indeed havens for terrorism. (quoted in Dobell, 2007: 85)

Instability had multiple connotations in a region where China was a still-limited presence, but it was the PICs that emphasised that they too valued security but in a more diversified form.

The collective diplomacy of the Boe Declaration on Regional Security of 2018, adopted by the PICs, extended the 2000 Biketawa Declaration into a less narrow view of security by including issues of human, food and environmental security, transnational crime, and cybersecurity. Its distinctive theme was recognition of the threat from climate change: 'We reaffirm that climate change remains the single greatest threat to the livelihoods, security and wellbeing of the peoples of the Pacific and our commitment to progress the implementation of the Paris Agreement' and that this was just part of an increasingly 'complex and dynamic geopolitical environment leading to an increasingly crowded and complex region' (Pacific Islands Forum, 2018). Geopolitics became a preoccupation of every state in the region and part of a more transactional politics (Wesley-Smith & Finin, 2021). As Transform Aqorau pointed out for Solomon Islands, 'leaders are *forced* to negotiate a path that, on the one hand, could secure much needed aid and investment and, on the other, may lead to external influence and the erosion of national sovereignty' (2023; our italics). Some metropolitan powers were quite willing to enable, even contribute to, such an erosion in their quest for strategic security.

Australia remained the largest source of development assistance in the region, yet Australia was less generous than most OECD countries and ranked poorly in aid allocated for climate change adaptation (Wood et al., 2021). Because of its failure to develop domestic policies to challenge climate change, it was criticised by island states that saw New Zealand as a more reliable and sympathetic partner, and China as a new, distinctive and potentially valuable partner (ignoring the military concerns of the metropolitan states). Australia thus sought greater involvement and a more direct engagement with PICs as equal partners. In 2019 it signed a Fiji-Australia Vuvale (Friendship) Partnership, involving joint peacekeeping and security forces training, with some Australian police embedded in Fiji's police force. It was the prototype and precursor of treaties and pacts to follow (and was upgraded to a treaty in 2025).

Australia's increased involvement took on commercial concerns. In 2021, Telstra, Australia's largest telecommunications company, partnered with, and was subsidised by, the Australian government to acquire Digicel Pacific, the biggest mobile phone operator in the Pacific region (and leading telco in PNG, Nauru, Samoa, Tonga and Vanuatu), a decision later criticised as 'a new benchmark in the retreat of economic principles from the policy domain, and their replacement by unspecified security concerns' (Howes & Liu, 2022). The Foreign Minister, Penny Wong, stated that 'the Australian Government's support for this transaction reflects our commitment to help build a stronger Pacific family through investment in high-quality infrastructure'. However, alongside climate change issues, trade agreements provoked PIC concern, notably over PACER-Plus and the trade imbalances between Australia, New Zealand and the PICs, where it was argued that Australia needed to move beyond official development assistance and transactions to enable equity in trading relationships.

Australia's stance on climate change and fossil fuels particularly troubled PICs. Climate change issues constantly raised concerns, focusing on the Australian Prime Minister, Scott Morrison, described by the Fijian Prime Minister, Frank Bainimarama, as having a 'dangerous addiction to coal', having tried to water down a regional climate declaration at the Pacific Islands Forum in Tuvalu in 2019. At the end of 2020, Bainimarama rejected the notion that 'Fijians and our Pacific Island sisters and brothers be some sacrificial canary for coal-burning countries' (quoted in SMH, 12 December 2020). Australia was deemed to have taken an inadequate response while continuing to mine, burn and export fossil fuels, contradicting international efforts to limit global warming. Approval in 2024 of three massive coal mine extensions and a major gas project the year after, by a Labor Party government, brought further opprobrium and deep disappointment, with Australia charged with 'moral hypocrisy'. Tuvalu pointed out that it was 'threatening our survival' while Vanuatu claimed that it 'severely undermines our national security and sabotages our future' and was potentially in breach of the ICJ advice on climate change (ABC, 29 May 2025, 11 September 2025). The urgency, ethics and moral overtones involved in climate change were an existential and 'human rights' issue that undercut Australia's step-up and its objective of strategic denial. By contrast, China was rarely taken to task in the Pacific—or the Caribbean—for its production of greenhouse gases and its contribution to climate change.

8.5.2 Soft Power and the 'Family'

In seeking to gain support in the region, Australia sought to place itself as part of a Pacific family, where common values were shared, and to use soft power particularly focused on sport, to suggest a broader 'kinship' beyond the gravity of geopolitics, and disguise and bypass a basic asymmetry of power within regional relationships. The broad notion of being a benevolent member of a family emerged from a speech by the then Prime Minister, Scott Morrison, in 2018: 'This is our part of the world … We are more than partners by choice. We are connected as members of a Pacific family'. The speech coincided with a China-Solomon Islands security pact and was part of what was intended to 'restore Australia's place as the partner of choice in the region' (quoted in Moore, 2024: 290). Subsequently, under different governments, the term was used repeatedly as Australia positioned itself as part of this 'Pacific family', despite tensions within the family over mining and climate change policy that threatened relationships, especially when Australia tried to water down agreements, including the Boe Declaration, and restrict action on climate change (Moore, 2024). It was in any case a lopsided family. Few 'other' family members were in Australia, especially in comparison with New Zealand, and migration policy discouraged Pacific islanders (Wallis, 2023). It was an improbable family, where seasonal workers would pick fruit and Australia would provide aid but without any modification of climate change or resource extraction policies. Moreover, seasonal employment was straining relationships in extended Pacific island families; exploitation was not unusual and scarcely the basis for strong family ties (Chap. 9). Relationships were ungracious and patronising as Australian politicians ignored agency on the part of PICs, usually regarding island states as small and vulnerable, so requiring and benefiting from a paternalistic family leader (Petrou & Connell, 2023b; Koro et al., 2023). In the context of migration and mobility particularly, the underlying dialectics of racism, capitalism and more broadly white possession, that had characterised a century of engagement with the region, more than suggested Australia's dominant position within the 'Pacific family' (Chacko, 2023; Moore, 2024). It had no evident resonance in Australia.

Few other countries took a condescending approach. New Zealand might have had a better claim to family membership, though the United States, under the Biden administration, did talk about the Pacific family (mainly referring to its own dependent territories) but in less personal terms. In the Caribbean the British Foreign Minister, Dominic Raab, suggested to Caribbean leaders that 'we are more than friends, we are family' (quoted in Cullum, 2021: 100), but the UK had limited involvement despite its several overseas territories there.

Beyond 'family' and the more 'conventional' relationships of trade and aid, Australia sought to advance its cause through soft power, recognising that effective ways to improve Pacific relations were not necessarily through budgetary assistance, infrastructure projects or capacity building, however welcome these might be, but through smaller-scale relationships between Australian and Pacific people, civil society groups (including churches), educational institutions (through scholarships), sporting codes and businesses. In most such contexts, notably in language and education, Australia had a distinct historical advantage (Wallis & Kemish, 2022; Wallis, 2023). Regional feelings and identity are perhaps most passionately expressed through sport. The Pacific Games, various football codes and other sporting events are where Pacific people meet, and affiliations shift from the local to the national and regional.

The Pacific Games, every four years, was held in Solomon Islands in 2023, mainly in the Chinese-constructed stadium, and supported by an Australian grant of A$17 million (drawn from the aid fund, over 10% of Australia's total aid to Solomon Islands), for accommodation, equipment and a temporary marine centre. In circumstances where competition with China was evident, that raised questions about how aid might be defined and whether that was a valuable or appropriate way of using aid finance in a poor country with weak education and health services. Australia had never previously used aid to support a Pacific Games (Howes, 2023). Financial support came from other countries, including New Zealand, India and Indonesia, but as one Solomon Islander wrote: 'while the funding was welcome for this important event, the dark undercurrent of financing for the Pacific Games is another example of the tug-of-war between major powers jockeying for regional influence and security positioning' where even the maintenance of the facilities would cost US$3 million annually (Liloqula, 2024). Whatever the merits of sport for fitness, health and pride in identity, criticism was inevitable.

Nonetheless, for Australia, sport proved a powerful political tool to gain support in the Pacific, however difficult that might be to measure. In mid-2022 Australia established a partnership with Nauru to develop Australian Rules football there (the only country in the world where it is the national sport), but soft sports power has mainly focused on rugby, by funding national island associations enabling island rugby teams, male and female, to compete with the best New Zealand and Australian teams. That incidentally had the potential to contribute to migration: almost half of all the players in Australia's National Rugby League in the 2022 season were migrants from PICs or of Pacific Island ancestry (SMH, 27 July 2022) and their remittances were considerable. The most significant and distinctive contribution was Australia pledging A$600 million in 2024 to support the entry of a PNG team into its National Rugby League (the most popular male sport in PNG but with a growing female component). The link was settled in the same week that Australia concluded a treaty with Nauru (see below), opened a training centre for Pacific police in Brisbane, while the Foreign Minister, Penny Wong, remarked on the 'permanent contest' between Australia and China across the blue Pacific (Dziedzic, 2024), a theme she repeated in 2025, despite PNG's Foreign Minister, Justin Tkatchenko, dubiously stating that its own treaty had 'nothing to do with China' (SMH, 5 December 2024). A police security clause was however inserted into the agreement, alongside an assurance that PNG would only partner with Australia or other PICs on security matters, a position emphasised in one more treaty between PNG and Australia late in 2025 that, like the Nakamal Treaty with Vanuatu, was delayed. As the PNG government suggested, when seeking inclusion in the Australian rugby league competition, 'the bid will celebrate and showcase PNG's passion for the game, sporting talent and events hosting capacity' (RNZ, 5 May 2022), with the Prime Minister, James Marape, observing that PNG was the only country in the world where rugby league was the national sport so that 'what the NRL is gifting us goes to the heart of uniting our country'. Inevitably, the agreement was criticised for everything from neo-colonial Australian bullying and patronising, demonising China, and sportswashing (to deflect attention from other issues where resources might better be used) and, in Fiji, Tonga and Samoa, where rugby was more dominant than rugby league, generating requests for similar financial support (Barrett, 2025). The arrangement had limited potential for expansion elsewhere.

PICs also had direct agency in developing their own soft power through sport. That had proved particularly successful for Fiji, which used its prowess in rugby, and its Sevens victory in the 2016 Olympic Games, to promote itself globally, strengthen its national image and support government claims to legitimacy, in the face of ethnic divisions and external concerns over repression of civil society (Connell, 2018). Fiji embarked on its own sports diplomacy with its 2024 Foreign Policy White Paper, claiming that sport is a 'bridge across cultural divides, fosters international goodwill, and projects Fiji as a vibrant and confident nation' that has benefited from rugby in developing collaborations with countries such as France and Germany (Fiji Ministry of Foreign Affairs, 2024: 11).

No obvious parallels exist for the Caribbean yet unquestionably individual athletes have played a role in positive images of Caribbean SIDS, not least in Jamaica, where Usain Bolt and many women athletes, from Merlene Ottey to Shelly-Ann Fraser-Pryce, have provided positive images. The winning of a gold medal for the first time ever by St. Lucia's Julien Alfred in the 2024 Olympic Games enabled her success to be used as a tourism marketing strategy (Chap. 4). Unlike in the Pacific, team games have attracted no real external support or great interest. The United States had little interest in funding sport in the region, and no interest in cricket, as it partly gave way to basketball in the region, supported by diasporic connections.

8.5.3 The United States: A Latter-Day Monroe Doctrine?

As Australia's regional involvement in the Pacific increased, so too did that of the United States, and in both regions. In November 2011, the US President, Barack Obama, addressed the Australian Parliament, symbolic of shared concerns, to announce a rebalancing in US foreign politics: a shift from the priority of the Middle East to the Pacific, to 'contain' China's growing power while also playing a greater role in South-east Asia. Security in the Indo-Pacific became a new theme, at a new scale, and the American presence in the Pacific expanded and intensified. New embassies were promised and built in Solomon Islands, Vanuatu and Tonga (but that in Kiribati was delayed, following Tarawa's switch of its one-China recognition from Taipei to Beijing). The United States offered in 2019 to send

US Peace Corps volunteers to Solomon Islands, but the Sogavare government did little to advance these proposals.

Unexpectedly, US commitments to the region expanded under the first President Trump administration, culminating in his meeting the leaders of the three North Pacific SIDS: Marshall Islands, FSM and Palau, the first such meeting in their history. The United States also upgraded and expanded facilities across multiple US Pacific territories, in each of the three SIDS, and also in Tinian (Marianas) and Guam, reinforcing both what defence planners sometimes called the 'first island chain' closest to the Chinese mainland (including American bases in Japan and the Philippines) and the smaller Pacific islands in the 'second island chain'.

Interest was maintained under President Biden, with the United States instigating its 'Pacific pledge', packaged into a Blue Pacific Act that focused on 'three essential pillars: security, development and shared values' and was intended to endorse the concept and vision of Pacific island leaders. The Secretary of State, Anthony Blinken, visited the Pacific in 2022 to discuss climate change, COVID-19 responses, coping with disasters and ways to 'further our shared commitment to democracy' (Pryke & Sora, 2022). The United States began to work with Australia, Japan and New Zealand to extend electrification in PNG and committed to upgrading Lombrum Naval Base (Manus, PNG). Pacific island leaders were twice hosted at the White House, US access to South Pacific tuna fisheries was renegotiated, and defence cooperation with both PNG and Fiji was initiated between 2022 and 2024. The latter carried strategic weight and was likely to mean a greater US military presence, such as troop rotations, joint exercises and coastguard missions, but no permanent base. It constituted recognition that Fiji was close to many maritime routes between Pacific ports used by both warships and merchant vessels, and that China now had a navy capable of challenging US power in the region.

US politicians have publicly declared that China seeks to set up military bases across the region. While access to sea lines was 'not being contested at the moment', the United States was concerned at China's increasing influence (*PacNews*, 27 November 2024). Bolstering its presence in Fiji, so important to the region, provided a physical and symbolic message that the Pacific region was of key strategic importance. However, the United States lacked the regional physical presence of Australia, in what was previously its backwater, and, like Australia, its failures on climate change proved an obstacle to more positive engagement. The early months of the second term of Donald Trump's presidency brought global uncertainty, not least in the Pacific, where aid cuts, tariffs, the claim that climate change was a hoax, ejection of 'illegal' residents and restrictions on migration, all suggested a deeply unreliable future partner, as the United States abandoned any sense of soft power.

8.5.4 Partners in the Blue Pacific

In mid-2022 the United States launched another attempt at strengthening regional ties—a Partners in the Blue Pacific (PBP) Initiative, alongside four other member states (Australia, Japan, New Zealand and the UK) and three 'partner countries' (South Korea, France and Germany). PBP was intended to be an open, informal group, bringing together several seemingly like-minded nations to establish a constructive, transparent and efficient development framework for the Pacific Islands that would support the PIF vision of a Blue Pacific Continent: a 'Sea of Islands' united by the ocean. It set out six arenas of cooperation ranging from building climate change resilience to 'people-centred' development and cybersecurity, and accorded the PICs a 'guiding' position in policy-making. However, direction from above was regarded as high-sounding rhetoric that was deceptive, controlling and hypocritical, by undermining the region's focus on a Blue Pacific that concerned self-empowerment and regional solidarity in the face of others' geopolitical objectives, effectively challenging the PIF structure centred on 'friends to all' rather than to a special self-selected group of five countries (Fry et al., 2022).

Since its inception, the PBP annually contributed more than US$2 billion in developmental assistance. At least US$22 million was committed to building a Pacific-owned research vessel to carry out research on oceans and fisheries, and PBP has collaborated with Pacific Islands and regional organisations such as the Forum Fisheries Agency to combat IUU Fishing in the region. It sought to pool resources and share best practices to enhance cybersecurity in the region. Without a secretariat it had no profile and was mainly indicative of how distant countries, such as Germany, could decide to take an interest in Pacific security and how countries closely involved in the region could fail to engage effectively with Pacific regionalism, by disrespecting regional sentiment and misjudging local priorities.

8.5.5 Regional Presences

Minimally, as the various countries involved in the PBP suggested, other countries had an interest in Pacific matters. France had long had a presence, mainly through its three territories (and in Vanuatu), that retained some of the military themes evident from the era of nuclear testing. Under his 'France 2030' industrial policy, French President Emmanuel Macron had announced considerable research funding and new initiatives around the exploration and potential exploitation of seabed minerals and strategic metals. With France claiming 7 million sq. kms of EEZ in the Pacific, the Blue Economy became a new 'frontier' for France, but the 2024 violence in New Caledonia reminded many PICs that France was an ongoing colonial power, yet ironically, through New Caledonia and French Polynesia, increasingly involved in regional agencies, including PIF. Other European states have exerted some interest and influence, responding to Chinese interests, but have experienced aid budget cuts (notably the UK and the EU), misaligned priorities (such as PBP) and lingering self-serving colonial attitudes that had limited local resonance (Smith & McNeill-Stowers, 2025).

New Zealand was committed through its three dependent territories and its former colony, Samoa: a mainly Polynesian presence. It was less concerned with China, either in terms of trade or security. Geography made a difference yet its 'Pacific reset' was largely in response to concerns over China's growing impact (Iati, 2021). New Zealand is also more evidently a Pacific island state, with an indigenous Maori population, and a founding document, the Treaty of Waitangi, that reflected principles of partnership and mutual respect, so increasingly framing itself as having a 'Pacific identity' (Powles, 2021; Powles & Wallis, 2024; McNeill, 2025). That was consolidated in the 2022 Duavata Partnership with Fiji, which opened by stressing that 'Fiji and New Zealand are natural partners' and covered a range of concerns from training in military and police matters to supporting the participation of women in Fijian society. Whereas Australia and the United States had largely ignored and discouraged migration from the Pacific islands (other than Micronesia for the United States), New Zealand recognised its own position, socially and geographically, as a Pacific island state and absorbed Pacific islanders into the nation and, visibly, into the political system. It otherwise shared perspectives with Australia, without the impediment of climate change policy. The Foreign Minister, Winston Peters, urged PICs to work together to 'form a cohesive approach, bolster regional development and security, and use our collective voice to hold bigger countries to account' (RNZ, 11 June 2025). The two most obvious 'bigger countries' were not specified.

Japan had become increasingly concerned over China's role and impact on security whether in east Asia or the Pacific, where it had long been a 'traditional' aid donor and where the Japanese International Cooperation Agency (JICA) had a valued presence in nine Pacific SIDS where its programme had shifted towards providing 'human security' to local communities by strengthening grassroots programmes, including education, health and WASH projects. As elsewhere, geopolitical issues took it further away from a focus on SDGs although it sought to increase engagement with regional countries in maritime security, and otherwise supported the 'established' rules-based order and its institutions through its Free and Open Indo-Pacific Strategy, which supported its own economic interests (mainly in fisheries), alongside continuity of supply chains and transport corridors (Koga, 2022). Japan sought to distinguish its role from that of China through 'quality' infrastructure, collaboration with local governments and the private sector, donating technology of environmental value, while stressing its lengthy presence in the region (Segawa, 2025) and its similar maritime interests and concerns, from piracy to natural hazards.

Other countries have become involved in the Pacific region, or stepped up their involvement, for different reasons, none being enamoured with China's growing role. South Korea scaled up its Pacific interaction, India and Indonesia (in defence of its presence in West Papua) boosted their engagement. Indonesia has stressed that in West Papua it has more Melanesians than in the Pacific islands (Wallis et al., 2022). Russia has maintained an interest, most visible in West Papua. Gulf states have taken new interest, with, for example, Solomon Islands reciprocating by opening an embassy in the UAE in 2024, 'to enhance, strengthen, and advance' its relationship with the UAE. Israel has been unusually well received in the region. Expansion of Pacific diplomatic interests was related to a search for climate funding and frustration that neither the United States nor Australia were sufficiently interested (Szadziewski, 2024). The key external players, other than China, remained Australia and the United States. The Anglophone countries had 'soft' advantages over China, in the primacy of English as the second language through most of the PICs, with English-language newspapers and radio and TV channels, and in the widespread Christianity infused into the region over two centuries. For most Pacific Island people identity is linked to Christian belief. Many churches have a clear regional or subregional identity, often with Australian or New Zealand roots, or occasionally with American roots (where Mormon islanders can migrate to the United States for education or employment). Historic ties of significance gave Australia, the United States and New Zealand soft power advantages (despite some legacies of colonialism and neo-colonialism, and inadequacy on climate change), but these were not necessarily able to combat the hard advantages of Chinese loans, infrastructure construction and vehicle deliveries.

8.6 Pacific Responses

Pacific SIDS were scarcely unaware that aid and infrastructure projects from any source were not altruistic but broadly retained the long-established position (as earlier articulated in the Caribbean) of 'friends to all and enemies to none' which provided scope to pick, choose or reject diplomatic partners and projects. Although PICs frequently emphasised that they had no wish to choose between China, Australia, the United States or anyone else, pressures increased and PICs could not simply be passive observers and expect benefits from some form of neutrality. Moreover, after around half a century of independence, SIDS had become adept at flexing more independent muscles, exerting self-interest and playing off larger states against each other, most evident in Solomon Islands but more difficult in smaller Palau. That had costs as both CARICOM and PIF were weakened by new and fluctuating divisiveness.

Solo Mara, the Fiji High Commissioner to the UK, suggested in 2013 that 'China had filled a vacuum left when the US and the UK withdrew and which Australia did not adequately fill'. The Samoan Prime Minister, Tuilaepa Sailele, observed that he welcomed Chinese assistance to provide the buildings and other infrastructure that Australia, New Zealand and the United States did not. He also accused the United States of showing a lack of interest in the South Pacific. Aid was valuable and a plethora of partners could be advantageous as great power competition intensified. Most Pacific SIDS were however circumspect about Chinese intentions, and several absorbed Australia's concern that China could establish military bases in the Pacific, recognised that its projects contributed little to employment, and were concerned that foreign aid, trade and investment might be 'weaponised' by China as it sought to enhance its own power and national security. Indeed, rather later, with the Australia-United Kingdom-United States (AUKUS) partnership initiated in 2021, through which Australia would purchase extremely expensive nuclear-powered submarines from the UK and United States, the initial Pacific response was concern over the contravention of regional commitments to keep the Pacific nuclear-free, and over the cost, that might have contributed to development in the region, and disappointment that Pacific leaders were not consulted (Vaha, 2025). As the Kiribati President, Taneti Maamau, said, 'we thought it would be a courtesy to raise it, to discuss it with your neighbours'. While Australia would have scarcely been willing to discuss sensitive military strategy, it added to the perception that the Pacific SIDS were still not prominent in Australian agendas (Fry, 2021). The 'family' was again found wanting. Defence had seemingly triumphed over foreign policy while it was hard to reconcile such enormous expenditure (over A$26 billion over 25 years) on militarism with that required to meet such basic SDG goals for the Pacific as the eradication of TB by 2030. It incidentally provided another bargaining chip for the region, while the weak Australian response on climate change had already provided PICs with leverage, which they sought to use collectively to assert greater agency. While Australia could find it difficult, however necessary, to match the largesse of China in the region, it retained soft power advantages in education, language, sport and Christianity, and a limited advantage in access to residence and the labour market in Australia. The latter expanded into the 2020s in what was ironically a 'reverse aid' programme where the principal beneficiaries were Australian farmers (Chap. 9).

While access to Chinese support, alongside that of 'traditional' partners, enabled PICs such as Fiji, Kiribati and Solomon Islands to be more assertive on the global stage, within these countries questions were being asked about the transparency, ownership and impact of projects led by Chinese companies, and their potential conflict with civil society, labour standards and environmental management (Szadziewski, 2022; Mausio, 2023). Fiji's 'China question' involved how the Fijian state and civil society would reconcile the benefits and challenges of a long-term Chinese presence, a question that was more evident than any 'United States' or 'Australian question', especially since China had taken a more obviously strategic approach rather than one centred around Blue Pacific narratives (Wallis et al., 2024). While the PICs were receptive to offers of loans, investment and infrastructure, they were much less receptive to China's narratives over the nature of development directions and retained their own narratives, more tied to a western canon, to retain their own agency and sovereignty (Rodd, 2020; Mausio, 2023; Waqavakatoga et al., 2025).

For all that the PICs could play off various would-be partners, and seek to be friends to all, that became increasingly difficult as geostrategic differences intensified, and taking sides between Taiwan and China became weaponised. The extent to which China was welcomed as a valuable contributor caused tensions within and between countries and between the great cultural divides. Balancing loyalties, heritage, ideals, philosophies and opportunities and taking sides came at a cost. When Nauru shifted its allegiance from Taiwan back to China, after a 20-year break, its Prime Minister, David Adeang, was suitably diplomatic: 'Nauru is proud of our strong and growing friendship with China … Our restored diplomatic relations with China are a testament that even the largest and the smallest of countries can find common ground and work toward mutual benefit'. Emphasising the 'friends to all' mantra, he went on, 'We must focus on building bridges, not walls, and work toward a more integrated, connected world' (quoted in *Pacific Island Times*, October 2024). Taiwan, however, had been disconnected.

Domestically, the expanding structures of defence and policing agreements across the Pacific sat uncomfortably with the rhetoric used by its leaders, most of whom invoked peace and non-alignment. The Fiji Prime Minister, Sitiveni Rabuka, championed the idea of an 'Ocean of Peace' in the Pacific, based around the old Pacific Way that celebrated consensus (*PacNews*, 27 November 2024). Rabuka and other Pacific leaders, including those with security arrangements with Beijing, were part of a 'tension' between rhetoric and military security. Minimally, China's presence offered new possibilities and an arena for asserting island power and influence. The Solomon Islands Prime Minister, Manasseh Sogavare, had spoken for many when he said:

> As long as our decision is consistent within international law, Solomon Islands will not allow itself to be used as a tool to satisfy the narrow geopolitical interest of foreign political powers. I will not allow that to happen. Solomon Islands is not a political football. (quoted in Waqavakatoga et al., 2025: 208)

Although Pacific leaders faced questions in an era of strategic competition, they were adept at exploiting that competition to get greater investment and new infrastructure, even as some buildings were badly constructed, others such as sports stadiums were scarcely necessary (and were costly to maintain), or environmental degradation followed, but other infrastructure was undeniably useful and gained through a new assertiveness. Leaders occasionally mentioned schools, hospitals and human security, but other visible infrastructure came first.

8.7 Caribbean Responses

The Pacific attracted widespread external involvement, but the Caribbean, for sometime, was relatively ignored by metropolitan powers. The Cold War had ended and Cuba posed no threats. Involvement from China, the

United States or other powers primarily related to trade and aid—although it seemed inevitable that China's presence would attract some concern, relating to SIDS 'selling out' or the establishment of ports and plausible military installations. Venezuela and its role in the region, illegal migration and possible repercussions from the plight of Haiti and most recently pressure from the US to expel Cuban doctors from the region have been at the centre of geopolitical tensions in the Caribbean. Most Caribbean SIDS have taken a pragmatic approach under Trump, seeking to appease the US and avoid consequences. Some have also articulated alternatives to the US alliance and the real potential of forging closer ties with the rest of the Commonwealth. Divisions in CARICOM have posed real challenges. Alongside CARICOM's attempt to nurture and preserve notions of the region as a zone of peace has been Trinidad's blatant support for the US' Operation Southern Spear, which has included airstrikes against fishing vessels in Caribbean waters.

At the end of the twentieth century, the United States remained the major regional superpower, but its interest in the region had declined, and it had lost moral authority through its perceived interference in local politics but responded to this lack of interest after 2005 with an activist foreign policy agenda centred on providing cheaper oil in the region. As Ralph Gonsalves, the Prime Minister of St. Vincent, pointed out, the savings from cheaper oil from Venezuela were 'used to fight poverty and to build houses and help with education and the health system' (quoted in *Searchlight*, 20 July 2021). Sanctions were unwelcome and contributed to divisions within CARICOM with some SIDS retaining support for Venezuelan initiatives. As the political situation in Venezuela worsened, with a disputed national election and US embargoes, the United States stressed that any nation assisting Venezuela in avoiding embargoes would face sweeping trade sanctions. CARICOM then called for the removal of US sanctions that prevented the purchase of cheaper Venezuelan oil. Somewhat later, the conflict between Venezuela and the United States and Trinidad's explicit and controversial support for Washington placed the Caribbean in the centre of a major geopolitical crisis, ostensibly aimed at regime change. The United States, under the second regime of Donald Trump, authorised and carried out multiple strikes on boats in the Caribbean, targeting and killing presumed Venezuelans, officially to protect the United States from drug trafficking, unofficially to pursue its security and economic interests. Trinidad has become the site of US military and radar facilities, threatening the perception of the Caribbean as a zone of peace and intensifying divisions in CARICOM.

In 2020 the US Secretary of State, Michael Pompeo, visited Jamaica and met with CARICOM member states to discuss 'expanding America's commitment to the Caribbean'. This visit and divisions it created within CARICOM reflected a continuation of ongoing geopolitical tensions between CARICOM and the United States and within CARICOM itself. The Prime Minister of Barbados, Mia Mottley, conceived of the meeting as a divide-and-rule tactic, a view that resonated with Trinidad, Grenada and Antigua. The antecedent of the problem lay in CARICOM's recognition of Nicolas Maduro as the president of Venezuela against US wishes. Through the Organisation of American States (OAS) the United States continued to exercise hegemony in the wider Latin America and Caribbean region through a divide-and-rule strategy, aimed at isolating Venezuela in exchange for aid. That was complicated further by refugee flows from Venezuela and Venezuela revitalising its claim to a large part of what had long been Guyana, prompting Guyana to seek diplomatic support for its sovereignty, amidst wider concern over Venezuela's aggressive stance.

Jamaica's Prime Minister, Andrew Holness, declared in 2020 that the United States and the Caribbean were 'natural allies and natural partners' and should 'move forward with even closer ties' to create more opportunities. However, 'opportunities' were not evident and Caribbean SIDS forged closer economic ties with China, despite US disapproval. The United States had lost regional status through acting both as a 'big brother' and with 'benign neglect' worsening under the perceived racism and autocracy of the first Trump administration while China continued to provide loans and construct infrastructure (MacDonald, 2022: 133). Yet the United States remained the main trading partner for most Caribbean SIDS and the dominant foreign actor in matters of national security, including gun and drug running, and no serious challenge to US hegemony in the region existed, even more so in 2025 as President Trump deployed an American naval buildup off the coast of Venezuela officially to combat drug cartels.

8.8 Free Association

Several Pacific island states are politically dependent on metropolitan states, notably France, the United States and New Zealand, and have unconstrained rights of migration to them. Nationals of Niue, Cook Islands and Tokelau may migrate freely to New Zealand; those of New Caledonia, French Polynesia and Wallis and Futuna may move to France; and those of Guam, American Samoa and the Northern Marianas have long been able to move to the United States. Broadly, the same circumstances exist in the Caribbean for the various British colonies, the Dutch territories and the US Virgin Islands and Puerto Rico (Connell & Aldrich, 2021). All such states have multiple powers but usually falling short of defence and foreign policy.

Unlike these states, most of which have sought to take on additional responsibilities from the colonial powers, from the mid-1980s, three PICS, the FSM, the Marshall Islands and Palau, signed Compacts of Free Association (COFA) with the United States. These granted the United States exclusive military rights over the 5.6 million sq kms EEZ of the three countries. In return the COFA states received financial aid and their citizens could freely live, work and be educated in the United States without a visa or green card. Before COFA was signed there was only limited migration from Palau, as lack of finance and support restricted migration, but as numbers grew in the United States, all three states experienced substantial emigration and declining populations. Palau lost more than half its population since the signing of COFA (Reklai, 2024; Abe & Wang, 2024), but overall losses have been even greater from the lone atoll state, the Marshall Islands, and mainly from the atolls in the other two states, FSM and Palau, such as Sonsorol (Palau) (Walda-Mandel, 2016; Chap. 9).

Though their political leaders simultaneously sought to develop more self-reliant and sustainable national economies, and they were able to exert significant autonomy, some saw the COFA as perpetuating neo-colonial relationships of dependency and the loss of identity, as in the Marshall Islands and FSM (Bordner et al., 2020; Puas & D'Arcy, 2021; Berta, 2025). Nonetheless, the relationship meant a high level of dependence on financial aid, eventually supported by remittances, and on 'security', as part of an extended military archipelago from Hawai'i to Guam. Negotiations between the countries determined the level of COFA funding, made easier for the PICs by the growing Chinese interest in the region. When COFA funding was delayed in mid-2024, the three SIDS argued to the US Senate that the delay 'resulted in undesirable opportunities for economic exploitation by competitive political actors active in the Pacific' (RNZ, 8 August 2024). Countries with relatively weak economies, but at the centre of intense geopolitical shifts, used this strategic location as a crucial bargaining chip in negotiating increased funding and support from their colonial partner.

The Micronesian COFA agreements incidentally became the model for a treaty, the Australia-Tuvalu Falepili Union Treaty, signed late in 2023, linking the largest and smallest independent states in the South Pacific. The treaty covered climate change adaptation, migration (offering Tuvalu migration opportunities in Australia, more generous than for any other Pacific SIDS) and providing a guarantee of security, as Tuvalu effectively ceded partial sovereignty to Australia. For a small atoll state of around 10,000 people, where a culture of migration had existed for a century, this was widely regarded as an attractive proposition, despite local concern over cultural identity, excessive migration and skill drain (Barnett et al., 2025). It came at a time when it was speculated that Tuvalu, strategically located in the Pacific and close to Kiribati and Nauru, both of which had recently switched allegiances from Taiwan to China, might also switch diplomatic ties from Taipei to Beijing following the pro-Taiwan Prime Minister Kausea Natano's defeat in the 2023 general elections.

A year later Australia signed a somewhat similar agreement with Nauru, focused on economic support and security but without any migration provisions (Nauru being the only state in the Pacific without significant emigration or seasonal labour migration). It too came at the request of the Nauru government. The two countries agreed to 'deepen and expand security cooperation' while Nauru would 'mutually agree with Australia any partnership, arrangement or engagement with any other State or entity on matters related to Nauru security'. Australia thus effectively acquired a veto over Nauru (and Tuvalu) achieving a security agreement with China (Wallis & Corbett, 2024) and ensured continuity of controversial offshore refugee processing there (while ignoring high-level corruption and the erosion of political rights, inconsistent with a democracy). Nauru secured over A$100 million in budget support over five years, at a time when its economy faced a deeply uncertain financial outlook after its only bank had closed. The agreement committed and co-opted the private Commonwealth Bank of Australia to operate in Nauru. A year later Australia concluded a further agreement with Nauru to dump unwanted 'undesirable immigrants' for further substantial budget support. The agreement was concluded in secrecy and was likely to cost Australia A$2.5 billion over 30 years, continuing a process of inflated contracts supporting dubious businesses linked to the President and with bikie involvement, a costly and injudicious outcome of Nauru becoming 'an absurdly outgrown feature of Australia's fixation with Chinese influence in the South Pacific' (Kampmark, 2025: 2). It said much about the morality of aid, desperate diplomacy, the economy of Nauru, especially the uncertainty of a 'refugee economy', and concern over climate change, that a country so despoiled by Australian interests should choose to become a quasi-Australian colony (and accept criminal deportees).

Australia signed a security agreement with Solomon Islands, three years after the latter had signed one with China, reflecting 'Australia's role as Solomon Islands' security partner of choice' although it was unclear whether the new agreement negated the Chinese one. Australia also signed a new security agreement with PNG (via rugby league) and sought to seal ties with the other crucial Melanesian partner in mid-2025 through the Nakamal Agreement (named after the prominent men's village meeting houses) that promised A$500 million in aid over a ten-year period. It was, however, rejected by Vanuatu since it was unwilling to make Australia, in the words of the Minister of Foreign Affairs, Penny Wong, its 'security partner of choice' to the exclusion of other relationships, principally with China, that might provide beneficial infrastructure. Vanuatu also sought ease of accessibility for its citizens to Australia but was unlikely to gain the equivalent of the freedoms awarded to Tuvalu. The 'arc of instability' was nonetheless becoming an arc of treaties and aid concessions.

With no push for independence (other than in New Caledonia) or for secession (other than in Bougainville) the smaller Pacific SIDS were benefiting from Chinese pressure as a largely welcome bargaining chip. It enabled strategically located PICs, as most were, to gain what were regarded there as generous concessions from Australia as they negotiated their own varied and informal versions of COFA. As one diplomatic observer commented, 'Nauru has been a tough negotiator and typically extracts a high price for signing up with Australia' (quoted in SMH, 4 December 2024). Remarkably, the United States had previously been encouraged to extend its Free Association arrangements for the Micronesian states to the trio of Nauru, Kiribati and Tuvalu (Gray, 2021). France had suggested in 2025 that New Caledonia might move towards a similar 'sovereignty in association with France' to resolve the question of independence, but it was quickly rejected. Australia alone took on new 'responsibilities' as it sought to renew, elaborate on and benefit from its presumed status as a trusted 'family member' or partner of PICs, in what was a costly and not particularly subtle recolonisation of the Pacific.

Predictable scepticism in the Pacific existed over the motivations underpinning Australia's actions, involving a break with tradition on migration alone and directly stemming from recent geopolitical rivalries. It was partly a bid to restore credibility, after its reputation had been undermined (almost literally) by inertia, the absence of effective action on climate change or on opportunities for mobility or migration (Chap. 9). The Falepili Union, despite references to some notions of climate justice, brought enigmatic responses and concerns over cultural identity, although PICs welcomed both the Pacific Australia Labour Mobility (PALM) scheme and the new Pacific Engagement Visa (PEV). Simultaneously, the PICs were more assertive but also, especially in the Melanesian arc, increasingly linked to their own much weaker version of free association that tied them closer to metropolitan powers.

8.9 Aid

SIDS have gained considerable overseas aid from metropolitan states, notably Australia, China, New Zealand and the United States, from the EU and through international agencies and multilateral organisations, including the UN family, the World Bank and the Asian Development Bank (ADB). Some of this—perhaps the majority from national donors—is the more formal version of geopolitical intrigue and diplomacy. Benefiting from 'small country bias', SIDS are the most heavily aid-assisted parts of the world on a per capita basis, characteristically receiving three times the aid of more impoverished sub-Saharan African states (Connell, 2013). Remarkably little analysis—and virtually none in the Caribbean—exists on the relationship between aid delivery and development. The role of national donors varies, and in this century China has become a substantial aid donor, alongside its trade and investment presence.

Aid delivery does not necessarily match development needs (nor is it always well-coordinated or effective), and needs are often those assumed relevant and appropriate by donors. Climate change has, however, received considerable attention from donors. The 'need' for climate change aid is so accepted by most donors that, in Tuvalu, for example, requests for aid in the 2020s for any project without a 'climate change' component were less likely to receive support. Funding for health in Pacific SIDS declined in the 2020s despite the health care problems and urgent needs revealed by COVID-19, a regional decline with Caribbean and global parallels (Dayant, 2020). That paralleled the widespread shift into large infrastructure projects. Global funding for water supply, sanitation and hygiene (WASH) projects and programmes reached its lowest level since the SDGs began, coinciding with the triple crises of COVID, rising costs of living and climate change, and despite WASH being essential for effective health systems, adequate nutrition, infection prevention and control, and economic development (Oza & Goff, 2023). That contributed to the Pacific being one of the worst world regions for meeting SDG6 on drinking water and sanitation.

Aid is most effective where it responds to effective local needs and practices, priorities and capacity, but the absorptive capacity of SIDS is limited. When new sources of funding for climate change mitigation and adaptation followed the Copenhagen climate summit in 2009, few SIDS had the capacity to respond, develop strategies and absorb new sources of funding, on top of already complex aid programmes whose management demands were considerable, challenging local bureaucrats with a different culture and language. SIDS may be overwhelmed and strategies and priorities distorted by external interventions (Webber, 2013; Dean et al.,

2017; Murray & Overton, 2011). In the Pacific, aid has been seen as 'boomerang aid', so tied to the interests of the donors that it flows back into purchases from them, for example for infrastructure, the advice and short-term 'expertise' of consultants, or scholarships (Koro & McNeill, 2024). Aid in the form of loans, such as much energy aid, tends to result in debts so that debt servicing similarly rebounds (Niles & Lloyd, 2013). Aid delivery is further complicated by the growing significance of aid from 'non-traditional' donors (such as China) that may be linked to natural resource use or global diplomacy, rather than local priorities, gender equality or poverty alleviation. That competition for influence, inherent in aid delivery, was evident in the symbolism of aid to Tonga after the 2022 volcanic eruption—the rush to get in first and with boxes, crates and other goods draped in the national flags of donors—far from any detached altruism. Typically, in this process, various countries (New Zealand, Australia, France, Fiji, the United States, Japan and the UK) coordinated content and delivery but China operated alone.

Aid exhibits and enhances urban bias and has spawned an urban bureaucracy arguably at some variance with local needs and national development strategies, contributing to some loss of control over national development strategies, policies and programmes, as these become politicised. Top-down approaches have inhibited independent modes of development, weakening local resilience and, in extreme circumstances, producing an 'inverse sovereignty effect' where key decisions are not taken nationally so distorting local needs, interests, values and aspirations (Murray & Overton, 2011; Overton et al., 2012, 2019; Webber, 2013; Bordner et al., 2020; Vunibola, 2023). Development practice in the SIDS typically overvalues external Eurocentric knowledge and undervalues local knowledge. Lack of capacity and political clarity, compounded by limited political will, has enabled international agencies and corporations to set the direction of development. From a donor perspective, discourses on aid can focus on financial statistics correlations; Pacific and Caribbean non-statistical and humanistic realities may be overlooked or discounted. Hierarchies are created that privilege outsiders and discriminate against islanders (Meki & Tarai, 2023). For a small state such as Kiribati, 'the scope and magnitude of donor requirements and conditions, relative to available capacity, can prevent projects from achieving even their intended objectives, and can negatively impact upon sense of empowerment, self-esteem and cultural identity, thus undermining the goals of both donors and recipients' (Dean et al., 2017: 63–64). Multiple barriers exist to the financing, development and sustainability of projects, including underdeveloped private sectors, limits to financial opportunity, weak bureaucratic capacity, land and cultural contexts and the social embeddedness of 'deals' (Anantharajah, 2019). As the Prime Minister of Tonga, Aisake Eke, has asserted:

> the limited time and resources of small administrations are expended on accessing climate finance and navigating the parameters of multiple funds…Even once finance is secured, the ability to deliver results is often compromised by the complexity of managing multiple agreements, delivery partners and external project implementation units… it is also clear that institutional capacity within Pacific administrations is a core driver of the problem. Yet, paradoxically, institutional capacity is further eroded as aid dependency increases. (Eke & Altangerel, 2025)

Partly in consequence, PICs have acquired few projects through the Green Climate Fund due to 'insufficient capacities to fulfil specific requirements established by donors and climate funds' (ESCAP, 2022: 8). A survey of 190 projects across 13 PICs, all financed by the Global Environment Facility Small Grants Programme, suggested that projects that were of short duration, received more co-financing and in-kind support (and thus more local support), were focused on a single adaptation approach, and in those SIDS where 'in-country managerial expertise' was more evident, were most likely to be successful (Hasan et al., 2021). Managing complex climate projects required an equally 'complex governance system with competencies and responsibilities diffused across different national actors and multiple governance scales' (Calliari & Vanhala, 2022). That was paralleled by a short-lived transition from large-scale infrastructure projects to softer more holistic approaches (Mohan, 2025a, b). Nonetheless, climate finance has had a greater impact on the reduction of carbon emissions in SIDS than in larger states (Lee et al., 2022). SIDS also faced power asymmetries where commercial interests and increasingly geopolitics distorted aid processes and ignored cultural contexts (thus often bypassing small local organisations).

Almost inevitably, local priorities conflict with those of donors, more so as donors have diversified. Heightened geopolitical tension between China and traditional partners coincided with an increased focus on hard infrastructure, rather than areas such as education and health (Duke & Ahsan, 2024; Rummery, 2025). That trend was already in place before the axing of USAID, and its accentuated opposition to 'wokeness' (and even empathy) that has meant that women, children and other vulnerable groups have been most likely to be disadvantaged (Mawdsley et al., 2025). Infrastructure aid may result in some success, but it usually fails to empower and respond to Pacific agency and reduce aid dependency. At a different scale, visible big-ticket infrastructure—ports, sports stadiums and hospitals—are symbolic and usually politically welcome. Yet infrastructure was often divorced from ongoing activities: equipment and medicines might be stranded on poorly managed wharfs and no provision made for training of workers.

China's aid has often been in the form of substantial loans, raising concerns over the ability of smaller states to repay these loans. By mid-2025 it was evident that several SIDS, including Tonga, Samoa and Vanuatu, were at some risk and facing substantial repayment burdens (RNZ, 5 June 2025). Otherwise in a competitive environment, China's aid has been characterised as being part of a process of accommodation and adaptation within the established order rather than a fundamental challenge to that order; growing competition between China and the west has not led to divergence but to convergence, while Australia and other donors have mirrored China's focus on infrastructure and debt financing (at the expense of health and education) (Skidmore, 2025; Hill, 2025b). That convergence has paralleled circumstances where SIDS recipients are less likely to differentiate between aid delivery from China and from the west, rejecting a binary division, especially one where 'northern' donors are the 'natural' architects and engineers of development, and favouring a diversity of sources. As in Antigua, SIDS nonetheless value Chinese aid as a form of interference but one that is better aligned with Antigua's development vision (Sou & Day, 2025). Such perspectives underpin the geopolitical complications discussed above.

Despite commitments to harmonisation, the proliferation of aid donors, including many NGOs, and projects (especially related to climate change) has sapped the attention, energy and resources of recipient governments, and reduced the capacity for aid management, policy development and practical implementation, which in turn has resulted in donors focusing on capacity building (e.g. of police forces) and 'good governance'. It has also given rise to the 'Pacific paradox' of high aid but low growth, although aid infrequently supports local private sector initiatives.

Changes in the context of aid have been typified, in the Pacific, by three clear shifts in Australian aid (the main donor) since the 2010s. Aid has become more politicised, deployed to support Australian interests and become increasingly privatised as much has been spent through just four large Australian and international corporations—with local Pacific companies marginalised—the boomerang at work, with the result that the content of aid has been increasingly contested by Pacific islander recipients concerned that development is not aid (Taylor & Middleby, 2023; Taylor et al., 2025; Stambolie & Hill, 2025). International corporations, with relatively conservative perceptions and strategies, place frequent demands for coordination and cooperation: the 'tyranny of experts' (Easterly, 2024). That has been accompanied by donor policy stasis and lack of interest in broadly radical directions that might threaten the political context where directions and strategies are largely determined (Day & Wells, 2025). A parallel transition has been from aid for economic and social development to the prevention of state fragility and failure, and enhancement of regional security. In this evolution aid could be seen as 'an unchecked donor-driven system, beset by excessive power and control to benefit the system and its agents' (Taylor & Middleby, 2023: 1). Such trends and processes emphasised that aid was distant from development, where development concerned local agency, self-determination and sovereignty, some congruence with the SDGs and the ability to restructure economies in favour of more sustainable and equitable development: a continued paradox in the decolonisation of aid (Taylor et al., 2025; Amankwa et al., 2025).

Power is always in the hands of those with capital. Accessing capital can be so complex that it creates an industry of its own, with scoping reports, concept designing, 'capacity building' and other types of technical assistance. A more holistic approach would mean shifting the locus of power (Prasad, 2024). Local organisations may be undervalued and overwhelmed by foreign aid organisations working in the same spaces and, sometimes, competing for attention. The COVID-19 crisis provided new opportunities for locally led programmes 'to shift our mindsets and decolonise' while overseas-based NGOs and staff were temporarily absent. The founder of the Tonga Women and Children's Crisis Centre, Ofa Rokotuibau, observed: 'We were always looking outside, valuing the outside approaches. But the answers have always been here … our forefathers, our ancestors had used them to build stability, social cohesion, development and progress' (quoted in Waide, 2024). There and elsewhere the absence of foreign staff led to increased opportunities for local staff, greater cultural sensitivity and relativity in the workplace, and successful ongoing operations of programmes by local people, distinct however from what were seen as broader issues of racism, neo-colonialism and the centring of whiteness in the development industry (Smith et al., 2024).

On accession early in 2025, the Prime Minister of Vanuatu, Jotham Napat, criticised delays in delivery, problematic after hazards: 'I will simply tell them that they must agree to do it our way—one that is feasible and efficient—so the funds can be utilised quickly. If they insist on prolonging the process, they can take their money back to their country'. Each donor imposes its own set of rules, complicating disbursement and management. Napat needed an integrated, holistic approach to ensure donor and government procedures were aligned for accountability and transparency (*Vanuatu Daily Post*, 19 February 2025). At the same time, Transform Aqorau, by then the Solomon Islands National University Vice-Chancellor, sought a greater diversity of donors, arguing that for decades traditional development partners had not delivered enough tangible benefits, yet PICs were being 'held to ransom by the geo-strategic interest of a narrow and smaller grouping of countries' (RNZ, 24 February 2025), hence the need to explore all options. Increasingly geopolitics, and therefore a bias towards urban infrastructure, emerged at the fore.

Formally, development aid 'has become increasingly inadequate, caught between elevated regional needs, economic fragility, and heightened geopolitical pressures'; rhetoric has outpaced action. Ironically, a surge of smaller donors has crowded and fragmented the aid landscape (Dayant et al., 2024: 4) while one large country, China, has complicated this landscape further, as it displaced the United States as the Pacific's second largest bilateral donor. As these processes evolved, that landscape became more detached from the agency and aspirations of local people and from the core elements of the SDGs, including gender equality and even climate action. This was further complicated by an increased number of national donors, with Fiji in 2024 having 36, more than doubling the number in 2008, a pattern typical of the PICs, and with varying motives from humanitarian and diplomatic strategies to simply securing UN votes (Dayant et al., 2024) and often without a long-term presence or strategies that would contribute to effective development outcomes. The Samoan Prime Minister, Fiame Naomi Mata'afa, observed that with 'every man and his dog coming to the Pacific', it was difficult for island administrations to cope with and navigate evolving geopolitical interests (Lewis, 2025). Nonetheless, despite widespread concerns and criticisms, aid has contributed to infrastructure improvements and improved education and health in many SIDS.

Overall global aid fell in 2024 for the first time in five years, a trend affecting recipients throughout the Caribbean and the Pacific. Late in 2025, the freeze and eventual closure of USAID, the foreign-aid programme that for decades underpinned influential US relationships, especially in the Caribbean, accelerated this decline to bring need and uncertainty. Other rich-world donors followed the US 'model' while aid shifted from notions of poverty alleviation to national self-interest. The Jamaican Prime Minister, Andrew Holness, acknowledged that the loss of USAID funding presented immediate financial challenges but suggested that Jamaica's long-term vision was of reducing reliance on aid while strengthening its economic foundation and independence: a somewhat obvious if optimistic response. US withdrawal left Caribbean public health systems, already under pressure from brain drain and underfunding, facing potential collapse (Mohamed, 2025). Aid fatigue, reduced metropolitan budgets, the costs of the Ukraine war, the USAID effect and a widespread surge in defence spending emphasised SIDS dependence on aid and the difficulties in making the kind of transition that Holness envisaged. As needs remained and climate change drove up the costs of response, donors effectively shifted from the soft power of aid to incipient hard power, but as the Falepili Union between Tuvalu and Australia demonstrated, both soft and hard power remain. Bilateral donors may retreat to regions more important to their interests, as the United States has done, and multilateral institutions may be unable to justify the diseconomies of supporting small island populations. The contours of future aid delivery are hard to envisage, in an era of changing global security and resource concerns: aid might simply be subordinated to a 'supplier of carrots and sticks', be more participatory and locally led or be much more selective (Hill, 2025a) but still unlikely to focus on the SDGs. However, the small size of SIDS—and the strategic location of many—means that relatively high levels of per capita aid, however organised and directed, may be maintained. Location is one partial guarantor of continued support.

8.10 It's the Geography (Stupid)

The possible strategic importance of SIDS to the west declined with the end of the Cold War and the collapse of the Soviet Union, especially in the Caribbean, with the Pacific largely a bystander to early power plays. Chinese expansionism, however, gave the SIDS much greater political and strategic relevance, this time particularly in the Pacific. China's focus on aid and infrastructure in the Pacific was increasingly regarded by metropolitan states, primarily the United States and Australia, as a means to a strategic end, providing a catalyst for widespread, expanded external interests. In the 2020s, the Pacific became of greater global and regional interest than at any time since the Second World War. The 'blue continent' was undeniably a contested space. As Surangel Whipps, the President of Palau, put it in mid-2025, 'we must help our people understand because of our location, we are under constant threat. I might venture to say we are already at war. The best way to combat this is through partnership with like-minded nations' (quoted in *Financial Times*, 12 September 2025). The region that Australia once regarded as least important had become the most important.

As Pacific SIDS switched allegiance from Taiwan to China, Chinese interests diversified with infrastructure and training seeming to have militaristic underpinnings, raising Australian anxiety about islands in the Melanesian arc potentially becoming military bases, witnessed particularly in the furore surrounding Solomon Islands' security agreement with China. Contestation and competition were rampant, a situation well summed up by Sitiveni Rabuka, the Prime Minister of Fiji (and the instigator of two 1987 military coups). Noting that sovereignty, security and prosperity were the traditional pillars of foreign policy, he observed:

> The context in which we must pursue them has profoundly changed. We are in the middle of what is likely to be a long transition from the US led liberal international order to a more multipolar world. The old liberal international order in which Fiji was a beneficiary and which saw the unprecedented level of global cooperation and economic interdependence is now slowly diminishing. In its place is a complicated competition for primacy between the United States and China played out more sharply in our Indo-Pacific region. At its most basic, China aspires to become the predominant power in the region and the US is also determined to hold on to its pre-eminence. How this competition will play out is an open question. The stakes are high. There are significant risks of miscalculation. (RNZ, 24 September 2024)

It was implicit that Australia and the United States were just as complicit, being seen by some as 'suzerain states' (Koga, 2022), while Fiji's size and strategic position were enabling it to benefit from multiple donors (Johnson & Young, 2025). As global powers competed for influence, so the SIDS acquired greater bargaining power, albeit with limits to how substantial and useful that might be. External interests and influence were no zero-sum game between foreign powers vying for hegemony, and being willing to pay for it, but provided a context that offered island leaders an opportunity 'to weave together and navigate multiple foreign influences in ways that frustrated colonial and neocolonial logics of international relations' (Davis et al., 2020: 13; cf. Mountz, 2011; Loyd et al., 2016). A vacillating Solomon Islands was seen as unusually frustrating.

All the past and present colonial and neo-colonial powers faced challenges in their engagement with the SIDS, as strategies too obviously accorded with their own interests. The United States was excoriated for its overreach in Partners of the Blue Pacific, the demise of its aid arm, USAID, departure from WHO and its effective denial of climate change (intensified after its exit from the Paris agreement). In Micronesia especially, there was resentment over US deportations of criminals, inadequate financial support and failure to act on the US nuclear weapons testing legacy. That was seen as part of an overall insensitivity to the priorities of the PICs themselves, in being designed primarily to counter China and developed without proper regional consultation, reflecting a lack of consideration over the abilities of the PICs to articulate and pursue their own interests (Puas & D'Arcy, 2021; Wesley-Smith & Finin, 2021; Liloqula, 2024). Criticisms and resentment over Cuban doctors, the demise of aid and attitudes to international law have all been evident in public debates in the Caribbean. President Trump's detachment and indifference, antagonism to climate finance and international law, support for extended fisheries and deep-sea mining, aid cuts and vacillations (notably over tariffs) brought unpredictability and distrust, reinforcing the need for SIDS to reconsider 'traditional' security interests and concerns.

Although Australia remained most visible as the largest provider of development assistance in the Pacific region, because of its failure to develop policies that challenged climate change, it too was strongly criticised by the PICs that saw New Zealand as a more reliable and sympathetic, if less affluent, partner. While Australia's specious claim to being a part of the 'Pacific family' was often ridiculed (as Manasseh Sogavare pointed out 'in all families there are issues to resolve'), it was nonetheless more involved in a range of ways, beyond its primary aid-delivery role, from sport and business to religion, in contrast to China's more narrow perspective. Language barriers, the lack of a 'common touch' and its unwillingness to hire local people weakened efforts to generate soft power, whether in the Pacific or the Caribbean.

Despite its contribution to climate change, China was rarely criticised directly, but concern was centred on the real value of Chinese aid and infrastructure and a basic uncertainty about what its objectives were as a new and unfamiliar player. China managed to escape Pacific criticism for its contribution to global warming, sought to remind rich states of 'their inescapable historical responsibilities over climate change' while subtly suggesting that 'development' was not so much an economic term but might be better thought of as a 'non-Western' departure from 'liberal values' (Chaguan, 2023). Even so, it was much harder for China to naturalise its presence in the two regions.

SIDS welcomed greater external engagement but not geopolitical rivalries that reduced development options. By 2024 the PIF Secretary-General had warned China and the United States to tone down their rhetoric against each other to reduce tensions in the region. When China test-launched an inter-continental ballistic missile into the Pacific Ocean (after 27 years of calm), the Secretary-General, Baron Waqa, simply described China's actions as a 'big no-no'. That both China and the United States were part of a world of autocracies, privileging power over rules, augured poorly for the SIDS. Dividing and ruling were rarely gentle. Chinese involvement in the Caribbean was similarly criticised and was as detached as in the Pacific, with little ability to utilise soft power. Initial expectations over the possibility of valuable and sensitive 'South–South' links with China were found wanting, given China's elitist business-centric approach to development, the avoidance of participatory approaches in China-Caribbean ventures and the passive incorporation of the Caribbean into China's global vision (Gonzalez-Vicente & Montoute, 2021). Equally a wider frustration existed that Chinese aid projects were rarely linked to long-term development issues but centred on physical infrastructure that generated little employment and no new skills. The BRI was no panacea but 'helped to perpetuate the [region's] structural constraints' (Gonzalez-Vicente, 2021: 190). China's engagement narrowed the scope for local social and economic participation, consolidating a model of top-down elite-led governance that eschewed transparency and participation in terms of decision-making and scrutiny and failed to solve longstanding development issues.

However, SIDS resented what were often seen in metropolitan states as worst-case scenarios over the impacts of China in the region and which ignored any sense of positive local agency. Solomon Islander, Transform Aqorau, noted: 'The fear that Pacific nations are merely passive pawns in a great power contest does not align with the historical and cultural realities of our region' (quoted in Futaiasi & Zhang, 2025). The Fijian academic, Steven Ratuva, stressed: 'the anxiety and fear of the Anglo-Western countries about a Chinese "takeover" is not just misplaced, but utterly irrational. It does not consider the agency of the BOSs [Big Ocean States] to wisely, strategically and imaginatively navigate their way through the treacherous geopolitical waters' (2022: 133). SIDS displayed the ability to mediate and balance external blandishments (Maggio, 2023) within a more nuanced and multifaceted security.

Criticism grew over the role and value of development assistance. Sports stadiums and conference centres proved white elephants, and heightened geopolitical tension between China and traditional partners brought an increased focus on hard infrastructure (some with environmental costs), rather than education and health, empowerment of women and youth, or slow capacity building (Duke & Ahsan, 2024). Poorly integrated and managed projects could overwhelm the absorptive capacity of the SIDS. The prominent Marshallese leader, Tony de Brum, commented: 'we have accepted prescriptions of our development partners as normal – that we must do what we are told to do, not what we want to do' (quoted in Tarte, 2015: 84). Programmes that might be valued could be double-edged: policing is of value to SIDS, but it was introduced as much to support the interests of donor nations such as Australia (to cut back the drug trade and gain insights into international security), rather than control local corruption or crime. In the Pacific, the Boe Declaration expressed a wider perspective on security, but in practice this had little resonance in the geopolitics of either region but was merely 'slow development' that failed to meet the needs of political leaders.

Amidst the jockeying for position, in a period of budget austerity, shifting alliances and partial commitments, SIDS sought to play a greater international role, especially in strengthening policies to mitigate and compensate for climate change, and determine their own economic partnerships and political alliances. Several territories still sought political emancipation. The Cook Islands, a self-governing Pacific state in 'free association' with New Zealand, with whom it had obligations to consult on matters of defence and security, was strongly criticised by New Zealand for agreeing to sign a comprehensive partnership agreement with China early in 2025 without revealing its details (Corlett, 2025). Otherwise, independence movements, some forged from painful colonial histories and socio-economic disparities, sought political sovereignty, but through referendums rather than violent struggles.

Seemingly paradoxically, while some territories sought greater independence, others were involved in compacts and treaties that tied them to long-term relationships with the United States or Australia, based on 'security' but involving migration and other rights, and increased development assistance. In some respects it constituted acquiescence to recolonisation. Pacific SIDS especially used their place in the world to 'sell' their sovereignty, escaping from being 'prisoners' of a geography that was once negatively seen as being that of small, isolated and vulnerable islands without prospects, to a new destiny where that location had become invaluable, enabling their being plausible beneficiaries of a new global geography. That geography remained uneven, evident in the multiple asymmetrical agreements between metropolitan states and PICs, whether the Falepili Union or Fiji's 2024 defence agreement with the United States in now more complex regions, where democratic institutions are occasionally volatile and US and Australia-China rivalry has weakened regional unity. Yet the significance of the Pacific region has meant that they also remained prisoners of a new form of international rivalry centred on securitisation.

A test for metropolitan powers was whether they could engage in the 'Pacific Way', an indefinable consensus politics centred on relationships with island nations linked by culture and heritage, the supposed theme of *falepili*, that values equality and neighbourliness, and where security is food security, accessible health services, schools and a means of generating an adequate and sustainable livelihood. The SDGs are key to island security, not the warped elitist grandeur of AUKUS or strategic alliances (and denials). Baron Waqa, the PIF Secretary-General (and former President of Nauru), in mid-2024 offered a reminder: 'Geopolitical manoeuvring means nothing to Pacific peoples who have cyclones coming over the horizon … who have water lapping at their doorsteps due to sea-level rise … who are focused on building resilience, peace and prosperity for our families, communities, nations and region' (RNZ, 24 June 2024). Treaties might help but health centres and primary schools influence elections. For several years, the daily *Fiji Sun* systematically included photographs of potholes in roads, basic problems to be fixed, comprehensible to local people, symbolic of a grounded development and a daily reminder that 'development' and security are inherently local. Unless otherwise prompted, ordinary people, unlike politicians, do not regard climate change as the most critical threat to existing livelihoods and security. More typically, on the Fijian and Vanuatu Islands of Mota Lava and Tongoa, 'climate-related issues did not figure prominently. Limited access to water and sanitation, land disputes, youth delinquency (e.g. teenage pregnancy, marijuana use, disrespect of elders and traditions), and food insecurity were the main issues' that emphasised family, spirituality and local livelihoods (Granderson, 2018: 485; Nand et al., 2024). That in turn suggests a structure of security shaped from the bottom up, which prioritises participation, agency and equity (Fabinyi et al., 2025). But metropolitan concepts of security have little time for SDGs and livelihoods, let alone questions of existential climate change (even when claiming to respond to it).

Remarkably, and significantly, the rise in strategic significance of island regions occurred while SIDS populations were declining. In some respects, this is a function of absent SDGs and livelihood opportunities, coinciding with, at a very different scale, emerging strategic realignment, as China and the United States contemplate a massive and valuable space, where fish, minerals and space itself are lures. For islanders this strategic space is becoming increasingly empty, and where treaties that enable migration to occur, greater emptiness may follow, an alarming trend especially in the blue continent. Asymmetry recurs in migration. At a time when the world's population has just reached 8 billion, two significant parts of the world are losing population. Margins are fading, an unusual twist on security, which we consider in the next chapter.

Bibliography

Abe, M., & Wang, A. (2024). Migration among the freely associated states in Micronesia: Trends, drivers, and implications. In E. Monaco & M. Abe (Eds.), *Sustainable development across Pacific Islands* (pp. 3–34). Springer.

Aldrich, R., & Connell, J. (1998). *The last colonies*. Cambridge University Press.

Allen, M., & Dinnen, S. (2017). *State-building and state formation in the Western Pacific. Solomon Islands in transition?* Routledge.

Amankwa, M., Murruka, G., Adusei-Asante, K., et al. (2025). Culture, community, and development: Unpacking the 'Socio-Cultural Milieu Dynamics' of Solomon Islands and their influence on Australia's aid localisation. *Public Administration and Development*.

Anantharajah, K. (2019). Governing climate finance in Fiji: Barriers, complexity and interconnectedness. *Sustainability, 11*(12), 3414.

Baker, K. (2023, April 18). Book review: Helpem Fren: Australia and the regional assistance mission to Solomon Islands. *Australian Outlook*.

Barnett, J., Farbotko, C., Kitara, T., & Aselu, B. (2025). Migration as adaptation? The Falepili Union Between Australia and Tuvalu. *Wiley Interdisciplinary Reviews: Climate Change, 16*(1), e924.

Barrett, C (2025, July 4). Pacific rugby leaders raise alarm about NRL's PNG deal. *Sydney Morning Herald*, 17.

Baud, M., Hogenboom, B., Gonzalez-Vicente, R., et al. (2022). *China's economic and political role in the Caribbean and Central America*. CEDLA-UvA.

Bernal, R. (2010). The Dragon in the Caribbean: China-CARICOM economic relations. *The Round Table, 99*(3), 281–302.

Bernal, R. (2015). The growing economic presence of China in the Caribbean. *The World Economy, 38*(9), 1409–1437.

Berta, O. (2025). Enduring ascriptions of dependence: Cultural autonomy and relational interdependence in the Marshall Islands. *Antipode, 57*, 1382–1401.

Bock, J., & Parilla, H. (2024, November 1). Hardly an inducement: Tourism from China gets up Palau's nose. *The Strategist*.

Boege, V. (2022). Rio Tinto and Bougainville. A fatal connection. A mine, a war, and an uncertain future. *Australian Journal of Politics and History, 68*(1), 18–35.

Bordner, A., Ferguson, C., & Ortolano, L. (2020). Colonial dynamics limit climate adaptation in Oceania: Perspectives from the Marshall Islands. *Global Environmental Change, 61*, 10205.

Brady, A.-M. (Ed.). (2022). *Looking north, looking south: China, Taiwan, and the South Pacific*. World Scientific.

Calliari, E., & Vanhala, L. (2022). The 'national turn' in climate change loss and damage governance research: Constructing the L&D policy landscape in Tuvalu. *Climate Policy, 22*(2), 184–197.

Caterson, J. (2024, October 1). How giant China looms over tiny Palau's economy. *The Strategist*.

Cavanough, E. (2020, September 2). Solomon Islands province announces independence vote amid China Tensions. *The Guardian*.

Cavanough, E. (2023). *Divided Isles: Solomon Islands and the China Switch*. La Trobe University Press.

Chacko, P. (2023). Racial capitalism and sphere of influence: Australian Assertions of White Possession in the Pacific. *Political Geography, 105*, 102923.

Chaguan. (2023, September 23). China, leader of the global south? *The Economist*, 28.

Cheung, K., & Day, J. (2025, June 22). Chinese investor perspectives on customary land disputes in Vanuatu. *Melbourne Asia Review*.

Clegg, P. (2012). Independence movements in the Caribbean: Withering on the vine? *Commonwealth and Comparative Politics, 50*(4), 422–438.

Connell, J. (2006). Nauru: The first Pacific failed state. *The Round Table, 95*, 47–63.

Connell, J. (2013). *Islands at risk: Environments, economies and contemporary change*. Edward Elgar.

Connell, J. (2018). Fiji, rugby and the geopolitics of soft power: Shaping national and international identity. *New Zealand Geographer, 74*(2), 92–100.

Connell, J. (2021a). The 2020 new Caledonia Referendum: The slow march to independence? *Journal of Pacific History, 56*(2), 144–160.

Connell, J. (2021b). Bougainville: A new Pacific Nation? *Small States & Territories, 3*(2), 375–396.

Connell, J., & Aldrich, R. (2021). *The ends of empire: The last colonies revisited*. Palgrave Macmillan.

Cooper, A. (2023). A small state taking on a Giant: Antigua and Barbuda 'getting the better' of the United States without overturning structural asymmetry. In G. Baldacchino (Ed.), *The success of small states in international relations* (pp. 96–17). Routledge.

Corbett, J. (2023). *Statehood à la carte in the Caribbean and the Pacific*. Oxford University Press.

Corbett, J., & Byron-Reid, J. (2024). Secessionism in Nevis: Why have tensions eased? *Island Studies Journal, 19*(1), 1–24.

Corlett, E. (2025, February 10). New Zealand raises 'concern' as Cook Islands prepares to sign China deal. *Guardian*.

Corva, D. (2008). Neoliberal globalization and the war on drugs: Transnationalizing illiberal governance in the Americas. *Political Geography, 27*(2), 176–193.

Crocombe, R. (2007). *The fourth wave: Chinese in the Pacific Islands in the twenty-first century*. CSCSD Occasional Paper Number 1.
Cullum, R. (2021). Comparing Australian/South Pacific and UK/Caribbean relations. *The Round Table, 110*(4), 492–509.
D'Arcy, P. (2014). The Chinese Pacifics: A brief historical review. *The Journal of Pacific History, 49*(4), 396–420.
Davis, S., Munger, L., & Legacy, H. (2020). Someone else's chain, someone else's road: US military strategy, China's Belt and Road Initiative, and island agency in the Pacific. *Island Studies Journal, 15*(2), 13–36.
Day, B., & Wells, T. (2025). The elite paradigm driving Australian aid policy: Learning to live with the 'cautious consensus'. *Asia & the Pacific Policy Studies, 12*(1), e70009.
Dayant, A. (2020, December). Health spending and foreign aid in the Pacific. *Pacific Economic Monitor*, 40–43.
Dayant, A., Duke, R., Ahsan, N., et al. (2024). *Pacific aid map. 2024 key findings report*. Lowy Institute.
Dean, A., Green, D., & Nunn, P. (2017). Too much sail for a small craft? Donor requirements, scale and capacity discourses in Kiribati. In E. Stratford (Ed.), *Island geographies: Essays and conversations* (pp. 54–77). Routledge.
Dobell, G. (2007). The 'Arc of Instability': The history of an idea. In R. Huisken & M. Thatcher (Eds.), *History as policy: Framing the debate on the future of Australia's Defence Policy* (pp. 85–104). ANU E Press.
Dornan, M., & Pryke, J. (2017). Foreign aid to the Pacific: Trends and developments in the twenty-first century. *Asia and the Pacific Policy Studies, 4*(3), 386–404.
Dorney, S. (2016). *The embarrassed Colonialist*. Lowy Institute and Penguin Random House.
Duke, R. & Ahsan, N (2024, November 18). Pacific education: Backsliding outcomes and a financing freefall. *Interpreter*.
Dziedzic, S. (2024, December 14). Friends and partners weigh up Canberra's 'relentless' Pacific push. *ABC*.
Easterly, W. (2024). *The Tyranny of experts*. Basic Books.
Eke, A., & Altangerel, T. (2025). Taking matters into our own hands: Aid dependency in the Pacific. *ANU DevPolicyBlog*, 3 July.
ESCAP. (2022). *Debt for climate swaps in the Pacific SIDS*. ESCAP (mimeo).
Fabinyi, M., Cvitanovic, C., Barclay, K., et al. (2025). Rethinking maritime security from the bottom up: Four principles to broaden perspectives and centre humans and ecosystems. *npj Ocean Sustainability, 4*(1), 29.

Feinberg, R. (1986). The 'Anuta Problem': Local sovereignty and national integration in the Solomon Islands. *Man, 21*(3), 438–452.

Fiji Ministry of Foreign Affairs. (2024). *Fiji's foreign policy white paper*. Ministry of Foreign Affairs.

Fry, G. (2021). AUKUS undermines Australia's "Pacific family". *ANU DevPolicyBlog*, 4 November.

Fry, G., Kabutaulaka, T., & Wesley-Smith, T. (2022). *'Partners in the Blue Pacific' initiative rides roughshod over established regional processes. ANU DevPolicyBlog*, 5 July.

Futaiasi, D., & Zhang, D. (2025). Solomon Islands' external relations amid great power competition. *ANU DevPolicyBlog*, 14 August.

Gonzalez-Vicente, R. (2021). Over hills and valleys too. China's Belt and Road Initiative in the Caribbean. In F. Schneider (Ed.), *Global perspectives on China's Belt and Road Initiative asserting agency through regional connectivity* (pp. 171–193). Amsterdam University Press.

Gonzalez-Vicente, R., & Montoute, A. (2021). A Caribbean perspective on China–Caribbean relations: Global IR, dependency and the postcolonial condition. *Third World Quarterly, 42*(2), 219–238.

Granderson, A. (2018). Value conflicts and the politics of risk: Challenges in assessing climate change impacts and risk priorities in rural Vanuatu. *Climate and Development, 10*(6), 481–494.

Gray, A. (2021, May 7). How the US can protect the sovereignty of the smallest Pacific Islands. *The Diplomat*.

Griffith, I. (2010). *Drugs and security in the Caribbean: Sovereignty under Siege*. Penn State University Press.

Griffith, I., & Munroe, T. (1995). Drugs and democracy in the Caribbean. *Journal of Commonwealth & Comparative Politics, 33*(3), 357–376.

Handa, S., & King, D. (1997). Structural adjustment policies, income distribution and poverty: A review of the Jamaican experience. *World Development, 25*(6), 915–930.

Hasan, S., Fleming, C., Smart, J., et al. (2021). Identifying and mitigating risks to completion of small grant climate change adaptation projects: Evidence from the Pacific. *Regional Environmental Change, 21*, 55.

Hill, C. (2025a). Realists, re-imaginers and reformers: Aid in the post-America multiverse. *ANU DevPolicyBlog*, 11 June.

Hill, C. (2025b). A new look at aid's 'great game'. *ANU DevPolicyBlog*, 12 November.

Howes, S. (2023, April 20). Questions about Australian aid to fund the Pacific Games. *Interpreter*.

Howes, S., & Liu, H. (2022). Australian government gifts US$190 million to Telstra to buy Digicel. *ANU DevPolicyBlog*, 21 April.

Iati, I. (2021). China's impact on New Zealand foreign policy in the Pacific: The Pacific reset. In G. Smith & T. Wesley-Smith (Eds.), *The China alternative. Changing regional order in the Pacific Islands* (pp. 143–166). ANU Press.

Johnson, B., & Young, A. (2025, July 21). Friends to all: Fiji prioritises shared value and regional leadership. *The Strategist*.

Jules, T., Arnold, R., & Smith, A. (2023). Caribbean exceptionalism and the rise of Sino-CARICOM relations in the post-bureaucratic era. *The Round Table, 112*(1), 57–72.

Kabutaulaka, T. (2006). Global capital and local ownership in Solomon Islands' Forestry Industry. In S. Firth (Ed.), *Globalisation and governance in the Pacific Islands* (pp. 239–257). ANU Press.

Kampmark, B. (2025). *Contract bikies and bribed politicians: Australia's Nauru refugee deal* (pp. 1–3). *Independent Australia*.

Koga, K. (2022). Japan's strategic approach toward Island States: Case of the Pacific Islands. *Journal of Indo-Pacific Affairs, 5*(7), 62–83.

Köllner, P. (2025). Explaining sender–receiver gaps in signalling: Australia's 'Pacific Step-up' and Solomon Islands' multi-alignment. *The British Journal of Politics and International Relations, 27*(1), 20–42.

Koro, M., & McNeill, H. (2024). Challenging colonial logics of habit in Australia's economic statecraft with Pacific Islands. *Asia & the Pacific Policy Studies, 11*(3), e398.

Koro, M., McNeill, H., Ivarature, H., & Wallis, J. (2023). Tā, Vā and Lā: Re-imagining the geopolitics of the Pacific Islands. *Political Geography, 105*, 102931.

Kowasch, M., & Lindenmann, P. (2014). New flags, upward forces and sheltered harbours: The new 'Great Game' in the Pacific Islands region. *Pacific Geographies, 41*, 4–9.

Kowasch, M., Batterbury, S., Bouard, S., & Wadrawane, E. (2022). The third independence referendum in New Caledonia – A fallback to colonialism? *Pacific Geographies, 57*, 11–15.

Lee, C., Li, X., Yu, C., & Zhao, J. (2022). The contribution of climate finance toward environmental sustainability: New global evidence. *Energy Economics, 111*, 106072.

Lewis, P. (1999). Revisiting the Grenada invasion: the OEC's role, and its impact on regional and international politics. *Social and Economic Studies, 48*, 85–120.

Lewis, L. (2025, May 22). 'Every man and his dog is coming': Samoan PM warns of growing geopolitical interest in Pacific. *Radio New Zealand.*
Liloqula, R. (2024). Behind the shrine of the Pacific Games lurks poor governance and corruption. *ANU DevPolicyBlog*, 29 January.
Loyd, J., Mitchell-Eaton, E., & Mountz, A. (2016). The militarization of islands and migration: Tracing human mobility through US bases in the Caribbean and the Pacific. *Political Geography, 53*, 65–75.
MacDonald, S. (2022). *The New Cold War, China, and the Caribbean*. Palgrave Macmillan.
McNeill, H. (2025). New Zealand's statecraft 'in and out of the Pacific'. In J. Wallis, H. McNeill, M. Rose, & A. Tidwell (Eds.), *Power and influence in the Pacific Islands. Understanding Statecraftiness* (pp. 140–154). Routledge.
Maggio, R. (2023). "this is not culture"/"yes, it is": Conceptualising Kiribati as an Interstitial Island in the Chinese Pacific. *Political Geography, 104*, 102907.
Mausio, A. (2023). A South-South Tango: China's soft power relations with Fiji since 1975. *Island Studies Journal, 18*(2).
Mawdsley, E., Banks, G., Sanyu, C., et al. (2025). Taking (anti-)'woke' seriously: The future of development cooperation and humanitarian aid. *Development Policy Review, 43*, e70031.
Meki, T., & Tarai, J. (2023). How can aid be decolonized and localized in the Pacific? Yielding and wielding power. *Development Policy Review, 41*, e12732.
Minto, J. (2019). The impact of Chinese outbound foreign direct investment in Jamaica, (2000-2017). In E. Peters (Ed.), *China's foreign direct investment in Latin America and the Caribbean* (pp. 287–308). Lynne Rienner.
Mohamed, A. (2025, February 18). Handouts are never free. The cruel US aid freeze is an opportunity for the Caribbean. *Guardian.*
Mohan, P. (2025a). Climate change adaptation in small island developing states: Evidence from the nationally determined contributions of Caribbean States. *Regional Environmental Change, 25*(1), 32.
Mohan, P. (2025b). Climate reparations for a just response to climate change: A review of historical responsibility and future implication. *Wiley Interdisciplinary Reviews: Climate Change, 16*, e70007.
Moore, L. (2024). A dysfunctional family: Australia's relationship with Pacific Island states and climate change. *Australian Journal of International Affairs, 78*(3), 286–305.
Morris, J. (2023). *Asylum and extraction in the Republic of Nauru*. Cornell University Press.
Mountz, A. (2011). The enforcement archipelago: Detention, haunting and asylum on islands. *Political Geography, 30*, 118–128.

Murray, W., & Overton, J. (2011). The inverse sovereignty effect: Aid, scale and neostructuralism in Oceania. *Asia Pacific Viewpoint, 52*(3), 272–284.

Nand, M., Clissold, R., McNamara, K., & Yee, M. (2024). Guiding our responses to climate change by what people value: Insights from Fiji. *Geoforum, 155*, 104080.

Niles, K., & Lloyd, B. (2013). Small island developing states (SIDS) & energy aid: Impacts on the energy sector in the Caribbean and Pacific. *Energy for Sustainable Development, 17*(5), 521–530.

Nobetau, O. (2025). *Bougainville's future: A roadmap for development*. Lowy Institute.

Overton, J., Murray, W., Prinsen, G., et al. (2019). *Aid, ownership and development. The inverse sovereignty effect in the Pacific Islands*. Routledge.

Overton, J., Prinsen, G., Murray, W., & Wrighton, N. (2012). Reversing the tide of aid: Investigating development policy sovereignty in the Pacific. *Journal de la Société des Océanistes, 135*, 229–242.

Oza, T., & Goff, F. (2023). Maximising the impact of Australia's scarce water and sanitation aid. *ANU DevPolicyBlog*, 6 July.

Pacific Islands Forum Secretariat. (2018). *Boe declaration action plan*. PIFS.

Peake, G., & Smith, G. (2022, January 27). Solomon Islands: Cops bearing gifts. *Interpreter*.

Petrou, K., & Connell, J. (2023a). *Pacific Islands guestworkers in Australia: The new blackbirds?* Palgrave Macmillan.

Petrou, K., & Connell, J. (2023b). Our 'Pacific family'. Heroes, guests, workers or a precariat? *Australian Geographer, 54*(2), 125–136.

Pouponneau, A., & Northrop, E. (2025, August 5). Big ocean: ICJ advisory opinion and marine boundaries. *Island Times*.

Powles, A. (2021). The 'Pacific Reset' and shaping New Zealand's relations with the Pacific. *Development Bulletin, 82*, 32–36.

Powles, A., & Wallis, J. (2024). Different nightmares, shared dreams? Australia and New Zealand's intuitive alliance. *Australian Journal of International Affairs, 78*(5), 536–560.

Prasad, J. (2024, October 23). Biases within capital flows to the Pacific. *Interpreter*.

Premdas, R., Steeves, J., & Larmour, P. (1984). The western breakaway movement in the Solomon Islands. *Pacific Studies, 7*, 34–34.

Pryke, J., & Sora, M. (2022, February 11). Kick-starting America's Pacific diplomacy. *Interpreter*.

Puas, G., & D'Arcy, P. (2021). Micronesia and the rise of China: Realpolitik meets the reef. *Journal of Pacific History, 56*(3), 274–295.

Puri, S. (2014). The Grenada Revolution in the Caribbean present: *Operation urgent memory*. Palgrave Macmillan.

Raine, S. (2017). *Regional disorder: The South China Sea disputes*. Routledge.

Ratuva, S. (2022, June 16). Pacific agency: A devalued political capital? *Australian Outlook*.

Regan, A. (2022, October 31). Overcoming suspicion: Reconciliation for Australia and Bougainville. *Interpreter*.

Regan, A., Baker, K., & Oppermann, T. (2022). The 2019 Bougainville Referendum and the question of independence: From conflict to consensus. *Journal of Pacific History, 57*(1), 58–88.

Reklai, L. (2024, June 28). Palau faces major challenges in halting outmigration. *Island Times*.

Ride, A., & Zhang, D. (2024). *Foreign actors, geopolitics and riots in the Pacific*. ANU Department of Pacific Affairs InBrief 2024/3.

Rodd, A. (2020). A road to island sovereignty and empowerment? Fiji's aims within the belt and road initiative. *Island Studies Journal, 15*(2), 93–118.

Rothe, D., Boas, I., Farbotko, C., & Kitara, T. (2024). Digital Tuvalu: State sovereignty in a world of climate loss. *International Affairs, 100*(4), 1491–1509.

Rummery, R. (2025). Buried beneath the bricks: A quiet retreat from health systems strengthening. *ANU DevPolicyBlog*, 23 July.

Segawa, N. (2025). Japan's FOIP-based diplomacy toward the Pacific Islands Region. *Pacific Affairs, 98*(2), 293–316.

Shah, K., & Chiorazzi, E. (2024, March 24). China's growing environmental footprint in the Caribbean. *New Security Beat*.

Skidmore, D. (2025). *China, the West and the global development finance regime. Competitive convergence*. Elgar.

Smith, N., & McNeill-Stowers, H. (2025, May 22) Europe's faltering statecraft in the Pacific. *East Asia Forum*.

Smith, Y., Craney, A., & Roche, C. (2024). Racism, colonialism and whiteness in development: Insights from Pacific professionals following repatriation of white staff during Covid-19. *Third World Quarterly, 45*(9), 1517–1535.

Song, L., Mosses, M., & Giraudeau, G. (2023). The ambiguous history of Matthew and Hunter Islands: Tracing the roots of Vanuatu and French claims. *Journal of Pacific History., 58*(3), 232–250.

Sora, M., Collins, J., & Keen, M. (2024). The great game in the Pacific Islands, Lowy Institute.

Sou, G., & Day, J. (2025). Postcolonial experiences of Chinese aid: Encountering and welcoming South–South aid from the middle. *Transactions of the Institute of British Geographers, 50*(2), e12693.

Stambolie, E., & Hill, C. (2025). 2025 Australian aid procurement update. *ANU DevPolicyBlog*, 21 November.

Steff, R. (2024). 'Our region is now a strategic theatre': New Zealand's balancing response to China. *The Pacific Review, 37*(6), 1120–1151.

Szadziewski, H. (2020). Converging anticipatory geographies in Oceania: The Belt and Road Initiative and Look North in Fiji. *Political Geography, 77*, 102119.

Szadziewski, H. (2022). Agents of the belt and road initiative or agents of agency? Fijian State and civil society perceptions of Chinese companies. In D. Pavlićević & N. Talmacs (Eds.), *The China question* (pp. 269–289). Palgrave Macmillan.

Szadziewski, H. (2024). *Going elsewhere: Oceania's globalism and the Gulf States.* ANU Department of Pacific Affairs Discussion Paper 2024/5.

Tarte, S. (2015). A new Pacific regional voice? The Pacific Islands development forum. In G. Fry & S. Tarte (Eds.), *The new Pacific Diplomacy* (pp. 79–88). ANU Press.

Taylor, M., & Middleby, S. (2023). Aid is not development: The true character of Pacific aid. *Development Policy Review, 41*(Suppl. 2), e12745.

Taylor, M., Middleby, S., & Vunibola, S. (2025). Aid in the age of Amazon: Imperial logics, Pacific resistance and an alternate paradigm. *Development Policy Review, 43*, e70030.

Vaha, M. (2025). Responsibility and anxiety in the 'Pacific family': AUKUS as a source of ontological insecurity. *Australian Journal of International Affairs, 79*(3), 406–422.

Valencia, M. (2007). The East China Sea dispute: Context, claims, issues, and possible solutions. *Asian Perspective, 31*(1), 127–167.

Vunibola, S. (2023). "Want to help someone? Shut up and listen": Foreign aid, maladaptation, and community development practices in the Pacific. *Development Policy Review, 41*, e12751.

Waide, S. (2024, August 13). The Pacific sees a lot of fly-in and fly-out experts-but what would happen if there was a shift to locally led aid? *ABC.*

Walda-Mandel, S. (2016). *'There is no place like home': Migration and cultural identity of the Sonsorolese, Micronesia.* Universitätsverlag Winter.

Wallis, J. (2023). The enclosure and exclusion of Australia's 'Pacific family'. *Political Geography, 106*, 102935.

Wallis, J., & Corbett, J. (2024, December 9). At first glance, Australia's new treaty with Nauru seems to be a win-win. But questions remain. *The Conversation.*

Wallis, J., & Kemish, I. (2022, August 10). Key to Pacific is education: Ours, not theirs. *Sydney Morning Herald*, 25.

Wallis, J., McNeill, H., Batley, J., & Powles, A. (2022). Security cooperation in the Pacific Islands: Architecture, complex, community, or something else? *International Relations of the Asia-Pacific, 23*(2), 263–296.

Wallis, J., Koro, M., & O'Dwyer, C. (2024). The 'Blue Pacific' strategic narrative: Rhetorical action, acceptance, entrapment, and appropriation? *The Pacific Review, 37*(4), 797–824.

Waqavakatoga, W., Habru, P., & Koro, M. (2025). How do Pacific Island countries respond to China's statecraft? In J. Wallis, H. McNeill, M. Rose, & A. Tidwell (Eds.), *Power and influence in the Pacific Islands. Understanding statecraftiness* (pp. 201–218). Routledge.

Webber, S. (2013). Performative vulnerability: Climate change adaptation policies and financing in Kiribati. *Environment and Planning A, 45*(11), 2717–2733.

Wesley-Smith, T. (2021). A new Cold War? Implications for the Pacific Islands. In G. Smith & T. Wesley-Smith (Eds.), *The China alternative. Changing regional order in the Pacific Islands* (pp. 71–106). ANU Press.

Wesley-Smith, T., & Finin, G. (2021). US-Pacific engagement and the Biden Presidency: The limits of a China-centred approach. *Journal of Pacific History, 56*(4), 437–458.

Wood, T. (2021). The cruel ironies of the 2021 Honiara riots. *ANU DevPolicyBlog*, 3 December.

Wood, T., Dornan, M., & Muller, S. (2021). *Change and continuity in Australian aid: What the aid flows show.* ANU Development Policy Centre.

Xie, G. (2025). China's strategic narratives in the Pacific. In J. Wallis, H. McNeill, M. Rose, & A. Tidwell (Eds.), *Power and influence in the Pacific Islands. Understanding statecraftiness* (pp. 190–200). Routledge.

Yamamoto, L., & Esteban, M. (2014). *Atoll Island States and international law.* Springer.

Yang, J. (2011). *The Pacific Islands in China's grand strategy.* Palgrave Macmillan.

Zhang, D. (2018). *China's growing interest in deep sea mining in the Pacific.* ANU Department of Pacific Affairs InBrief 2018/11.

9

Moving On: Migration, Remittances and Brain Drains

The SIDS have become dynamic and coveted geopolitical regions, where 'great' and lesser powers have sought footholds and productive political ties, 'moving in' in subtle and less subtle ways. By contrast, islanders themselves have long been 'moving out', migrating away from the regions, movements that have intensified and diversified in this century. In the Caribbean that movement has been long established, dating back to the late eighteenth century, whereas in the Pacific it only began a century later. But contemporary migration patterns have become similar in both regions, pointing to a new dispersed, demographic future. Islanders have moved in response to environmental stresses, the quest for resources, social tensions and even adventure, so that Epeli Hau'ofa (1993) regarded the Pacific not as a vast ocean of isolated islands but as a region bound together through oceanic networks, loosely integrated through trade and culture in webs of relationships, contacts and exchanges (Nero, 1997), intricately related to ecologies. Much later, that became one basis for some notions of blue economies. Such linkages and seafaring traditions have been best documented in the atolls of Micronesia (e.g. D'Arcy, 2006), where livelihoods were hard-won amidst hazards, and with

J. Connell, Y. Campbell, *Island Challenges*,
https://doi.org/10.1007/978-981-95-8177-1_9

contact with other atolls not infrequently violent: 'movement and migration were about survival' (Hanlon, 2018: 92) and were inherently risky.

Migration and mobility became more widely significant in the nineteenth century. The forced migration of slavery—the infamous middle passage—and colonialism shaped societies and economies in the Caribbean and instigated structures that proved driving forces in the dehumanisation, lack of citizenship rights, racial hierarchies and underdevelopment of the Caribbean, but which shaped new identities and resistance to such hierarchies. Slaves, forced to migrate from Africa to the Caribbean, made a significant contribution to modern capitalism and European development. With the abolition of slavery in 1833, Indian and Chinese workers were used to fill gaps, as former slaves refused to work on plantations. Labour migration from South Asia to the Caribbean and Fiji produced ethnically diverse countries, distinct in their cultural and livelihood complexities.

In the Pacific, labour migration came much later, with nineteenth-century 'blackbirding' taking workers to plantations scattered from Samoa to Australia and to mines from Chile to Palau. This era of mobility followed European expansion of commercial and colonial empires—as traders sought everything profitable, from whales and bêche-de-mer to copra, phosphate and sandalwood. In the second half of the century Melanesians went from the New Hebrides (Vanuatu) and Solomon Islands primarily to work as labourers on Australian sugar cane plantations. Caribbean islanders moved to Central and North America, and within the Caribbean, being a major force in constructing the Panama Canal. Migration has long been a feature of Pacific and Caribbean life.

The regions entered the era of independence with different structures of population origin, ethnicity, experience of migration and, in the Caribbean, the legacy of a slave trade and exploitative plantation system, more muted in the blackbirding of Pacific islands. While Caribbean islanders were disadvantaged by colonialism, English was universally spoken, whereas it was only ever a second language in the Pacific SIDS. Post-independence population growth brought greater convergence, with rural-urban migration and urbanisation within the SIDS (Chap. 5), but in the twenty-first century, population growth slowed, and outer islands and urban centres were left behind or bypassed as growing numbers of islanders moved overseas temporarily or more permanently. Long true of the smaller Caribbean SIDS, where more islanders lived overseas than at home, such a pattern was emerging even in relatively large Pacific SIDS like Tonga and Samoa. Few island states have growing populations (Fig. 1). While both regions face similar challenges in terms of the consequential effects of migration, qualitative and quantitative differences remain in numerical flows, destinations and impacts.

9.1 New Backgrounds

The contemporary rationale for migration is like that of past times. Seeking worthwhile employment provides an economic underpinning, partly a function of limited national development. However, many real and potential migrants are without the skills to access good jobs elsewhere. The selective migration of relatively few people, by number and proportion, with tertiary education and training, has left shortages of workers in skilled positions. Migration also has substantial social undertones: aspirations to superior education, health care and other services. Despite the primacy of a loosely economic rationale, and growing emphasis on environmental factors, culture often plays a role, as in the FSM island of Satowan, where islanders view mobility as related to 'the life stage (age, sex, marital status, childbirth), gender specific activities, religion, economic sustenance and cultural ceremonies marking transitions' (Bautista, 2010: 5). Indo-Fijians have left because of perceptions of political insecurity rather than economics. Despite diverse influences, multiple analyses conclude with a ritual incantation of the trilogy of 'jobs, education and healthcare not available' in the home country (e.g. Bordner et al., 2020: 3). Migration has been a response to perceived socio-economic inequalities, accentuated in this century by both a rise in expectations of incomes and living conditions and greater familiarity with overseas circumstances.

While environmental changes now influence migration (Chap. 2), they are rarely principal factors in international migration, though hazards are certainly catalysts. Serious hazards usually necessitate some migration—inland, between islands or to the capital city—but often temporarily. Most of those who left Ambae (Vanuatu) after serious volcanic eruptions eventually returned home (Chap. 2). When tens of thousands of Bahamians were displaced in the immediate aftermath of Hurricane Dorian in 2019, most of those from the two most affected islands migrated within the Bahamas, although many opted to migrate internationally, with at least 1500 going to the United States, an outcome with many precedents (Cooper, 2020; Robinson, 2024). Here too, many eventually returned to their home islands.

Even in the atoll states, international migration remains a primarily economic phenomenon. Despite climate change and environmental degradation being highly politicised, the rationale for migration from Tuvalu remained one of superior access to employment opportunities and the social and economic benefits derived from them, such as access to housing, health and education services (Mortreux & Barnett, 2009; Paton & Fairbairn-Dunlop, 2010). Indeed, on many atolls immediate concerns over resource scarcity, freshwater supplies and waste disposal were more urgent than climate change (Yarina & Takemoto, 2017). The notion of there already being 'environmental' refugees was dismissed in a New Zealand court case, while the Kiribati government has declined to support applications by its citizens for climate refugee status. People displaced by environmental change, should they wish to be potential migrants, remain subject to the same rules and legislation as all intending migrants. The distinction between 'voluntary' and 'forced' migration has involved exceptionally complicated and unresolved legal issues (Neef & Benge, 2022). Ironically, preferred destinations in all the atoll states are countries that are amongst the sources of environmental and climate change problems (however difficult to locate causality). SIDS have largely resisted the notion of 'environmental refugees' or 'climate change refugees' as unpleasant epithets that reduce their status to that of helpless victims fleeing disappearing islands and nations. Most islanders reject the use of the word 'refugees' as a stereotype diminishing their own agency and potential to adapt, whereas it is popular in the media as an easily understood and dramatic encapsulation for wider audiences (Farbotko & Lazrus, 2012; Dreher & Voyer, 2015). Islanders have no wish to be seen as tragic victims however much that elicits sympathy elsewhere.

In the face of climate change, for many it has become an article of faith that islanders have no wish to leave their homes, where lands and lagoons are precious, and ontological security exists, a situation repeatedly emphasised by island politicians (e.g. McNamara & Gibson, 2009). Yet many migrate freely where and when they can. Even in the Marshall Islands, where leaders insist that 'People want sea walls. They want to stay. Migration is the last option' (quoted in Coloma, 2025), the population is falling faster than in any other SIDS. Changing aspirations, a preference for regular employment, and distaste for agricultural work have gradually given migration a more obvious economic rationale: a movement in search of wage and salary employment, scarce, absent or poorly paid at home. While urban-rural migration and urban growth remain significant, especially in Melanesia (Chap. 6), aspirations have become international. Most islands with fewer than a thousand people are losing population, as in Malaita (Solomon Islands) where the 'saltwater people' of Malaita are leaving their small artificial lagoon islands to relocate to the mainland for better access to schools, land, markets and various other services (van der Ploeg et al., 2020). While such processes have been accompanied and superseded by international migration, they have contributed to rural and remote depopulation and a tightening of the margins of settlement.

9.2 International Migration

Contemporary international migration is an extension and elaboration of both mobility in colonial times and rural-urban migration. In the post-war years international migration began in the Pacific SIDS and took on a new dynamism in the Caribbean, as migration to the United Kingdom boomed, only slowing when the Commonwealth Immigrants Act of 1962 halted the flow and diverted it to the United States and Canada. As the largest SIDS, Jamaica dominated such flows. Transatlantic passages were well established and well documented, notably the dismal experience of the Windrush generation in the United Kingdom, the racist legacies of which remain (e.g. Thomas-Hope, 1993; Western, 1992; Gilroy, 2013).

A decade later, broadly similar international migration from the Pacific effectively began in the 1960s, particularly after Samoan independence in 1962, spurred by the global long boom, demand for employment and the arrival of jet transport, and mostly to New Zealand and the United States. Tonga soon followed. That relieved population pressure and initiated a modern phase of remitting to the islands, much as in the Caribbean (Philpott, 1973). Intendedly temporary, migration gradually became more permanent. With rare exceptions, a lack of intra-regional migration in the Pacific was indicative of limited island economic development and cultural differences. By contrast, Caribbean islanders moved from the smaller eastern Caribbean SIDS, particularly St. Vincent and Dominica, to Trinidad, amidst a complexity of intra-regional flows. Only Fiji has exerted any comparable pull in the Pacific. Governments increasingly intervened to develop and regulate distant employment opportunities, and the demographic balance began to shift away from the SIDS, with migration becoming more widespread, more diversified, more permanent, more selective and, through remittances, as valuable to those who remained at home.

Deteriorating economic conditions in the Caribbean, and the growth of beach-head diaspora populations able to provide initial support, brought accelerated migration rates from the 1980s. By the end of the twentieth century, it was said of Nevis, which was argued to typify the Caribbean, that 'every Nevisian has been, or is, either a migrant or a close relative of a migrant… Few do not include migration in their range of possible survival or life improvement strategies' (Byron, 1999: 117). That could equally have been said of most SIDS. Several SIDS have experienced population decline, notably some of the smaller eastern Caribbean states, while in the Pacific, Tonga, Tuvalu, FSM and especially the Marshall Islands are losing population (Table 1.1). Just as American Samoa is a Pacific stepping stone to the United States, the US Virgin Islands has played a similar role in the Caribbean. International migration is least from the culturally distinct Melanesian states where populations are growing most rapidly.

So significant has international migration become that, in Polynesia at least, households created small-scale metaphorical 'transnational corporations of kin', allocating family members to different countries, while migration became so familiar that it was part of a 'culture of migration' where migration is normative and reflects economic imperatives (Connell, 2013). That has led to some stagnation of national human resource development and a skill drain, not necessarily compensated for by remittances. As international migration has become of greater significance, so remittance flows are larger, especially in comparison with aid flows, and more stable than other sources of national and household finance, while remittances compensate in the aftermath of crises and disasters. Whether the immediate benefits from remittances translate into sustainable long-term economic and social gains remains in question.

Remittances are such a substantial component of most small SIDS that island states have been described as having MIRAB (Migration, Remittances, Aid and Bureaucracy) economies, an acronym first devised for Kiribati, Tuvalu (and Tokelau and the Cook Islands) (Bertram & Watters, 1985), but now more broadly relevant in both regions. Tonga and Samoa are in the top 20 countries in the world in terms of the contribution of remittances to GDP. In the Caribbean, only Jamaica comes close to this contribution (Table 1.1). Remittances became crucial to national development, and the MIRAB acronym shifted towards MURAB, as the growing bureaucracy became more obviously urban, with strong similarities between Caribbean and Pacific SIDS (Connell & Conway, 2000). MIRAB proved not so much a derogatory acronym suggesting a handout mentality but an exercise of valuable agency, at an international scale and with continued cultural overtones (Barcham et al., 2009). Yet economies could become 'mired in MIRAB', with domestic policies being driven by external influences. SIDS moved towards a 'remittance trap' with economies 'stuck on a lower growth, higher emigration treadmill', with remittances contributing to the training of skilled labour for export (Abbott & Pollard, 2020).

Most migration has been of individuals and later families, moving to metropolitan states, with many becoming permanent migrants. Several migrant generations now exist in the United States, Canada and New Zealand. Typically, migration has followed 'colonial' lines, notably from much of northern Micronesia to the United States (enabled in their Compacts of Free Association). However, Caribbean 'allegiances' have shifted from the United Kingdom to Canada and nearby United States, the destination of about 70% of all migrants. The three Compact states chose not to gain full independence, partly fearing that migration opportunities might be lost, just as Niue and the Cook Islands developed the same relationship with New Zealand. Migration was too attractive to give up.

Structural circumstances have changed. At least in the Pacific, several SIDS now benefit from some preferential and institutional arrangements, notably in New Zealand, whether permanent migration opportunities (the lotteries of New Zealand and the United States) or schemes for temporary employment (invariably in agriculture) and/or for training in particular activities (see below). Since 2002, New Zealand has offered a Pacific Access Category, which allocated a varied number of places to a range of PICs. It offered 1100 places a year to Samoa, 250 places a year to Tonga and Fiji, and 75 to migrants from both Kiribati and Tuvalu, but applicants were only granted a visa if they received a job offer from New Zealand, a significant problem in the atoll states. In 2024 Australia introduced a Pacific Engagement Visa (PEV) for all the PICs (and Timor-Leste) that enabled islanders to participate in a lottery for about 3000 migration positions in Australia, much like the United States' green card scheme. That was positively received by most PICs, although Samoa declined to participate arguing that it was already losing too many skilled migrants. In its first year, 56,000 people entered the ballot and three quarters of the 300 ballot winners applied for visas. The ballot applicants were dominated by Fiji, whose more skilled applicants were most likely to get visas (Turia & Howes, 2025). In the second year, applicant numbers substantially increased, with Samoa now being involved (Howes & Turia, 2025). Soon after the first PEV round, Australia agreed to the Falepili Union with Tuvalu, hitherto without any migration corridor, which gave the country migration opportunities in Australia, amounting to 280 visas a year, a concession available to no other PIC (Chap. 8). When the ballot was opened for these visas in mid-2025 more than 8500 Tuvaluans applied, two-thirds of the nation's population. While such schemes might have met New Zealand and Australian needs for labour, all these arrangements, particularly the Falepili Union, had strong geopolitical connotations. Such schemes had no Caribbean parallels.

9.3 National Strategies? Seasonal Schemes

Most migration from SIDS, other than rare resettlement schemes (Chap. 2), has been independent of government involvement, intervention (or even facilitation), but as migration became more obviously a development strategy, governments intervened. That was spearheaded by Kiribati and Tuvalu, first in developing mining opportunities in Nauru, for most of the twentieth century, and from the 1960s, with the establishment of national Marine Training Schools to enable workers, men and later women, to acquire the skills needed for employment as seafarers on merchant shipping lines.

Greater interest followed in a wider process of training people for migration to achieve what the former Kiribati President, Anote Tong, called 'migration with dignity', involving employment in occupations which enabled some degree of self-esteem and higher incomes, rather than low-status, low-income, entry-level positions, and which were also likely to offer some prospects of permanency and enable a more substantial flow of remittances. That resulted in upgrading of vocational education with, in the Pacific, the multinational Australia-Pacific Technical College (APTC) that enabled Pacific islanders to gain Australian skills and qualifications in a variety of trade areas, including caring, that would enable better access to overseas labour markets. It also brought recognition that language skills were needed, with Kiribati making a deliberate effort after 2013 to teach English so that migrants would be more successful in acquiring better jobs and adapting overseas (Ohi & Ingram, 2021). Migration became part of national development agendas. The greatest impact of government involvement in migration has been through the development of seasonal employment schemes, which have flourished in this century in the Pacific especially, but also in Jamaica. They began somewhat tentatively between Caribbean states and the United States, notably Florida, in the immediate post-war years, but grew and extended into the Canadian Seasonal Agricultural Worker Program

(SAWP) involving Caribbean SIDS and then into New Zealand and Australia.

9.3.1 Caribbean

Caribbean seasonal employment began half a century before that in the Pacific. An agricultural work programme to the United States started in 1943 and saw Barbados, the Bahamas and Jamaica sending farm workers to fill labour shortages. New inter-governmental programmes followed with source countries extending to include St. Lucia, St. Vincent and Dominica, but Jamaica numerically dominated the programme, with 33,000 working in the United States in 1945. As Mexicans began to dominate the workforce (and were regarded as more pliable), and tourism generated employment in the Caribbean SIDS, numbers never again approached early post-war levels.

The separate but similar SAWP programme began in Canada in 1966, again centred on Jamaica and subsequently extending to other countries, including all but the Bahamas of the present SIDS. As in the United States, workers were recruited to fill labour shortages in the domestic market during planting and harvesting seasons. After 2017 the number of Caribbean workers entering this programme declined, leaving in 2019 only Jamaica (8874 workers), Trinidad and Tobago (865) and St. Vincent (315). As in the United States, the vast majority came from Mexico (IOM, 2012). Eligible countries, like St. Kitts and Dominica, have chosen not to send workers since the 2010s, quite different from growing demand in the Pacific.

Early workers from Jamaica experienced precarity. Their remittances and savings were oriented to housing (building, modification and repair), children's education and consumer goods (such as clothes and appliances). Little investment meant that no entrepreneurial activity followed, as basic needs were first satisfied (Cecil & Ebanks, 1992). Rather later, seasonal migration from Jamaica brought benefits to home communities through greater capitalisation of farms, the transfer of new seeds and skills, and the ability to hire labour, enabling agricultural expansion. In considerable contrast, on the much smaller island of St. Vincent, where remittances took the same pattern as in the Pacific (Connell & Conway, 2000), purchase of land was of greater significance, enabling people to establish better housing, focus on less labour-intensive crops or simply abandon agriculture (Thomas-Hope, 2017). With few differences, these circumstances mirrored accounts of this century's programmes, whether in Australia, New Zealand or Canada. Poor working conditions, dissatisfaction with pay, perceptions of exploitation, a culture of fear and intimidation, job insecurity and a feeling of exclusion and alienation characterised life and working conditions (Beckford, 2016). However, the 'farmworker programme' for Jamaicans, whether in Canada or the United States, had 'long been the best—and often the only—way to earn the cash that made a difference between a decent life and desperation' (Hahamovich, 2011: 12). Workers returned repeatedly to try and achieve their goals; few succeeded in moving out of poverty and the main beneficiaries were American farmers.

By the 2020s, about 5000 Jamaican agricultural workers travelled to the United States each year, facilitated by the government and increasingly by private employment agencies, dominating the contemporary Caribbean market. In 2022, approximately 12,000 Jamaicans obtained employment through the US H-2B programme (increasingly focused on hospitality and construction employment rather than agriculture). Unlike other SIDS, Jamaica remained anxious to participate in the programmes. As a 2024 media release proclaimed,

> In recognizing the value of the programme … Minister [of Economic Growth and Job Creation] Matthew Samuda said "We want to ensure that Jamaica's presence in the US benefits both Jamaica and the United States". He noted that "many Jamaicans have been able to provide opportunities for their children, because of the programme, and many US farmers owe their success to the contribution of the Jamaican worker" [hence] the government was taking definitive steps to ensure that participating workers are properly oriented, noting that "they are our ambassadors, and so we must ensure that they know what is expected of them, what they are to expect, and how to conduct themselves". Minister Samuda praised the US Government for its support of the programme, noting that the relationship reflects a "good partnership, based on friendship and mutual respect". He however empha-

> sized that Jamaica stands ready to supply significantly more workers. The head of the US delegation, Deputy Assistant Secretary for Visa Services, Julie Stufft, confirmed that Jamaican workers were highly valued by the US government … noting that "Jamaica has a product which everyone wants".

The language and the reduction of workers to a 'product' were analogous to parallel contexts in Pacific SIDS. Migrants were expected and urged to work hard and be respectful representatives of the sending countries, to ensure that opportunities remained for others in later years. Despite continued and still growing Jamaican interest, seasonal migration was numerically much less than in the Pacific and of lesser importance in development.

9.3.2 Pacific

Seasonal worker schemes were introduced in New Zealand in the 1960s, but New Zealand faced substantial overstayer problems and the schemes were short-lived. Matching global trends, a new Recognised Seasonal Employer (RSE) scheme began in 2007 and was followed in 2008 by a similar Australian Seasonal Worker Programme (SWP) scheme, both centred on the employment of agricultural workers for up to seven months in a year and both modelled on SAWP.

New Zealand allowed up to 9500 migrant workers in a single year, and although Australia had no cap, numbers were consistently lower. Most Pacific SIDS have been involved. Samoa, Tonga and Vanuatu have taken the greatest advantage of them, whereas Tuvalu and Kiribati have only been marginally involved, disadvantaged by distance and transport costs, unfamiliarity and the difficulty of obtaining visas and passports, with potential workers lacking some basic skills, English language proficiency and the ability to develop workplace links necessary to smooth migration (Bedford et al., 2010, 2017). After initially low participation rates, numbers grew steadily until 2020, when COVID-19 produced a temporary halt.

After the pandemic, Australia's SWP scheme extended labour mobility opportunities into longer-term, non-seasonal employment in semi-skilled industries including hospitality, tourism, aged care, labour hire and meat processing, with the scheme renamed the Pacific Australia Labour Mobility (PALM) scheme. Contracts in these industries expanded to up to four years in length, enabling substantially increased earnings (Petrou & Connell, 2023). Within a year, workers had been deployed in tourism (especially in Great Barrier Reef hotels), caring, forestry, fisheries and meatpacking. This new diversity of occupations and substantial shortages in both New Zealand and Australia contributed to a skill drain of some consequence (see below) as the extension of seasonal agricultural labour mobility programmes into industry (especially abattoirs) and various service sectors (especially caregiving) significantly increased interest in longer-term overseas employment. Vanuatu became the largest supplier of PALM workers, while the expanded scheme continued to favour Australian employers, leaving questions over equity and uneven development.

In the Pacific that was accompanied by the growing significance of the Australia-Pacific Technical College later, significantly, renamed the Australia-Pacific Training Coalition, through which islanders acquired mainly trade skills of use both in the PICs and in Australia. That renaming, indicative of the demand for migration opportunities in the PICs, became further evident when demand within the PALM scheme fell in 2024 (as less regulated backpackers—working-holiday makers—returned in significant numbers) so that many workers applied for asylum in order to remain in Australia. After several Asian states, Vanuatu and Tonga were major sources of claimants, with some 10% of PALM workers taking this step. Though unlikely to be successful, it enabled applicants to remain in Australia while their claims were processed and be employed outside the regulated PALM system (SMH, 22 September 2024). It was one unusual measure of the considerable worker interest in the scheme as a means to further migration.

9.4 Socio-Economic Costs and Benefits

Migration has had massive economic significance, whether permanent or temporary. Seasonal agricultural work programmes benefitted individuals, communities and national economies—the widely touted 'triple win'—but not without costs and disadvantages. Benefits are often seen to be centred on remittances. In Jamaica, remittances from seasonal employment amount to over J$1 billion (US$6.2 million) annually. Men make up 92% of the participants in US agriculture whereas women accounted for 52% of those in the US hospitality programme (Government of Jamaica 2022). Men initially dominated the Australian and New Zealand agricultural schemes until women entered the packing sheds, and the caring and hospitality industries, moving towards a greater gender balance. In the Pacific, women's involvement was welcome since they were considered to remit more frequently and generously.

Almost immediately, seasonal migrants became well placed to contribute financially to development at home through their familiarity with needs. Mobile phones resulted in greater pressure on migrants to remit, to the extent that some became overwhelmed with requests and developed phone-call avoidance strategies (Hobbis, 2017). Both unskilled and skilled migrants sustained remittances at high levels, over long time periods, often more than 20 years, while skilled migrants also provided advice ('social remittances') on education, health and other matters (Connell & Brown, 2005). In many SIDS these proved basic adjuncts to household livelihood strategies (Quashie, 2019). Seasonal workers' remittances especially are a significant component of household incomes, often invaluable in meeting daily needs and expenses, including school fees, food and health care (World Bank, 2021): consumption rather than investment. Remittances have contributed to significant welfare gains—improved housing and water supplies, and better access to health and education—and to investments in land, stores and transport, and such small-scale services as bars, petrol stations and taxis. Substantial goods shipments, especially technology, solar panels and machinery, are also valued remittances. Exceptionally, as in Jamaica, remittances have contributed to food security for the relatively poor (Thomas-Hope, 2025).

Remittances settle debts, pay for airfares and school fees, make donations to churches or build homes and support businesses while fulfilling kinship obligations. They support local organisations and local needs, such as recovery from hazards. They have enabled some diversification out of agriculture, an increased flexibility of livelihood choices and over time a reduction in both poverty and inequality (Brown et al., 2014). Family remittances tend to have a greater impact on poverty reduction and welfare gains than other income flows as they reach a greater share of the population and more poor households, and remittances may constitute a valuable form of household insurance by diversifying livelihoods. They are not necessarily altruistic, but a form of insurance or social capital, maintaining land rights, to facilitate a return to the homeland in retirement or to enhance individual and household family prestige and status. Those who remitted most were those most likely to return. In remote Fijian villages like Nawairuku, these tended to be military personnel or professional rugby players—two groups who travelled overseas but were particularly likely to return—and, later, the experience of being in group contexts outside the village enabled them to become capable village leaders (Currenti et al., 2019). With now widespread smartphones, requests for remittances can be instantaneous—valuable after hazards, but not always otherwise welcome. However, despite benefiting the SDGs, migration and remittances alone are unlikely to be effective strategies for sustained economic development (Deonanan & Ramkissoon, 2018).

Most seasonal workers have specific goals that typically include investment in housing or land and establishing small businesses, though few come to fruition. Many experience too much competition and pressure from kin, so few survive. Business skills are not acquired from agricultural work, and new agricultural skills are not usually transferable to different ecological contexts. In both regions many returned workers have struggled to purchase or build a home, or establish a business, and many have returned several times to come close to their goals. Some experienced jealousy and faced land tenure challenges from other residents (Bailey et al., 2025). For many years neither Caribbean nor Pacific SIDS had programmes to assist in reintegrating seasonal workers to take advantage of any skills or capital they might have acquired. The obvious economic gains from participation in guestworker schemes were stressed in positive appraisals in exit surveys of workers, anxious to return, compared with more engaged discussions in workers' own languages which often revealed ambivalence, disappointment or frustration with living and working conditions overseas.

Social costs within the SIDS became more evident in the form of disrupted households, and a skill drain became more pronounced. Concern was attached to the impacts on households and families including substance abuse, infidelity, extramarital affairs, family breakdown and domestic violence: recurrent themes in popular accounts in the PICs (Petrou & Connell, 2025; World Bank, 2023). Repeated parental absence affected children, who may drop out of school or misbehave in the absence of parental guidance. Women who remained at home undertook more unpaid work but gained opportunities for more financial autonomy and empowerment (Petrou & Withers, 2024). Tonga benefited financially in the short term by having in 2024 more than 5000 workers overseas on seasonal schemes, about 5% of the national population, mostly between 18 and 30, but the 'exodus of its prime workforce' led to a critical shortage of labour on farms (resulting in higher food prices), in businesses and in government services, while school drop-outs, family breakups, affairs and divorces had increased. Crucially, the main income of a quarter of national households came from remittances (Siasau, 2024). In the Caribbean too, many children—traditionally known as 'barrel children'—are raised by extended family and suffer from parental absence, in a complex trade-off between income generation, female autonomy, education and social wellbeing (Crawford, 2003; Baldwin & Mortley, 2016). Economic benefits are troubled by social costs.

Seasonal migration also involved ongoing problems of unsafe working conditions, poor and expensive accommodation, social exclusion and wage theft. Other concerns surrounded labour rights, access to health care and insurance, and retaliation against workers who speak up about problems (Ornelas et al., 2021). Since labour shortages followed an inability to recruit and retain labour, due to a declining or unwilling local rural workforce, uncompetitive wages and demanding working conditions, that was unsurprising. For Jamaicans in Canada (exactly as for Pacific islanders in Australia and New Zealand),

> The migrant farmworkers participating in the farm work programmes have shared their own experiences and perspectives. These include poor living and working conditions, abuse and harassment from employers, injuries and deaths, lack of proper healthcare, and retaliation against workers who speak up about problems. These injured migrant farmworkers have been

speaking out about their real, lived experiences, not just 'negative publicity'. (*Jamaica Gleaner*, 26 July 2024)

A general perception among Jamaican workers was that they were often treated unfairly and given the hardest manual labour, believing that they were perceived as physically strong but also as resistant and so constantly under scrutiny, in comparison with Mexicans. As in Canadian and US destinations, commonly reported issues in Australia included unsuitable and overcrowded accommodation, wage theft, fluctuating work hours and overcharging of deductions for accommodation, transport and health insurance. Some workers—particularly women—experienced safety and wellbeing issues, including sexual assault, gender-based violence, illness, injuries and mental health issues (World Bank, 2023). In remote rural areas remedies were inaccessible. Opportunities and risks are unevenly distributed and, despite the needs of relatively poor countries and households, metropolitan farmers are the main beneficiaries, even as the Pacific SIDS have increasingly put pressure on both Australia and New Zealand to improve welfare provisions (Petrou & Connell, 2023, 2025). Temporary agricultural workers were denied permanency but encouraged to return every year, to take advantage of their experience, becoming 'permanently temporary', a situation that does nothing for equity and challenges the etymology of 'guest' workers. This may enhance domestic problems and raise questions over whether people are more valued as migrants or citizens and more basic questions about the use of remittances (and the supposedly more 'personal' engagement with the Pacific 'family') (Chap. 8). By contrast, brokers and agencies who selected workers and arranged placements, some of whom were local people, were key beneficiaries (Breman & van der Linden, 2024; Connell & Stilwell, 2006). Migration provides no simple development panacea, and a judicious combination of policies is needed to effectively maximise the benefits for households and countries.

As international migration became of greater significance, so remittance flows became larger, especially in comparison with aid flows, and more stable than other sources of national and household finance. Remittances compensate in the aftermath of crises and disasters, whether physical or economic, being counter-cyclical, in response to economic and environmental shocks, even increasing during COVID-19 (Le Dé et al., 2015; Le Dé & Jackson-Becerra, 2021). Growing recognition of their value as effective, informal, family-based systems of social protection for migrants' families brought institutional interest in managing flows and 'harnessing' remittances for financing investment and economic growth. At the same time, they may help to sustain a 'migration syndrome', while imposing sacrifices on migrants. The benefits have been effusively supported by PICS' governments and by Transform Aqorau, Vice-Chancellor of Solomon Islands National University, as providing, in the case of guestworkers from Solomon Islands:

> An exemplary model of mutually advantageous international collaboration. This initiative offers more than employment opportunities; it provides cultural exposure, skill enhancement and networking. For Solomon Islands, dealing with elevated unemployment, this scheme offers an immediate practical solution. It mitigates domestic joblessness by providing an external employment route and enables workers to remit substantial earnings back home, stimulating local economies and reducing poverty. Skills and experiences acquired abroad can be deployed to tackle local challenges, yielding wider societal advantages. (2023)

In practice no detailed study had been undertaken of the impact of guestworking in Solomon Islands and such an unusually positive perspective was rarely evident elsewhere. Encouraging migration reduces pressures on both source and destination governments.

9.5 Towards a Brain Drain

Over time, acquiring marketable skills has enabled migrants to gain better jobs, and migration has been of more evidently skilled migrants, such as teachers, health workers and sportsmen (and increasing numbers of women). Losses of skilled labour have become serious in several PICs, with small states losing disproportionately more skills than larger states. That has raised questions over whether migration constitutes both a brain drain of unacceptable proportions that hampers national development and a brain waste as some migrants fail to gain commensurate employment in destinations. Limited return migration has not remedied this problem, and remittances may not compensate for skill drains. Yet migration from SIDS gradually expanded, diversified and became more skilled.

Selectivity by skills and shortages elsewhere have resulted in the health and education sectors in the SIDS being particularly characterised by losses and by being the two sectors where global codes of practice have been introduced to prevent excessive and selective migration (see below). Thus in 2004, Grenada, St. Lucia, St. Kitts, Antigua, Jamaica and Fiji all had more than 50% of locally trained doctors working elsewhere (Docquier & Marfouk, 2005). More generally high proportions of graduates have migrated from the SIDS; more than 75% of graduates have left Grenada, Jamaica, St. Vincent, Trinidad, St. Kitts, Samoa and Tonga, with that proportion reaching over 85% in Jamaica (Lewis & Kirton, 2015), an indication that tertiary degrees are 'passports' to migration.

Alongside guestworker mobility, a potentially more permanent migration of others continued, made easier by growing diaspora populations that provided support and information via social media. Fiji, Vanuatu and other PICs have experienced a skill drain from public and private sectors: engineers, nurses, computer programmers, radio announcers and teachers migrated, resigned or took leave from skilled positions to pick fruit or be caregivers in Australia and New Zealand (Petrou & Connell, 2023). This skill drain from the formal sector became more important under PALM, as contracts were longer, counteracting some of the welfare gains made by households, and in its earliest days was emerging under PEV. Moreover, by 2025, many Fijians were using student visas as a migration route to Australia (FBC, 1 November 2025). By the 2020s, there was emerging evidence that because 'relatively "unskilled" people can go to work in Australia and New Zealand' and earn substantial wages, motivation for study in secondary schools had declined (Naidu, 2024). In Vanuatu at least, some youth deliberately dropped out or got expelled from schools so that they could work in the PALM and RSE schemes, a prospect more lucrative than completing education.

National shortages of workers in health (including dentistry), education, engineering, tourism and other formal sectors have emerged in many SIDS, accelerating after COVID-19, especially away from the health sector, itself a victim of COVID-19 (Connell, 2024b). That hampered the processes of governance and took a toll on both the private and public sectors, reducing the ability to achieve, manage and direct national development. As contract wages became much superior in metropolitan states, this brought a shift from earlier years, when the national goal of the scheme in the PICs was to provide opportunities for those in disadvantaged rural areas and remote islands, to an emerging bias in favour of relatively skilled urban workers. When the brain drain from Samoa became excessive after PALM, the government gave priority to unemployed individuals, with over six months verified unemployment, to reduce skilled worker losses, but some skilled workers deliberately left their jobs and waited six months, incidentally further enhancing local skilled worker shortages.

Simultaneously, international agencies were encouraging labour source countries to invest in training. As the Asian Development Bank noted, 'It's good that the governments are focusing on upskilling, making sure that the labour force has the proper education, the proper training' (quoted in *Islands Business,* January 2023, 12). However, Fiji lost more than 120 police officers to international migration in 2023 with the ratio of people to police officers increasing at a time of rising crime (*Fiji Times*, 18 December 2023). A Vanuatu Chamber of Commerce and Industry survey (Orozco & Spencer, 2023) found that half of all businesses had lost at least one employee to the labour mobility schemes in the previous 12 months, which threatened post-COVID revival. In both the Pacific and the Caribbean, workers laid off during the pandemic took advantage of the 'opportunity' to seek jobs overseas and pursue the certification required to get those jobs. Simultaneously, the most common reason for resignation was to work overseas. That was evident in the tourism industry of both regions, where wages were usually below national averages. Consequently, from Barbados to Vanuatu, many components of the tourism industry experienced labour shortages. Urban resorts in Vanuatu alone were losing skilled chefs and others (at a rate for some resorts in mid-2023, of one or two resignations per month) and replacing them at additional cost with migrants mainly from south Asia and the Philippines. The World Travel and Tourism Council observed:

> This outflow of skilled tourism workers to other countries often leads to these workers taking lower-tier jobs abroad, draining vital human capital from their home nations. Furthermore, many domestic tourism workers who shifted to other sectors during the pandemic have been reluctant to return to tourism, perceiving the industry as unstable, and too vulnerable to downturns. As a result, local tourism businesses struggle to fill vacancies with adequately skilled staff. For instance, the Fiji National Provident Fund, which operates several Fijian hotels, reported in mid-2022 that it was unable to fill approximately 30% of available positions with qualified, experienced workers. This shortage forces companies to either raise wages in an increasingly competitive market, potentially hiring from abroad, or to recruit less skilled staff, which can reduce productivity. (2024: ix)

Vanuatu created an emergency visa aimed at bringing in 1500 people, and its health care system brought nurses from Solomon Islands (Petrou & Connell, 2025). The most difficult positions to fill locally were trade occupations (including builders, electricians, plumbers and welders), retail jobs (e.g. salespeople, shopkeepers, cashiers and retail assistants) and restaurant staff (e.g. chefs, bakers, bartenders and waitstaff): some where the required skills took several years of training and workplace experience (Orozco & Spencer, 2023) and also required mentoring, management and continuity. Individual and household benefits from migration were at some cost to national development.

Skills disappear in every context—and for every occupation—but in the SIDS are often substantial and thus seen (and best documented) in education and health. Most problematic perhaps are skilled occupations in considerable demand overseas—such as air traffic controllers—where very few are needed; hence, attracting and training replacements provides a challenge. Occasionally this creates unusual problems; after the Vanuatu earthquake of 2024, there was a shortage of people with the ability to assess structural damage, slowing the task of reconstruction (RNZ, 8 January 2025). Three years after the Tongan capital was badly damaged from the 2022 tsunami, basic repairs and restoration were still incomplete since overseas migration had created a shortage of construction workers. But it is in education and health where numerical losses are often greatest, where training is crucial and where the implications for future generations and the achievement of the SDGs are severe. Losses from island bureaucracies, not always easy to classify by skillset, may also be demoralising where valuable managerial expertise and institutional knowledge disappear.

9.5.1 Health

Nurses constitute the largest distinctive skilled group in migration flows and, as in Jamaica, tend to be the more highly trained and experienced nurses rather than new Registered Nurses (Lewis & Kirton, 2015). Skilled migration has consequently become selective by gender, so that in Jamaica, once male-dominated migration has become female-dominated. WHO has listed 47 countries that are especially vulnerable where the number of nurses and midwives falls below the global average (48.6 per 10,000) and that also have low universal health care service coverage. Most are in sub-Saharan Africa, but six Pacific SIDS were listed: FSM, PNG, Samoa, Solomon Islands, Tuvalu and Vanuatu (WHO, 2023). Most SIDS have difficult and worsening health problems (Chap. 6); hence, ideally, human resources would be above the global average. That is rarely so and, despite the existence of a Global Code of Practice, intended to discourage migration and recruiting, developed countries have still recruited from SIDS (alongside ongoing voluntary migration).

The migration of nurses in the Caribbean has proved a constant concern, with the majority of Caribbean SIDS, including St. Kitts, St. Lucia, St. Vincent and Trinidad and Tobago, losing a considerable number of their professionals in health and education. At the start of the century, the Pan American Health Organization (PAHO) and CARICOM sought to encourage nurses to stay and encourage those in the diaspora to return. Identical processes went on simultaneously in the Pacific, following excessive migration from several PICs (Connell, 2009a). Most Caribbean nurses move to the United States, Canada and the United Kingdom, with some estimates suggesting that, by the 2010s, the Anglophone Caribbean region had an overall nursing vacancy rate of 40% (Jacobson, 2015). It has also been claimed that as many as 85% of Jamaica's trained nurses migrate soon after completing their training (*Jamaica Observer*, 29

November 2021): another 'passport' to migration, without even a short-term contribution. The Minister of Health and Wellness, Christopher Tufton, noted that every year Jamaica lost 500 or more nurses, most of them specialists, to larger countries:

> So beyond the registered nurses, we need nurses in critical care, intensive care unit (ICU), nephrology, post-natal care, operating theatre, and other specialised areas—those are the ones that are in really dire short supply because they get very attractive offers. In recent times those nurses have been offered significant incentives to migrate to North America [and] the United Kingdom. (*Jamaica Observer,* 23 June 2024)

Too rapid nurse turnover in the Caribbean has resulted in continuous shortages and chronic understaffing, jeopardising the ability of health systems to provide essential services and worsening patient care and outcomes (Rolle-Sands & Ingraham, 2024). Several Caribbean SIDS have waived qualifying exam requirements and quickly processed work permits for replacement migrant nurses by developing recruitment agreements with countries such as the Philippines and Nigeria. Losses were broadly similar in the Pacific, with both specialist and management losses, but attracting replacements from elsewhere was difficult, so that provision of adequate health care became inadequate (Yamamoto et al., 2012).

Nurses migrate in search of better socio-economic opportunities away from difficult local conditions: infrastructure, equipment, the availability of beds and intensive care facilities, income differentials and scarce opportunities for training and promotion. As in the Pacific, the rationale is dominated by the search for better working conditions (45%), better financial/economic opportunities (25%), limited education/training opportunities (15%) and too much bureaucracy (8%) (PAHO, 2019). While numerical losses are greatest for larger states such as Jamaica, Trinidad and Fiji, the impacts of migration of nurses are broadly similar in most SIDS (Connell, 2009a, b; Murphy et al., 2016; Sands et al., 2020), but migration can be more of a problem in smaller states, such as Dominica, St. Kitts and Tuvalu, where replacement is difficult. It is much less problematic in the more affluent Bahamas (Sands et al., 2020) which has benefited from nurse migration from other Caribbean SIDS.

Some of the strongest influences on employment and migration are only incidentally related to health care, but have much to do with attempts to improve the long-term welfare and status of families. Most skilled health workers entered the health professions less out of general altruism or a particular interest in medicine and health, but through recognition that this might be a means of maximising, or at least improving, family incomes and welfare. The predictability, availability, security and stability of jobs in health have long been strong influences on choice. For Fiji, Tonga and Samoa at least, where the propensity for international migration is considerable, half of those nurses who had migrated overseas joined the profession for income reasons compared with only a quarter (27%) of those still working in their home country. Those who claimed to have joined the profession because of its income potential were most likely to have migrated and least likely to have stayed (Brown & Connell, 2004, 2006).

For doctors and auxiliary health workers, such as radiologists and pharmacists, these 'general' reasons for migration extended into seeking more specialised knowledge, expertise and competence in modern technology and techniques, effectively constituting a dual culture of migration (Connell, 2014). Through education, and the social context of education, a more evidently medical culture of migration (extending into nursing, dentistry and allied activities) overlays and has become intertwined with an existing, more pervasive culture of migration. Both cultures are nurtured by the presence of overseas kin, social expectations of migration, training possibilities and sometimes the lure of citizenship. In SIDS, much migration and mobility is for further education and training, posing 'training conundrums': firstly, the more that skills have been developed the more they are of limited utility in small island states (Connell, 2006a, b), since workers need to be generalist and multi-skilled; secondly, the salary differential is greater for skilled workers (hence potentially increasing the propensity to migrate); thirdly, skilled workers are more likely to be bored and dissatisfied with local employment; and, fourthly, the more that training occurs overseas (as often it must, and at key times in life cycles), the greater the propensity to remain there afterwards. This is particularly true of the smallest SIDS, where training has gone beyond the needs of the relatively small-scale, flexible and generalist health care systems of SIDS. Ironically, part of the training of such skilled workers as teachers and nurses is conducted in facilities and with trainers funded by aid donations from the countries to which these graduates may eventually migrate. That is even more likely for specialist positions (Oman et al., 2006).

The migration of nurses has had a similar impact throughout the SIDS. Typically, it includes 'nurse shortages, lower quality of care, a greater burden of work on existing staff, lower staff morale, higher levels of industrial action, cancelled surgeries, merged patient care units and reduced government subsidy to train nurses' (Lewis & Kirton, 2015: 203). To that list might be added, in Fiji, less adequate supervision of and support for nurses, reduction in staff numbers (especially at the lone mental health hospital) and the closure of some remote health centres. In Trinidad at least, the youngest nurses are most likely to experience problems, less likely to cope and more likely to migrate (Tourigny et al., 2016). In Jamaica, migration has slowed the establishment of palliative health care, a relatively new and increasingly necessary branch of health care, but 'aggressive recruiting' has reduced numbers (Edwards et al., 2021).

Multiple policies have been used to discourage migration and encourage return. With health workers' education and training overseas often paid for, or subsidised by, the state, bonding has been commonly used by many SIDS to secure return. Various alternative management possibilities exist but few have been effective (Connell, 2009a, b; Blouin & Debnath, 2011; Lewis & Kirton, 2015) as health workers migrate freely. In the Caribbean the steady migration of nurses has subtly shifted towards a process of managed migration and 'training for export' rather than futile efforts to prevent excessive emigration (Lewis, 2011). International solutions, such as codes of practice, have had limited effect. As long as demand exists in the global north, where salaries and working conditions are superior, migration and critical shortages of health workers will characterise the south. Some solutions lie outside the health sector, involving whole of government responses. Although migration may be more managed, structured and slowed, it is likely to continue to be damaging in its outcomes.

9.5.2 Education

While teachers are less differentiated than health workers, the structure of migration is broadly similar, but data on migrant teachers is even more scarce. As in the health system, losses for SIDS in both regions are considerable, involving lost investment in their training and education across the entire education system. Minimally, the quality of island education has suffered. The Caribbean SIDS are major sources of migrant teachers in Canada and the United States. As in health care, the larger SIDS stand out numerically, with Jamaica among the top 13 source countries of teachers in the United States. In 2022 alone over 1500 teachers resigned their positions in Jamaica (most to migrate), leaving a vacuum and an appeal to retired teachers to help fill the gaps. A considerable number of UK work-permits for teachers are regularly issued for Jamaica, Trinidad, Barbados, St. Lucia and St. Vincent (all of which are in the top 30 sources of overseas teachers in the United Kingdom) (Higton et al., 2025).

The migration of teachers from Pacific SIDS is numerically less significant, but Fiji is a significant provider to Australia. Teachers, especially Indo-Fijians, have chosen teaching to get employment overseas, but to a lesser extent than nurses. Otherwise, the rationale for migration includes opportunities for their children, good employment and better wages, and access to better health care. Those who stayed were those with good jobs, good incomes and supportive extended families, usually in urban areas, whereas migration has created a deficit in outlying Fijian island schools. Science and computer science teachers have been most prone to migrate since they can more easily find employment overseas.

Education, and to a lesser extent health, is one sector where fluency in English is not just desirable but essential; hence, most SIDS have a comparative advantage over other potential source countries. However, PICs with minimal migration of teachers, such as Vanuatu, have poorly qualified teachers who would find immediate overseas employment impossible

(Iredale et al., 2015). The rise of PALM employment has brought migration of many teachers to take up manual employment. By 2023, Samoa had a 'chronic shortage' with teachers finding seasonal work in Australia and New Zealand, and finding better-paid jobs in other ministries or taking up theological studies (RNZ, 12 April 2023). Fiji had a teacher shortage, especially in the sciences, that had adversely affected the quality of education (RNZ, 21 February 2023). Since most are tertiary graduates, the value of loss is evident.

Teachers were more likely to be influenced by superior salaries overseas than by working conditions in their home countries, and salaries were several times those at home (George et al., 2020). As in other spheres of migration, COVID-19 provided a catalyst for more extensive migration from both the Pacific and the Caribbean (Thompson et al., 2023). Jamaican teachers in North Carolina, a popular destination, valued greater classroom resources, 'good leaders', training opportunities and the increased salary (Whittingham, 2022). Inevitably, teachers found fault with national education systems including the stress of a transition to online learning. Like nursing, voluntary migration was compounded by active recruitment efforts from the United Kingdom and the United States. Facing shortages in public schools in the early 2000s, the New York City Board of Education increased international recruitment, attracting hundreds of teachers from the Caribbean, particularly Jamaica and Barbados (Rudder, 2012). For small island states, even relatively few emigrating teachers create significant shortages (Bense, 2016). After COVID-19, more extensive migration became more damaging.

Barbados is one of the few SIDS to have developed policies designed to encourage teachers to stay, with strategies including direct engagement between the Ministry of Education and the recruiting governments, promotion of high standards and provision of incentives for local teachers, and co-operation with teachers' unions (Rudder, 2012). By contrast, in Jamaica, where teachers lost to recruitment and migration were generally the better teachers, it was concluded that stemming the flow was improbable so that international teacher recruitment might be seen rather as an 'opportunity for their people and to manage the process as efficiently as possible' (Appleton et al., 2006: 785). However, that migration has left less effective and more inequitable education systems (Lewis & Kirton, 2015). Managed migration, as in health care, has limitations.

Much as WHO developed protocols to discourage migration and recruiting, Caribbean governments mandated the Commonwealth Secretariat to develop a protocol for the ethical recruitment of teachers. The Commonwealth Teacher Recruitment Protocol responded to small Caribbean countries' concern that they were losing their teaching force to targeted recruitment drives. Commonwealth governments adopted the protocol to 'balance the rights of teachers to migrate internationally against the need to protect the integrity of national education systems, and prevent the exploitation of the scarce human resources of poor countries' (Commonwealth Secretariat, 2003). Once again, as with the migration of health workers, it was a non-binding code of conduct for countries and did not constrain individual teachers who wished to migrate. Skilled migration continues.

9.5.3 Further Skills

While teachers and nurses are numerically significant within migration streams, and their valuable role is lost to local development, many other smaller skilled groups have migrated, with flows also seemingly increasing after COVID-19 and, in the Pacific, with the institution of PALM. Public servants are prone to migrate. Many academics and administrators, who initially moved overseas on scholarships, have since moved away (Brissett, 2023). The relative lack of alternative employment in the SIDS means that innovation is not encouraged, whereas reliance on routine administration exists, with limited local mobility and promotion, and consequent low job satisfaction and frequent absenteeism (Sutton, 2007; Chap. 5). Interest in migration is the outcome of frustration and disappointment. Computer technologists have moved from Samoa, and military personnel have left Fiji, for the Gulf and Africa (Kanemasu & Molnar, 2017). Some smaller groups became locally significant when high proportions left, such as bus drivers from Fiji in 2023, disrupting public transport. An estimated 30% of all Fijian engineers left for employment in Australia and New Zealand over a 12-month period

(RNZ, 18 July 2023). Police officers have left Jamaica for the Cayman Islands.

Fijians work overseas as private security officers. Such opportunities for employment have expanded considerably in this century, and Fijians, with a long history of serving in the British Army and as peacekeepers in various UN missions, have made use of these opportunities. By mid-2005, there were over 1000 Fijians working in Iraq and Kuwait as soldiers, security guards, drivers and labourers. Another 2000 Fijian soldiers were in the British Army in 2006 (Maclellan, 2007). A proposal by the Australian government to recruit Pacific islanders into the national defence forces was greeted with interest in Fiji where the Prime Minister suggested that Fiji could provide the 5000 personnel required since their remittances would be invaluable (RNZ, 3 July 2025).

Sportsmen, notably rugby players (Stewart-Withers et al., 2017), make up a key component of the 'muscle trade' to the extent that Samoans, Fijians and Tongans make up more than a quarter of the world's professional rugby union players, giving them a distinctive ethnic presence in such countries as Japan (Schieder, 2014). In 2022 as many as 150 players of Fijian descent were playing rugby in Europe, not only in the professional leagues in the United Kingdom and France, but in lesser rugby nations, including Romania, Georgia and beyond in the United States (Robinson, 2022). Polynesians likewise dominated rugby league in England, Australia and New Zealand and are regarded as particularly prominent sources of remittances (Chap. 7).

Management of ports and shipping services in the Pacific has been disrupted where skilled workers, such as civil engineers, have migrated for better-paying jobs picking fruit, exacerbating an existing labour shortage (ADB, 2024). Construction workers including carpenters, plumbers and electricians have been lost from Jamaica, increasing the cost of construction, negatively affecting quality (Lewis & Kirton, 2015) and probably slowing progress on infrastructure. Other than in Fiji, PICs without domestic human capital struggle to attract and retain people, and local contractors cannot find skilled workers from overseas to deliver projects on time. The human resource challenge has extended to areas once thought of as adequate. In Jamaica, where murders are frequent, forensic pathologists (and police officers and prison warders) have often been in short supply. Workers in the Fijian garment industry have been particularly prone to migrate as that industry rapidly declined, but the rate was so substantial that industry representatives argued that the government both needed 'to support retention of our skilled workers rather than assisting their migration via the PALM (Pacific Labour Mobility) scheme' and needed to import 'skilled labour from other countries … to bring the industry to previous levels' (RNZ, 20 March 2025) as it was already doing for the sugar and construction industries. Skilled human resources are scarce, for example, in urban planning and management, since relevant courses do not exist in most SIDS, and local staff are not necessarily trained in relevant technical matters. Migration of workers from bureaucracies has removed experience and management expertise. Second-tier bureaucrats, sometimes described as 'foot soldiers', have limited skills or knowledge of the significance of evolving global strategies as they trickle down, and an unwillingness to take critical decisions. Tasks are not completed. Problems emanating from skill losses are usually greatest in the smallest SIDS despite fewer skilled workers being needed.

Island agriculture has been one victim of migration. Many workers have migrated from agricultural contexts to work in agriculture elsewhere, with the outcome being agricultural decline within the SIDS. By the end of the twentieth century, agriculture had virtually disappeared from dry islands like Barbuda, and from Nevis, where remittances enabled imported foods but concerns emerged over food security (Byron, 1999). In both the Caribbean and the Pacific seasonal migration has weakened domestic agriculture, since labour is lost, remittance-receiving households aspire to something 'better' and invest outside agriculture, and agricultural skills acquired overseas are rarely transferable to quite different environments (Ishemo & Semple, 2006). In Vanuatu, the large number of Ni-Vanuatu workers recruited for seasonal jobs abroad reduced the local workforce available for profitable kava cultivation, although demand for kava had reached record levels (RNZ, 2 December 2024). Similar losses have occurred in Samoa and Tonga, both prominent sources for RSE and PALM. Quite remarkably, in 2025, Fiji implemented a 'new recruitment model' to address severe labour shortages in the sugar cane industry, by bringing in workers from Bangladesh to ensure cane farming and mill operations continued without disruption, since local labour could not meet demand due to 'seasonal work abroad, internal migration, and the physically demanding nature of farm jobs' (FBC, 5 December 2025). As island agriculture declined or stagnated, migration supported its survival and development elsewhere.

9.6 Return Migration

Seasonal workers return (unless they abscond or seek asylum) but skilled workers, more easily able to get good employment overseas, are less likely to do so. As they become 'embedded' in destination societies, especially as their children attend school, and cultural memories and contacts decline, the idea of return migration is more likely to be rhetoric rather than reality. At the turn of the century as few as 10% of Samoan migrants to Australia had planned to return permanently but far fewer actually returned (Shankman, 2018). At no time during the past half century has there been substantial return migration to most islands and island states, other than of seasonal workers, with national populations increasingly concentrated on mainlands, in urban centres and in larger states. Outward flows have always exceeded return flows. The return migration of those trained and educated overseas is initially common, as those who have acquired tertiary education are bonded to return, but it is not necessarily long term, and often not beyond the duration of the bond—usually equivalent to the number of years spent away. In many islands, there is a belief that locally trained islanders, who had demonstrated commitment to local systems, should receive priority in job opportunities, hence returnees have sometimes experienced resistance or been required to start towards the bottom of the system, an obvious disincentive to return especially when additional skills and expertise have been acquired overseas. Social remittances have, however, been valuable in several contexts, whether in Trinidad or Polynesia (Conway et al., 2012; Connell, 2009a, b). Where the factors that encouraged migration remain, and social worlds are narrow, return is unattractive. Much return has been of households or individuals who have returned to care for ageing parents, or for other social reasons, contrasting with the more economic factors that encouraged migration (Gibson & McKenzie, 2011), rather than to more directly contribute to productive national development.

Nonetheless, PIC migrants especially, including skilled migrants, have sought security by remaining linked to 'anchor populations' at home, as a form of 'homeland security' for potential return migration (Rauchholz, 2012; Kolshus, 2017), where island livelihoods and cultural identity never lost significance. As COVID-19 demonstrated, where the virtues of an older order were retained and revived, they illustrated the value of an extended and culturally hybrid economy. Yet for many of those who delayed return into later years or after retirement, reintegration proves difficult.

Some returns are particularly problematic. Tongan youth are often sent from the diaspora to connect or reconnect with 'real' Tongan culture but face a challenging and often unwelcome task (Lee, 2016, 2017). In this century, miscreants have been deported from destination countries: a longstanding problem in some SIDS including Tonga, the Marshall Islands and Jamaica (Thomas-Hope, 2002; Johnson, 2012). An annual average of 29 Marshallese were deported from the United States between 2013 and 2019: there, as elsewhere, almost all were men (McNeill & Mackenzie, 2024). Their 'skills' are unwelcome, especially where they retain them in the SIDS (Chap. 6), and many have not retained cultural knowledge, language skills or family connections that would enable them to reintegrate. Many also experience social stigma. Forced deportations in the early months of Trump's second presidency added to this problem. In mid-2025, Australia signed an agreement with Nauru to deport around 280 people, who had failed to gain visas for Australia, a group that contained some serious criminals, to live amongst the local population. No SIDS had systems for suddenly dealing with numerous difficult deportees.

Skilled workers are least likely to return despite various global and national schemes encouraging this. Typically, when St. Kitts launched a new development strategy for Sustainable Economic Expansion and Diversification (SEED), its Prime Minister, Dr Terrance Dew, appealed to overseas residents:

> We need you to commit to diligent work and productivity, to embrace innovation and entrepreneurship. We need our diaspora to come home to

> bring their skills, experience, and passion back to the land of your ancestors. We need all hands on deck to seize the opportunities coming our way in construction, tourism, agriculture, and sustainability. I invite every citizen who is committed to seeing St Kitts and Nevis rise to join us in nurturing the SEED of our future. (quoted in *Jamaica Observer,* 18 October 2024)

Universally, it has proved difficult to get valuable people back, including nurses and teachers, particularly in their more productive years when they would be of most value. Yet, in an internet era, a few SIDS have gained through diaspora communities mobilising capital for small business investment, so moving beyond remittances to more productive and long-term economic contributions that invest in local ventures offer professional expertise and jointly develop projects across sectors. That has yet to become the norm.

9.7 Shifting Populations

Despite the growing significance of government intervention and regulation, throughout the SIDS migration involves individuals and households making their own decisions, usually moving either into already overstretched urban centres or to farms and cities overseas. Such moves are more permanent than the circular mobility of guestworkers in agriculture, yet under new PALM regimes, they too may be more permanent. Faltering national economies, aid fatigue and the immediate value and utility of remittances have shaped a substantial outward urge, which has drawn in SIDS hitherto less involved with international migration. Technological change has enabled islanders to be more mobile than ever, travelling over greater distances and more frequently during their lifetimes. Women have taken their place as independent migrants, but also join partners abroad, as households shift base, while women who remain in the islands have acquired skills to replace men (Boxall, 2025), challenging gendered workplace stereotypes.

People have used multiple strategies to achieve mobility: kinship, lotteries, national agreements, temporary work schemes and simply illegal overstaying—multiple ever-diversifying migration strands.

Simultaneously government interventions in favour of migration (not always with dignity) have increased, despite social costs and despite the loss of human resources. In theory remittances compensate for the skill drain (Chand et al., 2022); in practice that is rarely so, as social costs have mounted, the brain drain has increased (emphasised by training and selectivity), contributing to weaker educational institutions, particularly in the advanced sciences, where talented students leave for the global north in search of higher degrees and career opportunities. Migration from smaller SIDS especially has hollowed out populations, selectively taking skilled workers and leaders (whose authority and knowledge are lost), creating hourglass population pyramids (even resulting in the loss of carers for ageing populations) and, unlike larger states, being not easily able to find adequate high school graduates (or others) for training.

Migration is shaped by a familiar trilogy of education, employment and health care as core factors. Environmental change has contributed only to local moves, mainly away from cyclone damage or tectonic change, with climate change yet to determine migration (Connell, 2015; Allgood & McNamara, 2017; Mortreux & Barnett, 2009; Jarillo & Barnett, 2022) but that is likely to change and contribute further to a deep ambivalence over migration. Migration is one part of a wide-ranging deterritorialisation (Connell & Corbett, 2016): a shifting demographic balance, in which skilled migrants are less likely to return than the unskilled, and where national populations are contracting towards the centre at cost to the outer islands. As aspirations increase, the margins of settlement have moved inwards while expectations have gradually shifted towards the possibilities of an international future. In Nauru, economic concerns have resulted in people making 'furtive plans' and maintaining properties and bank accounts overseas for if and when migration was necessary (Morris, 2023: 251). In atoll states the possibility of migration is never far away, emotions are complex and bidirectional, and any hazard can prove a catalyst to migration.

International migration has deferred and mitigated, but not resolved, issues of poverty and development. The combination of scarce resources, weak economies, overburdened bureaucracies, urban unemployment, fractured social networks and uneven development challenges notions of sustainable development. In conventional economic terms most SIDS are likely to remain economically weak and become increasingly dependent on the wider world and require some degree of external support, partly evident in the proliferation of treaties that, mired in secrecy, suggested obsession with China and insecurity rather than security (Chap. 8). Migration thus constitutes one increasingly less hesitant solution to achieving development: an expanding and unsatisfied outward urge, a necessary bottom-up globalisation that will always be both uneven and somewhat unsatisfying. Demand for opportunities is unlikely to diminish but will continue to exceed their supply. Without substantial and improbable metropolitan changes, demand for cheap and pliable labour from SIDS will also continue, even as real and metaphorical walls are constructed. What was true previously is no less valid: 'as long as considerable economic challenges face island states, as their population growth rates remain above world averages, as development prospects are few, as the possibility of declining aid becomes more apparent, and as expectations rise, the ability to migrate will be crucial' (Connell & Brown, 2005: 55). People leave even as the new geopolitics makes these islands more important strategically: household livelihoods have more meaning than international strategies.

The outcome of international migration flows has been a substantial population of Pacific and Caribbean islanders living in metropolitan states where second, third and later generations swell these numbers: a continuing shift of demographic destiny beyond the islands. Migration is selective by age and by skill, but less so by gender: a now more obvious brain drain as skilled health workers, teachers, managers and public servants find jobs overseas, and island and institutional knowledge disappears.

Migration in search of superior livelihoods is a continuation of established past practices (Breman & van der Linden, 2024). Despite social costs, residents and governments are constantly seeking migration options, whether places, economic sectors or new categories of workers, typified by the rising demand for and supply of carers, as the world ages. Antigua and Barbuda has forged partnerships with other countries including a three-year programme with the British Crown Dependency of Jersey, for workers with hospitality experience. St. Lucia and St. Vincent are supporting recruitment drives by the British government for the enlistment of young men into the British Armed Forces. Programmes have been established by the Jamaican Ministry of Education to train hotel workers for employment in international labour markets. Despite massive skill losses, in mid-2024 Samoa was supporting the possibility of labour opportunities in American Samoa, because of its proximity and access to American dollars. Simultaneously, the Tongan ambassador in Japan was pointing to the need for foreign workers there as the population aged: 'We would support any opportunities for our people to earn money' (*Samoa Observer*, 13 July 2024). No migration programme that offered opportunities has been ignored. In various ways, migration involves transnational processes that shed light on the limited authority available to small states, and their people, in determining the rules of the mobility game. While the diversity of migration possibilities has increased, the legal rights and entitlements of migrants, especially seasonal workers, have scarcely changed, exemplifying the stratification, inequality and precarity within migration. Temporary and seasonal migrants are all too likely to be seen as foreign workers on temporary visas, easily able to be replaced, removed and denied citizenship and residential status: a new precarity. Migration is always fraught with uncertainty, notably as metropolitan living costs and housing shortages have fuelled mounting scepticism over immigration policies.

Whatever forms of migration develop, an attractive political, social and economic environment will encourage migrants to remain in contact and be involved in their home countries and islands, to return and to remit and, if feasible, to make investments there. National strategies in the PICs are slowly evolving in this direction. Migration has never been easy; there has rarely been a good time to be a migrant and global opposition to migration seems unlikely to decline imminently. Where, in colonial times, migrants were resettled within the Pacific, from Kiribati to Solomon Islands and from Tuvalu to Fiji (Connell, 2012), such choices are now improbable. Borders are hardening and legal restrictions increasing, at the same time as both greater demand and support for temporary migration, where flexible labour markets offer neither stable nor well-paid jobs, and as the politics of migration becomes increasingly sensitive and critical. By default, by design and by strategy, migration will continue to characterise SIDS.

A culture of migration, initially evident more than a century ago in localised mobility in search of improved livelihoods, suggests that the PICs especially may therefore be destined to be a form of permanent labour reserve (MacWilliam, 2022) as climate change becomes a greater threat. It is entirely fitting that this last substantive chapter ends the book where the first chapter started—with the recognition of environmental change and anticipation of its greater role in the future.

Bibliography

Abbott, K., & Pollard, S. (2020). Mired? Development planning in the Pacific. *World Development Perspectives, 20*, 100261.

Allgood, L., & McNamara, K. (2017). Climate-induced migration: Exploring perspectives in Kiribati. *Singapore Journal of Tropical Geography, 38*, 370–385.

Appleton, S., Morgan, W., & Sives, A. (2006). Should teachers stay at home? The impact of international teacher mobility. *Journal of International Development, 18*(6), 771–786.

Asian Development Bank. (2024). *Assessing and addressing risks in Pacific construction supply chains*. ADB.

Bailey, R., Bumseng, P., & Carlot, G. (2025). Development for whom? Land ownership for Vanuatu's returning migrants. *ANU DevPolicyBlog*, 10 September.

Baldwin, A., & Mortley, N. (2016). Reassessing Caribbean migration: Love, power and (re)building in the diaspora. *Journal of International Women's Studies, 17*(3), 164–176.

Barcham, M., Scheyvens, R., & Overton, J. (2009). New Polynesian triangle: Rethinking Polynesian migration and development in the Pacific. *Asia Pacific Viewpoint, 50*(3), 322–337.

Bautista, L. (2010). *Steadfast movement around Micronesia: Satowan enlargements beyond Micronesia*. Lexington.

Beckford, C. (2016). The experiences of Caribbean migrant farmworkers in Ontario, Canada. *Social and Economic Studies, 65*(1), 153–168.

Bedford, C., Bedford, R., & Ho, E. (2010). Engaging with the Recognised Seasonal Employer (RSE) work policy: The case of Tuvalu. *Asian and Pacific Migration Journal, 19*, 421–445.

Bedford, R., Bedford, C., Wall, J., & Young, M. (2017). Managed temporary labour migration of Pacific Islanders to Australia and New Zealand in the early twenty-first century. *Australian Geographer, 48*(1), 37–57.

Bense, K. (2016). International teacher mobility and migration: A review and synthesis of the current empirical research and literature. *Educational Research Review, 17*, 37–49.

Bertram, G., & Watters, R. (1985). The MIRAB economy in South Pacific microstates. *Pacific Viewpoint, 26*, 497–520.

Blouin, C., & Debnath, P. (2011). CARICOM return migration and brain circulation: Case study of Caribbean born nurses. *Canadian Foreign Policy Journal, 17*(2), 101–114.

Bordner, A., Ferguson, C., & Ortolano, L. (2020). Colonial dynamics limit climate adaptation in Oceania: Perspectives from the Marshall Islands. *Global Environmental Change, 61*, 10205.

Boxall, S. (2025). What about women? An overlooked solution to Pacific labour shortages. *ANU DevPolicyBlog*, 6 May.

Breman, J., & van der Linden, M. (2024). Migration: A view from below. *New Left Review, 148*, 145–159.

Brissett, N. (2023). 'I left because…': Caribbean high-skilled emigrants' reasons to migrate to the US. *Compare, 53*(2), 235–252.

Brown, R., & Connell, J. (2004). The migration of doctors and nurses from South Pacific island nations. *Social Science and Medicine, 58*, 2193–2210.

Brown, R., & Connell, J. (2006). Occupation-specific analysis of migration and remittance behaviour: Pacific island nurses in Australia and New Zealand. *Asia-Pacific Viewpoint, 47*, 133–148.

Brown, R., Connell, J., & Jimenez-Soto, E. (2014). Migrants' remittances, poverty and social protection in the South Pacific: Fiji and Tonga. *Population Space and Place, 20*(5), 434–454.

Byron, M. (1999). Migration as a way of life: Nevis and the post-war labour movement to Britain. In R. King & J. Connell (Eds.), *Small worlds, global lives. Islands and migration* (pp. 115–136). Pinter.

Cecil, R., & Ebanks, G. (1992). The Caribbean migrant farm worker programme in Ontario: Seasonal expansion of West Indian economic spaces. *International Migration, 30*(1), 19–37.

Chand, S., Clemens, M., & Dempster, H. (2022). Wage arbitrage through skilled emigration: Evidence from the Pacific Islands. *Asia & the Pacific Policy Studies, 9*(3), 430–446.

Coloma, R. (2025, June). 'No one wants to leave': Marshallese communities push back against the migration narrative. *Pacific Island Times.*
Commonwealth Secretariat. (2003). *A protocol for the recruitment of commonwealth teachers*. Commonwealth Secretariat.
Connell, J. (2006a). Nauru: The first Pacific failed state. *The Round Table, 95*, 47–63.
Connell, J. (2006b). "The taste of paradise": Selling Fiji and FIJI Water. *Asia Pacific Viewpoint, 47*, 342–350.
Connell, J. (2009a). *The global care chain. From the Pacific to the world*. Routledge.
Connell, J. (2009b). Bittersweet home? Return migration and health work in Polynesia. In D. Conway & R. Potter (Eds.), *Return migration of the next generation. 21st century transnational mobility* (pp. 139–160). Ashgate.
Connell, J. (2012). Population resettlement in the Pacific: Lessons from a hazardous history. *Australian Geographer, 43*(2), 127–142.
Connell, J. (2013). *Islands at risk: Environments, economies and contemporary change*. Edward Elgar.
Connell, J. (2014). The two cultures of health worker migration: A Pacific perspective. *Social Science and Medicine, 116*, 73–81.
Connell, J. (2015). Vulnerable islands: Climate change, tectonic change, and changing livelihoods in the Western Pacific. *The Contemporary Pacific, 27*(1), 1–36.
Connell, J. (2024). Culture, COVID-19, cyclones and supply chains: The contraction of tourism in Vanuatu. In G. Baldacchino (Ed.), *Archipelago tourism after the pandemic* (pp. 137–152). Routledge.
Connell, J., & Brown, R. (2005). *Remittances in the Pacific: An overview*. Asian Development Bank.
Connell, J., & Conway, D. (2000). Migration and remittances in island microstates: A comparative perspective on the South Pacific and the Caribbean. *International Journal of Urban and Regional Research, 24*, 52–78.
Connell, J., & Corbett, J. (2016). Deterritorialisation: Reconceptualising development in the Pacific Islands. *Global Society, 30*(4), 583–604.
Connell, J., & Stilwell, B. (2006). Merchants of medical care: Recruiting agencies in the global health care chain. In C. Kuptsch (Ed.), *Merchants of labour* (pp. 239–253). ILO.
Conway, D., Potter, R., & St Bernard, G. (2012). Diaspora return of transnational migrants to Trinidad and Tobago: The additional contributions of social remittances. *International Development Planning Review, 34*(2), 189–209.

Cooper, T. (2020). Hurricane Dorian: A case for building comprehensive climate change resilience frameworks for small island developing states in the Caribbean. In S. Rolle, J. Minnis, & I. Bethell-Bennett (Eds.), *Tourism development, governance and sustainability in the Bahamas* (pp. 201–215). Routledge.

Crawford, C. (2003). Sending love in a Barrel: The making of transnational families in Canada. *Canadian Woman Studies/les cahiers de la femme, 22*(3–4), 104–109.

Currenti, R., Pearce, T., Salabogi, T., Vuli, L., et al. (2019). Adaptation to climate change in an interior Pacific Island village: A case study of Nawairuku, Ra, Fiji. *Human Ecology, 47*, 65–80.

D'Arcy, P. (2006). *The people of the Sea: Environment, identity and history in Oceania*. University of Hawai'i Press.

Deonanan, R., & Ramkissoon, B. (2018). Remittances and economic development: Evidence from the Caribbean. *Social and Economic Studies, 67*(2/3), 95–132.

Docquier, F., & Marfouk, A. (2005). International migration by educational attainment 1990-2000. In C. Ozden & M. Schiff (Eds.), *International migration, brain drain and remittances* (pp. 151–199). Palgrave Macmillan.

Dreher, T., & Voyer, M. (2015). Climate refugees or migrants? Contesting media frames on climate justice in the Pacific. *Environmental Communication, 9*(1), 58–76.

Edwards, R., Patrician, P., Bakitas, M., & Markaki, A. (2021). Palliative care integration: A critical review of nurse migration effects in Jamaica. *BMC Palliative Care, 20*, 155.

Farbotko, C., & Lazrus, H. (2012). The first climate refugees? Contesting global narratives of climate change in Tuvalu. *Global Environmental Change, 22*(2), 382–390.

George, G., Rhodes, B., & Laptiste, C. (2020). Estimating the financial incentive for Caribbean teachers to migrate: An analysis of salary differentials using Purchasing Power Parity (PPP). *Journal of Education & Development in the Caribbean, 19*(2), *1–19*.

Gibson, J., & McKenzie, D. (2011). The microeconomic determinants of emigration and return migration of the best and brightest: Evidence from the Pacific. *Journal of Development Economics, 95*(1), 18–29.

Gilroy, P. (2013). *There ain't no black in the union jack*. Routledge.

Hahamovich, C. (2011). *No man's land. Jamaican guestworkers in America and the global history of deportable labor*. Princeton University Press.

Hanlon, D. (2018). A different historiography for "A Handful of Chickpeas Flung over the Sea". Approaching the federated states of Micronesia's Deeper Past. In W. Anderson, M. Johnson, & B. Brookes (Eds.), *Pacific futures. Past and present* (pp. 81–104). University of Hawai'i Press.

Hau'ofa, E. (1993). Our sea of islands. In E. Waddell, V. Naidu, & E. Hau'ofa (Eds.), *A new Oceania: Rediscovering our sea of islands* (pp. 2–16). USP.

Higton, J., White, Y., Nicoletti, R., et al. (2025). *International teacher recruitment*. Department of Education.

Hobbis, G. (2017). The microSDs of Solomon Islands. An offline remittance economy of digital multi-media. *Digital Culture & Society, 3*(2), 21–40.

Howes, S., & Turia, N. (2025). PEV popularity grows as Timor-Leste takes top slot. *ANU DevPolicyBlog*, 6 November.

IOM. (2012). *Mechanisms for labour migration in the Caribbean*. IOM.

Iredale, R., Voigt-Graf, C., & Khoo, S. (2015). Trends in international and internal teacher mobility in three Pacific Island Countries. *International Migration, 53*(1), 97–114.

Ishemo, A., & Semple, H. (2006). Population mobility and the survival of small farming in the Rio Grande Valley, Jamaica. *Geographical Journal, 172*(4), 318–330.

Jacobson, J. (2015). The complexities of nurse migration. *American Journal of Nursing, 115*(12), 22–23.

Jarillo, S., & Barnett, J. (2022). Repositioning the (is)land: Climate change adaptation and the atoll assemblage. *Antipode, 54*(3), 872–892.

Johnson, G (2012) Increased numbers of Marshall Islanders deported from U.S. *Marianas Variety*, 25 June.

Kanemasu, Y., & Molnar, G. (2017). Private military and security labour migration: The case of Fiji. *International Migration, 55*(4), 154–170.

Kolshus, T. (2017). Urban castaways: The precarious living of marooned islanders. In J. Taylor & H. Lee (Eds.), *Mobilities of Return: Pacific Perspectives* (pp. 171–188). ANU Press.

Le Dé, L., Gaillard, J.-C., Friesen, W., & Matautia Smith, F. (2015). Remittances in the face of disasters: A case study of rural Samoa. *Environment, Development and Sustainability, 17*(3), 653–672.

Le Dé, L., & Jackson-Becerra, J. (2021). COVID-19 and transnational remittances in Samoa: Maintaining family ties in the face of crisis. In Y. Campbell & J. Connell (Eds.), *COVID in the Islands: A comparative perspective on the Caribbean and the Pacific* (pp. 279–297). Palgrave Macmillan.

Lee, H. (2016). 'I was forced here': Perceptions of agency in second generation 'return' migration to Tonga. *Journal of Ethnic and Migration Studies, 42*(15), 2573–2588.

Lee, H. (2017). Overseas-born youth in Tongan high schools: Learning the hard life. In J. Taylor & H. Lee (Eds.), *Mobilities of return: Pacific perspectives* (pp. 75–98). ANU Press.

Lewis, P. (2011). Training nurses for export: A viable development strategy? *Social and Economic Studies, 60*, 67–104.

Lewis, P., & Kirton, C. (2015). Migration and remittances in development: A study of Jamaica. In W. Khonje (Ed.), *Migration and development. Perspectives from small states* (pp. 186–243). Commonwealth Secretariat.

Maclellan, N. (2007). Fiji, Iraq and Pacific island security. *Race and Class, 48*(3), 47–62.

McNamara, K., & Gibson, C. (2009). We do not want to leave our land': Pacific ambassadors at the United Nations resist the category of 'climate refugees. *Geoforum, 40*(3), 475–483.

McNeill, H., & Mackenzie, M. (2024). Sasa'a le fafao?: Approaches to return and reintegration of criminal deportees (returnees) into Samoa. *International Journal for Crime, Justice and Social Democracy, 13*(3), 59–70.

MacWilliam, S. (2022). "A road with no end": Making the South Pacific a permanent labour reserve. *Journal of Contemporary Asia., 52*(5), 830–854.

Morris, J. (2023). *Asylum and extraction in the Republic of Nauru*. Cornell University Press.

Mortreux, C., & Barnett, J. (2009). Climate change, migration and adaptation in Funafuti, Tuvalu. *Global Environmental Change, 19*, 105–112.

Murphy, G., MacKenzie, A., Waysome, B., et al. (2016). A mixed-methods study of health worker migration from Jamaica. *Human Resources for Health, 14*(Suppl 1), 36.

Naidu, R. (2024). The long, hard road to climate finance. *Islands Business, 50*(11), 10–14.

Neef, A., & Benge, L. (2022). Shifting responsibility and denying justice: New Zealand's contentious approach to Pacific climate mobilities. *Regional Environmental Change, 22*(1), 1–14.

Nero, K. (1997). The end of insularity. In D. Denoon & M. Meleisea (Eds.), *The Cambridge history of the Pacific Islanders* (pp. 439–467). Cambridge University Press.

Ohi, S., & Ingram, P. (2021). Preparing I-Kiribati for the future: Probing the English language research-policy nexus in Kiribati education in the South Pacific. *Asia Pacific Journal of Education, 42*(3), 478–496.

Oman, K., Bale, S., Baravilala, W., et al. (2006). Donor countries may unintentionally become major beneficiaries of their own aid: The case of postgraduate specialist training at the Fiji School of Medicine. *Fiji Medical Journal, 27*(2), 11–18.

Ornelas, I., Fung, W., Gabbard, S., & Carroll, D. (2021). *Findings from the national agricultural workers survey (NAWS) 2017–2018: A demographic and employment profile of United States farmworkers.* U.S. Department of Labor, Employment and Training Administration Research Report No. 14.

Orozco, L., & Spencer, J. (2023). *Vanuatu skills needs industry survey report.* Vanuatu Chamber of Commerce and Industry.

PAHO. (2019). *Health workers perception and migration in the Caribbean Region*. PAHO.

Paton, K., & Fairbairn-Dunlop, P. (2010). Listening to local voices: Tuvaluans respond to climate change. *Local Environment, 15*, 687–698.

Petrou, K., & Connell, J. (2023). *Pacific Islands guestworkers in Australia: The new blackbirds?* Palgrave Macmillan.

Petrou, K., & Connell, J. (2025). Guestworker schemes in Pacific island countries: Triple wins but social costs? *International Migration, 63*(5), e70085.

Petrou, K., & Withers, M. (2024). 'Sometimes men cannot do what women can': Pacific labour mobility, gender norms and social reproduction. *Global Networks, 24*, e12463.

Philpott, S. (1973). *West Indian migration. The Montserrat case*. Routledge.

Quashie, N. (2019). Social welfare and remittances to older adults in two Caribbean cities. *International Migration, 57*(5), 71–88.

Rauchholz, M. (2012). Discourses on Chuukese customary adoption, migration, and the laws of state(s). *Pacific Studies, 35*(1–2), 119–143.

Robinson, G. (2022, February 18). Why Drua will be a game-changer for Fijian rugby. *Sydney Morning Herald.*

Robinson, S. (2024). Patterns of hurricane induced displacement in The Bahamas: Building equitable resilience in small island developing states. *Climate Risk Management, 45*, 100634.

Rolle-Sands, S., & Ingraham, K. (2024). Caribbean nurse turnover: A scoping review. *International Journal of Bahamian Studies, 30*(1), 81–94.

Rudder, R. (2012). Managing teacher recruitment and migration: A case study of the Barbados experience. In J. Penson & A. Yonemura (Eds.), *Next steps in managing teacher migration* (pp. 43–45). Commonwealth Secretariat.
Sands, S., Ingraham, K., & Salami, B. (2020). Caribbean nurse migration – A scoping review. *Human Resources for Health, 18*, 19.
Schieder, D. (2014). Fiji Islander Rugby Players in Japan: Corporate rationales and migration routes. *Asia Pacific Journal of Sport and Social Science, 3*(3), 250–267.
Shankman, P. (2018). Samoan journeys: Migration, remittances and traditional gift exchange. In J. Connell & H. Lee (Eds.), *Change and continuity in the Pacific. Revisiting the region* (pp. 87–101). Routledge.
Siasau, K. (2024, October 9). Pacific labour mobility: Benefits and challenges for Tonga. *Matangi Tonga*.
Stewart-Withers, R., Sewabu, K., & Richardson, S. (2017). Rugby union driven migration as a means for sustainable livelihoods creation: A case study of iTaukei, Indigenous Fijians. *Journal of Sport for Development, 5*(9), 1–20.
Sutton, P. (2007). Public sector reform in small states: Cases from the Commonwealth Caribbean. In E. Kisanga & S. Danchie (Eds.), *Commonwealth small states. Issues and prospects* (pp. 218–235). Commonwealth Secretariat.
Thomas-Hope, E. (1993). *Explanations in Caribbean migration*. Macmillan.
Thomas-Hope, E. M. (2002). *Caribbean Migration*. University of the West Indies Press.
Thomas-Hope, E. (2017). Migration, small farming and food security in the Caribbean: Jamaica and St. Vincent and the Grenadines. *International Migration, 55*(4), 35–47.
Thomas-Hope, E. (2025). Migrant remittances, food security and inclusive social development in Jamaica. In J. Crush & S. Ramachandran (Eds.), *New directions in South-South migration* (pp. 63–82). Palgrave Macmillan.
Thompson, C., Cole, S., Moore, S., et al. (2023). Teacher migration and the Jamaican education system: Exploring push and pull factors and policy options. *Journal of Thought, 57*(3/4), 55–81.
Tourigny, L., Baba, V., & Lituchy, T. (2016). On the retention of younger nurses. *Journal of Nursing and Care, 5*, 3.
Turia, N., & Howes, S. (2025). The 2025 Pacific Engagement Visa is open- how did it go in 2014? *ANU DevPolicyBlog*, 18 August.

van der Ploeg, J., Sukulu, M., Govan, H., Minter, T., & Eriksson, H. (2020). Sinking islands, drowned logic: Climate change and community-based adaptation discourses in Solomon Islands. *Sustainability, 12*, 7225.

Western, J. (1992). *A passage to England. Barbadian Londoners Speak of Home*. UCL Press.

Whittingham, N. (2022). *Key motivating factors of worker migration: The migration of experienced Jamaican teachers*. PhD thesis, Northcentral University.

World Bank. (2021). *Pacific labor mobility, migration and remittances in times of COVID-19*. World Bank.

World Bank. (2023). *Improving outcomes of Pacific labor mobility for women, families and communities*. World Bank.

World Health Organisation. (2023). *WHO health workforce support and safeguards list*. WHO.

Yamamoto, T., Sunguya, B., Shiao, L., et al. (2012). Migration of health workers in the Pacific Islands: A bottleneck to health development. *Asia-Pacific Journal of Public Health, 24*(4), 697–709.

Yarina, E., & Takemoto, S. (2017). Interrupted atolls: Riskscapes and edge imaginaries in Tuvalu. *The Plan Journal, 2*(2), 461–495.

10

The Warm Winds of Change

> No national leader in the history of humanity has ever faced this question. Will we survive or will we disappear under the sea? I ask you all to think what it is like to be in my shoes. Stop and pause for a moment. If you were faced with the threat of the disappearance of your nation what would you do? (Tuvalu Prime Minister, Enele Sopoaga, quoted in Yarina & Takemoto, 2017: 465)

It is entirely appropriate that the last substantive chapter of this book concerned migration, and how in recent decades many residents of SIDS have migrated away, while far fewer returned. In the smallest islands, and in several SIDS, populations are falling: islanders are voting with their feet, perceiving a future beyond their home countries. Meanwhile, climate change adds extra pressures. The SIDS are surrounded by seas and oceans—whether the now inexorably rising waters of climate change, or the tuna-rich seas that are the core of a notional blue economy, at the interface between land and sea—the place of tourists, fishers and contemporary coastal squeeze: 'The marine environment is increasingly busy, changing, and a site of degradation, marginalization, injustice, contestation and conflict over declining resources and occupied spaces at local to global scales' (Bennett, 2019: 67). Power and political ecology, the weight

J. Connell, Y. Campbell, *Island Challenges*,
https://doi.org/10.1007/978-981-95-8177-1_10

of geography and history, environmental justice and equity are all salient themes in SIDS, shaping these contested seascapes and landscapes.

SIDS are no homogeneous group despite mostly sharing pressures related to precarious economies and environments, which provided the loose rationale for their initial grouping, exacerbated and emphasised as climate change took centre stage. Caribbean and Pacific SIDS differ, not necessarily between the two groups, but according to other factors such as location (and intervening opportunities), size, resource base and history. The duration and depth of colonialism, and human and physical resource bases, have shaped different pasts, presents and futures, yet, collectively, SIDS are in a marginal position in 'the uneven geographies of eco-imperial crisis' (Perry & Sealey-Huggins, 2023). When the economist-turned-politician Biman Prasad wondered, 'Why Fiji is not the "Mauritius" of the Pacific?' and suggested that this was partly the result of a 'lack of political stability and inclusive economic and political institutions as the major causes of poor economic performance for the last 25 years' (2014: 468), that pointed to critical institutional factors, which have acquired renewed prescience (Acemoglu et al., 2014). Management capacity and expertise could be defining factors, but they did not, however, explain why such variations had occurred. Implausible dreams remain. In 2017, the incoming President of Kiribati, Taneti Maamau, outlined his vision, based on fisheries and ecotourism, for a 'quantum leap' in development that would make the country 'the Singapore or Dubai of the Pacific' (RNZ, 15 March 2017). But Singapore's colonial history, the oil wealth of Dubai and the strategic location of both were vital. Some SIDS, notably archipelagic Tuvalu, whose remoteness and isolation had once seemed to exclude modernist development, belatedly discovered that these factors had acquired geopolitical value, and smallness might even be a position of strength (Bishop et al., 2025). Manifestly, differences are considerable, and 'a rising tide that lifts all boats' is an inaccurate and unfortunate metaphor. But the SIDS have become a formal group, coming together for perceived and real collective benefits, primarily to tackle environmental problems. SIDS bear a disproportionate burden of the hazardous consequences of the climate crisis, despite producing a tiny fraction of global emissions, but that crisis has given them unity.

Both the Caribbean and Pacific regions are changing rapidly: new patterns of temporary or permanent migration, more beneficial fishing agreements, debilitating hazards and ambitious treaties. But metropolitan interests fluctuate. Aid fatigue has negative impacts, just as the USAID closure has provoked a 'silent emergency' and a cascading crisis, stark reminders of the fragility of global health programmes (Bhattarai & Sigdel, 2025). A widespread shift away from agriculture has occurred despite pockets of intensification. Fisheries and deep-sea mining are real or potential components of an anticipated blue economy that has gained popularity in national, regional and donor circles. In the Pacific, Forum leaders launched *The Blue Pacific* in 2017 'as a new narrative that calls for inspired leadership and a long-term Forum foreign policy commitment to act as one "Blue Continent"' (Keen et al., 2018). That suggested a vision, if not always agency or strategy, for overcoming structural challenges, while cooperation, technological change and intensified geopolitics meant that SIDS were no longer isolated 'islands in a far sea'.

Achieving sustainable development, and the SDGs, is as challenging now as at any time in the past. Grand conclusions are impossible. The Caribbean scholar, Michelle Scobie, suggested that to assess environmental governance alone requires reflections on 'issues of social justice, equity, burden sharing, environmental norms, complexity, transformative change, power, scale, resilience etc' (2019: 6). The etc. is particularly daunting. A wider perspective on development would require even more variables. The SIDS, more than most, have experienced a global litany of difficulties, not least a cavalcade of unusually severe storms. Ocean temperatures and acidity are increasing. Fish are moving to cooler waters. Transnational crime, cyberwarfare, drugs and geopolitical (in)security add to future challenges. Ongoing outbreaks of dengue fever and zoonotic diseases, like bird flu, are reminders that pandemics may never be far away. These, too, should be assessed.

It is ominous that COVID-19 supposedly challenged the world to rethink development (Leach et al., 2021; Connell & Campbell, 2021; Granderson & Leotaud, 2021; Foley et al., 2022; DeShong et al., 2023) but that it changed remarkably little, whereas President Trump's inchoate insurgent populism demanded a quite different rethinking, with the abandonment of the rules-based order for one of strategic chaos where old certainties no longer applied. Trump's retreat from the Paris Accord, despite warming being more evidently heating, fostered distrust, cost the United States influence in the Caribbean and the Pacific, frustrated global ambitions to meet the existential problem of climate change and raised uncertainty about the future of multilateralism and international cooperation. Here SIDS are poorly placed, since the United States is the only great power bordering both the Caribbean and the Pacific and with dependent territories in both.

Meanwhile, new challenges have generated a mixture of division, denial and delay. COP29 achieved too little over any new climate finance goal, as developed countries procrastinated, much as they did later over global limits to plastic production and marine emissions. The 2025 COP30 in Brazil brought further frustration to the SIDS, with no breakthroughs on curbing emissions or fossil fuels, merely promises of more income for climate action. Pacific SIDS were particularly disappointed when Türkiye was chosen to host the next COP rather than Australia, which would have focused on their circumstances and put pressure on Australia to reduce fossil fuel mining and exporting. More positively COP30 provided a continued context for multilateralism and for the Global South to demand reparations and force wealthy states to negotiate openly (albeit with the absence of the United States). It again focused attention on how securing appropriate regional and international interdependence has always been difficult for SIDS. As the Boe Declaration had suggested for the Pacific, the notion of militarised national security was downplayed, to focus on a more local scale involving non-military threats such as global pandemics, climate change, food, water and energy insecurity, and 'natural' disasters, which threaten the livelihoods and environments of communities, so moving the focus from the survival of the state to the empowerment and protection of people and their environment. Flooded islands and cyclone damage inevitably capture headlines, so that, at a global scale, debates and discussions centre on climate change: arguably, other problems from overfishing to NCDs, drug abuse and food security affect ordinary livelihoods and ordinary security and, underpinned by fluctuating modes of income generation, are more imminent if less devastating.

10.1 A Climate of Concern

Any conclusion to a book addressing the future of SIDS must therefore return to climate change, which has propelled and provoked such small states as Tuvalu into global consciousness, partly through manipulated 'extinction narratives', in a way that could not have occurred with more obviously economic concerns. According to the *Lancet*, the greatest 'global health threat of the twenty-first century' is climate change, from the combined effects of malnutrition, floods, droughts and other disasters, rising seas and civil conflicts. Climate change claims the moral high ground. Indeed, it is a frequent rallying cry, as expressed at CHOGM in Samoa 2024 by Ralph Regenvanu, then Vanuatu's Special Envoy for Climate Change and Environment:

> The science and the position of Pacific governments is clear: we need to end fossil fuel expansion and support [for] us in transitioning to a Fossil Fuel-Free Pacific. This is why Vanuatu was the first country to call for a Fossil Fuel Non-Proliferation Treaty, so that we can manage a fast and fair global transition away from fossil fuels. … As a Commonwealth family, we look to those that dominate fossil fuel production in the Commonwealth to stop the expansion of fossil fuels in order to protect what we love and hold dear here in the Pacific.

Or even more pointedly by Fatumanava Pa'olelei Luteru, the Chair of AOSIS that first proposed the 1.5 °C target, who said in 2024: 'Our islands are quite literally sinking as the temperatures rise. The lack of ambition on climate change from bigger countries is consigning our states to a reality of devastating loss'. Climate change is increasingly evident; heat islands, flooded islands and drought-stricken islands are consequences.

Climate impacts are unevenly distributed. Poorer SIDS, and poorer people in settlements within the growing cities, are more exposed and less able to adapt, contributing to poverty and insecurity and, in turn, reducing the resilience of vulnerable populations, including women, children, older persons and people with disabilities. Environmental justice is least evident here. As Mia Mottley, the Prime Minister of Barbados, argued in response to Donald Trump's denial of climate change:

> If I can't live because I can't farm because I don't have access to water, or floods are now coming with an intensity and a regularity that makes it impossible for me to sustain my way of life, I'm going to shift where I'm living from. Or if I have no ability to access insurance, and insurance is critical to my ability to get a loan, I'm going to have to move from where I'm operating. So the volume of climate migration hopefully will wake up those who have been slow to see that this must be a win-win. (quoted in *The Guardian*, 14 November 2024)

Response has been limited: as long as the world burns oil, gas and coal, weather extremes will intensify. As Antonio Guterres, the UN Secretary-General, stated at COP29: 'climate finance is not charity, it's an investment. Climate action is not optional, it's imperative'. Yet, as 2024 became the hottest year ever, geopolitical tensions and conflicts crowded out the international agenda, undermining the chance of cooperation between key players.

From the pollination that supports agriculture to the natural flood protection and defence against erosion provided by mangroves, biodiversity provides a foundation for the global economy, yet is easily ignored. Not all environmental changes are consequences of climate change: mismanaged urbanisation and land clearance destroy biodiversity, plastics (often drifting from elsewhere) degrade landscapes and seascapes, and ghost nets entangle marine species. Nowhere in SIDS, or elsewhere, is environmental management a priority, despite the threat to biodiversity, health and tourist-oriented economies. Adaptive capacity varies considerably, incorporating traditional and modern elements in the face of external shocks. Despite disappointments, SIDS have managed to mitigate loss and damage through success on the world stage, working together and speaking with one voice, and have been at the forefront of action on climate change, forcing UN interest and action, gaining a reputation for emphasising the science and being instrumental in securing the Kyoto Protocol, the Paris Agreement and the High Seas Treaty. SIDS have made strategic use of a rhetoric of vulnerability, pointing to cyclones, devastating floods, rising seas, dying reefs and ocean acidification as tangible and immediate threats, alongside a reality of resilience, mitigation and preparation.

10.2 Beyond a Blue Economy

An emergent economic focus, partly rhetorical, proclaims a blue economy: a marine future at the interface between land and sea. Fisheries undoubtedly hold out hope and the PNA has been rightly celebrated for the sustainable management of tuna resources. Pacific states especially have untapped marine resources, in their EEZs and beyond in the high seas, where it is crucial they be properly managed, but the Caribbean faces greater challenges from illegal fishing and overfishing. Literally and metaphorically trawling the Pacific seabed for valuable nodules requires careful surveillance in the alluring 'final frontier'.

Meanwhile SDG15 (Life on Land) has taken a back seat in the rush to a blue economy. Food security requires diversification of food supply, and increased local production, to make rural communities more resilient to climate change and other shocks. Few in the SIDS, whether households or governments, see agriculture—especially subsistence agriculture—as part of a vibrant future. Commercial agriculture has similarly suffered, notably through loss of preferential access. Land-based economic strategies are crucial however much economies of scale and lost trade protection have discouraged local production ('legacy' exports have fast disappeared: bananas have gone, copra has almost gone and once mighty sugar struggles to survive). Niches, like kava, threatened by intervening opportunities, nonetheless have potential. However, many islanders believe that the only reliable route out of poverty is off the land and into better-paying urban jobs. COVID-19 forcibly reminded people of the virtues of home gardens and local foods, but food preferences, commercial marketing and convenience have perhaps irrevocably shifted ideas and attitudes away from what is burdensome. Land shortages, rigid land tenure, disdain, global images and ideologies have usurped local perspectives and possibilities. Soil erosion and land degradation, biodiversity loss, water shortages and lack of government priority and support for an 'old-fashioned' sector are disadvantageous. Greater self-reliance will be difficult, where a century has moved in the opposite direction.

Across all these sectors, sustainability is sometimes little more than a word with a powerful ideological presence, hinting at sensitive and evolving strategies towards an always distant endpoint and being a potent slogan in branding. It can no more be achieved in SIDS than anywhere else. The challenges and uncertainties of implementation are exemplified in the tourism sector where successes might seem to be easiest and necessary. The blue economy has come ashore, symbolised by inclusive coastal resorts centred around terrestrial swimming pools, and gone offshore into cruises: both resorts and cruise liners more detached from local economies and from 'small is beautiful' sentiments. Tourism, dependent on the aspirations, practices and strategies of distant companies and travellers, can never be entirely secure. The ecology of deep-sea mining remains uncertain.

The global shift towards the SDGs paralleled a livelihoods approach to development that left more scope for recognition of local agency and participation. Yet the basic livelihoods approach was criticised for failure 'to engage with processes of economic globalisation, debates about politics and governance, the challenges of environmental sustainability and fundamental shifts in rural economies' (Scoones, 2009:182). Its value and challenge lay in considering the distinctiveness of localities, the complexity of social life and social processes, the role of land tenure, power and cultural values (Fabinyi et al., 2014). Economic development strategies often bypassed or ignored sustainability, yet islanders have valued local knowledge and practices, concerning health, hazard response or local production, mobilising 'traditional' and new knowledge, and respecting traditional leaders and established churches (Mudaliar et al., 2025). Decent livelihoods and incomes are vital, however distant from grand strategic visions or the rhetoric of a blue economy.

When the Fijian Prime Minister, Sitiveni Rabuka, articulated his vision of the Pacific region as an 'Ocean of Peace', however much this aligned with regional aspirations for unity, non-alignment and security, it had little to do with equity, inclusivity, sustainability and grassroots development practices (Kant et al., 2025). At the grassroots, people have different needs and interests. At the Solomon Islands General Election in mid-2024, as two Solomon Islanders observed:

> Despite all the fanfare leading up to polling day, the primary concern of ordinary Solomon Islanders was not political wrangling, but the dire state of services in the country. The healthcare system is dilapidated, road conditions and infrastructure are poor and power cuts are constant. The increased cost of living and lack of educational and job opportunities have only made daily life more difficult for residents.. … Solomon Islanders have other pressing issues to worry about, such as how to pay school fees, how to feed their families, how to get their kids to school when the river floods and how to get fuel to take an expecting mother to the nearest health centre. This is what matters most to people's lives, not diplomatic tussles between global powers. (Habru & Habru, 2024)

Much as in other countries, SIDS islanders are concerned with their children's futures, health, education, housing, water and roads, local phenomena that require politicians and development planners to respond and ensure that basic services function and are distributed equitably. But urban and elite bias, incipient class formation, scarce skilled human resources, inertia and inadequate resources can frustrate delivery. States need resilient health systems, modern education sectors, sustainable food systems and employment for school-leavers or face instability and fragility. The strategic role of education in development is crucial but often ignored, perhaps by being so obvious, yet it is a force for reducing poverty and improving health, gender equality and social stability, enabling greater equality of opportunities. Intuitively this is well understood by so many households who prioritise this in their use of remittances. Quick 'solutions', such as constituency funds, are immediate 'solutions' where slow development is needed.

While geopolitical concerns are no longer distant, they are not the concerns of the urban poor in unfashionable urban areas, settlements and garrisons, where 'non-traditional security' is least evident, but precarity is greatest. Access to such resources as education, water and energy is uneven, reasonable employment a dream, with frustrations occasionally culminating in violence. Amidst leaking water pumps, incomplete electrical wiring, costly fish markets, blackboard jungles, potholed roads and pricey petrol pumps, the future of the SIDS may be written. That depends on political systems that encourage more equitable policies, including taxation. But politicians, sometimes 'big shots' detached from their origins, self-interested and bereft of moral obligations, infrequently visit constituencies, and local power and participation are too easily muted. Apathy is one outcome, corruption another, of the disconnect between local voices and deaf and distant politicians. Trust can be weak, the result of poor governance, political processes fraught with limited transparency and accountability, undelivered promises and inconsistent performance. Yet no SIDS are without social divisions, traditional or new, cultural or economic. Violent pasts have contemporary repercussions in divisiveness. Inclusion and equity, especially across social, gender and age divides, are far from inherent in small islands (McDougall, 2005, 2016). Different 'publics' have different ideas of public interest or 'common good' and make different demands on governments. Engagement and cooperation with others, in shortening time frames, is a more demanding element of politics at any scale.

10.3 Demographic Destinies

Despite youth bulges, SIDS populations are ageing. Continued urbanisation and outmigration, temporary and permanent, are products of both aspirations and uncertainty. Outer islands have consistently lost populations. Ageing populations strain social services, health care and pension systems, especially as carers and nurses migrate to the global north (and are even educated and trained to do so). Remittances soften the blow, but selective migration puts further pressure on those at 'home', where limited skilled human resources are a constraint on development. However, transformative social protection systems that emphasise equity, provide safety nets, tackle the underlying causes of poverty and reduce vulnerability to climate change impacts are becoming more evident (Prasad, 2023). Hazards and crises of any kind divert resources from longer-term development strategies and local will to act can be stymied by migration. Development challenges require an educated population, promoting local initiatives and development, but skills and expertise have been lost overseas producing calls for capacity building and consciousness raising. Practical policies have gone on amidst perennial capacity building, drawing attention to a 'capacity chasm in climate and public policy' (Cordonnier Segger, 2022), where SIDS priorities can be overwhelmed and distorted by external interventions. Meanwhile, the culture of migration has gathered pace.

Yet, as migratory numbers and aspirations have grown, global reaction to migration has become particularly strident, highlighting the political groundswell against globalisation and immigration, marked by draconian policies towards refugees. Should these trends persist, migration to higher-income countries will decline, trapping people in deteriorating situations and reducing adaptive capacity within SIDS (McLeman, 2019). However, in this century, in a new geopolitical order, Australia and New Zealand have opened doors to seasonal workers, and Australia has offered a distinctive pathway for migrants from Tuvalu, for sound economic and strategic reasons, partly shaped by soft power. Otherwise, migration pathways and possibilities are narrowing and becoming more selective.

10.4 The Elusive SDGs?

Remarkably few studies have examined sustainability, even for particular sectors in SIDS, and multiple variables have rarely enhanced analytical clarity (Connell, 2018). Countries agreed to 17 general goals, comprising 231 unique indicators and 169 targets to be achieved by 2030. No country will meet all these objectives, just as all failed to meet their MDG predecessors, partly because some SDGs conflict, some (such as the ending of hunger) are impossibly ambitious, and some are quite vague. It is difficult to measure such indicators as inequality. By 2025, the world was meeting or making 'moderate' progress on just 35% of the 137 targets (where data were available), insufficient progress on 47% of targets and 18% of targets had regressed from the 2015 baseline. Pacific SIDS were likely to miss 90% of the 116 measurable targets and unlikely to meet the 2030 goals (Allen et al., 2021). Even so, priorities and trade-offs must be made so that, for example, reduced poverty and greater equity might legitimise environmentally destructive processes.

Data relevant to the SIDS are riddled with gaps, especially related to poverty, hunger, gender equity and, more prosaically, waste management (Min et al., 2024; Graham, 2024; Diaz-Barrera et al., 2025). Less than a third of targets have clear numerical endpoints. Fewer than half the world's countries (and very few SIDS) have been able to report internationally comparable data since 2015 on key goals such as climate action, gender equality and governance. Geographic coverage is patchy, information collection often slow, and crucial data about marginalised groups—by ethnicity or socio-economic status—are missing. Absent libraries, digital divides, language and literacy are further constraints. Quite probably inequality is on the rise, but we cannot generalise (Archer & Gibson, 2023). It is an easy truism that development is difficult to achieve without adequate data, symbolised in the title of Elgie et al.'s (2021) study of waste management in Grenada: 'You can't manage what you can't measure'. Even with good data, monitoring and evaluating sustainable development is extraordinarily difficult (Uitto et al., 2017; Persaud & Dagher, 2021; Shah & Kelman, 2025). Human resource shortages encapsulate planning and management problems, where a local scientific base to policy formation and 'evidence-based planning' may be absent, in a context where complexity of choice and disjuncture between beliefs, intent and action is considerable. Most SIDS have limited institutional capacity and a weak 'culture of governance', hindered by differences over policies, strategies and priorities, compounded by the task of moving forward on wide-ranging goals simultaneously (Scobie, 2016; Eppinga et al., 2022; Archer & Gibson, 2023), resulting in unintended outcomes and slow progress towards development objectives. As in the global north, integrating and involving the private sector and civil society, with their different values and priorities, posed further problems. Realistically, the main value of the SDGs has been their focus on positive values and objectives, affirming people's rights to self-determination, participation and agency.

Permeating multiple studies of the SIDS, including some referred to in the previous paragraphs, are pervasive arguments that 'management' (sometimes 'integrated management' or 'adaptive management') is inadequate and therefore requires 'capacity building' and superior 'governance' at various interlocking scales. While such disadvantages are far from unique to SIDS and are probably universal, despite the mantra that small states face unique challenges, they have ironically become more pointed as increases in climate financing have placed a greater burden on SIDS governments (Wilkinson et al., 2025). Typically, as in marine spatial planning, SIDS struggle with 'issues of fragmented ocean governance, insufficient funding, data limitations, and ad hoc stakeholder engagement' that require, amongst other things, addressing 'capacity challenges' including 'leadership, human and technical resource deficits, data constraints and institutional inertia' (Singh et al., 2025). That then requires 'capacity building', a term arguably with neo-colonial connections, to the exclusion of existing indigenous capacity (Wilkinson et al., 2025). Otherwise, that amounts to a hollow and vacuous ritual of blaming the victim, in SIDS where the political economy is often rooted in postcolonial inheritances such as government structures, public institutions, legal systems and forms of access to land and resources.

However, to tackle climate change adaptation and disaster risk reduction, for example, requires a 'robust governance framework' usually at three regional levels (international, national and local) and involving policy planning and regulations; transparency, accountability and equity; and democratic processes and political commitment (Oh-Seng et al., 2025), the last of which stresses the widely recognised absence of political will and therefore trust. Such invocations apply to particular sectors, from food security to solid waste management, or for necessary water management policies that

> need to be implemented as an array of complementary measures that are simultaneously dynamic, flexible and relevant to the circumstances of each island... Ultimately the operationalisation of policies, practices and tools for good governance... will depend on available financial and skilled human resources, administrative reform across a multitude of agencies, robust enforcement of legislation and the use of innovative technological tools. (Mycoo, 2018: 67)

That is no easy task, and, as Michelle Mycoo cautions, to achieve water resource development, 'inordinate political courage' may be required of political leaders to take appropriate decisions, complicated in small states by complex patron-client relationships, competing policies and programmes, and scarce human and financial resources. That consequently suggests 'capacity strengthening' rather than more hierarchical building (sometimes from scratch) that recognises, as some accounts do, that government agencies are under-resourced and continue to face a colonial legacy of inappropriate structures that are not directed to existing strengths (e.g. Bennett et al., 2025; Mauli et al., 2023). One outcome is that policy formation, implementation and regulation may lack technical support and be rolled out only slowly. Bureaucracies need to be 'equipped with the necessary tools, skills, and knowledge [so] they can perform effectively' but that requires 'more than just training' (Mauli et al., 2023). It may mean parting with remnants of colonial pasts that inhibit development.

In little more than a year, culminating in 2025, various global initiatives, in a single sector, including the Ocean Rise and Coastal Resilience Coalition, the High Ambition Coalition for a Quiet Ocean, the Ocean Tourism Pact and the One Ocean Finance Facility, were all established. While virtuous and linked to the SDGs, they put a considerable burden on SIDS, distracting declining numbers of bureaucrats from other ongoing national and local activities and initiatives. Consultants, international bureaucrats and academics (including ourselves) add to this complexity by formulating other suggestions and directions; mercifully, most will be unread and ignored. What seem important in the cool confines of Geneva, Washington or Canberra may be irrelevant in tropical Suva or Basseterre. Building institutional capacity requires sustained long-term approaches and investments of time, purpose, enthusiasm and capital, and trust in government leadership and institutions.

Reversing long-established processes is difficult and tipping points, where change becomes irreversible, are closer on islands in the face of accelerated climate change (Singh et al., 2022; McKay et al., 2022). Beyond the knowledge and information required to make local and national changes (such as the transition to solar energy), governments and households require financial support, including more effective loss and damage payments. Yet the challenge of achieving sustainability is scarcely any easier, visibly so in atoll states, as the former President of the Marshall Islands, Hilda Heine, has said:

> Continuing sea-level rise intensifies salinity levels in our water lenses limiting types of food crops that can be grown to sustain us. There is no question that the solutions are few and far between and are not coming fast enough to ease the lives and livelihoods of the vulnerable in our communities. (2024: vii)

At COP28 in 2023, the Marshall Islands representative emphatically stated, 'We will not go silently to our watery graves'. SIDS face difficult and uncertain times—as often peripheral parts of a changing world order with limited ability to influence the courses of change despite remarkable legal and legislative successes. No surprise then that Naomi Klein (2014) described the climate crisis as an 'epic case of bad historical timing' since its emergence as a political issue that demands empathy and enlightened state intervention coincided with the 'ascendancy of free market ideology' where state intervention is weak and unwelcome and empathy vanquished.

A key element of SDGs is gender equity, limiting the achievement of other SDGs where women are marginalised and unable to reach their potential, barely represented in politics and disproportionate victims of domestic violence. Endemic violence against women is seen as the most significant human-rights issue in many SIDS and shows no sign of diminishing. Here particularly, motherhood statements are prone to dominate advisory documents and are again indicative of the tasks that face small states: thus a group of UN agencies concluded:

> Countries should establish targeted recruitment programs, including gender quotas, or enhance existing ones to increase the number of women in public service roles, aiming for equitable representation relative to their proportion in the working-age population. Furthermore, countries should prioritize public administration reforms that are gender responsive, including reforms of workplace culture, tackling gender stereotypes in the workplace through awareness-raising campaigns, implementing inclusive human resource policies, including work-life balance, flexible work arrangements and childcare policies, equal pay, gender responsive recruitment, retention and promotion, etc. (UNDP, UNODC, & OHCHR, 2024: 24)

No doubt they should, just as one academic scholar suggested that to enhance adaptation to climate change, 'Caribbean SIDS should prioritize data collection, integrate social and economic considerations, emphasize adaptation co-benefits, develop robust monitoring systems, and refine cost estimation practices' (Mohan, 2025a, b: 1; see also Scobie, 2016). Indeed so, but to be confronted with such lists begs obvious questions on how small states might respond, devise and implement policies and priorities, in every arena. Once again management is tested. The history of environmental governance and women's political representation in the SIDS emphasises that progress is not guaranteed: it must be continually fought for and zealously guarded. Persistent under-representation of women in government has made the task of implementing policies that protect women and their reproductive rights and equal pay more difficult. Such constraints occurred amid challenges to broader inclusiveness of minority groups, including youth, and a global backlash including increased authoritarianism and populism.

Development policy and practice has scarcely built more resilient and equitable societies. As Stephen Howes and Sadhana Sen (2022) have pointed out: 'The Pacific is awash with regional meetings, acronyms and declarations dealing with gender equity' that have brought little real change in over a decade. Widespread perceptions of men's legitimacy as political leaders, and leaders of bureaucracies, have created 'intractable barriers to women's election' as in Tonga (Taunisila & Palmieri, 2022). As the Vanuatu leader, Jennifer Kalpokas Doan, has asked: 'How do we shift from a transactional short-termist view of leadership to a desire for truly transformative governance?' (2022), avoiding votes of no-confidence and including women. Women steered villages and households through the problems of COVID-19 when 'traditional' skills proved invaluable, even where domestic violence increased, a familiar pattern of women taking the lead in coping with hazards (McNamara et al., 2020) and the aftermaths of conflict, as in post-war Bougainville, despite their widespread exclusion from formal politics itself. There is scope for optimism here, yet reconciling politics, economics, ecology and their gender implications, and linking these to the SDGs, is far from easy.

10.5 Externalities

The speed, scale and complexity of challenges, whether climate action or geopolitics, mean that no country can go it alone: shared policies and practices offer opportunities for building on regional and global cooperation. However, it is all very well to say that '[i]ntegrating climate policies, including adaptation, disaster risk reduction and decarbonisation strategies, with an inclusive social development agenda can help governments secure investments and stage interventions that tackle climate change in a way that also reduces poverty, hunger and inequality' (ESCAP, 2024: ii) but extremely difficult to specify exactly how that might be done, by whom and with what support. In response to climate change especially, small states have proved able to work individually and collectively to exercise authority, challenge global economic and legal regimes and gain concessions, policy shifts and at least a measure of environmental justice from much larger countries (Ruwet et al., 2023). Vision and focus on the existential threat of climate change, and on the conservation of oceans, and the visible performance of vulnerability, played key parts. Moral imperatives and negotiating cohesion in COP meetings, and ultimately through the International Court of Justice, have provoked changes of policy and practice in large countries and global bodies.

NGOs have supported SIDS through providing technical and legal advice and visibility. Cooperation has been less successful in trade negotiations and other matters such as security, enforcement of labour laws, health, human rights and gender equality. That is partly a function of how the global liberal order has positioned these issues, but also because of SIDS' own limitations stemming from their socio-cultural differences and priorities (Lindsay, 2019). While climate change commands the attention of some of the most powerful political leaders in SIDS, making it easier to secure coordination, social development is often left to public agencies and bureaucrats who are expected to advocate mainly at local and national level, but rarely on the global or regional stage.

Critics have chastised development that is dependent on aid, as others have seen aid as mere neo-colonialism: 'the Western aid-industrial complex is inherently fragile, self-serving and unpredictable' and now 'transactional, politicised and dictated by ideological swings' (Middleby, 2025) as 'the long-standing colonial grammars of development are being blurred by the rise of Chinese aid' (Sou & Day, 2025). Decades of aid and top-down approaches have weakened local resilience. International aspirations and goals—whether of companies, airlines, banks, NGOs or aid donors—have shaped many aspects of development and change, with neo-colonialism both endured and invited in. Aid itself is increasingly conditional and transactional, a diplomatic tool for pursuing donor interests rather than a means of poverty reduction or rebalancing equity, further emphasising inverse sovereignty, and a denial of agency, even more critical as it shifts towards infrastructure provision rather than service delivery. Shorn of ideology, aid has however been extremely beneficial, especially for service provision; hence, its sudden phasing out by the United States, and lesser reductions by other donors, posed problems for some SIDS.

A growing intensity of geopolitical contestation involves the digital world; the ocean's surface, depths and seabed; and the more established sectors of island economies. Too often impetuses have come from beyond and are of little or no value to island states. What form these take can never be certain: 'black swans' may arrive in the form of avian flu, global computer hacking or ever-changing American and Chinese global strategic ambitions and counter-responses. The growing complexity of external influences, from biosecurity (and seaweed infestations), through drug and gun running, to conditional aid, challenges national politics and development. Conventional development is inherently more difficult, in the Pacific SIDS especially, because of the tyranny of distance and isolation. Yet in an age of satellites, drones, digital technology and the internet, that can quickly change.

Occasionally distance is advantageous—cultures and biodiversity are preserved (both of which may benefit tourism), wars are elsewhere, pandemics are slow to arrive and plastic waste takes longer to disfigure coasts. Small states surrounded by water may seem apart from the great global shifts, but of course they are not. Their twenty-first-century economies connect them to global trends and tribulations. Islanders have shaped their own relationships with wider worlds in a multiplicity of ways. SIDS may be compromised by globalisation, evident in the exploitative incursions of distant nations into Pacific fishing stocks, in oceans where responsibilities are poorly defined and 'ocean grabbing' is rife (van Tatenhove, 2016; Mallin, 2018). While gains have been achieved, evident in the extended control of tuna through the PNA, heightened geopolitical tension between China and 'traditional' partners has coincided with an increased focus on hard infrastructure, reducing support for such areas as education and health. Progress can never be assured.

10.6 Mice that Roar, or Dolphins that Leap?

Geopolitical tensions, changes in the global economic order and most of all climate change, have led to a focus on the SIDS as critical sites of vulnerability. External perceptions have suggested the seeming inevitability of what appears 'a path dependency of poverty, high illiteracy rates, lack of expertise and limited institutional capacity' consequent on a legacy of colonialism that has constructed vulnerability and constrained adaptation (Dagher, 2019: 1041; Barnett, 2020; Perry & Sealey-Huggins, 2023; Weatherill, 2025). Small states certainly face challenges, but SIDS have been able to escape and emerge from vulnerability, despite wide-ranging asymmetries. As metropolitan states have abandoned soft power (Otis, 2025), enforced and freely chosen opportunities have led SIDS to rethink security 'from the bottom up', enhancing more equitable and sustainable local and national cultures and livelihoods (Fabinyi et al., 2025) and enabling an 'embedded security' within the deeper fabric and structure of the SIDS (Naupa, 2023; Barandon, 2025) at least partly divorced from the geopolitical conflicts of others.

Negativity has usually been the outcome of a focus on climate change, to the extent that Godfrey Baldacchino critically queried the simultaneous focus on climate change and vulnerability:

> SIDS continue to peddle a discourse of vulnerability [putting] climate change at the front and centre of international concern [but] tangible progress has been elusive and exasperating. [Yet] it might well be the single issue that holds so many SIDS together and offers them one powerful voice, in spite of glaring differences in size, natural resources and developmental

> stages. The danger here is that small state development gets associated exclusively with tackling climate change. (2022a: 240)

While that is hypercritical, it reflects both the value of unity and the need for diversity amidst climate change. Although climate change has certainly not been thwarted, a focus on its impacts obscures achievements that have been made through persistent pressure and cooperation, challenging vested interests and large and powerful countries, gaining moral and financial support, far in advance of what might have been achieved individually.

By combining forces at regional and international levels, diverse SIDS have succeeded by stressing similarities associated with vulnerability. The United Nations Convention on the Law of the Sea, ratified in 1994, turned the SIDS through AOSIS into 'large ocean states' that gave them extensive sovereign rights (Baldacchino, 2022b). The first UN Conference on the Sustainable Development of Small Island Developing States paved the way for a greater international presence and gains in other areas. Concerted pressure, focused on climate change, has brought improved, if imperfect, loss and damage compensation and finance, greater authority in the high seas and superior management and sustainable exploitation of migratory tuna resources. The Marshall Islands played a crucial role in negotiating the Paris Agreement and emissions reductions at the International Maritime Organisation, the latter through developing a particular capacity in maritime law nationally and within the Pacific region (Corbett et al., 2020). It also instigated the High Ambition Coalition and played a key role in negotiating the Paris Agreement, while Tuvalu consistently took a lead on climate matters. Most remarkably, a committed group of university students in Vanuatu took their claim for an advisory opinion on reparations and climate justice to the International Court of Justice, achieving improbable success through an innovatory approach in international law (Akhavan & Holst, 2024), so pushing forward the agenda on loss, damage and responsibility.

SIDS also challenged outdated international financial systems and aid architecture, developing and trialling innovative financial instruments, such as green bonds, that could be used to reduce, retain and transfer climate risks, so 'helping donors and international financial institutions develop better products and services for highly indebted, climate vulnerable nations' (Wilkinson, 2025: 90). Multilateralism across the Caribbean and Pacific was succeeding, with focused interventions highlighting the commitment, vision and capacity of small states, individually inhibited by size, to use informal networks and regional organisations to advance their interests and influence international affairs (Cumberbatch et al., 2024). SIDS have avoided being mere geopolitical pawns, neither 'backyards' nor 'battlegrounds'. Selectivity and focus were crucial.

Converting international political gains into a blue economy, let alone prosperous and sustainable livelihoods, faces difficulties, but the practical success of the PNA showed that working together was not merely valuable but vital. Regionalism largely failed to create more formidable blocs that might protect regional interests and achieve concessions, largely because countries had their own distinct interests, but the PNA demonstrated that capitalist incursions could be regulated and tamed by a small group of states to protect and manage the single most important renewable resource in the SIDS. It was unusual in triumphing economic matters against outsiders, suggesting a model for an expanded commitment to regional economic cooperation.

Ultimately SIDS may well be places of hope, where islanders own land, social safety nets exist and cultural identity more positive than patriarchy: 'in an age where new grand narratives of change—climate change, globalisation, and geo-political realignments—are again populating the research agendas of the region[s], it is critical that we make space and listen to local … narratives and understandings of change' (Banks & Overton, 2025: 41). Greater self-reliance can be valuable and virtuous (even if not necessarily freely chosen), land tenure and therefore 'homeland security' are invaluable, and 'traditional knowledge' contributes to coping in hazards. Even so, it is easy to say 'draw upon past knowledge and practice' (Mercer et al., 2014: 5586) but more difficult to do so in a technological age. Aspirations to greater affluence are universal; few welcome a 'return' to subsistence, and migration beyond the SIDS is attractive socially and economically. Livelihood diversification through migration and modernity has always made sense.

All economies face difficult times, within the already 'brutal tectonics of neo-liberal globalization' (Davis, 2006: 174), as the semblances of a rules-based order fade fast. While early chapters point to a litany of both disadvantage and difficulty, many challenges (not least adequate finance, management capacity and human resources) are well-nigh universal. SIDS do not face them alone. Frustrations over inequality and access have occasionally boiled over in SIDS, but respect and community are valued and the ignored private sector and civil society have a role. The final, often ignored, SDG17, 'Partnerships', is key. Successful development requires partnerships, at global, regional, national and local levels. At local level, however that is conceived, simple virtues, like fixing potholes, providing piped water and retaining biodiversity, unremarkable and unexciting, enable safety, security of livelihoods, a moral economy, safety and public confidence, inclusion and some degree of equity and economic justice.

The future of SIDS will continue to be a mix of cooperation, struggle, resistance and progress. How to sustain gains, amidst factors beyond the control of SIDs, will always remain difficult. The dominant global powers and hegemonic forces continue to write, (re)write and tear up the rules of the game, playing on ideological and cultural differences to divide and, more obviously, rule regionally and internationally. There can be no false optimism, but SIDS have coped, and can work towards achieving basic needs and service delivery, conducive to enabling people to stay in their homelands, although mobility will always be a feature of SIDS. More than ever before, SIDS demand to be treated as sovereign equals, mobilising island agency amidst increasingly dynamic—and troubling—geopolitics. Securing a future is not beyond the imagination and scope of SIDS that have shown their willingness and tenacity to work through the post-liberal order. Contemporary trends in the global north have made it clear that SIDS cannot rely on international organisations for their survival or economic and social viability in the international arena. As the state withers in transactional times, SIDS still face unusual difficulties, but the growing diversity of external interests has provided opportunities to exert, and benefit from, their strategic value. While sharks are always circling, tuna have been captured and dolphins display their potential to leap gracefully.

Bibliography

Acemoglu, D., Gallego, F., & Robinson, J. (2014). Institutions, human capital and development. *Annual Review of Economics, 6*(1), 875–912.

Akhavan, P., & Holst, R. (2024). *What are the legally binding obligations of states in respect of climate change?* IISD SDG Knowledge Hub.

Allen, C., Metternicht, G., Wiedmann, T., & Pedercini, M. (2021). Modelling national transformations to achieve the SDGs within planetary boundaries in small island developing states. *Global Sustainability, 4*, e15.

Archer, L., & Gibson, J. (2023). Policy dialogue on SDG implementation in an archipelagic nation. In W. Filho, N. Aguilar-Rivera, B. Borsari, et al. (Eds.), *SDGs in the Americas and Caribbean Region* (pp. 1301–1328). Springer Nature.

Baldacchino, G. (2022a). Review of J. Corbett, X. Yi-Chong & P. Weller (2021). *International organizations and small states: Participation, legitimacy and vulnerability* (Bristol University Press). *Small States and Territories, 5*(1), 239–240.

Baldacchino, G. (2022b). UNCLOS at 40: Small island developing states and new forays into positive sovereignty. *The Round Table, 111*(5), 598–611.

Banks, G., & Overton, J. (2025). The Pacific in transition: 50 Years on. *Asia Pacific Viewpoint, 66*, 34–42.

Barandon, E. (2025). Yumi talem, yumi mekem: How Vanuatu is shifting gears on security. *ANU DevPolicyBlog*, 28 August.

Barnett, J. (2020). Global environmental change II: Political economies of vulnerability to climate change. *Progress in Human Geography, 44*(6), 1172–1184.

Bennett, N. (2019). In political seas: Engaging with political ecology in the ocean and coastal environment. *Coastal Management, 47*(1), 67–87.

Bennett, M., March, A., Greer, R., & Failler, P. (2025). Value chain opportunities for Pacific Coastal Resources. *Sustainability, 17*, 1103.

Bhattarai, A., & Sigdel, A. (2025). The silent emergency: Awakening political decisions and a call to action for the future of global health. *Discover Health Systems, 4*, 123.

Bishop, M., Bouhia, R., Carter, G., et al. (2025). *Sustaining development in Small Islands: Climate change, geopolitical security and the permissive liberal order*. Cambridge University Press.

Connell, J. (2018). Islands: Balancing development and sustainability? *Environmental Conservation, 45*(2), 111–124.

Connell, J., & Campbell, Y. (2021). Aftermath: Towards a 'new normal'. In Y. Campbell & J. Connell (Eds.), *COVID in the islands: A comparative perspective on the Caribbean and the Pacific* (pp. 517–528). Palgrave Macmillan.

Corbett, J., Ruwet, M., Xu, Y., & Weller, P. (2020). Climate governance, policy entrepreneurs and small states: explaining policy change at the International Maritime Organisation, *Environmental Politics, 29*(5), 825–844.

Cordonnier Segger, M.-C. (2022). *Climate law and public policy innovation for the sustainable development goals: Accelerating pre-CoP27 Paris agreement compliance.* Bennett Institute for Public Policy.

Cumberbatch, J., Hinds, C., McConney, P., Speede, R., et al. (2024). Adaptive capacity, governance and small island developing states: A case study of sargassum management in the Eastern Caribbean. *Coastal Management, 52*(6), 340–358.

Dagher, R. (2019). Policy space under a constraining combination- open economies, austerity and small island states. *Third World Quarterly, 40*(6), 1040–1063.

DeShong, H., Devonish, D., Grenade, W., & Roberts, S. (2023). Afterword: Re-imagining Caribbean futures post-COVID-19. In S. Roberts, H. DeShong, W. Grenade, & D. Devonish (Eds.), *Interdisciplinary perspectives on Covid-19 and the Caribbean* (pp. 551–561). Palgrave Macmillan.

Davis, M. (2006). City of Quartz: *Excavating the future in Los Angeles*. Verso Books.

Diaz-Barrera, M., Alfaro-Aucca, C., Pacheco-Mendoza, J., et al. (2025). Bibliometric analysis of prominent topics in global scientific production on sustainable development goals in Scopus (2013–2022). *Discover Sustainability, 6*, 74.

Doan, J. (2022). Barriers to women's participation in Vanuatu. *ANU DevPolicyBlog*, 27 July.

Elgie, A., Singh, S., & Telesford, J. (2021). You can't manage what you can't measure: The potential for circularity in Grenada's waste management system. *Resources, Conservation and Recycling, 164*, 105170.

Eppinga, M., Mijts, E., & Santos, M. (2022). Ranking the sustainable development goals: Perceived sustainability priorities in small island states. *Sustainability Science, 17*, 1537–1556.

ESCAP. (2024). *People and planet: Addressing the interlinked challenges of climate change, poverty and hunger in Asia and the Pacific*. United Nations.

Fabinyi, M., Cvitanovic, C., Barclay, K., et al. (2025). Rethinking maritime security from the bottom up: Four principles to broaden perspectives and centre humans and ecosystems. *npj Ocean Sustainability, 4*(1), 29.

Fabinyi, M., Evans, L., & Foale, S. (2014). Social-ecological systems, social diversity, and power: Insights from anthropology and political ecology. *Ecology and Society, 19*(4), 28.

Foley, A., Moncado, S., Mycoo, M., et al. (2022). Small island developing states in a post-pandemic world: Challenges and opportunities for climate action. *WIREs Climate Change, 13*(3), e769.

Graham, R. (2024). The marine litter issue in the Windward Islands- a pathway to responses using the DPSIR framework. *Frontiers in Environmental Science, 11*, 1150722.

Granderson, A., & Leotaud, N. (2021). Towards a resilient, inclusive and green recovery in the Caribbean. In J. Connell & Y. Campbell (Eds.), *COVID in the Islands: A comparative perspective on the Caribbean and the Pacific* (pp. 495–515). Springer Nature.

Habru, P., & Habru, C. (2024). Will Solomon Islands' new leader stay close to China? *ANU DevPolicyBlog*, 2 May.

Heine, H. (2024). Foreword. In A. Dansie, H. Alleway, & B. Böer (Eds.), *The water, energy, and food security nexus in Asia and the Pacific: The Pacific* (pp. vii–viii). Springer.

Howes, S., & Sen, S. (2022). Time for action at the Forum on female political representation in the Pacific. *ANU DevPolicyBlog*, 8 July.

Kant, R., Wheeler, V., & Chung, M. (2025). The Pacific Ocean of peace: A promise or a paradox? *Australian Journal of International Affairs, 79*(3), 520–530.

Keen, M., Schwarz, A., & Wini-Simeon, L. (2018). Towards defining the blue economy: Practical lessons from Pacific ocean governance. *Marine Policy, 88*, 333–341.

Klein, N. (2014). *This changes everything. Capitalism vs. the climate.* Simon and Schuster.

Leach, M., MacGregor, H., Scoones, I., & Wilkinson, A. (2021). Post-pandemic transformations: How and why COVID-19 requires us to rethink development. *World Development, 138*, 105233.

Lindsay, C. (2019). Norm rejection: Why small states fail to secure special treatment in global trade politics. *Small States & Territories, 2*(1), 104–124.

McDougall, D. (2005). The unintended consequences of clarification: Development, disputing, and the dynamics of community in Ranongga, Solomon Islands. *Ethnohistory, 52*(1), 81–109.

McDougall, D. (2016). *Engaging with strangers: Love and violence in the rural Solomon Islands.* Berghahn.

McKay, D., Staal, A., Abrams, J., et al. (2022). Exceeding 1.5°C global warming could trigger multiple climate tipping points. *Science, 377*(6611), 1171–1180.

McLeman, R. (2019). International migration and climate adaptation in an era of hardening borders. *Nature Climate Change, 9*, 911–918.
McNamara, K., Clissold, R., & Westoby, R. (2020). Women's capabilities in disaster recovery and resilience must be acknowledged, utilized and supported. *Journal of Gender Studies, 30*(1), 119–125.
Mallin, F. (2018). From sea-level rise to seabed grabbing: The political economy of climate change in Kiribati. *Marine Policy, 97*, 244–252.
Mauli, S., Maelaua, J., Reeve, E., et al. (2023). Systemic capacity in food system governance in the Solomon Islands: "It's more than just training". *Sustainability, 15*(13), 10710.
Mercer, J., Kurvits, T., Kelman, I., & Mavrogenis, S. (2014). Ecosystem-based adaptation for food security in the AIMS SIDS: Integrating external and local knowledge. *Sustainability, 6*, 5566–5597.
Middleby, S. (2025, March 6). Is this the end of aid as we know it? *Interpreter.*
Min, Y., Chen, H., & Perucci, F. (2024). Data on SDGs are riddled with gaps. Citizens can help. *Nature, 633*, 279–281.
Mohan, P. (2025a). Climate change adaptation in small island developing states: Evidence from the nationally determined contributions of Caribbean States. *Regional Environmental Change, 25*(1), 32.
Mohan, P. (2025b). Climate reparations for a just response to climate change: A review of historical responsibility and future implication. *Wiley Interdisciplinary Reviews: Climate Change, 16*, e70007.
Mudaliar, C., Leach, M., Baker, K., et al. (2025). Popular political attitudes in Vanuatu: Findings of the Pacific attitudes survey. *Asia & the Pacific Policy Studies, 12*(3), e70028.
Mycoo, M. (2018). Achieving SDG 6: Water resources sustainability in Caribbean small island developing states through improved water governance. *Natural Resources Forum, 42*(1), 54–68.
Naupa, A. (2023, August 31). Cultural security in the Pacific: Why it matters for regional security. *Interpreter.*
Oh-Seng, J., Klöck, C., & Deenapanray, P. (2025). Conceptualising a governance framework for climate change adaptation and disaster risk reduction in small island developing states through a systematic review. *Sustainability, 17*(22), 9965.
Otis, G. (2025). Abandonment of soft power, neglect of climate change, and the shifting geopolitical landscape of the Pacific. *Visions and Voices, 2*(2).
Perry, K., & Sealey-Huggins, L. (2023). Racial capitalism and climate justice: White redemptive power and the uneven geographies of eco-imperial crisis. *Geoforum, 145*, 103772.

Persaud, N., & Dagher, R. (2021). *The role of monitoring and evaluation in the UN 2030 SDGs agenda*. Palgrave Macmillan.
Prasad, B. (2014). Why Fiji is not the "Mauritius" of the Pacific? Lessons for small island nations in the Pacific. *International Journal of Social Economics, 41*(6), 467–481.
Prasad, N. (2023). Past, current, and future social transformation in Pacific Island Countries. In K. Baehler (Ed.), *The Oxford handbook of governance and public management for social policy* (pp. 366–384). Oxford University Press.
Ruwet, M., Corbett, J., Yi-Chong, X., & Weller, P. (2023). Climate governance, multilateralism and small states: How the Marshall Islands rocked the boat at the International Maritime Organisation (IMO). In G. Baldacchino (Ed.), *The success of small states in international relations. Mice that roar?* (pp. 184–198). Routledge.
Scobie, M. (2016). Policy coherence in climate governance in Caribbean small island developing states. *Environmental Science & Policy, 58*, 16–28.
Scobie, M. (2019). *Global environmental governance and small states*. Edward Elgar.
Scoones, I. (2009). Livelihood perspectives and rural development. *Journal of Peasant Studies, 36*(1), 171–196.
Shah, K., & Kelman, I. (2025). Data management for sustainable development: Case analyses of small island developing states (SIDS). *Development Policy Review, 43*, e70017.
Singh, S., Huang, T., Nagabhatla, N., et al. (2022). Socio-metabolic risk and tipping points on islands. *Environmental Research Letters, 17*, 065009.
Singh, P., Linnér, B., & Singh, A. (2025). Marine spatial planning and ocean governance in small island developing states. *Regional Environmental Change, 25*(3), 91.
Sou, G., & Day, J. (2025). Postcolonial experiences of Chinese aid: Encountering and welcoming South–South aid from the middle. *Transactions of the Institute of British Geographers, 50*(2), e12693.
Taunisila, L., & Palmieri, S. (2022). *Intractable barriers to women's election in Tonga*. ANU Department of Pacific Affairs InBrief 2022/4.
Uitto, J., Kohlitz, J., & Todd, D. (2017). Evaluating sustainable development in SIDS-lessons from the Pacific and the Caribbean. In R. van den Berg, I. Naidoo, & S. Tamondong (Eds.), *Evaluation for agenda 2030: Providing evidence on progress and sustainability* (pp. 119–133). UNDP.
UNDP, UNODC, and OHCHR. (2024). *Global progress report on sustainable development goal 16 indicators. At the crossroads: Breakdown or breakthrough for peace, justice and strong institutions*. UNDP.

van Tatenhove, J. (2016). The environmental state at sea. *Environmental Politics, 25*(1), 160–179.

Weatherill, C. (2025). Colonial fantasies of invulnerability to climate change. *International Feminist Journal of Politics, 27*(1), 34–55.

Wilkinson, E. (2025). Small state innovations to strengthen financial resilience in an era of accelerating climate change. *Small States & Territories, 8*(1), 90–94.

Wilkinson, E., Manley, M., Lindsay, C., et al. (2025). *Fit for Size; rethinking capacity strengthening in small island developing states.* ODI Global.

Yarina, E., & Takemoto, S. (2017). Interrupted atolls: Riskscapes and edge imaginaries in Tuvalu. *The Plan Journal, 2*(2), 461–495.

Bibliography

Abbott, K., & Pollard, S. (2020). Mired? Development planning in the Pacific. *World Development Perspectives, 20*, 100261.

Abdulkadri, A., Cunningham-Myrie, S., Mkrtchyan, I., et al. (2021). *Addressing the adverse impacts of non-communicable diseases on the sustainable development of Caribbean countries*. ECLAC Studies No. 1000.

Abe, M., & Wang, A. (2024). Migration among the freely associated states in Micronesia: Trends, drivers, and implications. In E. Monaco & M. Abe (Eds.), *Sustainable development across Pacific Islands* (pp. 3–34). Springer.

Acemoglu, D., Gallego, F., & Robinson, J. (2014). Institutions, human capital and development. *Annual Review of Economics, 6*(1), 875–912.

Ackman, M., Hammond, D., Cooney, C., & Liu, L. (2018). *SDG16 in the Pacific. Strengthening and legitimising institutions to achieve sustainable development*. Institute for Economics and Peace.

ADB. (2023). *Invisible Hands: Enabling Pacific Women in the Informal Economy*. ADB.

Addinsall, C., Weiler, B., Scherrer, P., & Glencross, K. (2017). Agroecological tourism: Bridging conservation, food security and tourism goals to enhance smallholders' livelihoods on South Pentecost, Vanuatu. *Journal of Sustainable Tourism, 25*(8), 1100–1116.

J. Connell, Y. Campbell, *Island Challenges*,
https://doi.org/10.1007/978-981-95-8177-1

Adeoti, T., Fantini, C., Morgan, G., et al. (2020). *Infrastructure for small island developing states*. UNOPS.

Ahlgren, I., Yamada, S., & Wong, A. (2014). Rising oceans, climate change, food aid, and human rights in the Marshall Islands. *Health and Human Rights Journal, 16*(1), 69–80.

Ahmed, M., & Baroy, N. (2022). The spiralling debt crisis and what to do about it. *ANU DevPolicyBlog*, 22 April.

Akhavan, P., & Holst, R. (2024). *What are the legally binding obligations of states in respect of climate change?* IISD SDG Knowledge Hub.

Albert, S., Leon, J., Grinham, A., et al. (2016). Interactions between sea-level rise and wave exposure on reef island dynamics in the Solomon Islands. *Environmental Research Letters, 11*(5), 054011.

Albert, S., Bronen, R., Tooler, N., et al. (2018). Heading for the hills: Climate-driven community relocations in the Solomon Islands and Alaska provide insight for a 1.5°C future. *Regional Environmental Change, 18*(8), 2261–2272.

Aldrich, R., & Connell, J. (1998). *The last colonies*. Cambridge University Press.

Alexeyeff, K. (2020). Cinderella of the south seas? Virtuous victims, empowerment and other fables of development feminism. *Women's Studies International Forum, 80*, 102368.

Allen, C., Metternicht, G., Wiedmann, T., & Pedercini, M. (2021). Modelling national transformations to achieve the SDGs within planetary boundaries in small island developing states. *Global Sustainability, 4*, e15.

Allen, M. (2012). Land, identity and conflict on Guadalcanal, Solomon Islands. *Australian Geographer, 43*(2), 163–180.

Allen, M. (2018). *Resource extraction and contentious states: Mining and the politics of scale in the Pacific Islands*. Springer.

Allen, M. (2020). Framing food security in the Pacific: Resilience in Malo, Vanuatu. In J. Connell & K. Lowitt (Eds.), *Food security in small island states* (pp. 91–110). Springer.

Allen, M., & Dinnen, S. (2016). Beyond life support? Reflections on Solomon Islands after the Regional Assistance Mission. *Asia & the Pacific Policy Studies, 3*(1), 6–15.

Allen, M., & Dinnen, S. (2017). *State-building and state formation in the Western Pacific. Solomon Islands in transition?* Routledge.

Alleyne, P., & Barrow-Giles, C. (2021). An examination of the institutional governance mechanisms to promote political integrity in Barbados. In L. Briguglio, J. Byron, S. Moncada, & W. Veenendaal (Eds.), *Handbook of governance in small states* (pp. 64–76). Routledge.

Alleyne, R., Alleyne, A., Cezair, N., & Charles-Soverall, W. (2024). Sport tourism in small island developing states: A case study of the Barbados Sport Tourism Industry. In D. Van Rheenen, O. Naria, R. Melo, & C. Sobry (Eds.), *Sport tourism, island territories and sustainable development. A comparative perspective* (pp. 223–238). Springer.

Allgood, L., & McNamara, K. (2017). Climate-induced migration: Exploring perspectives in Kiribati. *Singapore Journal of Tropical Geography, 38*, 370–385.

Alofa, P., & Costello, C. (2025). Reshaping reality: Climate change's impact on Pasifika girls' lives. *ANU DevPolicyBlog*, 20 February.

Alofaituli, B. (2023). Religion and development in Samoa: Time to draw on the strength of local culture? *Development Policy Review, 41*(Suppl. 2), e12746.

Alston, M., Fuller, S., & Kwarney, N. (2025). Women and climate change in Vanuatu, Pacific Islands Region. *Gender, Place & Culture, 32*(1), 83–104.

Amankwa, M., Murruka, G., Adusei-Asante, K., et al. (2025). Culture, community, and development: Unpacking the 'Socio-Cultural Milieu Dynamics' of Solomon Islands and their influence on Australia's aid localisation. *Public Administration and Development.*

Ambrose, K., Box, C., Boxall, J., et al. (2019). Spatial trends and drivers of marine debris accumulation on shorelines in South Eleuthera, The Bahamas using citizen science. *Marine Pollution Bulletin, 142*, 145–154.

Amon, D., Palacios-Abrantes, J., Drazen, J., et al. (2023). Climate change to drive increasing overlap between Pacific tuna fisheries and emerging deep-sea mining industry. *npj Ocean Sustainability, 2*, 9. https://doi.org/10.1038/s44183-023-00016-8

Anantharajah, K. (2019). Governing climate finance in Fiji: Barriers, complexity and interconnectedness. *Sustainability, 11*(12), 3414.

Anderson, B. (2023). The costs of logging in Solomon Islands. *ANU DevPolicyBlog*, 3 August.

Anderson, I., & Irava, W. (2017). The implications of aging on the health systems of the Pacific Islands: Changes and opportunities. *Health Systems and Reform, 3*(3), 191–202.

Andrew, N., Bright, P., & de la Rua, L., et al. (2019). Coastal proximity of populations in 22 Pacific Island Countries and Territories. *PLoS One, 14*(9), e0223249.

Anthonj, C., Tracy, J., Fleming, L., et al. (2020). Geographical inequalities in drinking water in the Solomon Islands. *Science of the Total Environment, 712*, 135241.

Aporosa, S. (2019). De-mythologizing and re-branding of kava as the new 'world drug' of choice. *Drug Science, Policy and Law, 5*, 1–13.

Appleton, S., Morgan, W., & Sives, A. (2006). Should teachers stay at home? The impact of international teacher mobility. *Journal of International Development, 18*(6), 771–786.

Aqorau, T. (2019). *Fishing for success: Lessons in Pacific regionalism*. ANU Department of Pacific Affairs.

Aqorau, T. (2023). Reflecting on the 20th anniversary of RAMSI in Solomon Islands. *ANU DevPolicyBlog*, 26 July.

Aranda, A., Oxenford, H., Medina, J., et al. (2022). Widespread microplastic pollution across the Caribbean Sea confirmed using queen conch. *Marine Pollution Bulletin, 178*, 113582.

Archer, L., & Gibson, J. (2023). Policy dialogue on SDG implementation in an archipelagic nation. In W. Filho, N. Aguilar-Rivera, B. Borsari, et al. (Eds.), *SDGs in the Americas and Caribbean Region* (pp. 1301–1328). Springer Nature.

Armstrong, H., & Read, R. (2018). The impact of the 2008 global crisis on small economies in the Caribbean. *Canadian Journal of Latin American and Caribbean Studies, 43*(3), 394–416.

Armstrong, H., & Read, R. (2021). Tourism and sustainable growth in small (island) economies. In J. Roberts et al. (Eds.), *Shaping the future of small Islands* (pp. 93–108). Springer.

Asch, R., Cheung, W., & Reygondeau, G. (2018). Future marine ecosystem drivers, biodiversity and fisheries maximum catch potential in Pacific Island countries and territories under climate change. *Marine Policy, 88*, 285–294.

Asian Development Bank. (2005). *Towards a new Pacific Regionalism*. ADB.

Asian Development Bank. (2023). *Women's economic empowerment in the Pacific Region*. ADB.

Asian Development Bank. (2024). *Assessing and addressing risks in Pacific construction supply chains*. ADB.

Aswani, S., Albert, S., & Love, M. (2017). One size does not fit all: Critical insights for effective community based resource management in Melanesia. *Marine Policy, 81*, 381–391.

Aswani, S., Basurto, X., Ferse, S., et al. (2018). Marine resource management and conservation in the Anthropocene. *Environmental Conservation, 48*(2), 192–202.

Atteridge, A., & Canales, N. (2017). *Climate finance in the Pacific: An overview of flows to the region's small island developing states*. Stockholm Environmental Institute Working Paper 2017-04.

Atteridge, A., Verkuijil, C., & Dzebo, A. (2020). Nationally determined contributions (NDCs) as instruments for promoting national development agendas? An analysis of small island developing states (SIDS). *Climate Policy, 20*(4), 485–498.

Attzs, M. (2009). Natural disasters and remittances: Poverty, gender and disaster vulnerability in Caribbean SIDS. In W. Naudé, A. Santos-Paulino, & M. McGillivray (Eds.), *Vulnerability in developing countries* (pp. 249–261). United Nations University Press.

Atzori, D., Sonneveld, B., Alfarra, A., et al. (2024). Nutrition fragility in isolation: Food insecurity in small island developing states. *Food Security., 16*, 437–453.

Austin, L. (2014). Faith-based community radio and development in the South Pacific Islands. *Media International Australia, 150*(1), 114–121.

Awa, A. (2025, June). Tackling systemic inequities in the Solomon Islands Healthcare System: The race-health-governance nexus. *Visions and Voices*.

Ayson, R. (2007). The 'arc of instability' and Australia's strategic policy. *Australian Journal of International Affairs, 61*(2), 215–231.

Babinard, J., Bennett, C., Hatziolos, M., et al. (2014). Sustainably managing natural resources and the need for construction materials in Pacific island countries: The example of South Tarawa, Kiribati. *Natural Resources Forum, 38*(1), 58–66.

Bacchus, M. (2008). The education challenges facing small nation states in the increasingly competitive global economy of the twenty-first century. *Comparative Education, 44*, 127–145.

Bach, J. (2017). *Perceptions of environmental change: Nikutoru, Tabiteuea Maiaki, Kiribati*. PhD thesis, University of Montana.

Badjeck, M., Allison, E., Hall, A., & Dulvy, N. (2010). Impacts of climate variability and change on fishery-based livelihoods. *Marine Policy, 34*, 375–383.

Bailey, A., Minto-Coy, I., & Thakur, D. (2017). IT governance in E-government Implementations in the Caribbean: Key characteristics and mechanisms. In L. Rusu & G. Viscusi (Eds.), *Information technology governance in public organizations: Theory and practice* (pp. 201–227). Springer.

Bailey, R., Bumseng, P., & Carlot, G. (2025). Development for whom? Land ownership for Vanuatu's returning migrants. *ANU DevPolicyBlog*, 10 September.

Bainton, N. (2010). *The Lihir destiny. Cultural responses to mining in Melanesia*. ANU Press.

Bainton, N., & Skrzypek, E. (2022, August 3). Pacific nations are extraordinarily rich in critical minerals. But mining them may take a terrible toll. *The Conversation*.

Baird, A., Bishop, M., & Kerrigan, D. (2023). Differentiating the local impact of global drugs and weapons trafficking: How do gangs mediate 'residual violence' to sustain Trinidad's homicide boom? *Political Geography, 106*, 102966.

Baker, K. (2022). 'Copper and solar': The gendered politics of service delivery in Solomon Islands. *Third World Quarterly, 43*(4), 864–878.

Baker, K. (2023, April 18). Book review: Helpem Fren: Australia and the regional assistance mission to Solomon Islands. *Australian Outlook.*

Baldacchino, G. (2022a). Review of J. Corbett, X. Yi-Chong & P. Weller (2021). *International organizations and small states: Participation, legitimacy and vulnerability* (Bristol University Press). *Small States and Territories, 5*(1), 239–240.

Baldacchino, G. (2022b). UNCLOS at 40: Small island developing states and new forays into positive sovereignty. *The Round Table, 111*(5), 598–611.

Baldacchino, G., & Antat, S. (2023). Degrees of success: The alliance of small Island States at the 2015 Paris Climate Change Conference. In G. Baldacchino (Ed.), *The success of small states in international relations. Mice that roar?* (pp. 124–135). Routledge.

Baldwin, A., & Mortley, N. (2016). Reassessing Caribbean migration: Love, power and (re)building in the diaspora. *Journal of International Women's Studies, 17*(3), 164–176.

Ballagh, A. (2024). *ASEAN and the Pacific Islands: Pathways for Collaboration on climate and sustainable development.* Visions and Voices.

Ballard, C., McDonnell, S., & Calandra, M. (2020). Confronting the naturalness of disaster in the Pacific. *Anthropological Forum, 30*(1–2), 1–14.

Banivanua Mar, T. (2019). *Decolonisation and the Pacific: Indigenous globalisation and the ends of empire.* Cambridge University Press.

Banks, G., & Overton, J. (2025). The Pacific in transition: 50 Years on. *Asia Pacific Viewpoint, 66*, 34–42.

Barandon, E. (2025). Yumi talem, yumi mekem: How Vanuatu is shifting gears on security. *ANU DevPolicyBlog*, 28 August.

Barbara, J., & Baker, K. (2020). Addressing collective action problems in Melanesia: The Northern Islands Market Vendors' Association in Vanuatu. *Development in Practice, 30*, 994–1004.

Barbara, J., Mudaliar, C., Leach, M., et al. (2023). Popular attitudes to climate change in the Pacific: The case of Samoa. *Regional Environmental Change, 23*, 81.

Barbesgaard, M. (2018). Blue growth: Savior or ocean grabbing? *Journal of Peasant Studies, 45*(1), 130–149.

Barcham, M., Scheyvens, R., & Overton, J. (2009). New Polynesian triangle: Rethinking Polynesian migration and development in the Pacific. *Asia Pacific Viewpoint, 50*(3), 322–337.

Bardouille, N. (2018). Caribbean regionalisms in a comparative-historical perspective: The making of four regional systems. *Canadian Journal of Latin American and Caribbean Studies, 43*(2), 171–211.

Bardouille, P. (2024, July 31). Disaster response isn't keeping up with the climate crisis. *Project Syndicate.*

Barker, D. (2014). Caribbean agriculture in a period of global change: Vulnerabilities and opportunities. *Caribbean Studies, 40*(2), 41–61.

Barnett, J. (2017). The dilemmas of normalising losses from climate change: Towards hope for Pacific atoll countries. *Asia Pacific Viewpoint, 58*(1), 3–13.

Barnett, J. (2020). Global environmental change II: Political economies of vulnerability to climate change. *Progress in Human Geography, 44*(6), 1172–1184.

Barnett, J., Jarillo, S., Swearer, S., Lovelock, C., et al. (2022). Nature-based solutions for atoll habitability. *Philosophical Transactions of the Royal Society B, 377*, 20210124.

Barnett, J., Farbotko, C., Kitara, T., & Aselu, B. (2025). Migration as adaptation? The Falepili Union Between Australia and Tuvalu. *Wiley Interdisciplinary Reviews: Climate Change, 16*(1), e924.

Barrett, C (2025, July 4). Pacific rugby leaders raise alarm about NRL's PNG deal. *Sydney Morning Herald*, 17.

Barrientos, A. (2010). Vulnerability and social protection in small island states: The case of Grenada. *Social and Economic Studies, 59*, 3–30.

Barrowclough, D., & Eugui, D. (2021). *Plastic production and trade in small states and SIDS: The shift towards a circular economy.* Commonwealth Secretariat.

Barrow-Giles, C. (2015). Beyond the status quo, centring women in the Westminster system in the Commonwealth Caribbean: A preliminary analysis. *Commonwealth & Comparative Politics, 53*(1), 49–70.

Barry, T., & Gahman, L. (2020). Agrarian struggle and food system injustice in the Anglo-Caribbean: Centering social reproduction by (re)turning to Creft and Fanon. *Human Geography, 13*(2), 174–178.

Barry, T., & Gahman, L. (2021). Food system and social reproduction realities for women in agriculture across the Caribbean: Evidence from Grenada, St. Lucia and St. Vincent and the Grenadines. *Journal of Agrarian Change, 21*, 815–833.

Basel, B., Goby, G., & Johnson, J. (2020). Community-based adaptation to climate change in villages of Western Province. *Solomon Islands. Marine Pollution Bulletin, 156*, 111266.

Basurto, X., Gutierrez, N., Franz, N., et al. (2025). Illuminating the multidimensional contributions of small-scale fisheries. *Nature, 637*, 875–884.

Batalofo, M., Pollard, A., Ride, A., et al. (2024). What can the experiences of rural women in Solomon Islands teach us about innovation in aquatic food systems? *Asia Pacific Viewpoint, 65*(1), 96–110.

Baud, M., Hogenboom, B., Gonzalez-Vicente, R., et al. (2022). *China's economic and political role in the Caribbean and Central America*. CEDLA-UvA.

Bautista, L. (2010). *Steadfast movement around Micronesia: Satowan enlargements beyond Micronesia*. Lexington.

Bayliss-Smith, T., & Christensen, A. (2024). From copra sequestration to outboard engine trawling: Changes in the carbon footprint of Ontong Java Atoll, Solomon Islands. *Journal of Pacific History, 59*(4), 468–490.

Beaumont, N., Aanesen, M., Austen, M., et al. (2019). Global ecological, social and economic impacts of marine plastic. *Marine Pollution Bulletin, 142*, 189–195.

Becken, S., & Hay, J. (2007). *Tourism and Climate Change: Risks and opportunities*. Multilingual Matters.

Beckford, G. (1987). The social economy of bauxite in the Jamaican man-space. *Social and Economic Studies*, 1–55.

Beckford, C. (2016). The experiences of Caribbean migrant farmworkers in Ontario, Canada. *Social and Economic Studies, 65*(1), 153–168.

Beckford, C., & Campbell, D. (2013). *Domestic food production and food security in the Caribbean*. Palgrave Macmillan.

Bedford, C., Bedford, R., & Ho, E. (2010). Engaging with the Recognised Seasonal Employer (RSE) work policy: The case of Tuvalu. *Asian and Pacific Migration Journal, 19*, 421–445.

Bedford, R., Bedford, C., Wall, J., & Young, M. (2017). Managed temporary labour migration of Pacific Islanders to Australia and New Zealand in the early twenty-first century. *Australian Geographer, 48*(1), 37–57.

Bell, J., Reid, C., Batty, Allison, E., et al. (2011). Implications of climate change for contributions by fisheries and aquaculture to Pacific Island economies and communities. In J. Bell, J. Johnson & A. Hobday (Eds.), *Vulnerability of tropical Pacific fisheries and aquaculture to Pacific Island economies and communities* (pp. 733–801), SPC.

Bell, J., Ganachaud, A., Gehrke, P., et al. (2013). Mixed responses of tropical Pacific fisheries and aquaculture to climate change. *Nature Climate Change, 3*, 591–599.

Bell, C., Latu, C., Na'ati, E., Snowdon, W., Moodie, M., & Waqa, G. (2021a). Barriers and facilitators to the introduction of import duties designed to prevent noncommunicable disease in Tonga: A case study. *Globalization and Health, 17*, 136.

Bell, J., Senina, I., Adams, T., et al. (2021b). Pathways to sustaining tuna-dependent Pacific Island economies during climate change. *Nature Sustainability, 4*, 900–910.

Bell-James, J., & Watson, J. (2025). Ambitions in national plans do not yet match bold international protection and restoration commitments. *Nature Ecology and Evolution, 9*, 417–424.

Belmar, Y., McNamara, K., & Morrison, T. (2016). Water security in small island developing states: The limited utility of evolving governance paradigms. *WIREs Water, 3*(2), 181–193.

Béné, C., Wood, R., Newsham, A., & Davies, M. (2012). *Resilience: New utopia or new tyranny? Reflection about the potentials and limits of the concept of resilience in relation to vulnerability reduction programmes.* IDS Working Papers, 2012(405), 1–61.

Benett, I. (2023). Space unmade: Hurricane Dorian, autoethnography and dislocation. *Caribbean Studies, 51*(1), 39–75.

Bennett, N. (2019). In political seas: Engaging with political ecology in the ocean and coastal environment. *Coastal Management, 47*(1), 67–87.

Bennett, N., Finkheiner, E., Ban, N., et al. (2020). The COVID-19 pandemic, small-scale fisheries and coastal fishing communities. *Coastal Management, 48*(4), 336–347.

Bennett, N., Alava, A., Ferguson, C., Blythe, J., et al. (2023). Environmental (in)justice in the Anthropocene ocean. *Marine Policy, 147*, 105383.

Bennett, S., Talouli, A., & Dixit, V. (2024). Towards a circular logistics partnership: Regional trade, waste, and inter-sectoral cooperation. In A. Dansie, H. Alleway, & B. Böer (Eds.), *The water, energy, and food security nexus in Asia and the Pacific: The Pacific* (pp. 197–199). Springer.

Bennett, M., March, A., Greer, R., & Failler, P. (2025). Value chain opportunities for Pacific Coastal Resources. *Sustainability, 17*, 1103.

Bense, K. (2016). International teacher mobility and migration: A review and synthesis of the current empirical research and literature. *Educational Research Review, 17*, 37–49.

Benwell, R. (2011). The Canaries in the Coalmine: Small states as climate change champions. *The Round Table, 100*(413), 199–211.

Bernal, R. (2010). The Dragon in the Caribbean: China-CARICOM economic relations. *The Round Table, 99*(3), 281–302.

Bernal, R. (2015). The growing economic presence of China in the Caribbean. *The World Economy, 38*(9), 1409–1437.

Berno, T. (2025). "Do you want fries with that?". Reflections on the relationship between Pacific cuisine, tourism and sustainability. In E. Fusté-Forné & E. Wolf (Eds.), *Food and beverage tourism on Islands* (pp. 133–146). Edward Elgar.

Berta, O. (2025). Enduring ascriptions of dependence: Cultural autonomy and relational interdependence in the Marshall Islands. *Antipode, 57*, 1382–1401.

Bertana, A., & Gaytán, M. (2025). Racialized downgrading and upgrading: Dis/articulation and the Fijian kava commodity chain. *Environment and Planning A: Economy and Space, 57*(1), 22–39.

Bertram, G., & Watters, R. (1985). The MIRAB economy in South Pacific microstates. *Pacific Viewpoint, 26*, 497–520.

Bevacqua, R., Hunter, S., Deshpande, S., & Morgan, P. (2025). *Supporting small-scale entrepreneurs in the Pacific Region*. ADB Institute Policy Brief No. 2025-10.

Beyerl, K., Mieg, H., & Weber, E. (2019). Comparing perceptions of climate-related environmental changes for Tuvalu, Samoa and Tonga. In C. Klöck & M. Fink (Eds.), *Dealing with climate change on Small Islands: Towards effective and sustainable adaptation?* (pp. 143–174). Göttingen University Press.

Bhattacharya, P., & Dash, A. (2021). Determinants of blue economy in Asia-Pacific island countries: A study of tourism and fisheries sector. *Ocean and Coastal Management, 211*, 105774.

Bhattarai, A., & Sigdel, A. (2025). The silent emergency: Awakening political decisions and a call to action for the future of global health. *Discover Health Systems, 4*, 123.

Binetti, U., Silburn, B., Russell, J., et al. (2020). First marine litter survey on beaches in Solomon Islands and Vanuatu, South Pacific. *Marine Pollution Bulletin, 161*, 111827.

Bird, Z., Wairiu, M., des Combes, H., & Iese, V. (2021). Religious and cultural-spiritual attributions of climate-driven changes on food production: A case study from North Malaita, Solomon Islands. In J. Luetz & P. Nunn (Eds.), *Beyond belief: Opportunities for faith-engaged approaches to climate-change adaptation in the Pacific Islands* (pp. 39–56). Springer Nature.

Bird, Z., Iese, V., des Combes, H., et al. (2022). Assessing the impacts of climate change on domestic crop production: Experience and perceptions of local farmers in North Malaita, Solomon Islands. *Pacific Dynamics, 6*(1), 77–98.

Bishop, M. (2010). Tourism as a small-state development strategy: Pier pressure in the Eastern Caribbean. *Progress in Development Studies, 10*(2), 99–114.

Bishop, M., Corbett, J., & Veenendaal, W. (2020). Labor movements and party system development: Why does the Caribbean have stable two-party systems, but the Pacific does not? *World Development, 126*, 104719.

Bishop, M., Bouhia, R., Carter, G., et al. (2024). *Sustaining development in Small Islands: Climate change, geopolitical security and the permissive liberal order*. Cambridge University Press.

Blair, A., & Momtaz, S. (2018). Climate change perception and response: Case studies of fishers from Antigua and Efate. *Ocean and Coastal Management, 157*, 86–94.

Bleeker, A., Escribano, P., Gonzales, C., et al. (2021). *Advancing gender equality in environmental migration and disaster displacement in the Caribbean*. ECLAC and IOM.

Block, A., & Krausner, P. (1987). Masters of Paradise Island: Organized crime, neo-colonialism and the Bahamas. *Dialectical Anthropology, 12*, 85–102.

Blouin, C., & Debnath, P. (2011). CARICOM return migration and brain circulation: Case study of Caribbean born nurses. *Canadian Foreign Policy Journal, 17*(2), 101–114.

Blumfield, T., Reverchon, F., & Vigulu, V. (2018). The importance of market access for timber growers in small island developing states. *Land Use Policy, 77*, 598–602.

Bock, J., & Parilla, H. (2024, November 1). Hardly an inducement: Tourism from China gets up Palau's nose. *The Strategist*.

Boege, V. (2022). Rio Tinto and Bougainville. A fatal connection. A mine, a war, and an uncertain future. *Australian Journal of Politics and History, 68*(1), 18–35.

Bola, A. (2016). Potential for sustainable sea transport: A case study of the Southern Lomaiviti, Fiji islands. *Marine Policy, 75*, 260–270.

Bomhauer-Beins, L., de Guttry, C., & Ratter, B. (2019). When culture materializes: Societal dynamics in resilience of social-ecological systems in the case of conch management on Abaco, the Bahamas. *Sustainability, 11*(4), 1080.

Bordner, A., Ferguson, C., & Ortolano, L. (2020). Colonial dynamics limit climate adaptation in Oceania: Perspectives from the Marshall Islands. *Global Environmental Change, 61*, 10205.

Bourne, C. (2009). Economic growth, poverty and income inequality. *Journal of Eastern Caribbean Studies, 34*, 21–48.

Bourne, P. A. (2025). Violence in western Jamaica: A secondary data analysis. *International Journal of Insights & Transformations in Law, Crime & Justice, 9*(2), 18–44.

Bowling, B. (2010). *Policing the Caribbean: Transnational security cooperation in practice*. Oxford University Press.

Bourrel, M., Swaddling, A., Atalifo, V., & Tawake, A. (2018). Building in-country capacity and expertise to ensure good governance of the deep sea mineral industry within the Pacific region. *Marine Policy, 95*, 372–379.

Bove, C., Mudge, L., & Bruno, J. (2022). A century of warming on Caribbean reefs. *PLOS Climate, 1*(3), e0000002.

Boxall, S. (2025). What about women? An overlooked solution to Pacific labour shortages. *ANU DevPolicyBlog*, 6 May.

Bradley, M., & Swaddling, A. (2018). Addressing environmental impact assessment challenges in Pacific island countries for effective management of deep sea minerals activities. *Marine Policy, 95*, 356–362.

Brady, A.-M. (Ed.). (2022). *Looking north, looking south: China, Taiwan, and the South Pacific*. World Scientific.

Braithwaite, S. (2022). Estimating the economic costs of noncommunicable diseases in CARICOM. In C. Cannonier & M. Burke (Eds.), *Contemporary issues within Caribbean economies* (pp. 65–82). Palgrave Macmillan.

Bratrud, T. (2024). Climate change in the Pacific and the question of relocation. *Human Organization, 83*(3), 209–215.

Breman, J., & van der Linden, M. (2024). Migration: A view from below. *New Left Review, 148*, 145–159.

Brent, Z., Barbesgaard, M., & Pedersen, C. (2020). The blue fix: What's driving blue growth? *Sustainability Science, 15*, 31–43.

Bridges, K., & McClatchey, W. (2014). Living on the margin: Ethnoecological insights from Marshall Islanders at Rongelap Atoll. *Global Environmental Change, 19*(2), 140–146.

Brissett, N. (2017). Sustainable development goals (SDGs) and the Caribbean: Unrealizable promises? *Progress in Development Studies, 18*(1), 18035.

Brissett, N. (2023). 'I left because…': Caribbean high-skilled emigrants' reasons to migrate to the US. *Compare, 53*(2), 235–252.

Britton, R. (1977). Making tourism more supportive of small state development: The case of St. Vincent. *Annals of Tourism Research, 4*(5), 268–278.

Britton, S. (1980). The spatial organisation of tourism in a neocolonial economy: A Fiji case study. *Pacific Viewpoint, 21*(2), 144–165.

Brodie, G., Holland, E., N'Yeurt, A., et al. (2020). Seagrasses and seagrass habitats in Pacific small island developing states: Potential loss of benefits via human disturbance and climate change. *Marine Pollution Bulletin, 160*, 111573.

Brooks, S. (2016). Informal settlements in Jamaica's tourism space: Urban spatial development in a small island developing state. *Urban Island Studies, 2*, 72–94.

Brooks, N., & Adger, N. (2004). Assessing and enhancing adaptive capacity. In B. Lim & E. Spanger-Siegfried (Eds.), *Adaptation policy framework for climate change* (pp. 165–182). Cambridge University Press.

Brown, P., Daigneault, A., & Gawith, D. (2017). Climate change and the economic impacts of flooding on Fiji. *Climate and Development, 9*(6), 493–504.

Brown, R., & Connell, J. (1993). The global flea market: Migration, remittances and the informal economy in Tonga. *Development and Change, 24*(4), 611–647.

Brown, R., & Connell, J. (2004). The migration of doctors and nurses from South Pacific island nations. *Social Science and Medicine, 58*, 2193–2210.

Brown, R., & Connell, J. (2006). Occupation-specific analysis of migration and remittance behaviour: Pacific island nurses in Australia and New Zealand. *Asia-Pacific Viewpoint, 47*, 133–148.

Brown, R., Connell, J., & Jimenez-Soto, E. (2014). Migrants' remittances, poverty and social protection in the South Pacific: Fiji and Tonga. *Population Space and Place, 20*(5), 434–454.

Bruce, L., & Marshall, L. (2020). What does neighbourhood level urban resilience look like in Honiara? In D. Sanderson & L. Bruce (Eds.), *Urbanisation at Risk in the Pacific and Asia* (pp. 74–89). Routledge.

Buggy, L., & McNamara, K. (2015). The need to reinterpret "community" for climate change adaptation: a case study of Pele Island, Vanuatu. *Climate and Development, 8*(3), 1–11.

Burch, E. (2020). A sea change for climate refugees in the South Pacific. How social media – not journalism – tells their real story. *Environmental Communication, 15*(2), 250–263.

Burnett, G., & Suluma, W. (2024). Formal education in Nauru: A Pacific development assistance story. *Asia Pacific Viewpoint, 65*(2), 276–287.

Butler, R. (1980). The concept of a tourist area cycle of evolution: Implications for management of resources. *Canadian Geographer, 24*(1), 5–12.

Byron, M. (1999). Migration as a way of life: Nevis and the post-war labour movement to Britain. In R. King & J. Connell (Eds.), *Small worlds, global lives. Islands and migration* (pp. 115–136). Pinter.

Byron, J., & Montoute, A. (2023). The role of churches in regional cooperation and governance in the Caribbean. In L. Briguglio, M. Briguglio, & C. Slatter (Eds.), *Handbook of civil society and social movements in small states* (pp. 189–203). Routledge.

Cabral-Ramírez, M., Niño-Barrero, Y., & DiBella, J. (2025). Lessons from the implementation of the Sendai framework for disaster risk reduction from Latin America and the Caribbean. *International Journal of Disaster Risk Science, 16*(1), 72–83.

Cahu, P., & Sondergaard, L. (2023). *Estimating interim learning poverty for Pacific Island countries*. World Bank.

Cain, T. (2015). Rebuild or reform; regional or sub-regional architecture in the Pacific island region. *Journal de la Société des Océanistes, 140*(1), 49–58.

Cain, T. (2022). *A drop in the ocean? Organised crime dynamics in Pacific Island countries*. Global Initiative.

Caldeira, M., Sekinairai, A., & Vierros, M. (2025). Weaving science and traditional knowledge: Toward sustainable solutions for ocean management. *Marine Policy, 174*, 106591.

Calliari, E., & Vanhala, L. (2022). The 'national turn' in climate change loss and damage governance research: Constructing the L&D policy landscape in Tuvalu. *Climate Policy, 22*(2), 184–197.

Calliari, E., Serdeczny, O., & Vanhala, L. (2020). Making sense of the politics in the climate change loss and damage debate. *Global Environmental Change, 64*, 102133.

Campbell, Y. (2018). Security cooperation in Latin America and the Caribbean: Threats, institutions and challenges. In E. Fiddian-Qasmiyeh & P. Daley (Eds.), *Routledge handbook of south-south relations* (pp. 309–313). Routledge.

Campbell, Y. (2020). *Citizenship on the margins: State power, security and precariousness in 21st-century Jamaica*. Springer Nature.

Campbell, Y. (2025). Colonial continuities in citizenship and the role of civil society organisations. In *Routledge handbook of Caribbean studies* (pp. 148–159). Routledge.

Campbell, Y., & Clarke, C. (2017). The garrison community in Kingston and its implications for violence, policing, de facto rights, and security in Jamaica. In L. Macdonald & T. Hilgers (Eds.), *Violence in Latin America and the*

Caribbean: Subnational structures, institutions, and clientelistic networks (pp. 93–111). Cambridge University Press.

Campbell, Y., & Connell, J. (Eds.). (2011). *COVID in the Islands: A comparative perspective on the Caribbean and the Pacific*. Palgrave Macmillan.

Campbell, Y., & Johnson-Myers, T. A. (2023). *Gender inequality and women's citizenship: Evidence from the Caribbean*. Routledge.

Campbell, Y., & Harriott, A. (2024). The resort to emergency policing to control gang violence in Jamaica: Making the exception the rule. *Journal of Latin American Studies, 56*(1), 115–136.

Campbell, D., & Lee, S. (2021). Vulnerability of Jamaica's South coast fishing communities to coastal erosion and flooding. In S. Moncada et al. (Eds.), *Small island developing states: Vulnerability and resilience under climate change* (pp. 211–231). Springer.

Campbell, B., Steenbergen, D., Li, O., et al. (2024). Characterising a diversity of coastal community fisheries in Kiribati and Vanuatu. *Fish and Fisheries, 25*, 837–857.

Campbell, Y., Harriott, A., Grey, F. A., & Blake, D. (2026). From the 'war on drugs' to the 'war on guns': South–South cooperation between Mexico and the Caribbean. *European Journal of International Security, 11*(1), 96–114.

Campling, L., & Colás, A. (2021). *Capitalism and the sea: The maritime factor in the making of the modern world.* Verso.

Carballo, D., & Dos Reis, L. (2018). *The agricultural insurance market in the Caribbean*. World Bank.

Carby, B. (2021). Integrating disaster risk reduction in national development planning: Experiences and challenges of Jamaica. *Environmental Hazards, 17*, 219–233.

CARE. (2022). *That's not new money*. CARE Denmark.

Carew-Reid, J. (1989). *Environment, aid and regionalism in the South Pacific*. National Centre for Development Studies, ANU.

Caribbean Disaster Emergency Management Authority. (2016). *The regional response mechanism*. CDEMA.

Carmichael, C., & Jia, G. (2022). Economic growth through trade liberalization for small island developing states in the Pacific: Regionalism versus globalization. In J. Randall (Ed.), *Annual report on Global Islands 2021* (pp. 135–148). Island Studies Press.

Carr, L., & Heyman, W. (2009). Jamaica bound? Marine resources and management at a crossroads in Antigua and Barbuda. *Geographical Journal, 175*(1), 17–38.

Casola, W., Oren, J., Register, M., et al. (2021). Modernization of artisanal fishing communities on Andros Island, The Bahamas, as a treadmill of production. *Ocean & Coastal Management, 201*, 105487.

Casola, W., Rehnberg, M., Peterson, M., et al. (2022). Drivers of long-term support for marine protected areas in The Bahamas. *Ocean & Coastal Management, 217*, 106000.

Cassells, R. (2019). Engaging with Churches to address development-related challenges in Solomon Islands. *Sites, 16*(1), 109–121.

Cassinat, J., Cassinat, K., Segi, T., et al. (2022). Traditional village roles and gender shape Samoan perceptions of climate change. *Current Research in Environmental Sustainability, 4*, 100173.

Caterson, J. (2024, October 1). How giant China looms over tiny Palau's economy. *The Strategist.*

Cavallo, E., & Noy, I. (2009). *The economics of natural disasters: A survey*. The Inter-American Development Bank Working Paper No. 124.

Cavanough, E. (2020, September 2). Solomon Islands province announces independence vote amid China Tensions. *The Guardian.*

Cavanough, E. (2023). *Divided Isles: Solomon Islands and the China Switch*. La Trobe University Press.

Cecil, R., & Ebanks, G. (1992). The Caribbean migrant farm worker programme in Ontario: Seasonal expansion of West Indian economic spaces. *International Migration, 30*(1), 19–37.

Chacko, P. (2023). Racial capitalism and sphere of influence: Australian Assertions of White Possession in the Pacific. *Political Geography, 105*, 102923.

Chaguan. (2023, September 23). China, leader of the global south? *The Economist,* 28.

Chambers, A., & Chambers, K. (2018). Engaging the world: Four decades of intensifying change in Tuvalu. In J. Connell & H. Lee (Eds.), *Change and continuity in the Pacific: Revisiting the region* (pp. 17–36). Routledge.

Chand, S., Sing, B., & Kumar, S. (2020). The economic burden of non-communicable disease mortality in the South Pacific: Evidence from Fiji. *Plos One, 15*(7), e0236068.

Chand, S., Clemens, M., & Dempster, H. (2022). Wage arbitrage through skilled emigration: Evidence from the Pacific Islands. *Asia & the Pacific Policy Studies, 9*(3), 430–446.

Chandra, A., & Gaganis, P. (2016). Deconstructing vulnerability and adaptation in a coastal river basin ecosystem: A participatory analysis of flood risk in Nadi, Fiji Islands. *Climate and Development, 8*(3), 256–269.

Chandra, A., McNamara, K., Clissold, R., et al. (2023). Climate-induced non-economic loss and damage: Understanding policy responses, challenges and future directions in Pacific small island developing states. *Climate, 11*, 74.

Charan, D., Kaur, M., & Singh, P. (2018). Customary land and climate change induced relocation - A case study of Vunidogoloa Village, Vanua Levu, Fiji. In W. Filho & J. Nalau (Eds.), *Limits to climate change adaptation* (pp. 345–358). Springer.

Charlson, F., Diminic, S., & Whiteford, H. (2015). The rising tide of mental disorders in the Pacific region. *Asia and the Pacific Policy Studies, 2*(2), 280–292.

Cheer, J., Cole, S., Keir, J., et al. (2017). Tourism and islandscapes: Cultural realignment, social-ecological resilience and change. *Shima, 11*(1), 40–54.

Cheng, Y., Tseng, Y., Iwaki, Y., & Huang, M. (2021). Sustainable food security in small island developing states (SIDS): A case study of horticulture project in Marshall Islands. *Marine Policy, 128*, 104378.

Chesneaux, J., & Maclellan, N. (1998). *After Moruroa: France in the South Pacific*. Ocean Press.

Cheung, K., & Day, J. (2025, June 22). Chinese investor perspectives on customary land disputes in Vanuatu. *Melbourne Asia Review*.

Christensen, A. (2011). Marine gold and atoll livelihoods: The rise and fall of the bêche-de-mer trade on Ontong Java, Solomon Islands. *Natural Resources Forum, 35*, 9–20.

Cinner, J., Adger, N., Allison, E., et al. (2018). Building adaptive capacity to climate change in tropical coastal communities. *Nature Climate Change, 8*(2), 117–123.

Ciplet, D., Falzon, D., Uri, I., et al. (2022). The unequal geographies of climate finance: Climate injustice and dependency in the world system. *Political Geography, 99*, 102769.

Clark, T., & Cisneros-Montemayor, P. (2024). Colonialism and the blue economy: Confronting historical legacies to enable equitable ocean development. *Ecology and Society, 29*(3), 4.

Clarke, M. (2013). Good works and God's work: A case study of Churches and community development in Vanuatu. *Asia Pacific Viewpoint, 54*(3), 340–351.

Clegg, P. (2012). Independence movements in the Caribbean: Withering on the vine? *Commonwealth and Comparative Politics, 50*(4), 422–438.

Clissold, R., Westoby, R., & McNamara, K. (2020). Women as recovery enablers in the face of disasters in Vanuatu. *Geoforum, 113*, 101–110.

Cloos, P., Belloiseau, M., McPherson, N., et al. (2023). Discussing linkages between climate change, human mobility and health in the Caribbean: The case of Dominica. A qualitative study. *The Journal of Climate Change and Health, 11*, 100237.

Coke-Hamilton, P. (2020). Impact of COVID-19 on tourism in small island developing states, UNCTAD.

Coloma, R. (2025, June). 'No one wants to leave': Marshallese communities push back against the migration narrative. *Pacific Island Times.*

Commonwealth Secretariat. (2003). *A protocol for the recruitment of commonwealth teachers*. Commonwealth Secretariat.

Comolli, V. (2024, March 14). Geopolitics, influence and crime in the Pacific islands. *The Strategist.*

Conley, J. (2024, September 10). Pacific Islands Push ICC to Recognise Ecocide as International Crime. *Common Dreams.*

Connell, J. (1983). The end ever nigh: Contemporary population change on Pitcairn Island. *GeoJournal, 16*(2), 193–200.

Connell, J. (2003a). Losing ground? Tuvalu, the greenhouse effect and the garbage can. *Asia Pacific Viewpoint, 44*(2), 89–107.

Connell, J. (2003b). Regulation of space in the contemporary postcolonial Pacific city: Port Moresby and Suva. *Asia Pacific Viewpoint, 44*, 243–258.

Connell, J. (2006a). Nauru: The first Pacific failed state. *The Round Table, 95*, 47–63.

Connell, J. (2006b). "The taste of paradise": Selling Fiji and FIJI Water. *Asia Pacific Viewpoint, 47*, 342–350.

Connell, J. (2007a). Towards free trade in the Pacific? The genesis of the 'Kava-Biscuit War' between Fiji and Vanuatu. *Geographical Research, 45*, 1–12.

Connell, J. (2007b). Local skills and global markets? The migration of health workers from Caribbean and Pacific Island States. *Social and Economic Studies, 56*(1–2), 67–95.

Connell, J. (2008). Niue: Embracing a culture of migration. *Journal of Ethnic and Migration Studies, 34*(6), 1021–1040.

Connell, J. (2009a). *The global care chain. From the Pacific to the world*. Routledge.

Connell, J. (2009b). Bittersweet home? Return migration and health work in Polynesia. In D. Conway & R. Potter (Eds.), *Return migration of the next generation. 21st century transnational mobility* (pp. 139–160). Ashgate.

Connell, J. (2010). *Migration and the globalisation of health care. The health worker exodus?* Edward Elgar.

Connell, J. (2011). Elephants in the Pacific? Pacific urbanisation and its discontents. *Asia Pacific Viewpoint, 52*(2), 121–135.

Connell, J. (2012). Population resettlement in the Pacific: Lessons from a hazardous history. *Australian Geographer, 43*(2), 127–142.

Connell, J. (2013a). *Islands at risk: Environments, economies and contemporary change*. Edward Elgar.

Connell, J. (2013b). Ants, rats, snakes and taro blight: Contemporary invasions in the Pacific Islands. In S. Larrue (Ed.), *Biodiversity and societies in the Pacific Islands* (pp. 51–70). PUP.

Connell, J. (2013c). Medical tourism in the Caribbean islands: A cure for economies in crisis? *Island Studies Journal, 8*(1), 115–130.

Connell, J. (2014). The two cultures of health worker migration: A Pacific perspective. *Social Science and Medicine, 116*, 73–81.

Connell, J. (2015a). Food security in the island Pacific: Is Micronesia as far away as ever? *Regional Environmental Change, 15*(7), 1299–1311.

Connell, J. (2015b). Vulnerable islands: Climate change, tectonic change, and changing livelihoods in the Western Pacific. *The Contemporary Pacific, 27*(1), 1–36.

Connell, J. (2015c). Competing Islands? The Mamanuca and Yasawa Islands, Fiji. In G. Baldacchino (Ed.), *Archipelago Tourism* (pp. 183–197). Ashgate.

Connell, J. (2016). Last days in the Carteret Islands? Climate change, livelihoods and migration on coral atolls. *Asia Pacific Viewpoint, 57*(1), 3–15.

Connell, J. (2018a). Islands: Balancing development and sustainability? *Environmental Conservation, 45*(2), 111–124.

Connell, J. (2018b). Fiji, rugby and the geopolitics of soft power: Shaping national and international identity. *New Zealand Geographer, 74*(2), 92–100.

Connell, J. (2018c). Nothing there Atoll? "Farewell to the Carteret Islands". In T. Crook & P. Rudiak-Gould (Eds.), *Pacific climate cultures* (pp. 73–87). DeGruyter.

Connell, J. (2021a). The 2020 new Caledonia Referendum: The slow march to independence? *Journal of Pacific History, 56*(2), 144–160.

Connell, J. (2021b). Bougainville: A new Pacific Nation? *Small States & Territories, 3*(2), 375–396.

Connell, J. (2021c). Double jeopardy: Distance and decentralisation in Tuvalu. In Y. Campbell & J. Connell (Eds.), *COVID in the Islands: A comparative perspective on the Caribbean and the Pacific* (pp. 163–175). Palgrave Macmillan.

Connell, J. (2023a). *Climate change and health: A security challenge in the Pacific Islands*. Toda Peace Institute Policy Brief No. 166.

Connell, J. (2023b). COVID-19 and tourism in the island Pacific: Gender tribulations and transformations in different seas. In A. Datta, J. Momsen, & A. Oberhauser (Eds.), *Bridging worlds- Building feminist geographies* (pp. 212–220). Routledge.

Connell, J. (2024). Culture, COVID-19, cyclones and supply chains: The contraction of tourism in Vanuatu. In G. Baldacchino (Ed.), *Archipelago tourism after the pandemic* (pp. 137–152). Routledge.
Connell, J., & Aldrich, R. (2021). *The ends of empire: The last colonies revisited.* Palgrave Macmillan.
Connell, J., & Brown, R. (2005). *Remittances in the Pacific: An overview*. Asian Development Bank.
Connell, J., & Campbell, Y. (2021). Aftermath: Towards a 'new normal'. In Y. Campbell & J. Connell (Eds.), *COVID in the islands: A comparative perspective on the Caribbean and the Pacific* (pp. 517–528). Palgrave Macmillan.
Connell, J., & Conway, D. (2000). Migration and remittances in island microstates: A comparative perspective on the South Pacific and the Caribbean. *International Journal of Urban and Regional Research, 24*, 52–78.
Connell, J., & Corbett, J. (2016). Deterritorialisation: Reconceptualising development in the Pacific Islands. *Global Society, 30*(4), 583–604.
Connell, J., & Lutkehaus, N. (2017). Escaping Zaria's fire? The volcano resettlement problem of Manam Island, Papua New Guinea. *Asia Pacific Viewpoint, 58*(1), 14–26.
Connell, J., & Negin, J. (2012). The malaise of modernity: Urbanization, mobility and HIV in the Pacific. *Pacific Studies, 35*(1–2), 1–35.
Connell, J., & Pritchard, W. (1990). Tax havens and global capitalism: Vanuatu and the Australian Connection. *Australian Geographical Studies, 28*, 38–50.
Connell, J., & Rugendyke, B. (2008). Tourism and local people in the Asia-Pacific region. In J. Connell & B. Rugendyke (Eds.), *Tourism at the grassroots. Villagers and visitors in the Asia-Pacific* (pp. 1–40). Routledge.
Connell, J., & Stanley-Niaah, S. (2008). Remembering the fire: Music tourism and the Bob Marley Heritage in Jamaica. *Caribbean Geography, 15*(1), 76–90.
Connell, J., & Stilwell, B. (2006). Merchants of medical care: Recruiting agencies in the global health care chain. In C. Kuptsch (Ed.), *Merchants of labour* (pp. 239–253). ILO.
Connell, J., & Tabucanon, G. (2016). From Banaba to Rabi: A pacific model for resettlement? In S. Price & J. Singer (Eds.), *Global implications of development, disasters and climate change* (pp. 142–158). Routledge.
Connell, J., & Taulealo, T. (2021). Island tourism and COVID-19 in Vanuatu and Samoa: An unfolding crisis. *Small States and Territories, 4*(1), 105–124.
Conway, D., Potter, R., & St Bernard, G. (2012). Diaspora return of transnational migrants to Trinidad and Tobago: The additional contributions of social remittances. *International Development Planning Review, 34*(2), 189–209.

Cooper, T. (2020). Hurricane Dorian: A case for building comprehensive climate change resilience frameworks for small island developing states in the Caribbean. In S. Rolle, J. Minnis, & I. Bethell-Bennett (Eds.), *Tourism development, governance and sustainability in the Bahamas* (pp. 201–215). Routledge.

Cooper, A. (2023). A small state taking on a Giant: Antigua and Barbuda 'getting the better' of the United States without overturning structural asymmetry. In G. Baldacchino (Ed.), *The success of small states in international relations* (pp. 96–17). Routledge.

Cooper, S., Cloos, P., Abraham, C., et al. (2025). What can be said about risks, vulnerabilities, and adaptation to climate change in Caribbean small island developing states (SIDS)? The case of Dominica. A qualitative study. *PLOS Climate, 4*(4), e0000275.

Corbett, J. (2015). "Everybody knows everybody": Practising politics in the Pacific Islands. *Democratization, 22*(1), 51–72.

Corbett, J. (2023). *Statehood à la carte in the Caribbean and the Pacific*. Oxford University Press.

Corbett, J. (2024). *Statehood à la carte in the Caribbean and the Pacific*. Oxford University Press.

Corbett, J., & Byron-Reid, J. (2024). Secessionism in Nevis: Why have tensions eased? *Island Studies Journal, 19*(1), 1–24.

Corbett, J., & Connell, J. (2015). All the world is a stage: Global governance, human resources and the 'problem' of smallness. *The Pacific Review, 28*(3), 435–459.

Corbett, J., & Veenendaal, W. (2016). Westminster in small states: Comparing the Caribbean and Pacific experience. *Contemporary Politics, 22*(4), 432–449.

Corbett, J., Yi-Chong, X., & Weller, P. (2018). Climate change and the active participation of small states in international organisations. *The Round Table, 107*(1), 103–105.

Corbett, J., Xu, Y., & Weller, P. (2019). Norm entrepreneurship and diffusion 'from below' in international organisations: How the competent performance of vulnerability generates benefits for small states. *Review of International Studies, 45*(4), 647–668.

Corbett, J., Ruwet, M., Xu, Y., & Weller, P. (2020). Climate governance, policy entrepreneurs and small states: Explaining policy change at the International Maritime Organisation. *Environmental Politics, 29*(5), 825–844.

Corbett, J., Veenendaal, W., & Connell, J. (2021). The core executive and small states: Is coordination the primary challenge? *Public Administration, 99*(1), 103–117.

Corbett, J., Yi-Chong, X., & Weller, P. (2021). *International Organizations and Small States: Participation, Legitimacy and Vulnerability*. Bristol University Press.

Corbin, C. (2023). The role of waste management in underpinning the blue economy. In P. Clegg, R. Mahon, P. McConney, & H. Oxenford (Eds.), *The Caribbean blue economy* (pp. 195–209). Routledge.

Cordonnier Segger, M.-C. (2022). *Climate law and public policy innovation for the sustainable development goals: Accelerating pre-CoP27 Paris agreement compliance*. Bennett Institute for Public Policy.

Corlett, E. (2025, February 10). New Zealand raises 'concern' as Cook Islands prepares to sign China deal. *Guardian*.

Corva, D. (2008). Neoliberal globalization and the war on drugs: Transnationalizing illiberal governance in the Americas. *Political Geography, 27*(2), 176–193.

Cowell, N. (1987). The impact of bauxite mining on peasant and community relations in Jamaica. *Social and Economic Studies*, 171–216.

Cox, J. (2009). Active citizenship or passive clientelism? Accountability and development in Solomon Islands. *Development in Practice, 19*(8), 964–980.

Cox, J., Finau, G., Kant, R., et al. (2018). Disaster, divine judgement, and original sin: Christian interpretations of tropical cyclone Winston and climate change in Fiji. *The Contemporary Pacific, 30*(2), 380–410.

Coxon, A., & Munce, K. (2008). The global education agenda and the delivery of aid to Pacific education. *Comparative Education, 44*, 147–165.

Cramb, R., & Curry, G. (2012). Oil palm and rural livelihoods in the Asia-Pacific region: An overview. *Asia Pacific Viewpoint, 53*, 223–239.

Craney, A. (2022). *Youth in Fiji and Solomon Islands: Livelihoods, leadership and civic engagement*. ANU Press.

Crawford, C. (2003). Sending love in a Barrel: The making of transnational families in Canada. *Canadian Woman Studies/les cahiers de la femme, 22*(3–4), 104–109.

Crisman, T., & Winters, Z. (2023). Caribbean small island developing states must incorporate water quality and quantity in adaptive management of the water-energy-food nexus. *Frontiers in Environmental Science, 11*, 1212552.

Crocombe, R. (2007). *The fourth wave: Chinese in the Pacific Islands in the twenty-first century*. CSCSD Occasional Paper Number 1.

Croes, R. (2006). A paradigm shift to a new strategy for small island economies: Embracing demand side economics for value enhancement and long-term economic stability. *Tourism Management, 27*, 453–465.

Cross, M. (1979). *Urbanization and urban growth in the Caribbean*. Cambridge University Press.

Cross, I., & Congreve, A. (2020). Teaching (Super) wicked problems: Authentic learning about climate change. *Journal of Geography in Higher Education, 45*(4), 491–516.

Crossley, M., Bray, M., & Packer, S. (2011). *Education in small states: Policies and priorities*. Commonwealth Secretariat.

Crump, L. (2024). *Unity and discord in the Pacific: Negotiating the Pacific Agreement on closer economic relations plus*. Griffith University Asia Institute.

Cullum, R. (2021). Comparing Australian/South Pacific and UK/Caribbean relations. *The Round Table, 110*(4), 492–509.

Culpin, A. (2017). Child health and nutrition in the Marshall Islands. *Development Bulletin, 78*, 83–86.

Cumberbatch, J., Drakes, C., Mackey, T., Nagdee, M., et al. (2020). Social vulnerability index- Barbados – A case study. *Coastal Management, 48*(5), 505–526.

Cumberbatch, J., Compton, S., & McConney, P. (2023). Managing change to implement the small-scale fisheries guidelines: Perspectives from a Caribbean fisheries policy process. *Maritime Studies, 22*, 23.

Cumberbatch, J., Hinds, C., McConney, P., Speede, R., et al. (2024). Adaptive capacity, governance and small island developing states: A case study of sargassum management in the Eastern Caribbean. *Coastal Management, 52*(6), 340–358.

Currenti, R., Pearce, T., Salabogi, T., Vuli, L., et al. (2019). Adaptation to climate change in an interior Pacific Island village: A case study of Nawairuku, Ra, Fiji. *Human Ecology, 47*, 65–80.

Curtain, R. (2025). Why do Australia's seasonal worker numbers continue to fall? *ANU DevPolicyBlog*, 16 January.

Dagher, R. (2019). Policy space under a constraining combination- open economies, austerity and small island states. *Third World Quarterly, 40*(6), 1040–1063.

D'Arcy, P. (2006). *The people of the Sea: Environment, identity and history in Oceania*. University of Hawai'i Press.

D'Arcy, P. (2014). The Chinese Pacifics: A brief historical review. *The Journal of Pacific History, 49*(4), 396–420.

Datta, A., & Chaffin, B. (2022). Evolving adaptive governance: Challenging assumptions through an examination of fisheries law in Solomon Islands. *Ecology and Society, 27*(2), 30.

Dauvergne, P., Allan, J., Beaudoin, S., et al. (2025). Competing axes of power in the global plastics treaty: Understanding the politics of progress and setbacks in negotiating a high-ambition agreement. *Marine Policy, 181*, 106820.

Davila, F., Bourke, R., McWilliam, A., Crimp, S., et al. (2021). COVID-19 and food systems in Pacific Island Countries, Papua New Guinea, and Timor-Leste: Opportunities for actions towards the sustainable development goals. *Agricultural Systems, 191*, 103137.

Davis, M. (2006). *City of Quartz: Excavating the future in Los Angeles.* Verso Books.

Davis, S., Munger, L., & Legacy, H. (2020). Someone else's chain, someone else's road: US military strategy, China's Belt and Road Initiative, and island agency in the Pacific. *Island Studies Journal, 15*(2), 13–36.

Davis-Mattis, L. (2023). The future of deep-seabed minerals and marine genetic resources in blue economies. In P. Clegg, R. Mahon, P. McConney, & H. Oxenford (Eds.), *The Caribbean blue economy* (pp. 182–194). Routledge.

Dawkins-Cabrazos, M. (2024). An overview of youth crime and violence in the Caribbean community (CARICOM). In S. Morris-Francis, C. Gibson, & L. Grant (Eds.), *Youth crime and violence in the Caribbean* (pp. 19–32). Lexington.

Day, B., & Wells, T. (2025). The elite paradigm driving Australian aid policy: Learning to live with the 'cautious consensus'. *Asia & the Pacific Policy Studies, 12*(1), e70009.

Dayant, A. (2020, December). Health spending and foreign aid in the Pacific. *Pacific Economic Monitor*, 40–43.

Dayant, A., Duke, R., Ahsan, N., et al. (2024). *Pacific aid map. 2024 key findings report.* Lowy Institute.

Dean, A., Green, D., & Nunn, P. (2017). Too much sail for a small craft? Donor requirements, scale and capacity discourses in Kiribati. In E. Stratford (Ed.), *Island geographies: Essays and conversations* (pp. 54–77). Routledge.

De Beukelaer, C., Taukäve, J., Sahib, M., et al. (2026). Sailing Through Time: Building Pacific Maritime Resilience. *Global Policy*, 1–12.

De Boer, J. (2016). Risk, resilience and the fragile city. In D. Sanderson, J. Kayden, & J. Leis (Eds.), *Urban disaster resilience: New dimensions from international practice in the built environment* (pp. 113–127). Routledge.

De Coteau, D. (2024). *Black fins white sharks: Unmasking the genealogy of Caribbean political corruption.* Palgrave Macmillan.

de Waegh, R., House, J., Duarte, A., et al. (2023). Social inclusion and resilience. In C. Ooi, R. de Waegh, C. Trifan, & Y. Zhang (Eds.), *Islands and resilience. Experiences from the pandemic era* (pp. 17–34). Springer.

Dedekorkut-Howes, A., Torabi, E., & Howes, M. (2020). When the tide gets high: A review of adaptive responses to sea level rise and coastal flooding. *Journal of Environmental Planning and Management, 63*(12), 2102–2143.

Dehm, J., Le Gendre, R., Lal, M., et al. (2025). Water quality within the greater Suva urban marine environment through spatial analysis of nutrients and water properties. *Marine Population Bulletin, 213*, 117601.

Del Campo, F., Singh, S., Fishman, T., et al. (2023). Can a small island nation build resilience? The significance of resource-use patterns and socio-metabolic risks in The Bahamas. *Journal of Industrial Ecology, 27*, 491–507.

Demerath, P. (1999). The cultural production of educational utility in Pere Village. *PapuaNew Guinea. Comparative Education Review, 43*(2), 162–192.

Deonanan, R., & Ramkissoon, B. (2018). Remittances and economic development: Evidence from the Caribbean. *Social and Economic Studies, 67*(2/3), 95–132.

DeShong, H., Devonish, D., Grenade, W., & Roberts, S. (2023). Afterword: Re-imagining Caribbean futures post-COVID-19. In S. Roberts, H. DeShong, W. Grenade, & D. Devonish (Eds.), *Interdisciplinary perspectives on Covid-19 and the Caribbean* (pp. 551–561). Palgrave Macmillan.

Dewey, S. (2011). Markets and women's market trading in the Pacific Islands: An overview of social contexts and ongoing challenges. *Asian Women, 27*(3), 1–23.

Dey, M., Gosh, K., Valmonte-Santos, R., et al. (2016). Economic impact of climate change and climate change adaptation strategies for fisheries sector in Solomon Islands: Implication for food security. *Marine Policy, 67*, 171–178.

Diaz-Barrera, M., Alfaro-Aucca, C., Pacheco-Mendoza, J., et al. (2025). Bibliometric analysis of prominent topics in global scientific production on sustainable development goals in Scopus (2013–2022). *Discover Sustainability, 6*, 74.

DeShong, H. (2024). Gendered violence and gender as a site of violence: Reflections from the Caribbean. *Frontiers: A Journal of Women Studies, 45*(2), 79–101.

Diedrich, A., & Aswani, S. (2016). Exploring the potential impacts of tourism development on social and ecological change in the Solomon Islands. *Ambio, 45*, 808–818.

Dinnen, S. (2017). The inexorable rise of private security in Papua New Guinea. *Development Bulletin, 78*, 63–66.

Dinnen, S., & Allen, M. (2012). Paradoxes of postcolonial police-building: Solomon Islands. *Policing and Society, 23*(2), 222–242.

Dinnen, S., & Allen, M. (2018). Reflections on hybridity as an analytical lens on state formation: The case of Solomon Islands. In J. Wallis et al. (Eds.), *Hybridity on the ground in peacebuilding and development: Critical conversations* (pp. 129–144). ANU Press.

Dinnen, S., & McLeod, A. (2009). Policing Melanesia – international expectations and local realities. *Policing and Society, 19*(4), 333–353.

Doan, J. (2022). Barriers to women's participation in Vanuatu. *ANU DevPolicyBlog*, 27 July.

Dobell, G. (2007). The 'Arc of Instability': The history of an idea. In R. Huisken & M. Thatcher (Eds.), *History as policy: Framing the debate on the future of Australia's Defence Policy* (pp. 85–104). ANU E Press.

Docquier, F., & Marfouk, A. (2005). International migration by educational attainment 1990-2000. In C. Ozden & M. Schiff (Eds.), *International migration, brain drain and remittances* (pp. 151–199). Palgrave Macmillan.

Dodman, D. (2009). Globalization, tourism and local living conditions on Jamaica's north coast. *Singapore Journal of Tropical Geography, 30*(2), 204–219.

Dodd, R., Shanthosh, J., Lung, T., et al. (2021). Gender, health and ageing in Fiji: A mixed methods analysis. *International Journal for Equity in Health, 20*, 205.

Doerr, A. (2021). Regulating for social and environmental sustainability: A stakeholder perspective from the Bahamian spiny lobster fishery. *Marine Policy, 124*, 104366.

Dommen, E., & Hein, P. (1985). *States, microstates and islands*. Croom Helm.

Donner, S., & Webber, S. (2014). Obstacles to climate change adaptation decisions: A case-study of sea-level rise and coastal protection measures in Kiribati. *Sustainability Science, 9*(3), 331–345.

Dornan, M. (2015). Renewable energy development in small island developing states of the Pacific. *Resources, 4*, 490–506.

Dornan, M., & Cain, T. (2014). Regional service delivery among Pacific Island Countries: A case study. *Asia and the Pacific Policy Studies, 1*(3), 541–560.

Dornan, M., & Pryke, J. (2017). Foreign aid to the Pacific: Trends and developments in the twenty-first century. *Asia and the Pacific Policy Studies, 4*(3), 386–404.

Dornan, M., & Shah, K. (2016). Energy policy, aid, and the development of renewable energy resources in small island developing states. *Energy Policy, 98*, 759–767.

Dornan, M., Morgan, W., Cain, T., & Tarte, S. (2018). What's in a term? 'Green growth' and the 'blue-green' economy in the Pacific islands. *Asia and the Pacific Policy Studies, 5*, 408–425.

Dorney, S. (2016). *The embarrassed Colonialist*. Lowy Institute and Penguin Random House.

Doyle, J., & Knox-Vydmanov, C. (2024). Financing social protection in the Pacific and Timor-Leste. *ANU DevPolicyBlog*, 12 August.

Drakes, C., & Moore, W. (2015). The success factors of manufacturing firms in Caribbean small island states: A look at Barbados. *Central Bank of Barbados, 15*(10).

Dreher, T., & Voyer, M. (2015). Climate refugees or migrants? Contesting media frames on climate justice in the Pacific. *Environmental Communication, 9*(1), 58–76.

Duany, J. (2022). Main historical trends in Caribbean migration. In A. Feldmann, X. Bada, J. Durand, & S. Schütze (Eds.), *The Routledge history of modern Latin American migration* (pp. 49–61). Routledge.

Duke, R. & Ahsan, N (2024, November 18). Pacific education: Backsliding outcomes and a financing freefall. *Interpreter*.

Dulal, H., Shah, K., & Ahmad, N. (2009). Social equity considerations in the implementation of Caribbean climate change adaptation policies. *Sustainability, 1*, 363–383.

Dumaru, P. (2010). Community-based adaptation: Enhancing community adaptive capacity in Druadrua Island, Fiji. *WIRES Climate Change, 1*, 751–763.

Duncan, N., & Mctair, D. (2025, January 5). The Rastafarians opening up Caribbean cannabis farms to promote the benefits of marijuana. *Guardian*.

Dunn, D., Stewart, K., Bjorkland, R., et al. (2010). A regional analysis of coastal and domestic fishing effort in the wider Caribbean. *Fisheries Research, 102*(1-2), 60–68.

Duval, D. (2003). When hosts become guests: Return visits and diasporic identities in a Commonwealth Eastern Caribbean Community. *Current Issues in Tourism, 6*(4), 267–308.

Duval, D. (2004). *Tourism in the Caribbean: Trends, development, prospects*. Routledge.

Duvat, V. (2013). Coastal protection structures in Tarawa Atoll, Republic of Kiribati. *Sustainability Science, 8*(3), 363–379.

Duvat, V., Magnan, A., & Pouget, F. (2013). Exposure of atoll population to coastal erosion and flooding: A South Tarawa assessment, Kiribati. *Sustainability Science, 8*(3), 423–440.

Dyson, J. (1982). *The south seas dream. Adventures in paradise*. Heinemann.

Dziedzic, A. (2023). Judicial mobilities. Travelling judges in the Pacific. In A. Dziedzic & S. Young (Eds.), *The Cambridge handbook of foreign judges in domestic courts* (pp. 300–322). Cambridge University Press.

Dziedzic, S. (2024, December 14). Friends and partners weigh up Canberra's 'relentless' Pacific push. *ABC*.

Easterly, W. (2024). *The Tyranny of experts*. Basic Books.

Edgar, G., Bates, A., Krijeck, N., et al. (2024). Stock assessment models overstate sustainability of the world's fisheries. *Science, 385*, 860–865.

Edmonds, C., & Noy, I. (2018). The economics of disaster risks and impacts in the Pacific. *Disaster Prevention and Management, 27*(5), 478–494.

Edwards, R., Patrician, P., Bakitas, M., & Markaki, A. (2021). Palliative care integration: A critical review of nurse migration effects in Jamaica. *BMC Palliative Care, 20*, 155.

Eke, A., & Altangerel, T. (2025). Taking matters into our own hands: Aid dependency in the Pacific. *ANU DevPolicyBlog*, 3 July.

Elgie, A., Singh, S., & Telesford, J. (2021). You can't manage what you can't measure: The potential for circularity in Grenada's waste management system. *Resources, Conservation and Recycling, 164*, 105170.

Elliott, M., MacDonald, M., Chan, T., et al. (2017). Multiple household water sources and their use in remote communities with evidence from Pacific Island Countries. *Water Resources Research, 53*(11), 9106–9117.

Ells, P. (2000). *Where the hell is Tuvalu?* Virgin.

Enayati, S., Campbell, J., & Li, H. (2023). Vaccine distribution with drones for less developed countries: A case study in Vanuatu. *Vaccine:X, 14*, 100312.

Eppinga, M., Mijts, E., & Santos, M. (2022). Ranking the sustainable development goals: Perceived sustainability priorities in small island states. *Sustainability Science, 17*, 1537–1556.

Eriksson, H., Sulu, R., Blythe, J., et al. (2020). Intangible links between household livelihoods and food security in Solomon Islands: Implications for rural development. *Ecology and Society, 25*(4), 18.

Eriksson, H., Blythe, J., Österblom, H., & Olsson, P. (2021). Beyond social-ecological traps: Fostering transformations towards sustainability. *Ecology and Society, 26*(1), 13.
Errington, F., Fujikura, T., & Gewertz, D. (2013). *The noodle narratives.* University of California Press.
ESCAP. (2022). *Debt for climate swaps in the Pacific SIDS.* ESCAP (mimeo).
ESCAP. (2024). *People and planet: Addressing the interlinked challenges of climate change, poverty and hunger in Asia and the Pacific.* United Nations.
Esquivel, V. (2016). Power and the sustainable development goals: A feminist analysis. *Gender and Development, 24*(1), 9–23.
Evans, D. (2019). Bad places turned good: Security in Melanesia's settlement communities. *ANU DevPolicyBlog*, 4 February.
Fabinyi, M., & Barclay, K. (2022). *Asia-Pacific fishing livelihoods.* Palgrave Macmillan.
Fabinyi, M., Cvitanovic, C., Barclay, K., et al. (2025). Rethinking maritime security from the bottom up: Four principles to broaden perspectives and centre humans and ecosystems. *npj Ocean Sustainability, 4*(1), 29.
Fabinyi, M., Evans, L., & Foale, S. (2014). Social-ecological systems, social diversity, and power: Insights from anthropology and political ecology. *Ecology and Society, 19*(4), 28.
Fache, E., & Fair, H. (2020). Turning away from wicked ways: Christian climate change politics in the Pacific Island Region. *Anthropological Forum, 30*(3), 233–253.
Fache, E., & Pauwels, S. (2020). Tackling coastal "overfishing" in Fiji: Advocating for indigenous worldview, knowledge, and values to be the backbone of fisheries management strategies. *Maritime Studies, 19*(1), 41–52.
Fache, E., Le Meur, P.-Y., & Rodary, E. (2021). The new scramble for the Pacific: A frontier approach. *Pacific Affairs, 94*(1), 57–75.
Fair, H. (2020). Their sea of islands? Pacific climate warriors, oceanic identities, and world enlargement. *The Contemporary Pacific, 32*(2), 341–369.
Fairbairn, T., & Worrell, D. (1996). *South Pacific and Caribbean Island Economies: A Comparative Study.* Foundation for Development Cooperation.
Fanning, A., & Raworth, K. (2025). Doughnut of social and planetary boundaries monitors a world out of balance. *Nature, 646*, 47–56.
FAO. (2014). *Youth and agriculture: Key challenges and concrete solutions.* FAO.
FAO. (2024). *The state of food security and nutrition in the world 2024.* FAO.

Farbotko, C. (2005). Tuvalu and climate change: Constructions of environmental displacements in the Sydney Morning Herald. *Geografiska Annaler B, 7*(4), 279–293.

Farbotko, C. (2010a). 'The global warming clock is ticking so see these places while you can': Voyeuristic Tourism and Model Environmental Citizens on Tuvalu's Disappearing Islands. *Singapore Journal of Tropical Geography, 31*, 224–238.

Farbotko, C. (2010b). Wishful sinking: Disappearing islands, climate refugees and cosmopolitan experimentation. *Asia Pacific Viewpoint, 51*(1), 47–60.

Farbotko, C., & Campbell, J. (2022). Who defines atoll 'uninhabitability'? *Environmental Science and Policy, 138*, 182–190.

Farbotko, C., & Kitara, K. (2021). The political ecology and visual culture of the pacific climate warriors. In T. Demos, E. Scott, & S. Banerjee (Eds.), *The Routledge companion to contemporary art, visual culture, and climate change* (pp. 399–408). Routledge.

Farbotko, C., & Lazrus, H. (2012). The first climate refugees? Contesting global narratives of climate change in Tuvalu. *Global Environmental Change, 22*(2), 382–390.

Farmer, P. (2010). The consumption of the poor: Tuberculosis in the twenty-first century. In H. Saussy (Ed.), *Partner to the poor: A Paul farmer reader* (pp. 222–247). University of California Press.

Farrelly, T., Borrelle, S., & Fuller, S. (2021). The strengths and weaknesses of Pacific Islands plastic pollution policy frameworks. *Sustainability, 13*, 1252.

Fazey, I., Pettorelli, N., Kenter, J., et al. (2011). Maladaptive trajectories of change in Makira, Solomon Islands. *Global Environmental Change, 21*, 1275–1289.

Feeny, S., & McDonald, L. (2016). Vulnerability to multidimensional poverty: Findings from households in Melanesia. *Journal of Development Studies, 52*(3), 447–464.

Feinberg, R. (1986). The 'Anuta Problem': Local sovereignty and national integration in the Solomon Islands. *Man, 21*(3), 438–452.

Ferguson, C., Bennett, N., Kostka, W., et al. (2022a). The tragedy of the commodity is not inevitable: Indigenous resistance prevents high-value fisheries collapse in the Pacific islands. *Global Environmental Change, 73*, 102477.

Ferguson, C., Tuxson, T., Manghubai, S., et al. (2022b). Local practices and production confer resilience to rural Pacific food systems during the COVID-19 pandemic. *Marine Policy, 137*, 104954.

Ferguson Irlanda, C., Mangubhai, S., Waqa, E., et al. (2024). Women's participation promotes effective and equitable fisheries management but is hindered by restrictive gender roles in Fiji. *SPC Women in Fisheries Information Bulletin, 40*, 1–5.

Ferguson Irlanda, C., Mangubhai, S., Waqa, E., Govan, H., et al. (2025). Insights on the roles of women in effective and procedurally just environmental governance from coastal fisheries management in Fiji. *Conservation Biology, 39*(6), e70121.

Ferreira, M., Thompson, J., Paris, A., et al. (2020). Presence of microplastics in water, sediments and fish species in an urban coastal environment of Fiji, a Pacific small island developing state. *Marine Pollution Bulletin, 153*, 110991.

Fielding, D. (2010). Aid and Dutch disease in the South Pacific and in other small island states. *Journal of Development Studies, 46*, 918–940.

Figueroa, E., & Rotarou, E. (2021). Island tourism-based sustainable development at a crossroads: Facing the challenges of the COVID-19 pandemic. *Sustainability, 13*, 10081.

Fiji Ministry of Foreign Affairs. (2024). *Fiji's foreign policy white paper*. Ministry of Foreign Affairs.

Fitzgerald, S., Moriarty, A., Basas, R., & Shagazatova, M. (2022). *Making Space: Women's participation in higher education in Solomon Islands*. ADB Policy Brief.

Fitzmaurice, M., & Rydberg, A. (2023). Using international law to address the effects of climate change: A matter for the international court of justice? *Yearbook of International Disaster Law Online, 4*(1), 281–305.

Fitz-Henry, E. (2022, July 12). Climate change is white colonisation of the atmosphere. It's time to tackle this entrenched racism. *The Conversation*.

Fleming, L., Anthonj, C., Thakkar, M., Tikoisuva, W. et al. (2019). Urban and rural sanitation in the Solomon Islands: How resilient are these to extreme weather events?. *Science of the Total Environment, 683*, 331–340.

Floerl, O., Atalah, J., Bugnot, A., et al. (2021). A global model to forecast coastal hardening and mitigate associated socioecological risks. *Nature Sustainability, 4*, 1060–1067.

Flores-Palacios, X. (2019). Climate-induced migration in Lotofaga village in Samoa. In C. Klöck & M. Fink (Eds.), *Dealing with climate change on small islands: Towards effective and sustainable adaptation?* (pp. 313–337). Göttingen University Press.

Foale, S. (2001). 'Where's our Development?' Landowner aspirations and environmentalist agendas in Western Solomon Islands. *Asia Pacific Journal of Anthropology, 2*, 44–67.

Foley, A., Moncado, S., Mycoo, M., et al. (2022). Small island developing states in a post-pandemic world: Challenges and opportunities for climate action. *WIREs Climate Change, 13*(3), e769.

Folkersen, M., Fleming, C., & Hasan, S. (2018). Deep sea mining's future effects on Fiji's tourism industry: A contingent behaviour study. *Marine Policy, 96*, 81–89.

Fonua, F. (2023, January 12). Tonga volcano eruption: Mango and 'Atata Islanders resettle nearly a year on from disaster. *Radio New Zealand.*

Forbes, K. (2021, April 27). Caribbean food security during COVID-19 can only be ensured through debt relief. *The Conversation.*

Forsyth, G., & Atkinson, Q. (2024). A brief history of political instability in Vanuatu. *Anthropological Forum, 34*(3), 253–278.

Foster, T., & Willetts, J. (2018). Multiple water source use in rural Vanuatu: Are households choosing the safest option for drinking? *International Journal of Environmental Health Research, 28*(6), 579–589.

Fraenkel, J., & Iati, I. (2022). Introduction: What Can We Learn from the 2021 Political Crisis About Long-run Stability and Democracy in Sāmoa?. *Journal of Pacific History, 57*(4), 393–406.

Frauts, M. (2019). Resilience and the creative economy in Kingston, Jamaica. *Cultural Studies, 33*(3), 391–413.

Fry, G. (1980). Regionalism and international politics of the South Pacific. *Pacific Affairs, 54*(3), 455–484.

Fry, G. (2015). *Recapturing the Spirit of 1971: Towards a new regional policy settlement in the Pacific*. ANU SSGM Discussion Paper 2015/3.

Fry, G. (2021). AUKUS undermines Australia's "Pacific family". *ANU DevPolicyBlog*, 4 November.

Fry, G., Kabutaulaka, T., & Wesley-Smith, T. (2022). *'Partners in the Blue Pacific' initiative rides roughshod over established regional processes. ANU DevPolicyBlog*, 5 July.

Fukuoka, T., Sakane, F., Kinoshita, C., Sato, K., et al. (2022). Covid-19-derived plastic debris contaminating marine ecosystem: Alert from a sea turtle. *Marine Pollution Bulletin, 175*, 113389.

Futaiasi, D., & Zhang, D. (2025). Solomon Islands' external relations amid great power competition. *ANU DevPolicyBlog*, 14 August.

Gagaeolo, F., Hemstock, S., & Price, C. (2020). Christianity, Kiribati and climate change-induced migration. In F. Gagaeolo, S. Hemstock, & C. Price (Eds.), *Science, faith and the climate crisis* (pp. 1–14). Emerald.

Gahman, L., & Thongs, G. (2020). Development justice, a proposal: Reckoning with disaster, catastrophe, and climate change in the Caribbean. *Transactions of the Institute of British Geographers, 45*(4), 763–778.

Gahman, L., Thongs, G., & Greenidge, A. (2021). Disaster, debt, and 'underdevelopment': The cunning of Colonial-Capitalism in the Caribbean. *Development, 64*(1–2), 112–118.

Galaitsi, S., Corbin, C., Cox, C., et al. (2024). Balancing climate resilience and adaptation for Caribbean small island developing states (SIDS): Building institutional capacity. *Integrated Environmental Assessment and Management, 20*(5), 1237–1255.

Galappaththi, E., Susarla, V., Loutet, S., et al. (2022). Climate change adaptation in fisheries. *Fish and Fisheries, 23*, 4–21.

Galloway, L. (2022). *Palau's world-first 'good traveller' incentive.* Retrieved from www.bbc.com/travel/article/20220517

Ganivatu, E. (2025). Fuel cost threat. Economic challenge for island aviation. *Islands Business, 51*(5), 14.

Garcia, C., Makhoul, N., Fox, M., & Tuxson, T. (2022). Looking at the impacts of COVID- 19 on coastal communities in the Pacific using a gender and social inclusion lens. *SPC Women in Fisheries Information Bulletin, 35*, 21–25.

Garces-Ordonez, O., Spence, R., Canals, M., & Thiel, M. (2025). Macrolitter contamination in beach, dune, and mangrove ecosystems on a Caribbean island: A comparative analysis. *Marine Pollution Bulletin, 213*, 117616.

Garonna, I., Fazey, I., Brown, M., & Pettorelli, N. (2009). Rapid primary productivity changes in one of the last coastal rainforests: the case of Kahua, Solomon Islands. *Environmental Conservation, 36*(3), 253–260.

Gayle, D., & Goodrich, J. (1993). Caribbean tourism marketing, management and development strategies. In D. Gayle & J. Goodrich (Eds.), *Tourism marketing and management in the Caribbean* (pp. 1–19). Routledge.

Genave, A., Blancard, S., & Garabedian, S. (2020). An assessment of energy vulnerability in small island developing states. *Ecological Economics, 171*, 106595.

George, N. (2016). Light, heat and shadows: Women's reflections on peacebuilding in post-conflict Bougainville. *Peacebuilding, 4*(2), 166–179.

George, E. (2024, May 29). Barbuda's unending disaster. *Caribbean Investigative Journalism Network.*

George, G., Rhodes, B., & Laptiste, C. (2020). Estimating the financial incentive for Caribbean teachers to migrate: An analysis of salary differentials using Purchasing Power Parity (PPP). *Journal of Education & Development in the Caribbean, 19*(2), *1–19.*

Georgeou, N., Hawksley, C., Wali, N., Lountain, S., et al. (2022). Food security and small holder farming in Pacific Island countries and territories: A scoping review. *PLoS Sustainability and Transformation, 1*(4), e0000009.

Gewertz, D., & Errington, F. (2021). Stingy galitarianism: Precarity and Jealousy at the Sisiak Settlement, Madang, Papua New Guinea. In N. Bainton, D. McDougall, K. Alexeyeff, & J. Cox (Eds.), *Unequal lives: Gender, race and class in the Western Pacific* (pp. 213–235). ANU Press.

Gheuens, J., Nagabhatla, N., & Perera, E. D. P. (2019). Disaster-risk, water security challenges and strategies in small island developing states (SIDS). *Water, 11*(4), 637.

Giardino, A., Nederhoff, K., & Vousdoukas, M. (2018). Coastal hazard risk assessment for small islands: Assessing the impact of climate change and disaster risk reduction measures on Ebeye (Marshall Islands). *Regional Environmental Change, 18*, 2237–2248.

Gibson, C. (2021). Theorising tourism in crisis: Writing and relating to place. *Tourist Studies, 21*(1), 84–95.

Gibson, D. (2025). Island delights: Exploring the culinary tapestry of Fiji through food and beverage tourism. In E. Fusté-Forné & E. Wolf (Eds.), *Food and beverage tourism on Islands* (pp. 4–14). Edward Elgar.

Gibson, J., & McKenzie, D. (2011). The microeconomic determinants of emigration and return migration of the best and brightest: Evidence from the Pacific. *Journal of Development Economics, 95*(1), 18–29.

Gibson, K., Barnett, J., Haslam, N., & Kaplan, I. (2020). The mental health impacts of climate change: Findings from a Pacific Island atoll nation. *Journal of Anxiety Disorders, 73*, 102237.

Giglio, S., Kelly, B., & Stroebel, J. (2021). Climate finance. *Annual Review of Financial Economics, 13*, 15–36.

Gilbert, N. (2022, April 28). Climate change will force new animal encounters - and boost viral outbreaks. *Nature.*

Gillett, R., & Fong, M. (2023). *Fisheries in the economies of Pacific Island countries and territories*. Pacific Community.

Gilroy, P. (2013). *There ain't no black in the union jack*. Routledge.

Gina. (2024, December 17). Solomons renews safe skies partnership with Australia. *In-depth Solomons.*

Glasgow, S., & Schrecker, T. (2016). The double burden of neoliberalism? Noncommunicable disease policies and the global political economy of risk. *Health & Place, 39*, 204–211.

Goldsmith, M. (2015). The big smallness of Tuvalu. *Global Environment, 8*(1), 134–151.

Gonzalez-Vicente, R. (2021). Over hills and valleys too. China's Belt and Road Initiative in the Caribbean. In F. Schneider (Ed.), *Global perspectives on China's Belt and Road Initiative asserting agency through regional connectivity* (pp. 171–193). Amsterdam University Press.

Gonzalez-Vicente, R., & Montoute, A. (2021). A Caribbean perspective on China–Caribbean relations: Global IR, dependency and the postcolonial condition. *Third World Quarterly, 42*(2), 219–238.

Gordon-Strachan, G. P., S, Harewood, H., et al. (2025). The 2024 small island developing states report of the *Lancet* Countdown on health and climate change. *The Lancet Global Health, 13*(1), e146–e166.

Gössling, S., & Higham, J. (2021). The low carbon imperative: Destination management under urgent climate change. *Journal of Travel Research, 60*(6), 1167–1179.

Gotschall, I. (2021). An overview of I-Kiribati women in fisheries. *SPC Women in Fisheries Information Bulletin, 33*, 31–36.

Goujon, M., & Wagner, L. (2023). Fragility of small island developing states. In D. Carment & Y. Samy (Eds.), *Handbook of fragile states* (pp. 299–315). Edward Elgar.

Goundar, A., Newell, A., Nuttall, P., et al. (2017). King Canute muses in the South Seas: Why aren't Pacific Islands transitioning to low carbon sea futures? *Marine Policy, 81*, 80–90.

Government of Jamaica. Ministry of Labour and Social Security. (2023). *Jamaican government redoubles efforts to increase overseas employment opportunities for Jamaicans*. Ministry of Labour and Social Security.

Government of Vanuatu. (2002). *Vanuatu poverty survey: Analysis report*. Vanuatu Statistics Office.

Graci, S., & van Vliet, L. (2020). Examining stakeholder perceptions towards sustainable tourism in an Island destination: The case of Savusavu, Fiji. *Tourism Planning Development, 17*, 62–81.

Graham, R. (2024). The marine litter issue in the Windward Islands- a pathway to responses using the DPSIR framework. *Frontiers in Environmental Science, 11*, 1150722.

Granderson, A. (2017). The role of traditional knowledge in building adaptive capacity for climate change: Perspectives from Vanuatu. *Weather Climate and Society, 9*(3), 545–561.

Granderson, A. (2018). Value conflicts and the politics of risk: Challenges in assessing climate change impacts and risk priorities in rural Vanuatu. *Climate and Development, 10*(6), 481–494.

Granderson, A., & Leotaud, N. (2021). Towards a resilient, inclusive and green recovery in the Caribbean. In J. Connell & Y. Campbell (Eds.), *COVID in the Islands: A comparative perspective on the Caribbean and the Pacific* (pp. 495–515). Springer Nature.

Grant, R., Roy, S., Zablah, C., et al. (2023). Climate adaptation and resilience indices for the Caribbean region: An assessment of four leading indices. *Climate and Development*, 1–11.

Graves, C., Powell, A., Stone, M., Redfern, et al. (2021). Marine water quality of a densely populated Pacific atoll (Tarawa, Kiribati): Cumulative pressures and resulting impacts on ecosystem and human health. *Marine Pollution Bulletin, 163*, 111951.

Gray, A. (2021, May 7). How the US can protect the sovereignty of the smallest Pacific Islands. *The Diplomat*.

Graycar, A. (2022, March 16). Sustainable catch: Navigating global fisheries crime. *Interpreter*.

Green, S., White, A., Christie, P., et al. (2011). Emerging marine protected area networks in the coral triangle: Lessons and way forward. *Conservation and Society, 9*(3), 173–188.

Grenade, W. (2011). *Regionalism and sub-regionalism in the Caribbean: Challenges and prospects – Any insights from Europe?* University of Miami Jean Monnet/Robert Schuman Paper Series 11/4.

Griffen, S., & Robinson, S. (2023). (Un)just post-disaster mobilities in small island developing states: Revisiting the patterns and outcomes of three major environmental disasters in the Caribbean. *International Journal of Disaster Risk Reduction, 97*, 104029.

Grilli, G., Tyllianakis, E., Luisetti, T., et al. (2021). Prospective tourist preferences for sustainable tourism development in small island developing states. *Tourism Management, 82*, 104178.

Griffith, I. (2010). *Drugs and security in the Caribbean: Sovereignty under Siege*. Penn State University Press.

Griffith, I., & Munroe, T. (1995). Drugs and democracy in the Caribbean. *Journal of Commonwealth & Comparative Politics, 33*(3), 357–376.

Griswold, D. (2021). Growing hope: Island agriculture and refusing catastrophe in climate change adaptation in Fiji. *Island Studies Journal, 16*(2), 136–155.

Grossman, L. (1998). *The political ecology of bananas*. University of North Carolina Press.

Gruby, R. (2017). Macropolitics of Micronesia: Toward a critical theory of regional environmental governance. *Global Environmental Politics, 17*(4), 9–27.

Gruby, R., & Campbell, L. (2013). Scalar politics and the region: Strategies for transcending Pacific Island smallness on a global environmental governance stage. *Environment and Planning A, 45*(9), 2046–2063.

Grydehøj, A., & Kelman, I. (2017). The eco-island trap: Climate change mitigation and conspicuous sustainability. *Area, 49*(1), 106–113.

Grynberg, R. (2010). The trade and health debate. *Islands Business, 36*(9), 34–35.

Grynberg, R., & Razzaque, M. (2004). *The trade performance of small states*. Commonwealth Secretariat.

Guell, C., Brown, C., Iese, V., et al. (2021). "We used to get food from the garden." Understanding changing practices of local food production and consumption in small island states. *Social Science & Medicine, 284*, 114214.

Guell, C., Brown, C., Navunicagi, O., et al. (2022). Perspectives on strengthening local food systems in small island developing states. *Food Security, 14*(5), 1227–1240.

Guell, C., Saint Ville, A., Anderson, S., et al. (2024). Small island developing states: Addressing the intersecting challenges of non-communicable diseases, food insecurity, and climate change. *The Lancet Diabetes & Endocrinology, 12*(6), 422–432.

Gupta, J., & Vegelin, C. (2016). Sustainable development goals and inclusive development. *International environmental agreements: Politics, Law and Economics, 16*, 433–448.

Guthrie, K. (2013). Aviation regionalism in the Pacific. *Journal of Pacific History, 48*(3), 294–308.

Ha'apio, M., Gonzalez, R., & Wairiu, M. (2019). Is there any chance for the poor to cope with extreme environmental events? Two case studies in the Solomon Islands. *World Development, 122*, 514–524.

Habru, P., & Habru, C. (2024). Will Solomon Islands' new leader stay close to China? *ANU DevPolicyBlog*, 2 May.

Hahamovich, C. (2011). *No man's land. Jamaican guestworkers in America and the global history of deportable labor*. Princeton University Press.

Hair, C., Foale, S., Kinch, J., et al. (2016). Beyond boom, bust and ban: The sandfish (*Holothuria scabra*) fishery in the Tigak Island, Papua New Guineas. *Regional Studies in Marine Science, 5*, 69–79.

Hall, C., & Higham, J. (Eds.). (2005). *Tourism, recreation and climate change*. Channel View Publications.

Hall, I., & McDonald, A. (2020). The Abandoned Cars of Pohnpei. Reflections on a small island economy and environment. *Shima, 14*(2), 121–130.

Hallegatte, S., Bangalore, M., Bonzanigo, L., et al. (2016). *Shock waves: Managing the impacts of climate change on poverty*. World Bank.

Halliday, C., Morrissey, K., Saint Ville, A., et al. (2023). Trends in food supply, diet, and the risk of non-communicable diseases in three small island developing states: Implications for policy and research. *Frontiers in Sustainable Food Systems, 7*, 1058540.

Hameiri, S. (2009a). Governing disorder: The Australian Federal Police and Australia's new regional frontier. *Pacific Review, 22*(5), 549–557.

Hameiri, S. (2009b). The region within: RAMSI, the Pacific Plan and new modes of governance in the Southwest Pacific. *Australian Journal of International Affairs, 63*, 348–360.

Handa, S., & King, D. (1997). Structural adjustment policies, income distribution and poverty: A review of the Jamaican experience. *World Development, 25*(6), 915–930.

Handmer, J., Monson, R., & Schinko, T. (2025). Addressing the diversity of loss and damage in Pacific Island countries to foster a just transition towards a climate-resilient future. *Climate and Development, 17*(7), 657–669.

Hanich, Q., Wabnitz, C., Ota, Y., Amos, M., et al. (2018). Small-scale fisheries under climate change in the Pacific Islands region. *Marine Policy, 88*, 279–284.

Hanich, Q., Jung, M., McDonald, A., et al. (2021). Tuna fisheries conservation and management in the Pacific Islands region: Implications for Korean distant water fisheries. *Asia-Pacific Journal of Ocean Law and Policy, 6*(2), 192–220.

Hanlon, D. (2018). A different historiography for "A Handful of Chickpeas Flung over the Sea". Approaching the federated states of Micronesia's Deeper Past. In W. Anderson, M. Johnson, & B. Brookes (Eds.), *Pacific futures. Past and present* (pp. 81–104). University of Hawai'i Press.

Haraksingh, I. (2023). Renewable energy: An emerging blue economy sector. In P. Clegg, R. Mahon, P. McConney, & H. Oxenford (Eds.), *The Caribbean blue economy* (pp. 156–168). Routledge.

Harohau, D., Blythe, J., Sheaves, M., & Diedrich, A. (2020). Limits of Tilapia aquaculture for rural livelihoods in Solomon Islands. *Sustainability, 12*, 4592.

Harohau, D., Allison, E., & Teoh, S. (2025). Nature-based solutions to aquaculture: A way forward for aquaculture development in Pacific Island countries. *SPC Fisheries Newsletter, 176*, 50–60.

Harrington, R., Harvey, N., Larkins, S., & Redman-MacLaren, M. (2021). Family planning in Pacific Island Countries and Territories (PICTs): A scoping review. *PLoS One, 16*(8), e0255080.

Harrison, A., James, C., & Pemberton, A. (2023). Human trafficking of adolescents: A Caribbean perspective. *Current Pediatrics Reports, 11*, 116–112.

Hasan, S., Fleming, C., Smart, J., et al. (2021). Identifying and mitigating risks to completion of small grant climate change adaptation projects: Evidence from the Pacific. *Regional Environmental Change, 21*, 55.

Hassall, G. (2020). *Government and public policy in the Pacific Islands*. Emerald.

Hassall, G., & Tipu, F. (2008). Local government in the South Pacific islands. *Commonwealth Journal of Local Governance, 1*, 6–29.

Hassanali, K. (2020). CARICOM and the blue economy- Multiple understandings and their implications for global engagement. *Marine Policy, 120*, 104137.

Hauck, V. (2010). The role of churches in creating social capital and improving governance in Papua New Guinea: Lessons for working in fragile situations. *Public Administration and Development, 30*(1), 49–65.

Hau'ofa, E. (1993). Our sea of islands. In E. Waddell, V. Naidu, & E. Hau'ofa (Eds.), *A new Oceania: Rediscovering our sea of islands* (pp. 2–16). USP.

Hau'ofa, E. (1998). The ocean in us. *The Contemporary Pacific, 10*(2), 392–410.

Havice, E., & Reed, K. (2012). Fishing for development? Tuna resource access and industrial change in Papua New Guinea. *Journal of Agrarian Change, 12*(2–3), 413–435.

Hawley, N., & McGarvey, S. (2015). Obesity and diabetes in Pacific Islanders: The current burden and the need for urgent action. *Current Diabetes Reports, 15*(6), 29.

Haworth, B., McKinnon, S., & Eriksen, C. (2022). Advancing disaster geographies: From marginalisation to inclusion of gender and sexual minorities. *Geography Compass, 16*(11), e12664.

Hayle, C., Singh, R., & Wright, A. (2010). Planning for tourism resilience in the Caribbean. *Journal of Eastern Caribbean Studies, 35*(1), 36–59.

Heck, N., Beck, M., Reguero, B., et al. (2023). Global climate change risk to fisheries – A multi-risk assessment. *Marine Policy, 148*, 105489.

Heffernan, O. (2024). *The high seas*. Greystone.

Hein, J., Klepp, S., & Bercht, A. (2024). Bringing a justice lens to ocean access. *Maritime Studies, 23*, 45.

Heine, H. (2024). Foreword. In A. Dansie, H. Alleway, & B. Böer (Eds.), *The water, energy, and food security nexus in Asia and the Pacific: The Pacific* (pp. vii–viii). Springer.

Hendriks, S. (2022). Sustainable small-scale fisheries can help people and the planet. *Nature, 606*(7915), 650–652.

Henry-Lee, A., Watson, P., Osei, P., & Trezelle, S. (2010). Social risk management in a small state: The case of St Lucia. *Social and Economic Studies, 59*, 61–89.

Hicks, R., McDougall, D., & Oakeshott, D. (2021). The promise of education: Schooling journeys in the Southwestern Pacific. *The Contemporary Pacific, 33*(2), 301–336.

Hidalgo, D., Witten, I., Nunn, P., et al. (2020). Sustaining healthy diets in times of change: Linking climate hazards, food systems and nutrition security in rural communities of the Fiji Islands. *Regional Environmental Change, 20*, 73.

Hidalgo, D., Nunn, P., & Beazley, H. (2021). Uncovering multilayered vulnerability and resilience in rural villages in the Pacific: A case study of Ono Island, Fiji. *Ecology and Society, 26*(1), 26.

Higgins-Desbiolles, F. (2020). Socialising tourism for social and ecological justice after Covid-19. *Tourism Geographies, 22*(3), 610–623.

Higham, J., Font, X., & Wu, J. (2022). Code red for sustainable tourism. *Journal of Sustainable Tourism, 30*(1), 1–13.

Higton, J., White, Y., Nicoletti, R., et al. (2025). *International teacher recruitment*. Department of Education.

Hill, C. (2025a). Realists, re-imaginers and reformers: Aid in the post-America multiverse. *ANU DevPolicyBlog*, 11 June.

Hill, C. (2025b). A new look at aid's 'great game'. *ANU DevPolicyBlog*, 12 November.

Hill, L. (2025c). Identities in disasters: Opportunities and challenges for developing inclusive disaster risk governance networks for gender and sexual minorities in Dominica and neighbouring small Caribbean islands. In D. Dominey-Howes, A. Rushton, W. Leonard, et al. (Eds.), *Queering disasters, climate change and humanitarian crises* (pp. 313–333). Springer Nature.

Hill, N., Peatman, T., Wakefield, C., et al. (2018). Improving guidelines for implementing harvest strategies in capacity-limited fisheries – Lessons from Tonga's deepwater line fishery. *Marine Policy, 98*, 85–91.

Hills, J., Michalena, E., & Chalvatzis, K. (2018). Innovative technology in the Pacific: Building resilience for vulnerable communities. *Technological Forecasting and Social Change, 129*, 16–26.

Hirano, S. (2008). The development of national sustainable development strategies in small island developing states. In J. Strachan & C. Vigilance (Eds.), *Sustainable development in small island developing states* (pp. 4–9). Commonwealth Secretariat.

Hitkari, C. (2024). *Revamping Partners in the Blue Pacific Initiative*. PacNet 13 Pacific Forum.

Ho, S., Dias Baptista, M., & McEvoy, D. (2023). Community profiling to support inclusive urban community-based climate adaptation: Experiences of a survey-based approach in urban informal settlements in Honiara, Solomon Islands. *Climate and Development, 15*(4), 325–339.

Hoarau-Heemstra, H., Wigger, K., Olsen, J., & James, L. (2023). Cruise tourism destinations: Practices, consequences and the road to sustainability. *Journal of Destination Marketing & Management, 30*, 100820.

Hobbis, G. (2017). The microSDs of Solomon Islands. An offline remittance economy of digital multi-media. *Digital Culture & Society, 3*(2), 21–40.

Hobbis, G. (2019a). New media, new Melanesia? In E. Hirsch & W. Rollason (Eds.), *The Melanesian world* (pp. 546–560). Routledge.

Hobbis, S. (2019b). A road to development? Rural perspectives on infrastructure maintenance in Solomon Islands. *Development in Practice, 29*(6), 748–759.

Hobbis, G. (2020). *The digitizing family. An ethnography of Melanesian smartphones*. Palgrave Macmillan.

Hofman, C., Stancioff, C., Richards, A., et al. (2021). Resilient Caribbean communities: A long-term perspective on sustainability and social adaptability to natural hazards in the Lesser Antilles. *Sustainability, 13*(17), 9807.

Hogarth, J., & Wójcik, D. (2016a). An evolutionary approach to adaptive capacity assessment: A case study of Whitehouse, Jamaica. *Journal of Rural Studies, 43*, 248–259.

Hogarth, J., & Wójcik, D. (2016b). An evolutionary approach to adaptive capacity assessment: A case study of Soufriere, Saint Lucia. *Sustainability, 8*, 228.

Holden, A., Jamal, T., & Burini, F. (2022). The future of tourism in the Anthropocene. *Annual Review of Environment and Resources, 47*(1), 423–447.

Horton, P. (2012). Pacific Islanders in global rugby: The changing currents of sports migration. *International Journal of the History of Sport, 29*(17), 2388–2404.

Hope, K. R. (1974). CARIFTA and Caribbean trade: An overview. *Caribbean Studies, 14*(1), 169–179.

Hosein, R. (2021). *Oil and gas in Trinidad and Tobago managing the resource curse in a small petroleum-exporting economy*. Palgrave Macmillan.

House, J., Delise, A., Malverus, V., et al. (2024). Towards inclusion of persons with disabilities in coastal fisheries in the Pacific: A review of current practice and approaches. *SPC Fisheries Newsletter, 174*, 28–35.

Howard, A. (2019). Diaspora no more? The role of Facebook in the development of a global Rotuman community. *Diaspora, 20*(2), 177–203.

Howden, M. (2022, July 5). Floods will continue to intensify so we must act. *Sydney Morning Herald.*

Howes, S. (2023, April 20). Questions about Australian aid to fund the Pacific Games. *Interpreter.*

Howes, S., & Chowdhury, R. (2025). GNDI: Anew approach to measuring economic performance in the Pacific. *ANU DevPolicyBlog*, 28 November.

Howes, S., & Liu, H. (2022). Australian government gifts US$190 million to Telstra to buy Digicel. *ANU DevPolicyBlog*, 21 April.

Howes, S., & Sen, S. (2022). Time for action at the Forum on female political representation in the Pacific. *ANU DevPolicyBlog*, 8 July.

Howes, S., & Turia, N. (2025). PEV popularity grows as Timor-Leste takes top slot. *ANU DevPolicyBlog*, 6 November.

Huebner, A. (2012). Public perceptions of destination vulnerability to climate change and implications for long-haul travel decisions to small island states. *Journal of Sustainable Tourism, 20*(7), 939–951.

Huff, A., & Naess, L. (2022). Introduction: Reframing climate and environmental justice. *IDS Bulletin, 53*(4), 1–11.

Huffer, E. (2006). Regionalism and cultural identity: Putting the Pacific back into the plan. In S. Firth (Ed.), *Globalisation and governance in the Pacific Islands* (pp. 25–48). ANU Press.

Hughes, A., Teece, R., Heard, D., & Hatayama, K. (2021). *Pacific skills partnerships: Addressing skills shortages in Fiji and other Pacific Island Countries.* Overseas Development Institute.

Hughes, C. (1958). Experiments towards closer union in the British West Indies. *The Journal of Negro History, 43*(2), 85–104.

Hughes, E. (2023). Tourism and sustainable development goals in small island states in the Pacific. In M. Stephenson (Ed.), *Routledge handbook on tourism and small states in the Pacific* (pp. 28–43). Routledge.

Hughes, T., Barnes, M., Bellwood, D., et al. (2017). Coral reefs in the anthropocene. *Nature, 546*(7656), 82–90.

Hukula, F. (2017). Making a home in the city: Ideas of employment and work. *Development Bulletin, 78*, 45–48.

Human Rights Watch. (2025). *'There's just no more land'. Community-led planned relocation as last-resort adaptation to sea-level rise in Solomon Islands.* Human Rights Watch.

Huntington, H., & Shenoy, A. (2021). Does insecure land tenure deter investment? Evidence from a randomized controlled trial. *Journal of Development Economics, 150*, 102632.

Hurley, G., Wilkinson, E., Weaver, J., & Kasanawaqa, V. (2025). *Turning the tide. Enhancing ocean equity for small island developing states*. ODI Global.

Iati, I. (2021). China's impact on New Zealand foreign policy in the Pacific: The Pacific reset. In G. Smith & T. Wesley-Smith (Eds.), *The China alternative. Changing regional order in the Pacific Islands* (pp. 143–166). ANU Press.

Idris, I. (2021). *LGBT rights and inclusion in small island developing states (SIDS)*. K4D Helpdesk Report 970. Institute of Development Studies.

Iese, V., Kiem, A., Mariner, A., Malsale, P., et al. (2021). Historical and future drought impacts in the Pacific islands and atolls. *Climatic Change, 116*, 19.

International Finance Corporation. (2018). *From the farm to the tourist's table: A study of fresh produce demand from Fiji's hotels and resorts*. IFC.

International Finance Corporation. (2019). *Assessment of the economic impact of cruise ship tourism in Fiji*. IFC.

Intergovernmental Panel on Climate Change [IPCC]. (2021). Summary for policymakers. In V. Masson-Delmotte et al. (Eds.), *Climate change 2021: The physical science basis. Contribution of Working Group I to the Sixth Assessment Report of the Intergovernmental Panel on Climate Change*. Cambridge University Press.

Intergovernmental Panel on Climate Change [IPCC]. (2022). *Climate change 2022: Impacts, adaptation and vulnerability (Working Group II contribution to the IPCC's Sixth Assessment Report)*. Cambridge University Press.

IOM. (2012). *Mechanisms for labour migration in the Caribbean*. IOM.

Iredale, R., Voigt-Graf, C., & Khoo, S. (2015). Trends in international and internal teacher mobility in three Pacific Island Countries. *International Migration, 53*(1), 97–114.

Ishemo, A., & Semple, H. (2006). Population mobility and the survival of small farming in the Rio Grande Valley, Jamaica. *Geographical Journal, 172*(4), 318–330.

Isley, C., & Taylor, M. (2018). Air quality management in the Pacific Islands: A review of past performance and implications for future directions. *Environmental Science & Policy, 84*, 26–33.

Jackson, G., & Piggott-McKellar, A. (2021). Gods, spirits and natural hazards: Ontologies and epistemologies of natural hazards and climate change in Kiribati and Papua New Guinea. In J. Luetz & P. Nunn (Eds.), *Beyond belief: Opportunities for faith-engaged approaches to climate-change adaptation in the Pacific Islands* (pp. 81–98). Springer Nature.

Jackson, J., Donovan, M., Cramer, K., & Lam, V. (Eds.). (2014). *Status and trends of Caribbean Coral Reefs: 1970–2012*. IUCN.

Jackson, G., N'Guetta, A., De Rosa, S., et al. (2023). An emerging governmentality of climate change loss and damage. *Progress in Environmental Geography, 2*(1–2), 33–57.

Jacobs, B. (1975). Administrative problems of small countries. In P. Selwyn (Ed.), *Development Policy in Small Countries* (pp. 134–143). Croom Helm.

Jacobson, J. (2015). The complexities of nurse migration. *American Journal of Nursing, 115*(12), 22–23.

Jaeckel, A., Harden-Davies, H., Amon, D., van der Grient, J., et al. (2023). Deep seabed mining lacks social legitimacy. *npj Ocean Sustainability, 2*(1), 1.

Jamaica Education Transformation Committee. (2021). *The Reform of Education in Jamaica, 2021*. Kingston.

James, K., & Clarke, J. (2025, June 11). Securing our blue future: Caribbean leadership for global marine protection. *IISD.*

James, P., Nadarajah, Y., Haive, K., & Stead, V. (2012). *Sustainable communities, sustainable development: Other paths for Papua New Guinea*. University of Hawai'i Press.

James, P., Tidd, A., & Kaitu, L. (2018). The impact of industrial tuna fishing on small-scale fishers and economies in the Pacific. *Marine Policy, 95*, 189–198.

Jansen, H., Stren, A., & Weiss, E. (2015). *Linking farmers and agroprocessors to the tourism industry in the Eastern Caribbean*. World Bank.

Jarillo, S., & Barnett, J. (2021). Contingent communality and community-based adaptation to climate change: Lessons from a Pacific rural atoll. *Journal of Rural Studies, 87*, 137–145.

Jarillo, S., & Barnett, J. (2022). Repositioning the (is)land: Climate change adaptation and the atoll assemblage. *Antipode, 54*(3), 872–892.

Jessop, D. (2019, September 15). Reassessing the role of cruise tourism. *Tourism Analytics.*

Jinnah, S. (2011). Climate change bandwagoning: The impacts of strategic linkages on regime design, maintenance and death. *Global Environmental Politics, 11*(2), 1–9.

Johansson-Fua, S. (2022). Wansolwara: Sustainable development, education, and regional collaboration in Oceania. *Comparative Education Review, 66*(3), 516–534.

Johnson, G (2012) Increased numbers of Marshall Islanders deported from U.S. *Marianas Variety*, 25 June.

Johnson, G. (2015). *Idyllic no More*. Create Space.

Johnson, G. (2021). *Our ocean's promise: From aspirations to inspirations: The Marshall Islands fishing story*. Create Space.

Johnson, B., & Young, A. (2025, July 21). Friends to all: Fiji prioritises shared value and regional leadership. *The Strategist.*

Johnson-Myers, T., & Campbell, Y. (2024). The cost of politics: How money acts as a barrier for women in politics in the Caribbean. *Canadian Journal of Latin American and Caribbean Studies, 49*(2), 220–239.

Johnston, R., Crooks, V., & Ormond, M. (2015). Policy implications of medical tourism development in destination countries: Revisiting and revising an existing framework by examining the case of Jamaica. *Global Health, 11*, 29.

Jolly, J., & O'Carroll, L. (2023, October 20). EU raises security concerns as it reveals five Caribbean states have sold 88,000 'golden passports'. *The Guardian.*

Jolly, M., Lee, H., Lepani, K., et al. (2015). *Falling through the net? Gender and Social Protection in the Pacific*. UNWomen.

Jones, P., & Lea, J. (2007). What has happened to urban reform in the island Pacific? Some lessons from Kiribati and Samoa. *Pacific Affairs, 80*(3), 473–491.

Jones, F., & Serieux-Lubin, L. (2018). *Disability, human rights and public policy in the Caribbean: A situation analysis*. ECLAC.

Jones, E., Deere-Birkbeck, C., & Woods, N. (2010). *Manoeuvring at the margins: Constraints faced by small states in international trade negotiations.* Commonwealth Secretariat.

Jones, C., Murray, W., & Overton, J. (2017). FIJI Water, water everywhere: Global brands and democratic and social injustice. *Asia Pacific Viewpoint, 58*(1), 112–123.

Jourdan, C. (2017). Haosgel: Parenté, classes sociales et transformations urbaines à Honiara. *Journal de la Société des Océanistes, 144–145*, 131–146.

Joshi, K. (2021, December 29). In Australia's climate wars, delay and deception are the new denial. *Guardian.*

Jules, D. (2008). Rethinking education for the Caribbean: A radical approach. *Comparative Education, 44*(2), 203–214.

Jules, T. (2024). Introduction: CARICOM @ 50- is it time to decolonise southern regionalism? *The Round Table, 113*(1), 1–14.

Jules, T., Arnold, R., & Smith, A. (2023). Caribbean exceptionalism and the rise of Sino-CARICOM relations in the post-bureaucratic era. *The Round Table, 112*(1), 57–72.

Jumo, M. (2022). Time to reprioritise girls' education. *ANU DevPolicy Blog*, 3 May.

Jupiter, S., Wenger, A., Klein, C., et al. (2017). Opportunities and constraints for implementing integrated land-sea management on islands. *Environmental Conservation, 44*(3), 254–266.

Kabutaulaka, T. (2005). Australian foreign policy and the RAMSI intervention in Solomon Islands. *The Contemporary Pacific*, 283–308.

Kabutaulaka, T. (2006). Global capital and local ownership in Solomon Islands' Forestry Industry. In S. Firth (Ed.), *Globalisation and governance in the Pacific Islands* (pp. 239–257). ANU Press.

Kabutaulaka, T. (2021). Mapping the blue Pacific in a changing regional order. In G. Smith & T. Wesley-Smith (Eds.), *The China alternative. Changing regional order in the Pacific Islands* (pp. 41–70). ANU Press.

Kagy, V., Wong, M., Vandenbroucke, H., et al. (2016). Traditional banana diversity in Oceania: An endangered heritage. *PloS One, 11*(3), e0151208.

Kampmark, B. (2025). *Contract bikies and bribed politicians: Australia's Nauru refugee deal* (pp. 1–3). *Independent Australia.*

Kanemasu, Y. (2008). Weapons of the weak: Employees in the Fiji hotel scene. In J. Connell & B. Rugendyke (Eds.), *Tourism at the grassroots* (pp. 114–130). Routledge.

Kanemasu, Y., & Molnar, G. (2017). Private military and security labour migration: The case of Fiji. *International Migration, 55*(4), 154–170.

Kant, R., Wheeler, V., & Chung, M. (2025). The Pacific Ocean of peace: A promise or a paradox? *Australian Journal of International Affairs, 79*(3), 520–530.

Karcher, D., Fache, E., Breckwoldt, A., Guilloux, B., et al. (2020). Trends in South Pacific fisheries management. *Marine Policy, 118*, 104021.

Kaotoke, J., Taukave, J., Wairiu, M., et al. (2024). *Holomui ki mu'a: Pacific catalytic action at IMO proves the power of small.* Micronesian Center for Sustainable Transport, Working Paper 2.

Keen, M., & Connell, J. (2019). Regionalism and resilience? Meeting urban challenges in Pacific Island states. *Urban Planning and Research, 37*(3), 322–337.

Keen, M., & Ride, A. (2019). Trading places: Inclusive cities and market vending in the Pacific Islands. *Asia Pacific Viewpoint, 60*(3), 239–251.

Keen, M., & Sora, M. (2024, January 18). Nauru's diplomatic switch to China – The rising stakes in Pacific geopolitics. *Interpreter.*

Keen, M., Schwarz, A., & Wini-Simeon, L. (2018). Towards defining the blue economy: Practical lessons from Pacific ocean governance. *Marine Policy, 88*, 333–341.

Keen, M., Sanderson, D., Osborne, K., Deo, R., & Pranaya, U. (2022). Area-based approaches and urban recovery in the Pacific: Lessons from Fiji, Solomon Islands and Vanuatu. *Environment and Urbanization, 34*(2), 489–507.

Keene, B. (2008). *Tribe wanted: My adventure on paradise or bust*. Ebury Press.
Kelly, S., Walsh, T., Delport, D., et al. (2023). Health and economic benefits of achieving contraceptive and maternal health targets in small island developing states in the Pacific and Caribbean. *BMJ Global Health, 8*, e010018.
Kelly, F., Wright, S., Woodward, G., & Fussell, J. (2025). Plastic pollution under the influence of climate change: Implications for the abundance, distribution, and hazards in terrestrial and aquatic ecosystems. *Frontiers in Science, 3*, 1636665.
Kelman, I. (2018). Islandness within climate change narratives of small island developing states (SIDS). *Island Studies Journal, 13*(1), 149–166.
Kelman, I. (2021). Words without meaning? Examining sustainable development terminology through small states and territories. *Island Studies Journal, 4*(2), 231–244.
Kelsey, J. (2004). World trade and small nations in the South Pacific region. *Kansas Journal of Law and Public Policy, 14*, 247–306.
Kempadoo, K. (2016). The war on humans: Anti-trafficking in the Caribbean. *Social and Economic Studies, 65*(4), 5–32.
Kempf, W. (2020). Introduction: Climate change and Pacific Christianities. *Anthropological Forum, 30*(3), 215–232.
Kench, P., Ford, M., & Owen, S. (2018). Patterns of island change and persistence offer alternate adaptation pathways for atoll nations. *Nature Communications, 9*, 605.
Kench, P., Sengupta, M., Ford, M., & Owen, S. (2024). Island change framework defines dominant modes of atoll island dynamics in response to environmental change. *Communications Earth & Environment, 5*, 585.
Keppel, G., Morrison, C., Watling, D., et al. (2012). Conservation in tropical Pacific Island countries: Why most current approaches are failing. *Conservation Letters, 5*(4), 256–265.
Kersey, K., Blechinger, P., & Shirley, R. (2021). A panel data analysis of policy effectiveness for renewable energy expansion on Caribbean islands. *Energy Policy, 155*, 112340.
Kharas, H. McArthur, J., & Onyechi, O. (2024). *How is the world doing on the SDGs? Four tests and eight findings*. Brookings Institute Center for Sustainable Development Working Paper No. 188.
Kiddle, L., McEvoy, D., Mitchell, D., et al. (2017). Unpacking the Pacific urban agenda: Resilience challenges and opportunities. *Sustainability, 9*(10), 1878.
Kiddle, L., Zari, M., Blaschke, P., et al. (2021). An Oceania urban design agenda linking ecosystem services, nature-based solutions, traditional ecological knowledge and wellbeing. *Sustainability, 13*(22), 12660.

Kilroy, J., McNamara, K., Witt, B., & Raynes, S. (2025). 'I don't know anything else besides gardening': Exploring sensitivities to climate change in Vanuatu. *Asia Pacific Viewpoint, 66*(1) (in press).

Kim, K. (2024, June 25). Navigating the intersection of climate change and ocean contamination: Addressing plastic waste issues for the future of the Pacific. *Visions and Voices*.

Kinch, J., Vitukawalu, B., Nalasi, U., et al. (2019). Socioeconomic aspects of oyster harvesting in the Rewa River delta area, Fiji. *SPC Fisheries Newsletter, 159*, 45–54.

King, J., Edwards, N., Watling, H., & Hair, S. (2019). Barriers to disability-inclusive disaster management in the Solomon Islands: Perspectives of people with disability. *International Journal of Disaster Risk Reduction, 34*, 459–466.

Kissling, C. (Ed.). (1984). *Transport and Communications for Pacific Microstates: Issues in Organisation and Management*. USP.

Klein, N. (2014). *This changes everything. Capitalism vs. the climate*. Simon and Schuster.

Klepp, S., & Fünfgeld, H. (2022). Tackling knowledge and power: An environmental justice perspective on climate change adaptation in Kiribati. *Climate and Development, 14*(8), 757–769.

Klöck, C., Duvat, V., & Nunn, P. (2022). Maladaptive diffusion? The spread of hard protection to adapt to coastal erosion and flooding along island coasts in the Pacific and Indian Ocean. *Regional Environmental Change, 22*, 136.

Klöck, C., Castro, P., Chin-Yee, S., et al. (2025). Beyond AOSIS: Small island states' presence and participation at COP27. *Climate and Development*, 1–15.

Koczberski, G., Curry, G., & Connell, J. (2001). Full circle or spiralling out of control? State violence and the control of urbanisation in Papua New Guinea. *Urban Studies, 38*(11), 2017–2036.

Koga, K. (2022). Japan's strategic approach toward Island States: Case of the Pacific Islands. *Journal of Indo-Pacific Affairs, 5*(7), 62–83.

Köllner, P. (2025). Explaining sender–receiver gaps in signalling: Australia's 'Pacific Step-up' and Solomon Islands' multi-alignment. *The British Journal of Politics and International Relations, 27*(1), 20–42.

Kolshus, T. (2017). Urban castaways: The precarious living of marooned islanders. In J. Taylor & H. Lee (Eds.), *Mobilities of Return: Pacific Perspectives* (pp. 171–188). ANU Press.

Koro, M., & McNeill, H. (2024). Challenging colonial logics of habit in Australia's economic statecraft with Pacific Islands. *Asia & the Pacific Policy Studies, 11*(3), e398.

Koro, M., McNeill, H., Ivarature, H., & Wallis, J. (2023). Tā, Vā and Lā: Re-imagining the geopolitics of the Pacific Islands. *Political Geography, 105*, 102931.

Korovulavula, I., Nunn, P., Kumar, R., & Fong, T. (2020). Peripherality as key to understanding opportunities and needs for effective and sustainable climate-change adaptation: A case study from Viti Levu Island, Fiji. *Climate and Development, 12*(10), 888–898.

Kowasch, M., & Lindenmann, P. (2014). New flags, upward forces and sheltered harbours: The new 'Great Game' in the Pacific Islands region. *Pacific Geographies, 41*, 4–9.

Kowasch, M., Batterbury, S., Bouard, S., & Wadrawane, E. (2022). The third independence referendum in New Caledonia – A fallback to colonialism? *Pacific Geographies, 57*, 11–15.

Kroodsma, D., Mayorsga, J., Hochberg, T., Miller, N., & Park, J. (2018). Tracking the global footprint of fisheries. *Science, 359*(6378), 904–908.

Kuang, Y., Zhang, B., Li, C., & Cao, Y. (2016). A review of renewable energy utilization in islands. *Renewable and Sustainable Energy Reviews, 59*, 504–513.

Kumar, L., & Taylor, S. (2015). Exposure of coastal built assets in the South Pacific to climate risks. *Nature Climate Change, 5*(11), 992–996.

Kuruppu, N., & Liverman, D. (2011). Mental preparation for climate adaptation: The role of cognition and culture in enhancing adaptive capacity of water management in Kiribati. *Global Environmental Change, 21*(2), 657–669.

Kuster, C., Vuki, V., & Zann, L. (2005). Long-term trends in subsistence fishing patterns and coral reef fisheries yield from a remote Fijian island. *Fisheries Research, 76*(2), 221–228.

Kwansing, R. (2025). Leave no-one behind: Social protection reducing poverty in Kiribati. *ANU DevPolicyBlog*, 5 August.

Kwiatkowska, B. (2007). The 2006 Barbados/Trinidad and Tobago Award: A landmark in compulsory jurisdiction and equitable maritime boundary delimitation. *International Journal of Marine and Coastal Law, 22*(1), 7–60.

Laalaai-Tuasa, C. (2022, November 15). Climate change is a harsh reality in Sāmoa – and the world's leaders need to face it. *The Spin-Off.*

Labh, N. (2024, May 18). Connectivity in crisis: The problem with shipping in the Pacific. *Visions and Voices.*

Labuinao, E. (2020). Kastom, gender and economic development: The case of the fish aggregating device in To'abaita, Solomon Islands. *SPC Women in Fisheries Information Bulletin, 32*, 13–18.

Lachmann, F., Almroth, B., Baumann, H., et al. (2017). *Marine plastic litter on small island developing states (SIDS): Impacts and measures.* Swedish Institute for the Marine Environment, University of Gothenburg.

Laffoley, D., Baxter, J., Amon, D., Claudet, J., et al. (2022). The forgotten ocean: Why COP26 must call for vastly greater ambition and urgency to address ocean change. *Aquatic Conservation: Marine and Freshwater Ecosystems, 32*(1), 217–228.

Lai, M., Robinson, S., Salas, E., et al. (2022). Climate justice for small island developing states: Identifying appropriate international financing mechanisms for loss and damage. *Climate Policy, 22*(9–10), 1213–1224.

Lal, P., Kinch, J., & Wickham, F. (2009). *Review of economic and livelihood impact assessments of, and adaptation to, climate change in Melanesia*. SPREP.

Lallani, S. (2024). Cultural globalization at sea: The rise of the modern Caribbean cruise industry. *Globalizations, 21*(7), 1290–1308.

Lam, C., Islam, M., Devi, R., et al. (2025). Examining wealth-related inequality in childhood vaccination in Fiji using the UNICEF Multiple Indicator Cluster Survey 2021. *International Health, 17*(5).

Lam, V., Allison, E., Bell, J., et al. (2020). Climate change, tropical fisheries and prospects for sustainable development. *Nature Reviews Earth and Environment, 1*(9), 440–454.

Lamb, V. (2023). Constructing the global sand crisis: Four reasons to interrogate crisis and scarcity in narrating extraction. *The Extractive Industries and Society, 15*, 101282.

Larmour, P. (2012). *Interpreting corruption. Culture and politics in the Pacific Islands*. University of Hawai'i Press.

Laslett, K. (2014). *State crimes on the margins of empire: Rio Tinto, the war on Bougainville and resistance to mining*. Pluto Press.

Lata, S., & Nunn, P. (2012). Misperceptions of climate-change risk as barriers to climate-change adaptation: A case study from the Rewa Delta, Fiji. *Climatic Change*, 169–186.

Latai-Niusulu, A., Binns, T., & Nel, E. (2020). Climate change and community resilience in Samoa. *Singapore Journal of Tropical Geography, 41*(1), 40–60.

Latai-Niusulu, A., Tsujita, M., & Neef, A. (2023). Climate micro-mobilities as adaptation practice in the Pacific: The case of Samoa. *Philosophical Transactions of the Royal Society B, 378*, 20220392.

Latai, A., Taulealo, A., & Connell, J. (2024). 'Burn like Hot Stones': Children's perceptions of climate change in Samoa. *Island Studies Journal, 19*(1).

Lau, J., & Sutcliffe, S. (2021). Of isolation and atolls: Coping with COVID-19 in Manus, Papua New Guinea. In Y. Campbell & J. Connell (Eds.), *COVID in the Islands: A comparative perspective on the Caribbean and the Pacific* (pp. 353–366). Palgrave Macmillan.

Lauer, M. (2016). Governing uncertainty: Resilience, dwelling and flexible resource management in Oceania. *Conservation and Society, 14*(1), 34–47.

Lauer, M. (2023). *Sensing disaster: Local knowledge and vulnerability in Oceania.* University of California Press.

Lauer, M., & Aswani, S. (2010). Indigenous knowledge and long-term ecological change: Detection, interpretation, and responses to changing ecological conditions in Pacific Island communities. *Environmental Management, 45*, 985–997.

Lauer, M., Atger, T., Holbrook, S., & Rassweiler, A. (2022). Who is perturbed by ecological perturbations? Marine scientists' and Polynesian fishers' understandings of a crown-of-thorns starfish outbreak. In S. Hoffman, T. Eriksen, & P. Mendes (Eds.), *Cooling down: Local responses to global climate change* (pp. 65–89). Berghahn.

Lazarus, L. (2020). Enacting citizenship, debating sex and sexuality: Conservative Christians' participation in legal processes in Jamaica and Belize. *Commonwealth & Comparative Politics, 58*(3), 366–386.

Leach, M., MacGregor, H., Scoones, I., & Wilkinson, A. (2021). Post-pandemic transformations: How and why COVID-19 requires us to rethink development. *World Development, 138*, 105233.

Le Dé, L., Gaillard, J.-C., Friesen, W., & Matautia Smith, F. (2015). Remittances in the face of disasters: A case study of rural Samoa. *Environment, Development and Sustainability, 17*(3), 653–672.

Le Dé, L., & Jackson-Becerra, J. (2021). COVID-19 and transnational remittances in Samoa: Maintaining family ties in the face of crisis. In Y. Campbell & J. Connell (Eds.), *COVID in the Islands: A comparative perspective on the Caribbean and the Pacific* (pp. 279–297). Palgrave Macmillan.

Lee, H. (2016). 'I was forced here': Perceptions of agency in second generation 'return' migration to Tonga. *Journal of Ethnic and Migration Studies, 42*(15), 2573–2588.

Lee, H. (2017). Overseas-born youth in Tongan high schools: Learning the hard life. In J. Taylor & H. Lee (Eds.), *Mobilities of return: Pacific perspectives* (pp. 75–98). ANU Press.

Lee, H. (2018). Becoming Tongan today. In J. Connell & H. Lee (Eds.), *Change and continuity in the Pacific: Revisiting the region* (pp. 118–135). Routledge.

Lee, C., Li, X., Yu, C., & Zhao, J. (2022). The contribution of climate finance toward environmental sustainability: New global evidence. *Energy Economics, 111*, 106072.

Lenderking, H., Robinson, S., & Carlson, G. (2021). Climate change and food security in Caribbean small island developing states: Challenges and strategies. *International Journal of Sustainable Development & World Ecology, 28*(3), 238–245.

Lenzen, M., Moran, D., Kanemoto, K., et al. (2012). International trade drives biodiversity threats in developing nations. *Nature, 486*(7401), 109–112.

Lenzen, M., Sun, Y., Faturay, F., et al. (2018). The carbon footprint of global tourism. *Nature Climate Change, 8*, 522–528.

Lester, S., & Rhiney, K. (2018). Going beyond basic access to improved water sources: Towards deriving a water accessibility index. *Habitat International, 73*, 129–140.

Levasseur, P. Rouly, S & Seetahul, S (2025) Water insecurity and purchases of soft and sweet beverages: A case study in the obesogenic context of the Pacific Islands, *Journal of the Asia Pacific Economy,* 30.

Levi, M. (2019). Employment of the weak: The role of a multinational factory in the life trajectory of early school leavers in Sāmoa. *The Contemporary Pacific, 31*(2), 417–446.

Lewis, P. (1999). Revisiting the Grenada invasion: the OEC's role, and its impact on regional and international politics. *Social and Economic Studies, 48*, 85–120.

Lewis, P. (2011). Training nurses for export: A viable development strategy? *Social and Economic Studies, 60*, 67–104.

Lewis, J. S. (2020). *Scammer's yard: The crime of black repair in Jamaica.* University of Minnesota Press.

Lewis, L. (2025, May 22). 'Every man and his dog is coming': Samoan PM warns of growing geopolitical interest in Pacific. *Radio New Zealand.*

Lewis, P., & Kirton, C. (2015). Migration and remittances in development: A study of Jamaica. In W. Khonje (Ed.), *Migration and development. Perspectives from small states* (pp. 186–243). Commonwealth Secretariat.

Lewis-Cameron, A. (2021). Understanding small island states and territories. In A. Lewis-Cameron, L.-A. Jordan, & S. Roberts (Eds.), *Managing crises in tourism. Resilience strategies from the Caribbean* (pp. 13–32). Palgrave Macmillan.

Lewis-Cameron, A., & Brown-Williams, T. (2023). Civil society and the pursuit of sustainable tourism in the small island jurisdiction of Tobago. In

L. Briguglio, M. Briguglio, & C. Slatter (Eds.), *Handbook of civil society and social movements in small states* (pp. 13–25). Routledge.

Li, T. (2009). Exit from agriculture: A step forward or step backward for the rural poor. *Journal of Peasant Studies, 36*(3), 629–636.

Liloqula, R. (2024). Behind the shrine of the Pacific Games lurks poor governance and corruption. *ANU DevPolicyBlog*, 29 January.

Lindley, J., & Lothian, S. (2024). Exposure to transnational maritime crime in the Pacific Islands Region. *Ocean and Society, 1*.

Lindsay, C. (2019). Norm rejection: Why small states fail to secure special treatment in global trade politics. *Small States & Territories, 2*(1), 104–124.

Lindstrom, L. (2018). Surprising times on Tanna, Vanuatu. In J. Connell & H. Lee (Eds.), *Change and Continuity in the Pacific* (pp. 55–68). Routledge.

Liu, X., Peerally, J., De Fuentes, C., et al. (2022). Who makes or breaks energy policymaking in the Caribbean small island jurisdictions? A study of stakeholders' perceptions. *Sustainability, 14*, 1902.

Llana, S. (2019, December 17). Rebuilding the Bahamas: How a hurricane blows up social divides. *Christian Science Monitor.*

Llivina, C. (2018). Small states and regional dispute settlement mechanisms: The Caribbean and Pacific experiences. In P. Butler, E. Lein, & R. Salim (Eds.), *Integration and international dispute resolution in small states* (pp. 27–59). Springer Nature.

Loehr, J., & Becken, S. (2025). Institutionalised tourism policy goals: Fit to address climate change. *Journal of Sustainable Tourism, 33*(10), 2075–2093.

Louey, P. (2022). The Pacific blue economy: An instrument of political maneuver. *Marine Policy, 135*, 104880.

Love, M., Beal, C., Gonzalez, D., & Hagabore, J. (2022). A livelihood analysis of resettlement outcomes: Lessons for climate-induced relocations. *Development Policy Review, 40*(4), e12597.

Lowe, R., Gasparrini, A., Van Meerbeeck, C., Lippi, C., et al. (2018). Nonlinear and delayed impacts of climate on dengue risk in Barbados: A modelling study. *PLoS medicine, 15*(7), e1002613.

Lowenthal, D. (1987). Social features. In C. Clarke & T. Payne (Eds.), *Politics, security and development in small states* (pp. 26–49). Allen and Unwin.

Loyd, J., Mitchell-Eaton, E., & Mountz, A. (2016). The militarization of islands and migration: Tracing human mobility through US bases in the Caribbean and the Pacific. *Political Geography, 53*, 65–75.

Lucas, H., Fifita, S., Talab, I., et al. (2017). Critical challenges and capacity building needs for renewable energy deployment in Pacific small island developing states (Pacific SIDS). *Renewable Energy, 107*, 42–52.

Lutz, C. (1988). *Unnatural emotions: Everyday sentiments on a Micronesian Atoll.* University of Chicago Press.

Lyons, K., Walters, P., & Riddell, E. (2015). The role of faith-based organizations in environmental governance: The case of forestry in Solomon Islands. *Journal of Environmental Policy & Planning, 18*(3), 342–360.

Malherbe, W., Sauer, W., & Aswani, S. (2020). Social capital reduces vulnerability in rural coastal communities of Solomon Islands. *Ocean & Coastal Management*, 191, 105186.

McAdam, J. (2014). Historical cross-border relocations in the Pacific: Lessons for planned relocations in the context of climate change. *Journal of Pacific History, 49*, 301–332.

McAdam, J. (2015, September 1). Secret refugee deal with Nauru just hides the problem. *Sydney Morning Herald.*

McAdoo, B., Moore, A., & Baumwoll, J. (2009). Indigenous knowledge and the near-field population response during the 2007 Solomon Islands tsunami. *Natural Hazards, 48*(1), 73–82.

MacCarthy, M. (2023). "It will be like a town here, things are really coming up!": Inequality in village-based cruise ship tourism in the Trobriand Islands. *The Contemporary Pacific, 35*(1), 61–93.

McConney, P., Cox, S.-A., & Parsram, K. (2020). Fisheries governance and food security in the Eastern Caribbean. In J. Connell & K. Lowitt (Eds.), *Food security in small island states* (pp. 199–218). Springer.

McCubbin, S., Smit, B., & Pearce, T. (2015). Where does climate fit? Vulnerability to climate change in the context of multiple stressors in Funafuti, Tuvalu. *Global Environmental Change, 30*, 43–55.

MacDonald, S. (2022). *The New Cold War, China, and the Caribbean.* Palgrave Macmillan.

MacDonald, M., Chan, T., Elliott, M., Kearton, L., et al. (2017). Temporal and thematic trends in water, sanitation and hygiene (WASH) research in Pacific Island countries: A systematic review. *Journal of Water Sanitation and Hygiene for Development, 7*(3), 352–368.

MacDonald, M., Elliott, M., Langidrik, D., et al. (2020). Mitigating drought impacts in remote island atolls with traditional water usage behaviors and modern technology. *Science of the Total Environment, 741*, 140230.

McDonnell, S. (2017). Urban land grabbing by political elites: Exploring the political economy of land and the challenges of regulation. In S. McDonnell et al. (Eds.), *Kastom, property and ideology: Land transformations in Melanesia* (pp. 283–304). ANU Press.

McDonnell, S. (2021). The importance of attention to customary tenure solutions: Slow onset risks and the limits of Vanuatu's climate change and resettlement policy. *Current Opinion in Environmental Sustainability, 50*, 281–288.

McDonnell, S. (2023). *My land, my life. Dispossession at the frontier of desire.* University of Hawai'i Press.

McDonnell, S. (2025). Renewed hope: Why the ICJ decision matters to us all. *ANU DevPolicyBlog*, 25 July.

McDougall, D. (2005). The unintended consequences of clarification: Development, disputing, and the dynamics of community in Ranongga, Solomon Islands. *Ethnohistory, 52*(1), 81–109.

McDougall, D. (2016). *Engaging with strangers: Love and violence in the rural Solomon Islands*. Berghahn.

McEvoy, D., Mitchell, D., & Trundle, A. (2020). Land tenure and urban climate resilience in the South Pacific. *Climate and Development, 12*(1), 1–11.

McEvoy, D., Tara, A., Vahanvati, M., et al. (2024). Localized nature-based solutions for enhanced climate resilience and community wellbeing in urban informal settlements. *Climate and Development, 16*(7), 600–612.

MacFeely, S. (2020). Measuring the sustainable development goal indicators: An unprecedented statistical challenge. *Journal of Official Statistics, 36*(2), 361–378.

MacFeely, S., Peltola, A., Barnat, N., et al. (2021). Constructing a criteria-based classification for small island developing states: An investigation. *Journal of Marine and Island Cultures, 10*(1), 155–195.

McGinn, A., & Solofa, A. (2020). Multi-level governance of climate change adaptation: A case study of country-wide adaptation projects in Samoa. In W. Filho (Ed.), *Managing climate change adaptation in the Pacific region* (pp. 151–170). Springer.

McIlgorm, A., Raubenheimer, K., McIlgorm, D., & Nichols, R. (2022). The cost of marine litter damage to the global marine economy: Insights from the Asia-Pacific into prevention and the cost of inaction. *Marine Pollution Bulletin, 174*, 113167.

McIntyre, W. (2014). *Winding up the British Empire in the Pacific Islands*. Oxford University Press.

McKay, D., Staal, A., Abrams, J., et al. (2022). Exceeding 1.5°C global warming could trigger multiple climate tipping points. *Science, 377*(6611), 1171–1180.

McKenzie, N. (2025, June 19). AFP probes firm's insured luxury. *Sydney Morning Herald.*

Maclellan, N. (2007). Fiji, Iraq and Pacific island security. *Race and Class, 48*(3), 47–62.

Maclellan, N. (2022). The region in review: International issues and events 2021. *The Contemporary Pacific, 34*(2), 422–446.

McLeman, R. (2019). International migration and climate adaptation in an era of hardening borders. *Nature Climate Change, 9*, 911–918.

McLennan, A., & Ulijaszek, S. (2015). Obesity emergence in the Pacific islands: Why understanding colonial history and social change is important. *Public Health Nutrition, 18*(8), 1499–1505.

MacLeod, M., Arp, H., Tekman, M., & Jahnke, A. (2021). The global threat from plastic pollution. *Science, 373*, 61–65.

McLeod, M. (2024). *Island tourism policy and sustainable development*. Routledge.

McLeod, M., Dodds, R., & Butler, R. (2024). *Island tourism sustainability and resiliency*. Routledge.

Mcleod, E., Bruton-Adams, M., Förster, J., et al. (2019). Lessons from the Pacific Islands – Adapting to climate change by supporting social and ecological resilience. *Frontiers in Marine Science, 6*, 289.

McMichael, C., Katonivualiku, M., & Powell, T. (2019). Planned relocation and everyday agency in low-lying coastal villages in Fiji. *Geographical Journal, 185*(3), 325–337.

McMichael, C., Powell, T., Piggott-McKellar, A., & Yee, M. (2025). Climate change and the planned relocation of people: A longitudinal analysis of Vunidogoloa. *Fiji, Ambio, 54*, 1043–1056.

McMillen, H., Ticktin, T., Friedlander, A., et al. (2014). Small islands, valuable insights: Systems of customary resource use and resilience to climate change in the Pacific. *Ecology and Society, 19*(4), 44.

McMurray, C., & Smith, R. (2001). *Diseases of globalization. Socioeconomic transitions and health*. Earthscan.

McNamara, K. E. (2013). Taking stock of community-based climate-change adaptation projects in the Pacific. *Asia Pacific Viewpoint, 54*(3), 398–405.

McNamara, K., & Gibson, C. (2009). We do not want to leave our land: Pacific ambassadors at the United Nations resist the category of climate refugees. *Geoforum, 40*(3), 475–483.

McNamara, K., & Farbotko, C. (2017). Resisting a 'doomed' fate: An analysis of the Pacific Climate Warriors. *Australian Geographer, 48*(1), 17–26.

McNamara, K., & Prasad, S. (2014). Coping with extreme weather: Communities in Fiji and Vanuatu share their experiences and knowledge. *Climate Change, 123*, 121–132.

McNamara, K., Clissold, R., & Westoby, R. (2020). Women's capabilities in disaster recovery and resilience must be acknowledged, utilized and supported. *Journal of Gender Studies, 30*(1), 119–125.

McNamara, K., Clissold, R., & Westoby, R. (2022). Teaching our children from books, not the sea': How climate change is eroding human right in Vanuatu. *The Conversation.*

Macnaught, T. (1982). Mass tourism and the dilemmas of modernization in Pacific island communities. *Annals of Tourism Research, 9*(3), 359–381.

McNeill, H. (2025). New Zealand's statecraft 'in and out of the Pacific'. In J. Wallis, H. McNeill, M. Rose, & A. Tidwell (Eds.), *Power and influence in the Pacific Islands. Understanding Statecraftiness* (pp. 140–154). Routledge.

McNeill, H., & Koro, M. (2024). Pacific states are setting the terms of diplomatic engagement. *ANU DevPolicyBlog*, 11 September.

McNeill, H., & Mackenzie, M. (2024). Sasa'a le fafao?: Approaches to return and reintegration of criminal deportees (returnees) into Samoa. *International Journal for Crime, Justice and Social Democracy, 13*(3), 59–70.

McNeill, H., & Walton, G. (2025). Passports, prosperity and problems: (Counter)geopolitical narratives of Vanuatu's citizenship by investment scheme. *Geopolitics, 30*(4), 1795–1825.

MacWilliam, S. (2022). "A road with no end": Making the South Pacific a permanent labour reserve. *Journal of Contemporary Asia., 52*(5), 830–854.

Maebuta, H., & Maebuta, J. (2009). Generating livelihoods: A study of urban squatter settlements in Solomon Islands. *Pacific Economic Bulletin, 24*(3), 118–131.

Maggio, R. (2023). "this is not culture"/"yes, it is": Conceptualising Kiribati as an Interstitial Island in the Chinese Pacific. *Political Geography, 104*, 102907.

Maharaj, S., Asmath, H., Ali, S., Agard, J., et al. (2019). Assessing protected area effectiveness within the Caribbean under changing climate conditions: A case study of the small island, Trinidad. *Land Use Policy, 81*, 185–193.

Mahon, R., Petrie, J., Trotman, A., et al. (2021). Climate services for tourism: Insights from Caribbean small island developing states. *Climate Services, 24*, 100262.

Mahoney, I., & Collins, V. (2020). The capitalist voyeur: Commodification, consumption and the spectacle of the cruise. *Leisure Studies, 39*(2), 280–293.

Mallin, F. (2018). From sea-level rise to seabed grabbing: The political economy of climate change in Kiribati. *Marine Policy, 97*, 244–252.

Mallin, F., & Barbesgaard, M. (2020). Awash with contradiction: Capital, ocean space and the logics of the Blue Economy Paradigm. *Geoforum, 113*, 121–132.

Manglou, M., Rocher, L., & Bahers, J.-B. (2022). Waste colonialism and metabolic flows in island territories. *Journal of Political Ecology, 29*(1), 1–19.

Mansur, A., Doyle, J., & Ivaschenko, O. (2017). *Social protection and humanitarian assistance nexus for disaster response: Lessons learnt from Fiji's Tropical Cyclone Winston*. World Bank.

Marre, J.-B., & Imhof, C. (2021). An assessment of socio-economic impacts due to COVID-19 on coastal fisheries in Tonga. *SPC Fisheries Newsletter, 165*, 28–32.

Marre, J.-B., & Hunt, A. (2021). Coastal fisheries and FAD monitoring project in Kadavu, Fiji: An assessment of data collected from 2017-2020. *SPC Fisheries Newsletter, 166*, 29–37.

Martin, P., Nunn, P., Leon, J., & Tindale, N. (2018). How sea-level and climate change affect island people: The example of Yadua Island, Fiji. *Climate Risk Management, 21*, 7–15.

Masta, M., & Jackson, E. (2024, July 31). Gender equality, the Pacific way. *Interpreter*.

Matagi, L., Richards, D., Yabar, H., et al. (2024). Enhancing solid waste management in Fiji: A comprehensive approach with LCA, GIS, and waste treatment strategies. *Resources, Conservation & Recycling Advances, 23*, 200228.

Mauli, S., Maelaua, J., Reeve, E., et al. (2023). Systemic capacity in food system governance in the Solomon Islands: "It's more than just training". *Sustainability, 15*(13), 10710.

Mausio, A. (2021). Spectre of neoliberal gentrification in Fiji: The price Suva's poor must pay. *Asia Pacific Viewpoint, 62*(1), 2–14.

Mausio, A. (2023). A South-South Tango: China's soft power relations with Fiji since 1975. *Island Studies Journal, 18*(2).

Massola, J. (2020, July 13). Tension on the high seas. *Sydney Morning Herald*, 14–15.

Mawdsley, E., Banks, G., Sanyu, C., et al. (2025). Taking (anti-)'woke' seriously: The future of development cooperation and humanitarian aid. *Development Policy Review, 43*, e70031.

Mecartney, S., & Connell, J. (2017). Urban Melanesia: The challenges of managing land, modernity and tradition. In S. McDonnell, M. Allen, & C. Filer (Eds.), *Kastom, property and ideology: Land transformations in Melanesia* (pp. 57–84). ANU Press.

Meessen, B., Ancia, A., Gill, D., et al. (2024). When one size does not fit all: Aid and health system strengthening for small island developing states. *Health Policy and Planning, 39*(Suppl.1), i4–i8.

Meki, T., & Tarai, J. (2023). How can aid be decolonized and localized in the Pacific? Yielding and wielding power. *Development Policy Review, 41*, e12732.
Mercer, J., Kelman, I., Alfthan, B., & Kurvits, T. (2012). Ecosystem-based adaptation to climate change in Caribbean small island developing states: Integrating local and external knowledge. *Sustainability, 4*, 1908–1932.
Mercer, J., Kurvits, T., Kelman, I., & Mavrogenis, S. (2014). Ecosystem-based adaptation for food security in the AIMS SIDS: Integrating external and local knowledge. *Sustainability, 6*, 5566–5597.
Meschede, H., Bertheau, P., Khalili, S., & Breyer, C. (2022). A review of 100% renewable energy scenarios on islands. *WIREs Energy and Environment, 11*, e450.
Michalena, E., & Hills, J. (2018). Paths of renewable energy development in small island developing states of the South Pacific. *Renewable and Sustainable Energy Reviews, 82*(1), 343–352.
Middelbeek, L., Kolle, K., & Verrest, H. (2014). Built to last? Local climate change adaptation and governance in the Caribbean: The case of an informal urban settlement in Trinidad and Tobago. *Urban Climate, 8*, 138–154.
Middleby, S. (2025, March 6). Is this the end of aid as we know it? *Interpreter*.
Mikhailovich, K., Mackenzie, E., & Smith, D. (2023). Enhancing coastal livelihoods through half-pearl aquaculture and handicrafts in Tonga and Fiji. *Development in Practice, 33*(1), 110–121.
Militz, T., & Foale, S. (2017). The "Nemo Effect": Perception and reality of Finding Nemo's impact on marine aquarium fisheries. *Fish and Fisheries, 18*(3), 596–606.
Miller, P. (2014). Education for all in the Caribbean: Promise, paradox and possibility. *Research in Comparative and International Education, 9*(1), 1–3.
Millette, S., Williams, W., & Hull, C. (2019). Materials flow analysis in support of circular economy development: Plastics in Trinidad and Tobago. *Resources, Conservation & Recycling, 150*, 104436.
Mills, M., Pressey, R., Weeks, R., Foale, S., & Ban, N. (2010). A mismatch of scales: Challenges in planning for implementation of marine protected areas in the Coral Triangle. *Conservation Letters, 3*, 291–303.
Mimura, N., Nurse, L., McLean, R., et al. (2007). Small islands. In M. Perry et al. (Eds.), *Climate change 2007. Impacts, adaptation and vulnerability* (pp. 687–716). Cambridge University Press.
Min, Y., Chen, H., & Perucci, F. (2024). Data on SDGs are riddled with gaps. Citizens can help. *Nature, 633*, 279–281.
Minnis, J., Rolle, S., & Bethell-Bennett, I. (2020). The impact of tourism development on small island communities in The Bahamas. The case of Abaco,

Bimini and Exuma. In S. Rolle, J. Minnis, & I. Bethell-Bennett (Eds.), *Tourism development, governance and sustainability in the Bahamas* (pp. 31–51). Routledge.

Minter, T., & van der Ploeg, J. (2023). 'Our happy hour became a hungry hour': Logging, subsistence and social relations in Solomon Islands. *International Forestry Review, 25*(1), 113–135.

Minto, J. (2019). The impact of Chinese outbound foreign direct investment in Jamaica, (2000-2017). In E. Peters (Ed.), *China's foreign direct investment in Latin America and the Caribbean* (pp. 287–308). Lynne Rienner.

Mishra, P. (2006). *Emigration and brain drain: Evidence from the Caribbean.* IMF Working Paper No. 06/25.

Mitchell, J. (2004). "Killing Time" in a postcolonial town: Young people and settlements in Port Vila, Vanuatu. In V. Lockwood (Ed.), *Globalisation and cultural change in the Pacific Islands* (pp. 338–376). Pearson.

Mitchell, J. (2011). "Operation restore public hope": Youth and the magic of modernity in Vanuatu. *Oceania, 81*(1), 36–30.

Mitchell, J., & McEvoy, D. (2019). *Land tenure and climate vulnerability.* UN Habitat.

Moala, K. (2024, October 23). The Pacific INA - at 50. *ANU DevPolicyBlog.*

Moesinger, A. (2019a). Influence of socio-economic stressors on interpretations of climate change on Takuu Atoll, Papua New Guinea. *Journal de le Société des Océanistes, 149*, 223–234.

Moesinger, A. (2019b). *Modifications to natural resource use in response to perceptions of changing weather conditions on Takuu Atoll, Papua New Guinea.* SPC Traditional Marine Resource Management and Knowledge Information Bulletin No. 40.

Moghal, Z., & O'Connell, E. (2018). Multiple stressors impacting a small island tourism destination-community: A nested vulnerability assessment of Oistins, Barbados. *Tourism Management Perspectives, 26*, 78–88.

Moglia, M., Perez, P., & Burn, S. (2008). Water troubles in a Pacific atoll town. *Water Policy, 10*, 613–637.

Mohammadi, E., Singh, S., & Habib, K. (2021a). Electronic waste in the Caribbean: An impending environmental disaster or an opportunity for a circular economy?, *Resources. Conservation and Recycling, 164*, 105106.

Mohammadi, E., Singh, S., & Habib, K. (2021b). How big is circular economy potential on Caribbean islands considering e-waste? *Journal of Cleaner Production, 317*, 128457.

Mohammadi, E., Singh, S., McCordic, C., & Pittman, J. (2022). Food security challenges and options in the Caribbean: Insights from a scoping review. *Anthropocene Science, 1*(1), 91–108.

Mohamed, A. (2025, February 18). Handouts are never free. The cruel US aid freeze is an opportunity for the Caribbean. *Guardian.*

Mohan, P. (2022a). Climate finance to support Caribbean small island developing states efforts in achieving their Nationally Determined Contributions in the energy sector. *Energy Policy, 169*, 113208.

Mohan, P. (2022b). Sustainable tourism and the sustainable development goals in sub-national island jurisdictions: The case of Tobago. *Island Studies Journal, 17*(2), 168–191.

Mohan, P. (2023). Financing needs to achieve nationally determined contributions under the Paris Agreement in Caribbean small island developing states. *Mitigation and Adaptation Strategies for Global Change, 28*, 26.

Mohan, P. (2025a). Climate change adaptation in small island developing states: Evidence from the nationally determined contributions of Caribbean States. *Regional Environmental Change, 25*(1), 32.

Mohan, P. (2025b). Climate reparations for a just response to climate change: A review of historical responsibility and future implication. *Wiley Interdisciplinary Reviews: Climate Change, 16*, e70007.

Mohan, P., & Strobl, E. (2025). The macroeconomic impact of disasters: A review of hurricanes in Caribbean small island developing states. *Climate and Development, 17*(7), 602–615.

Mohee, R., Mauthoor, S., Bundhoo, Z. M., Somaroo, G., et al. (2015). Current status of solid waste management in small island developing states: A review. *Waste Management, 43*, 539–549.

Molu, K., Pratt, S., & Tarte, S. (2024). Sport tourism and local sustainable development in small island territories: The case of Kiribati. In D. Van Rheenen, O. Naria, R. Melo, & C. Sobry (Eds.), *Sport tourism, island territories and sustainable development. A comparative perspective* (pp. 285–295). Springer.

Monnereau, I., Mahon, R., McConney, P., Nurse, L., et al. (2021). Fisheries sector vulnerabilities to climate change in small island developing states. In S. Moncada et al. (Eds.), *Small island developing states: Vulnerability and resilience under climate change* (pp. 233–255). Springer.

Moore, A. (2019). *Destination anthropocene: Science and tourism in The Bahamas.* University of California Press.

Moore, L. (2024). A dysfunctional family: Australia's relationship with Pacific Island states and climate change. *Australian Journal of International Affairs, 78*(3), 286–305.

Moorhead, A. (2021). Small steps for sustainable tourism in Fiji. ANUDevPolicyBlog, 13 December.

Morgan, W., Carter, S., & Manoa, F. (2024). Leading from the frontline: A history of Pacific climate diplomacy. *Journal of Pacific History, 59*(3), 353–374.

Morris, J. (2021). Colonial afterlives of infrastructure: From phosphate to refugee processing in the Republic of Nauru. *Mobilities, 16*(5), 688–706.

Morris, J. (2023). *Asylum and extraction in the Republic of Nauru.* Cornell University Press.

Morrison, R., & Munro, J. (1999). Waste management in the small island developing states of the South Pacific: An overview. *Australian Journal of Environmental Management, 6*(4), 232–246.

Mortimer, N., Campbell, H., Tulloch, A., et al. (2017). Zealandia: Earth's hidden continent. *GSA Today, 27*(3), 27–35.

Mortley, N. (2011). Strategic opportunities from diaspora tourism: The Jamaican perspective. *Canadian Foreign Policy Journal, 17*(2), 171–185.

Mortreux, C., & Barnett, J. (2009). Climate change, migration and adaptation in Funafuti, Tuvalu. *Global Environmental Change, 19*, 105–112.

Mortreux, C., Jarillo, S., Barnett, J., & Waters, E. (2023). Climate change and migration from atolls? No evidence yet. *Current Opinion in Environmental Sustainability, 60*, 101234.

Mosley, L., Sharp, D., & Singh, S. (2004). Effect of a tropical cyclone on the drinking water quality of a remote Pacific island. *Disasters, 38*, 405–417.

Moulton, A. (2024). Modernity's Antillean ecologies: Dispossession, disasters, justice and repair across the Caribbean archipelago. *Progress in Environmental Geography, 3*(1), 3–16.

Moulton, A., & Machado, M. (2019). Bouncing forward after Irma and Maria: Acknowledging colonialism, problematizing resilience and thinking climate justice. *Journal of Extreme Events, 6*(1), 1940003.

Mountz, A. (2011). The enforcement archipelago: Detention, haunting and asylum on islands. *Political Geography, 30*, 118–128.

Movono, A., & Hughes, E. (2022). Tourism partnerships: Localizing the SDG agenda in Fiji. *Journal of Sustainable Tourism, 30*(10), 2318–2332.

Movono, A., & Scheyvens, R. (2021). Tourism in a world of disorder: A return to the Vanua and kinship with nature in Fiji. In Y. Campbell & J. Connell

(Eds.), *COVID in the islands: A comparative perspective on the Caribbean and the Pacific* (pp. 265–278). Palgrave Macmillan.

Movono, A., Scheyvens, R., & Auckram, S. (2022). Silver linings around dark clouds: Tourism, Covid-19 and a return to traditional values, villages and the *vanua*. *Asia Pacific Viewpoint, 63*(2), 164–179.

Mudaliar, C., Leach, M., Baker, K., et al. (2025). Popular political attitudes in Vanuatu: Findings of the Pacific attitudes survey. *Asia & the Pacific Policy Studies, 12*(3), e70028.

Mullings, B. (2009). Neoliberalization, social reproduction and the limits to labour in Jamaica. *Singapore Journal of Tropical Geography, 30*(2), 174–188.

Munroe, M. (2023). Jamaica's transnational violence: When geography matters the most. *The Round Table, 112*(6), 613–628.

Murphy, J., & Gard, R. (2025). The deep-sea mining commodity frontier: Blue economy agenda in the Pacific Ocean. *Academia Environmental Sciences and Sustainability, 2*(1).

Murphy, G., MacKenzie, A., Waysome, B., et al. (2016). A mixed-methods study of health worker migration from Jamaica. *Human Resources for Health, 14*(Suppl 1), 36.

Murray, W. (2000). Neoliberal globalisation, "exotic" agro-exports and local change in the Pacific islands: A study of the Fijian kava sector. *Singapore Journal of Tropical Geography, 21*, 355–373.

Murray, W., & Overton, J. (2011). The inverse sovereignty effect: Aid, scale and neostructuralism in Oceania. *Asia Pacific Viewpoint, 52*(3), 272–284.

Mycoo, M. (2014). Sustainable tourism, climate change and sea level rise adaptation policies in Barbados. *Natural Resources Forum, 38*, 47–45.

Mycoo, M. (2018a). Urban sustainability in Caribbean small island developing states: A conceptual framework for urban planning using a case study of Trinidad. *International Development Planning Review, 40*(2), 143–174.

Mycoo, M. (2018b). Achieving SDG 6: Water resources sustainability in Caribbean small island developing states through improved water governance. *Natural Resources Forum, 42*(1), 54–68.

Mycoo, M. (2021). Vulnerabilities to climate change and enhancing resilience in Caribbean small island developing states: A spatial planning framework. In S. Moncada et al. (Eds.), *Small Island developing states: Vulnerability and resilience under climate change* (pp. 259–283). Springer.

Mycoo, M. (2022). Caribbean island cities: Urban issues, urbanization processes and opportunities for transformation. In J. González-Pérez, C. Irazábal, &

R. Lois-González (Eds.), *The Routledge handbook of urban studies in Latin America and the Caribbean* (pp. 579–602). Routledge.

Mycoo, M., & Roopnarine, R. (2024). Water resource sustainability: Challenges, opportunities and research gaps in the English-speaking Caribbean small island developing states. *PLOS Water, 3*(1), e0000222.

Mycoo, M., Robinson, S.-A., Nguyen, C., et al. (2021). Human adaptation to coastal hazards in Greater Bridgetown, Barbados. *Frontiers in Environmental Science, 9*, 647788.

Naidu, R. (2024a). Dropouts declining, but education challenges remain. *Islands Business, 50*(2), 25–26.

Naidu, R. (2024b). The long, hard road to climate finance. *Islands Business, 50*(11), 10–14.

Nakamura, N., & Kanemasu, Y. (2022). A minority group's response to a severe climatic event: A case study of rural Indo-Fijians after Tropical Cyclone Winston in 2016. *Disasters, 46*(1), 1–20.

Nand, M., Clissold, R., McNamara, K., & Yee, M. (2024a). Guiding our responses to climate change by what people value: Insights from Fiji. *Geoforum, 155*, 104080.

Nand, M., Bardsley, D., & Suh, J. (2024b). Climate change loss and damage from droughts: Key insights from Fiji's sugar industry. *Local Environment, 28*(6), 768–783.

Napper, I., & Thompson, R. (2020). Plastic debris in the marine environment: History and future challenges. *Global Challenges, 4*(6), e1900081.

Narsey, W. (2023). *Beyond 33%. The economic empowerment of Fiji women and girls*. Fiji Women's Rights Movement.

Nasim, N., Anthony, S., Daurewa, T., et al. (2023). Understanding on-site sanitation in rural Fiji: Where definitions of sanitation back-ends differ. *Environmental Science: Water Research & Technology, 9*(7), 1913–1931.

Naupa, A. (2023, August 31). Cultural security in the Pacific: Why it matters for regional security. *Interpreter*.

Naupa, A. (2025). How can Vanuatu's new parliament rebuild public trust? *ANU DevPolicyBlog*, 28 January.

Naupa, A., Mecartney, S., Pechan, L., & Howlett, N. (2021). An industry in crisis: How Vanuatu's tourism sector is seeking economic recovery. In Y. Campbell & J. Connell (Eds.), *COVID in the Islands: A comparative perspective on the Caribbean and the Pacific* (pp. 231–252). Palgrave Macmillan.

Neate, R. (2022, February 12). Bitcoin paradise? Briton creates 'crypto utopia' in South Pacific. *Guardian*.

Neef, A. (2021). The contentious role of tourism in disaster response and recovery. *Frontiers in Earth Science, 9*, 771345.

Neef, A., & Benge, L. (2022). Shifting responsibility and denying justice: New Zealand's contentious approach to Pacific climate mobilities. *Regional Environmental Change, 22*(1), 1–14.

Neef, A., Benge, L., Boruff, B., Pauli, N., Weber, E., & Varea, R. (2018). Climate adaptation strategies in Fiji: The role of social norms. *World Development, 107*, 125–137.

Neemia, U. (1986). *Cooperation and conflict: Costs, benefits and national interests in Pacific regional cooperation*. Institute of Pacific Studies, USP.

Nef, D., Kotra, K., Stauffacher, M., et al. (2022). Vulnerability to food insecurity in a telecoupled world: Insights from Vanuatu. *Frontiers in Sustainable Food Systems, 6*, 855272.

Nero, K. (1997). The end of insularity. In D. Denoon & M. Meleisea (Eds.), *The Cambridge history of the Pacific Islanders* (pp. 439–467). Cambridge University Press.

Neufville, Z. (2022). *Overharvesting and poaching devastate Jamaica's fisheries*. InSight Crime Investigation Report.

Newell, A., Nuttall, P., Prasad, B., & Veitayaki, J. (2017). Turning the tide: The need for sustainable sea transport in the Pacific. *Marine Policy, 75*, 249–259.

Newell, A., Nuttall, P., & Newell, P. (2025). *Cruise liners – are they good for Aotearoa and the Pacific and what more do we need to know?* Micronesian Center for Sustainable Transport, Working Paper 5.

Ngcamu, B. (2023). Climate change effects on vulnerable populations in the Global South: A systematic review. *Natural Hazards, 118*, 977–991.

Nguyen, C., & Hopkins, W. (2022). *Corruption and money laundering in the Pacific: Intertwined challenges and interlinked responses*. Transparency International New Zealand.

Nibbs, F. (2024, May 13). Thirsty in paradise: Water crises are a growing problem across the Caribbean islands. *The Conversation*.

Nichols, A. (2019). Climate change, natural hazards, and relocation: Insights from Nabukadra and Navuniivi villages in Fiji. *Climatic Change, 156*, 255–271.

Niles, K., & Lloyd, B. (2013). Small island developing states (SIDS) & energy aid: Impacts on the energy sector in the Caribbean and Pacific. *Energy for Sustainable Development, 17*(5), 521–530.

Niles, K., & Lloyd, B. (2014). Using power sector reform as an opportunity to increase the uptake of renewable energy in the power sector: Responding to peak oil and climate change in Caribbean and Pacific small island developing States, between 1970-2010. *Natural Resources Forum, 38*, 14–26.

Nilles, E., Manaia, A., Ruaia, B., et al. (2018). Re-emergence of thiamine deficiency disease in the Pacific islands (2014-15): A case-control study. *PLoS One, 13*(6), e0198590.

Nobetau, O. (2025). *Bougainville's future: A roadmap for development*. Lowy Institute.

Noble, C., Pereira, N., & Saune, N. (2011). *Urban youth in the Pacific: Increasing resilience and reducing risk for involvement in crime and violence*. UNDP Pacific Centre and Pacific Island Forum Secretariat.

Nolet, E. (2016). 'Are you prepared?' Representations and management of floods in Lomainikoro, Rewa (Fiji). *Disasters, 40*(4), 720–739.

Nomura, K., Woodill, A., Sweeney, J., Harte, M., & Watson, J. (2024). Emergent geopolitical risks from fishing activities and past conflicts in the Pacific Ocean. *Marine Policy, 166*, 106234.

Nunn, P. (2009). Responding to the challenges of climate change in the Pacific Islands: Management and technological imperatives. *Climate Research, 40*(2-3), 211–231.

Nunn, P., Runman, J., Falanruw, M., & Kumar, R. (2017). Culturally grounded responses to coastal change on islands in the Federated States of Micronesia, northwest Pacific Ocean. *Regional Environmental Change, 17*(4), 959–971.

Nunn, P., Klöck, C., & Duvat, V. (2021a). Seawalls as maladaptations along island coasts. *Ocean and Coastal Management, 205*, 105554.

Nunn, P., Smith, T., & Elrick-Barr, C. (2021b). Path dependency and future adaptation of coastal cities: Examples from the Asia-Pacific. *Frontiers in Environmental Science, 9*, 642385.

Nunn, R., Itinteang, T., Ioteba, N., & Tan, W. (2024). *Child mortality in Kiribati: A wake-up call for urgent action*. ANU Devpolicy Blog, 1 November.

Nurse, L., McLean, R., Agard, J., et al. (2014). Small islands. In V. Barros et al. (Eds.), *Climate change 2014* (pp. 1613–1654). Cambridge University Press.

OECS. (2024). *Power below the surface. Harnessing geothermal energy in the OECS*. OECS Commission.

O'Garra, T. (2012). Economic valuation of a traditional fishing ground on the coral coast of Fiji. *Ocean and Coastal Management, 56*, 44–55.

Ohi, S., & Ingram, P. (2021). Preparing I-Kiribati for the future: Probing the English language research-policy nexus in Kiribati education in the South Pacific. *Asia Pacific Journal of Education, 42*(3), 478–496.

Oh-Seng, J., Klöck, C., & Deenapanray, P. (2025). Conceptualising a governance framework for climate change adaptation and disaster risk reduction in small island developing states through a systematic review. *Sustainability, 17*(22), 9965.

O'Keefe, M. (2024, August 30). 'Hot mic' moment aside, the Albanese government's Pacific policing deal is a masterstroke of diplomacy. *The Conversation.*

Olszowy, K., Pomer, A., Dancause, K., et al. (2015). Impact of modernization on adult body composition on five islands of varying economic development in Vanuatu. *American Journal of Human Biology, 27*(6), 832–844.

Oman, K., Bale, S., Baravilala, W., et al. (2006). Donor countries may unintentionally become major beneficiaries of their own aid: The case of postgraduate specialist training at the Fiji School of Medicine. *Fiji Medical Journal, 27*(2), 11–18.

Oral, L. (2025). *ICJ climate change advisory opinion: Noteworthy trends.* IISD Knowledge Hub.

Orams, M. (2013). Economic activity derived from whale-based tourism in Vava'u, Tonga. *Coastal Management, 41*(6), 481–500.

Ornelas, I., Fung, W., Gabbard, S., & Carroll, D. (2021). *Findings from the national agricultural workers survey (NAWS) 2017–2018: A demographic and employment profile of United States farmworkers.* U.S. Department of Labor, Employment and Training Administration Research Report No. 14.

Orozco, L., & Spencer, J. (2023). *Vanuatu skills needs industry survey report.* Vanuatu Chamber of Commerce and Industry.

Ostrom, E. (2000). Collective action and the evolution of social norms. *Journal of Economic Perspectives, 14*(3), 137–158.

O'Sullivan, D. (2022, June 30). Police violence in Fiji: An endemic problem with no simple end in sight. *Australian Institute of International Affairs.*

Otis, G. (2025). Abandonment of soft power, neglect of climate change, and the shifting geopolitical landscape of the Pacific. *Visions and Voices, 2*(2).

Overton, J., & Murray, W. (2018). Migration, education and marginality: Networks and strategies in the Pacific Islands. In S. Pelc & M. Koderman (Eds.), *Nature, tourism and ethnicity as drivers of (de)marginalization. Perspectives on geographical marginality* (pp. 215–233). Springer.

Overton, J., Murray, W., Prinsen, G., et al. (2019). *Aid, ownership and development. The inverse sovereignty effect in the Pacific Islands.* Routledge.

Overton, J., Prinsen, G., Murray, W., & Wrighton, N. (2012). Reversing the tide of aid: Investigating development policy sovereignty in the Pacific. *Journal de la Société des Océanistes, 135*, 229–242.

Owens, E., Zhang, Q., & Mihelcic, J. (2011). Material flow analysis applied to household solid waste and marine litter on a small island developing state. *Journal of Environmental Engineering, 137*, 937–944.
Oza, T., & Goff, F. (2023). Maximising the impact of Australia's scarce water and sanitation aid. *ANU DevPolicyBlog*, 6 July.
Pacific Freedom Forum. (2024). *For us, by us, about us*. Suva.
Pacific Islands Forum Secretariat. (2013). *The Pacific plan review*. PIFS.
Pacific Islands Forum Secretariat. (2014). *Framework for Pacific regionalism*. PIFS.
Pacific Islands Forum Secretariat. (2017). *State of Pacific regionalism*. PIFS.
Pacific Islands Forum Secretariat. (2018). *Boe declaration action plan*. PIFS.
Pacific Islands Forum Secretariat. (2022). *2050 strategy for the blue Pacific continent*. PIFS.
Pacific Private Sector Development Initiative. (2021). *Looking forward: Evaluating the challenges for Pacific tourism after COVID-19*. PPSDI.
Paeniu, L., Holland, E., Miller, C., & Anderson, G. (2017). Rainfall trends, drought frequency and La Niña in Tuvalu: A small equatorial island state in the Pacific Ocean. *Journal of Environmental and Analytical Toxicology, 7*, 501.
PAHO. (2019). *Health workers perception and migration in the Caribbean Region*. PAHO.
Pakaang, A., & Gurung, O. (2025). *Bridging Fiji's digital divide: A holistic approach to digital and financial inclusion*. Griffth Asia Institute.
Palmieri, S. (2025, July 10). Remarks reveal enduring sexist norms in Samoan politics. *East Asia Forum*.
Panke, D. (2012). Dwarfs in international negotiations: How small states make their voices heard. *Cambridge Review of International Affairs, 25*(3), 313–328.
Papalexiou, S., & Montanari, A. (2019). Global and regional increase of precipitation extremes under global warming. *Water Resources Research, 55*(6), 4901–4914.
Pareti, K. (2024). Addressing Kiribati's 'alarming' water situation. *Islands Business, 50*(8), 48–50.
Paris, A., Kwaoga, A., & Hewavitharane, C. (2022). An assessment of floating marine debris within the breakwaters of the University of the South Pacific, Marine Studies Campus at Laucala Bay. *Marine Pollution Bulletin, 174*, 113290.
Patagan, R. (2022, December, 13–16). Social protection in the Federated States of Micronesia: Spotlight on subsistence and food security. *Pacific Economic Monitor*.

Patay, D., Schram, A., & Friel, S. (2023). The challenges in protecting public health interests in multisectoral governance in the context of small island developing states: The case of tobacco control in Fiji and Vanuatu. *Globalization and Health, 19*(1), 31.

Pathirana, A. (2025). Small islands: Living laboratories revealing climate and sustainable development challenges. *Frontiers in Climate, 6*, 1445378.

Patil, P., Virdin, J., & Colgan, C. (2023). The blue economy as a global initiative. In P. Clegg, R. Mahon, P. McConney, & H. Oxenford (Eds.), *The Caribbean blue economy* (pp. 1–8). Routledge.

Paton, K., & Fairbairn-Dunlop, P. (2010). Listening to local voices: Tuvaluans respond to climate change. *Local Environment, 15*, 687–698.

Payne, A., & Sutton, P. (2007). *Repositioning the Caribbean within globalisation*. The Centre for International Governance Innovation, Caribbean Papers No. 1.

Peake, G., & Smith, G. (2022, January 27). Solomon Islands: Cops bearing gifts. *Interpreter*.

Pearce, T., Currenti, R., Mateiwai, A., & Doran, B. (2018). Adaptation to climate change and freshwater resources in Vusama village, Viti Levu, Fiji. *Regional Environmental Change, 18*(2), 501–510.

Pena, M., McConney, P., Perch, L., & Phillips, T. (2022). Capacities and gaps in Caribbean fisherfolk organization leadership. *Gender, Technology and Development, 26*(3), 503–521.

Pendeverana, L. (1987). Rethinking Honiara's water supply: Voices from the settlements. *Development Bulletin, 78*, 59–61.

Perch, L., & Roy, R. (2010). *Social policy in the post-crisis context of small island developing states: A synthesis*. International Policy Centre for Inclusive Growth.

Perry, K. (2024). Epistemic silences, subversive politics: Post-disaster economic assessments as technologies of persistent coloniality and route to an emancipatory climate justice agenda in the Caribbean. *Climate and Development, 16*(9), 798–810.

Perry, K., & Sealey-Huggins, L. (2023). Racial capitalism and climate justice: White redemptive power and the uneven geographies of eco-imperial crisis. *Geoforum, 145*, 103772.

Persaud, N., & Dagher, R. (2021). *The role of monitoring and evaluation in the UN 2030 SDGs agenda*. Palgrave Macmillan.

Persson, K., Zampoukos, K., & Ljunggren, I. (2022). No (wo)man is an island – socio-cultural context and women's empowerment in Samoa. *Gender, Place and Culture, 29*(4), 482–501.

Perumal, N. (2018). "The place where I live is where I belong": Community perspectives on climate change and climate-related migration in the Pacific island nation of Vanuatu. *Island Studies Journal, 13*(1), 45–64.

Peterson, R., DiPietro, R., & Harrill, R. (2020). In search of inclusive tourism in the Caribbean: Insights from Aruba. *Worldwide Hospitality and Tourism Themes, 12*(3), 225–243.

Peteru, S., Regan, S., & Klak, T. (2010). Local vibrancy in a globalizing world: Evidence from Dominica. *Focus, 53*(4), 125–133.

Petrou, K. (2020). *If everyone returned, the island would sink: Urbanisation and migration in Vanuatu*. Berghahn.

Petrou, K., & Connell, J. (2017). Rural-urban migrants, translocal communities and the myth of return migration in Vanuatu: The case of Paama. *Journal de la Société des Océanistes, 144–145*, 51–62.

Petrou, K., & Connell, J. (2020). Overcoming precarity? Social media, agency and Ni-Vanuatu seasonal workers in Australia. *Journal of Australian Political Economy, 84*, 116–146.

Petrou, K., & Connell, J. (2023a). *Pacific Islands guestworkers in Australia: The new blackbirds?* Palgrave Macmillan.

Petrou, K., & Connell, J. (2023b). Our 'Pacific family'. Heroes, guests, workers or a precariat? *Australian Geographer, 54*(2), 125–136.

Petrou, K., & Connell, J. (2025). Guestworker schemes in Pacific island countries: Triple wins but social costs? *International Migration, 63*(5), e70085.

Petrou, K., & Withers, M. (2024). 'Sometimes men cannot do what women can': Pacific labour mobility, gender norms and social reproduction. *Global Networks, 24*, e12463.

Pettinotti, L., Colenbrander, S., Kamninga, T., et al. (2025). *A fair share of climate finance? Assessing quantity, quality and alignment with gender goals*. ODI Global.

Petzold, J., & Magnan, A. (2019). Climate change: Thinking small islands beyond small island developing states (SIDS). *Climatic Change, 152*, 145–165.

Petzold, J., & Scheffran, J. (2024). Climate change and human security in coastal regions. *Cambridge Prisms: Coastal Futures, 2*(e5), 1–14.

Petzold, J., Ratter, B., & Holdschlag, A. (2018). Competing knowledge systems and adaptability to sea-level rise in The Bahamas. *Area, 50*(1), 91–100.

Pew Charitable Trusts. (2016). *Netting billions. A global valuation of Tuna*. Pew Charitable Trusts.

Phillips, T., & Keen, M. (2016). *Sharing the City: Urban growth and governance in Suva, Fiji*. ANU State, Society and Governance in Melanesia Discussion Paper 2016/6.

Phillips, T., & Narayan, A. (2017). Meeting the health care challenges posed by rapid urbanisation in the Pacific: The view from Fiji. *Development Bulletin, 78*, 79–82.

Philpott, S. (1973). *West Indian migration. The Montserrat case*. Routledge.

Pickering, K., Pearce, T., Manuel, L., Doran, B., & Smith, T. (2023). Socio-ecological challenges and food security in the 'salad bowl' of Fiji, Sigatoka Valley. *Regional Environmental Change, 23*(2), 61.

Piguet, E. (2022). Linking climate change, environmental degradation, and migration: An update after 10 years. *Wiley Interdisciplinary Reviews: Climate Change, 13*(1), e746.

Piggott-McKellar, A., Pearson, J., McNamara, K., & Nunn, P. (2020). A livelihood analysis of resettlement outcomes: Lessons for climate-induced relocations. *Ambio, 49*(9), 1474–1489.

Pilling, G., Harley, S. J., Nicol, S., Williams, P., & Hampton, J. (2020). Can the tropical Western and Central Pacific tuna purse seine fishery contribute to Pacific Island population food security?. In J. Connell & K. Lowitt (Eds.), *Food Security in Small Island States*. Springer.

PINA. (2024, October 1). *Pacific and Caribbean SIDS demand global shipping emissions levy at MEPC82 to protect survival.*

Plange, N., & Alam, M. (2024). Averting the old age crisis in Fiji: Policy, providence and the crisis of livelihood. *Sage Open, 14*(2).

Pomponio, A. (1993). Education is development on a ten-acre island. In V. Lockwood et al. (Eds.), *Contemporary Pacific societies* (pp. 197–215). Prentice Hall.

Pondorfer, A. (2019). The perception of climate change: Comparative evidence from the small-island societies of Bougainville and Palawan. *Environmental Development, 30*, 21–34.

Potter, R., Barker, D., Conway, D., & Klak, T. (2004). *The contemporary Caribbean*. Pearson.

Pouponneau, A., & Northrop, E. (2025, August 5). Big ocean: ICJ advisory opinion and marine boundaries. *Island Times*.

Powles, A. (2021). The 'Pacific Reset' and shaping New Zealand's relations with the Pacific. *Development Bulletin, 82*, 32–36.

Powles, A., & Wallis, J. (2024). Different nightmares, shared dreams? Australia and New Zealand's intuitive alliance. *Australian Journal of International Affairs, 78*(5), 536–560.

Praene, J., Payet, M., & Bénard-Sora, F. (2018). Sustainable transition in small island developing states: Assessing the current situation. *Utilities Policy, 54*, 86–91.

Prakash, R., & Nakamura, N. (2024). Sugarcane farmers' perceptions of climate change and their "adaptation" methods in Fiji: adaptation for the industry or for their livelihood?. *Climate and Development, 16*(8), 665–672.

Prasad, B. (2014). Why Fiji is not the "Mauritius" of the Pacific? Lessons for small island nations in the Pacific. *International Journal of Social Economics, 41*(6), 467–481.

Prasad, N. (2023). Past, current, and future social transformation in Pacific Island Countries. In K. Baehler (Ed.), *The Oxford handbook of governance and public management for social policy* (pp. 366–384). Oxford University Press.

Prasad, J. (2024, October 23). Biases within capital flows to the Pacific. *Interpreter.*

Prasad, B., & Kausimae, P. (2012). *Social policies in Solomon Islands and Vanuatu.* Commonwealth Secretariat and United Nations Research Institute for Social Development.

Pratap, A., Mani, F., & Prasad, S. (2020). Heavy metals contamination and risk assessment in sediments of Laucala Bay, Suva, Fiji. *Marine Pollution Bulletin, 156*, 111238.

Pratt, S., Magbalot-Fernandez, A., & Ohe, Y. (2022). Motivations and constraints of developing agritourism under the challenges of climate change: The case of Samoa. *International Journal of Tourism Research, 24*(4), 610–622.

Premdas, R., Steeves, J., & Larmour, P. (1984). The western breakaway movement in the Solomon Islands. *Pacific Studies, 7*, 34–34.

Prentice, C., Kundra, S., Alam, M., et al. (2023). Utopia or dystopia–deterrents to ecotourism development in Fiji. *Tourism Geographies, 25*(2–3), 843–864.

Pryke, J., & Sora, M. (2022, February 11). Kick-starting America's Pacific diplomacy. *Interpreter.*

Puas, G. (2023). Shaping the legal landscape. Foreign judges in the federated states of Micronesia. In A. Dziedzic & S. Young (Eds.), *The Cambridge handbook of foreign judges in domestic courts* (pp. 447–462). Cambridge University Press.

Puas, G., & D'Arcy, P. (2021). Micronesia and the rise of China: Realpolitik meets the reef. *Journal of Pacific History, 56*(3), 274–295.

Puas, G., & Oliver, N. (2022). *Electing the first woman to the Congress of the Federated States of Micronesia.* ANU Department of Pacific Affairs InBrief 2022/1.

Pueyo, A., & Haas, J. (2025). Energy transitions in the Pacific, one island at a time: The case of Efate island, Vanuatu. *Energy Policy, 206*, 114751.

Pugh, J. (2005). Social transformation and participatory planning in St Lucia. *Area, 37*(4), 384–392.

Purcell, S., Tagliafico, A., Cullis, B., & Cocks, N. (2024). Wicked problem of improving fishery livelihoods through capacity building. *Marine Policy, 163*, 106108.

Puri, S. (2014). *The Grenada Revolution in the Caribbean present: Operation urgent memory*. Palgrave Macmillan.

Quashie, N. (2019). Social welfare and remittances to older adults in two Caribbean cities. *International Migration, 57*(5), 71–88.

Rabone, M., Wiethase, J., Simon-Lledo, E., Emery, A., et al. (2023). How many metazoan species live in the world's largest mineral exploration region? *Current Biology, 33*, 2383–2396.

Raddatz, C. (2009). *The Wrath of God: Macroeconomic costs of natural disasters*. World Bank Policy Research Working Paper 5039.

Raghoo, P., Surroop, D., Wolf, F., et al. (2018). Dimensions of energy security in small island developing state. *Utilities Policy, 53*, 94–101.

Raine, S. (2017). *Regional disorder: The South China Sea disputes*. Routledge.

Rajan, J. (2005). Gilt-edged packet or economic straight jacket? A case study of cannery workers in Levuka, Fiji Islands. In I. Novaczek, J. Mitchell, & J. Veitayaki (Eds.), *Pacific voices: Equity and sustainability in Pacific Island fisheries* (pp. 153–166). Institute of Pacific Studies, USP.

Randin, G., & Turagabaleti, P. (2021). COVID-19 and food security in Fiji's urban areas: The rise of urban farming and home gardening as a 'new way of life'. In Y. Campbell & J. Connell (Eds.), *COVID in the islands: A comparative perspective on the Caribbean and the Pacific* (pp. 353–366). Palgrave Macmillan.

Rampersad, I. (2024). Contemporary dynamics of drug and gang activism in Trinidad and Tobago. In S. Morris-Francis, C. Gibson, & L. Grant (Eds.), *Youth crime and violence in the Caribbean* (pp. 231–250). Lexington.

Rarai, A., Parsons, M., Nursey-Bray, M., & Crease, R. (2022). Situating climate change adaptation within plural worlds: The role of Indigenous and local knowledge in Pentecost Island, Vanuatu. *Environment and Planning E: Nature and Space, 5*(4), 2240–2282.

Rashid, A., Nakib, T., Shahriar, T., et al. (2024). Energy and economic analysis of an ocean thermal energy conversion plant for Bangladesh: A case study. *Ocean Engineering, 293*, 116625.

Raturi, A., Singh, A., & Prasad, R. (2016). Grid-connected PV systems in the Pacific Island Countries. *Renewable and Sustainable Energy Reviews, 58*, 419–428.

Ratuva, S. (2022, June 16). Pacific agency: A devalued political capital? *Australian Outlook*.

Rauchholz, M. (2012). Discourses on Chuukese customary adoption, migration, and the laws of state(s). *Pacific Studies, 35*(1–2), 119–143.

Rauchholz, M. (2023). Family and gender violence in Chuuk, Micronesia. In A. Kelly-Hanku, P. Aggleton, & A. Malcolm (Eds.), *Sex and gender in the Pacific. Contemporary perspectives on sexuality, gender and health* (pp. 181–191). Routledge.

Ravulo, J., Hollier, J., Waqa, M., et al. (2024). A queer resilience: Reviving indigenous-Pacific perspectives and practices. *ETropic: Electronic Journal of Studies in the Tropics, 23*(2), 213–234.

Read, R. (2006). Sustainable natural resource use and economic development in small states: The tuna fisheries in Fiji and Samoa. *Sustainable Development, 14*, 93–103.

Reaser, J., Meyerson, L., Cronk, Q., et al. (2007). Ecological and socioeconomic impacts of invasive alien species in island ecosystems. *Environmental Conservation, 34*, 98–111.

Reddy, M., & Wilkes, K. (2012). *Tourism, Climate Change and Sustainability*. Routledge.

Reed, K., Wehner, M., & Zarzycki, C. (2022). Attribution of 2020 hurricane season extreme rainfall to human-induced climate change. *Nature Communication, 13*, 1905.

Reeve, E., Thow, A.-M., Huse, O., et al. (2021). Policy-makers' perspectives on implementation of cross-sectoral nutrition policies, Western Pacific Region. *Bulletin of the WHO, 99*, 865–873.

Regan, A. (2017). Bougainville: Origins of the conflict, and debating the future of large-scale mining. In C. Filer & P.-Y. Le Meur (Eds.), *Large scale mines and local-level politics. Between new Caledonia and Papua New Guinea* (pp. 353–414). ANU Press.

Regan, A. (2022, October 31). Overcoming suspicion: Reconciliation for Australia and Bougainville. *Interpreter*.

Regan, A., Baker, K., & Oppermann, T. (2022). The 2019 Bougainville Referendum and the question of independence: From conflict to consensus. *Journal of Pacific History, 57*(1), 58–88.

Reklai, L. (2022, August 17). Palau inaugurates its first digital residence. *Island Times*.

Reklai, L. (2024, June 28). Palau faces major challenges in halting outmigration. *Island Times*.

Remling, E., & Veiteyaki, J. (2016). Community-based action in Fiji's Gau Island: A model for the Pacific? *International Journal of Climate Change Strategies and Management, 8*(3), 375–398.

Rey, T., Le Dé, L., Leone, F., & Gilbert, D. (2017). An integrative approach to understand vulnerability and resilience post-disaster: The 2015 Cyclone Pam in urban Vanuatu as case study. *Disaster Prevention and Management, 26*(3), 259–275.

Rhiney, K. (2011). Agri-tourism linkages in Jamaica: Case study of the Negril all-inclusive hotel sub-sector. In R. Momsen (Ed.), *Tourism and agriculture: New geographies of consumption, production and rural restructuring* (pp. 117–138). Routledge.

Rhiney, K. (2012). The Negril tourism industry: Growth, challenges and future prospects. *Caribbean Journal of Earth Science, 43*, 25–34.

Rhiney, K. (2015). Geographies of Caribbean vulnerability in a changing climate: Issues and trends. *Geography Compass, 9*(3), 97–114.

Rhiney, K. (2020). Dispossession, disaster capitalism and the post-hurricane context in the Caribbean. *Political Geography, 78*, 102171.

Rhiney, K., & Baptiste, K. (2019). Adapting to climate change in the Caribbean: Existential threat or development crossroads? *Caribbean Studies, 47*(2), 59–80.

Rhoden-Trader, J. (2024). Human trafficking of children and youth in the Caribbean. In S. Morris-Francis, C. Gibson, & L. Grant (Eds.), *Youth crime and violence in the Caribbean* (pp. 189–208). Lexington.

Rhodes, K., Warren-Rhodes, K., Houk, P., Cuetos-Bueno, J., & Fong, Q. (2011). *An interdisciplinary study of market forces and nearshore fisheries management in Micronesia*. Nature Conservancy Asia Pacific Conservation Region Marine Program Report No 6/11.

Rhodes, K., Warren-Rhodes, K., Sweet, S., et al. (2015). Marine ecological footprint indicates unsustainability of the Pohnpei (Micronesia) coral reef fishery. *Environmental Conservation, 42*(2), 182–190.

Rich, R., Hambly, L., & Morgan, M. (Eds.). (2008). *Political parties in the Pacific Islands*. ANU Press.

Richardson-Ngwenya, P. (2010). The EU sugar reform and the responses of Caribbean sugar producers. *Geography, 95*(2), 70–79.

Ride, A. (2022). *Riots in the Pacific: Control and change.* Canberra: Australia Pacific Security College Discussion Paper.
Ride, A., & Zhang, D. (2024). *Foreign actors, geopolitics and riots in the Pacific.* ANU Department of Pacific Affairs InBrief 2024/3.
Ride, A., Eriksson, H., Tutuo, J., et al. (2021). Catching fish in COVID-19 currents: Food security and governance in rural communities in Solomon Islands. In Y. Campbell & J. Connell (Eds.), *COVID in the islands: A comparative perspective on the Caribbean and the Pacific* (pp. 367–384). Palgrave Macmillan.
Ride, A., Ballard, T., & Smith, G. (2025, March 13). Foreign state-sponsored disinformation in the Pacific Islands. *Australian Outlook.*
Rikimani, B. (2024). Climate justice and Pacific Island countries – A case study on grassroots advocacy. *The Round Table, 113*(4), 374–384.
Rise, N., Oura, C., & Drewry, J. (2022). Climate change and health in the Caribbean: A review highlighting research gaps and priorities. *The Journal of Climate Change and Health, 8*, 100126.
Roach, C., Sabharwal, M., Abraham, R., & Charles-Soverall, W. (2022). Does the public service motivation model hold in the Caribbean? *Review of Public Personnel Administration, 42*(3), 514–536.
Robinson, T. (2008). Gender, nation and the common law constitution. *Oxford Journal of Legal Studies, 28*(4), 735–762.
Robinson, G. (2022, February 18). Why Drua will be a game-changer for Fijian rugby. *Sydney Morning Herald.*
Robinson, T. (2023). Foreign judging and securing judicial independence in the Anglo Caribbean. In A. Dziedzic & S. Young (Eds.), *The Cambridge handbook of foreign judges in domestic courts* (pp. 324–344). Cambridge University Press.
Robinson, S. (2024). Patterns of hurricane induced displacement in The Bahamas: Building equitable resilience in small island developing states. *Climate Risk Management, 45*, 100634.
Robinson, S.-A., & Wren, C. (2020). Geographies of vulnerability: A research note on human system adaptations to climate change in the Caribbean. *Geografisk Tidsskrift-Danish Journal of Geography, 120*(1), 79–86.
Robinson, R., Martins, A., Solnet, D., & Baum, T. (2019). Sustaining precarity: Critically examining tourism and employment. *Journal of Sustainable Tourism, 27*(7), 1008–1025.
Rodd, A. (2020). A road to island sovereignty and empowerment? Fiji's aims within the belt and road initiative. *Island Studies Journal, 15*(2), 93–118.

Rolle, S. (2020). Marijuana agro-tourism habitat. In S. Rolle, J. Minnis, & I. Bethell-Bennett (Eds.), *Tourism development, governance and sustainability in the Bahamas* (pp. 177–188). Routledge.

Rolle-Sands, S., & Ingraham, K. (2024). Caribbean nurse turnover: A scoping review. *International Journal of Bahamian Studies, 30*(1), 81–94.

Rooney, M. (2021). "We want development": Land and water (dis)connections in Port Moresby, Urban Papua New Guinea. *The Contemporary Pacific, 33*(1), 1–30.

Roscher, M., Eriksson, H., Harohau, D., et al. (2022). Unpacking pathways to diversified livelihoods from projects in Pacific Island coastal fisheries. *Ambio, 51*(10), 2107–2117.

Rothe, D., Boas, I., Farbotko, C., & Kitara, T. (2024). Digital Tuvalu: State sovereignty in a world of climate loss. *International Affairs, 100*(4), 1491–1509.

Rouse, J. (2013). Sustainability of wastewater treatment and excess sludge handling practices in the Federated States of Micronesia. *Sustainability, 5*, 4183–4194.

Rousso, B., Sanderson, R., Love, M., et al. (2014). A multi-country survey on sanitation systems in underserved urban settlements in the Melanesian Pacific region. *Clean Water, 7*(93).

Rudiak-Gould, P. (2012). *Surviving paradise. One year on a disappearing Island.* Sterling.

Rudiak-Gould, P. (2014). *Climate change and tradition in a small island state: The rising tide.* Routledge.

Ruehr, S. (2022). Beyond the vulnerability/resilience dichotomy: Perceptions of and responses to the climate crisis on Emau. *Vanuatu. Island Studies Journal, 17*(1), 157–176.

Rummery, R. (2025). Buried beneath the bricks: A quiet retreat from health systems strengthening. *ANU DevPolicyBlog*, 23 July.

Russell, L. (2009). *Stayin' alive. A report on urban livelihoods in Honiara, Solomon Islands.* APHEDA.

Russell, J., & Kueffer, C. (2019). Island biodiversity in the Anthropocene. *Annual Review of Environment and Resources, 44*, 31–60.

Ryan, F. (2024). A quantitative goal with no quantum – three years of climate finance talks. *ANU DevPolicyBlog*, 15 October.

Rudder, R. (2012). Managing teacher recruitment and migration: A case study of the Barbados experience. In J. Penson & A. Yonemura (Eds.), *Next steps in managing teacher migration* (pp. 43–45). Commonwealth Secretariat.

Ruwet, M., Corbett, J., Yi-Chong, X., & Weller, P. (2023). Climate governance, multilateralism and small states: How the Marshall Islands rocked the boat at the International Maritime Organisation (IMO). In G. Baldacchino (Ed.), *The success of small states in international relations. Mice that roar?* (pp. 184–198). Routledge.

Saati, A. (2022). The political consequences of smallness: The case of St Lucia. *Commonwealth and Comparative Politics, 60*(4), 339–356.

Saint Ville, A., Hickey, G., & Phillip, L. (2017). Institutional analysis of food and agriculture policy in the Caribbean: The case of Saint Lucia. *Journal of Rural Studies, 51*, 198–210.

Samuwai, J., Fihaki, E., & Underhill-Sem, Y. (2020). Demystifying climate finance impacts in small island developing states: Pacific women's perspectives from Funafuti and Weno. *Small States and Territories, 3*(2), 283–302.

Sands, S., Ingraham, K., & Salami, B. (2020). Caribbean nurse migration – A scoping review. *Human Resources for Health, 18*, 19.

Sanmarti, M. (2025, December 4). Astropolitics of Fiji: An introduction. *Australian Outlook*.

Santagata, E., Regenauer-Lieb, K., & Corkish, R. (2024). Geothermal energy for PICTs. In A. Dansie, H. Alleway, & B. Boer (Eds.), *The water, energy and food security Nexus in Asia and the Pacific: The Pacific* (pp. 416–444). Springer.

Sasingian, S. (2025, November 10). Pacific Islands: Without literacy, climate change is nigh on impossible. *Informer*.

Satnarine-Singh, N., Hosein, R., & Saridakis, G. (2023). Structural change and export diversification: A comparison of CARICOM's position. *The Journal of International Trade & Economic Development, 33*(6), 1163–1189.

Savage, A., Bambrick, H., & Gallegos, D. (2021a). From garden to store: Local perspectives of changing food and nutrition security in a Pacific island country. *Food Security, 13*, 819–847.

Savage, A., Bambrick, H., & Gallegos, D. (2021b). Climate change and socio-economic determinants are structural constraints to agency in diet-related non-communicable disease prevention in Vanuatu: A qualitative study. *BMC Public Health, 21*, 1231.

SCF. (2025). *Regional child protection situational analysis – Pacific*. SCF.

Scheyvens, R. (2003). Church women's groups and the empowerment of women in Solomon Islands. *Oceania, 74*(1–2), 24–43.

Scheyvens, R., & Cheer, J. (2021). Tourism, the SDGs and partnerships. *Journal of Sustainable Tourism, 30*(10), 2271–2281.

Scheyvens, R., & Laeis, G. (2021). Linkages between tourist resorts, local food production and the sustainable development goals. *Tourism Geographies, 23*(4), 787–809.

Scheyvens, R., & Momsen, J. (2007). Tourism in small island states: From vulnerability to strengths. *Journal of Sustainable Tourism, 16*(5), 491–510.

Scheyvens, R., & Russell, M. (2012). Tourism, land tenure and poverty alleviation in Fiji. *Tourism Geographies, 14*, 1–25.

Schieder, D. (2014). Fiji Islander Rugby Players in Japan: Corporate rationales and migration routes. *Asia Pacific Journal of Sport and Social Science, 3*(3), 250–267.

Schlosberg, D., & Collins, L. (2014). From environmental to climate justice: Climate change and the discourse of environmental justice. *WIREs Climate Change, 5*(3), 359–374.

Schoburgh, E., & Martin, J. (2016). From developmental local government to developmental local governance. In E. Schoburgh, J. Martin, & S. Gatchair (Eds.), *Developmental local governance: A critical discourse in 'alternative development'* (pp. 227–243). Palgrave Macmillan.

Schrader, J., Weigelt, J., Cai, L., et al. (2024). Islands are key for protecting the world's plant endemism. *Nature, 634*(8035), 868–874.

Schrecongost, A., & Wong, K. (2015). *Unsettled: Water and sanitation in urban settlements communities of the Pacific*. Pacific Regional Infrastructure Facility.

Schrecongost, A., Wong, K., Dutton, P., & Blackett, I. (2015). Delivering water and sanitation to Melanesian informal settlements. *Water: Journal of the Australian Water Association, 42*(3), 40–43.

Schuermann, S., & Lauer, M. (2018). Disaster recovery in the Western Pacific: Scale vulnerability and traditional exchange practices. *Natural Hazards, 84*(2), 1287–1306.

Schurman, R. (1998). Tuna dreams: Resource nationalism and the Pacific Islands. *Development and Change, 29*, 107–136.

Schwarz, A., Béné, C., Bennett, G., et al. (2011). Vulnerability and resilience of remote rural communities to shocks and global changes: Empirical analysis from Solomon Islands. *Global Environmental Change, 21*(3), 1128–1140.

Scobie, M. (2016). Policy coherence in climate governance in Caribbean small island developing states. *Environmental Science & Policy, 58*, 16–28.

Scobie, M. (2019). *Global environmental governance and small states.* Edward Elgar.

Scobie, M., Gopinathan, U., Kelman, I., & Myhre, S. (2023). The agency of community groups in health and climate change adaptation governance and policy in SIDS: The case of Toco, Trinidad and Tobago. *Environmental Science & Policy, 147*, 116–125.

Scoones, I. (2009). Livelihood perspectives and rural development. *Journal of Peasant Studies, 36*(1), 171–196.

Scorlazzi, C., Gingerich, S., van Dongeren, A., Cheriton, O., et al. (2018). Most atolls will be uninhabitable by the mid-21st century because of sea-level rise exacerbating wave-driven flooding. *Science Advances, 4*, eaap9471.

Scott, K. (2022, June 5). Cruising set to launch as new rules come into place. *Sun-Herald*, 24.

Scott, D., Hall, C., & Stefan, G. (2012a). *Tourism and climate change: Impacts, adaptation and mitigation*. Routledge.

Scott, D., Simpson, M., & Sim, R. (2012b). The vulnerability of Caribbean coastal tourism to scenarios of climate change related sea level rise. *Journal of Sustainable Tourism, 20*(6), 883–898.

Sealey, K., & Smith, J. (2014). Recycling for small island tourism developments: Food waste composting at Sandals Emerald Bay, Exuma, Bahamas. *Resources, Conservation and Recycling, 92*, 25–37.

Seebens, H., Meyerson, L., Richardson, D., et al. (2025). Biological invasions: A global assessment of geographic distributions, long-term trends and data gaps. *Biological Reviews*.

Segawa, N. (2025). Japan's FOIP-based diplomacy toward the Pacific Islands Region. *Pacific Affairs, 98*(2), 293–316.

Sengupta, M., Ford, M., & Kench, P. (2021). Shoreline changes in coral reef islands of the Federated States of Micronesia since the mid-20th century. *Geomorphology, 377*, 107584.

Sengupta, S. (2022, December 8). Tiny Vanuatu uses its 'unimportance' to launch big climate ideas. *The New York Times*.

Seto, K., Miller, N., Young, M., & Hanich, Q. (2022). Toward transparent governance of transboundary fisheries: The case of Pacific tuna transshipment. *Marine Policy, 136*, 104200.

Shah, K. (2021). Air pollution in small island developing states: A growing but avoidable challenge requiring multilevel interventions. *Ramphal Institute Policy Brief, 2*(3), 1–7.

Shah, K., & Chiorazzi, E. (2024, March 24). China's growing environmental footprint in the Caribbean. *New Security Beat*.

Shah, K., & Kelman, I. (2025). Data management for sustainable development: Case analyses of small island developing states (SIDS). *Development Policy Review, 43*, e70017.

Shah, S., Moroca, A., & Bhat, J. (2018). Neo-traditional approaches for ensuring food security in Fiji Islands. *Environmental Development, 28*, 83–100.

Shah, K., Niles, K., Ali, S., Surroop, D., & Jaggeshar, D. (2019). Plastics waste metabolism in a petro-island state: Towards solving a "wicked problem" in Trinidad and Tobago. *Sustainability, 11*, 6580.

Shah, K., Awojobi, M., & Soomauroo, Z. (2022). Electric vehicle adoption in small island economies: Review from a technology transition perspective. *Wiley Interdisciplinary Reviews: Energy and Environment, 11*(4), e432.

Shankman, P. (2018). Samoan journeys: Migration, remittances and traditional gift exchange. In J. Connell & H. Lee (Eds.), *Change and continuity in the Pacific. Revisiting the region* (pp. 87–101). Routledge.

Shantz, A., Ladd, M., & Burkepile, D. (2020). Overfishing and the ecological impacts of extirpating large parrotfish from Caribbean coral reefs. *Ecological Monographs, 90*(2), e01403.

Sharman, J. (2011). *The money laundry. Regulating criminal finance in the global economy*. Cornell University Press.

Sheller, M. (2021). Reconstructing tourism in the Caribbean: Connecting pandemic recovery, climate resilience and sustainable tourism through mobility justice. *Journal of Sustainable Tourism, 29*(9), 1436–1449.

Sheller, M. (2025). Caribbean Islands and the coloniality of climate change. In P. Noxolo, K. Rhiney, & R. Cummings (Eds.), *Routledge handbook of Caribbean studies* (pp. 35–46). Routledge.

Shirley, R., & Kammen, D. (2013). Renewable energy sector development in the Caribbean: Current trends and lessons from history. *Energy Policy, 57*, 244–252.

Shultz, J., Sands, D., Kossin, J., & Galea, S. (2020). Double environmental injustice—climate change, Hurricane Dorian, and the Bahamas. *New England Journal of Medicine, 382*(1), 1–3.

Siasau, K. (2024, October 9). Pacific labour mobility: Benefits and challenges for Tonga. *Matangi Tonga*.

Sievert, K., Lawrence, M., Naika, A., & Baker, P. (2019). Processed foods and nutrition transition in the Pacific: Regional trends, patterns and food system drivers. *Nutrients, 11*, 1328.

Silas, E., Makikon, J., & Kinsella, S. (2022). Service delivery reform- lessons from Vanuatu. *ANU DevPolicyBlog*, 15 March.

Simonelli, A., & Novalski, K. (2023). Rainbows of comfort in rising seas: How literalist Bible interpretations impact climate change communication in the Marshall Islands. *Weather, Climate and Society, 15*, 773–785.

Singh, S. (2017). The media and journalism challenges in Melanesia: Addressing the impacts of external and internal threats in Fiji, Papua New Guinea, the Solomon Islands and Vanuatu. *Pacific Journalism Review, 26*(1), 48–62.

Singh, S., Huang, T., Nagabhatla, N., et al. (2022). Socio-metabolic risk and tipping points on islands. *Environmental Research Letters, 17*, 065009.

Singh, P., Linnér, B., & Singh, A. (2025). Marine spatial planning and ocean governance in small island developing states. *Regional Environmental Change, 25*(3), 91.

Skidmore, D. (2025). *China, the West and the global development finance regime. Competitive convergence*. Elgar.

Smith, S. (2020). Structural vulnerability and marital sexual violence in Chuuk, Micronesia. In M. Torres & K. Yllö (Eds.), *Sexual violence in intimacy* (pp. 44–58). Routledge.

Smith, G. (2024, October 15). The US isn't the only country voting on Nov 5. This small nation is also holding an election – and China is watching. *The Conversation*.

Smith, R. (2025). Indigenous vulnerability, disaster governance, and environmental justice. Case study of the 2021 La Soufrière Eruptions in St. Vincent and the Grenadines. In P. Noxolo, K. Rhiney, & R. Cummings (Eds.), *Routledge handbook of Caribbean Studies* (pp. 75–98). Routledge.

Smith, A., & Dalzell, P. (1991). *Fisheries resources and management investigations in Woleai Atoll, Yap State*. Federated States of Micronesia, Noumea: South Pacific Commission.

Smith, N., & McNeill-Stowers, H. (2025, May 22) Europe's faltering statecraft in the Pacific. *East Asia Forum*.

Smith, S., & Katzman, F. (2020). The collective power of women's organisations in Chuuk, FSM. *Global Public Health, 15*(8), 1144–1156.

Smith, R., & Rhiney, K. (2015). Climate change, vulnerability, land and livelihoods: The case of the Black Caribs in northeastern St Vincent. *Geoforum, 73*, 22–31.

Smith, Y., Craney, A., & Roche, C. (2024). Racism, colonialism and whiteness in development: Insights from Pacific professionals following repatriation of white staff during Covid-19. *Third World Quarterly, 45*(9), 1517–1535.

Snowdon, W., Moodie, M., Schultz, J., & Swinburn, B. (2011). Modelling of potential food policy interventions in Fiji and Tonga and their impacts on noncommunicable disease mortality. *Food Policy, 36*, 597–605.

Soanes, L., Pike, S., Armstrong, S., et al. (2021). Reducing the vulnerability of coastal communities in the Caribbean through sustainable mangrove management. *Ocean and Coastal Management, 210*, 105702.

Sobion, J. (2025). *ICJ advisory opinion on climate change: The case for the Caribbean*. IISD SDG Knowledge Hub.

Sofer, M. (2018). Revisiting the Fijian periphery: A 30-year perspective on Kadavu Island. In J. Connell & H. Lee (Eds.), *Change and continuity in the Pacific: Revisiting the Region* (pp. 166–186). Routledge.

Song, L., Mosses, M., & Giraudeau, G. (2023). The ambiguous history of Matthew and Hunter Islands: Tracing the roots of Vanuatu and French claims. *Journal of Pacific History., 58*(3), 232–250.

So'o, A., & Jackson, L. (2025). Fiame Naomi Mata'afa's tumultuous tenure has ended – what's next? *ANU DevPolicyBlog*, 30 May.

Sora, M., Collins, J., & Keen, M. (2024). The great game in the Pacific Islands, Lowy Institute.

Sou, G., & Day, J. (2025). Postcolonial experiences of Chinese aid: Encountering and welcoming South–South aid from the middle. *Transactions of the Institute of British Geographers, 50*(2), e12693.

Sousa-Santos, J. (2019, June 26). The Pacific is in danger of becoming a semi-narco region. *The Guardian*.

Sousa-Santos, J. (2022). *Drug trafficking in the Pacific Islands: Impact of transnational crime*. Lowy Institute.

Sousa-Santos, J., & Howes, L. (2022). Policing illicit drugs in the Pacific: The role of culture and community on the frontline. *Journal of Contemporary Criminal Justice, 38*(3), 364–379.

Spalding, M. (2016). The new blue economy: The future of sustainability. *Journal of Ocean and Coastal Economics, 2*(2), 8.

Spann, M., & Maetala, R. (2024). Economic empowerment programs for women and culturally embedded entrepreneurism: A viewpoint from Solomon Islands. *Development in Practice, 34*(4), 527–531.

SPEC. (1982). *Pacific energy programme mission report: Vanuatu 1982*. Pacific Islands Development Program, East West Center.

Spencer, N., Strobl, E., & Campbell, A. (2022). Sea level rise under climate change: Implications for beach tourism in the Caribbean. *Ocean & Coastal Management, 225*, 106207.

Spennemann, D. (1996). Nontraditional settlement patterns and typhoon hazard on contemporary Majuro Atoll, Republic of the Marshall Islands. *Environmental Management, 20*, 337–348.

Spennemann, D. (2021). Groundwater, graves and golf: Layers of heritage tourism on a Fiji resort island. *Sustainability, 13*, 5863.

Sprague, J. (2015). From international to transnational mining: The industry's shifting political economy and the Caribbean. *Caribbean Studies*, 73–112.

Spriggs, R. (2010). Bougainville women's role in conflict resolution in the Bougainville Peace Process. In S. Dinnen (Ed.), *A kind of mending: Restorative justice in the Pacific Islands* (pp. 195–253). ANU Press.

Srinivasan, P. (2023, October 3). 'End the licences': Vanuatu oil rig registry sparks concern amid climate advocacy. *Guardian.*

Stambolie, E. (2024). Facebook use in the Pacific: All over the place. *ANU DevPolicyBlog*, 19 June.

Stambolie, E., & Hill, C. (2025). 2025 Australian aid procurement update. *ANU DevPolicyBlog*, 21 November.

Stanberry, D., & Fletcher-Paul, L. (2024). *Food security in the Caribbean.* Palgrave Macmillan.

Staunton, M. (2024, August 1). 'Government is really working' to fix poverty, the main driver of Fiji's drug crisis - minister. *Radio New Zealand.*

Steenbergen, D., Neihapi, P., Koran, D., et al. (2020). COVID-19 restrictions amidst cyclones and volcanoes: A rapid assessment of early impacts on livelihoods and food security in coastal communities in Vanuatu. *Marine Policy, 121*, 104199.

Steff, R. (2024). 'Our region is now a strategic theatre': New Zealand's balancing response to China. *The Pacific Review, 37*(6), 1120–1151.

Stege, M. (2018). Atoll habitability thresholds. In W. Filho & J. Nalau (Eds.), *Limits to climate change adaptation* (pp. 381–399). Springer.

Steibl, S., Kench, P., Young, H., Wegmann, A., et al. (2024). Rethinking atoll futures: Local resilience to global challenges. *Trends in Ecology & Evolution, 39*(3), 258–266.

Steinberg, P. (2001). *The social construction of the ocean.* Cambridge University Press.

Steiner, C. (2015). A sea of warriors: Performing an identity of resilience and empowerment in the face of climate change in the Pacific. *The Contemporary Pacific, 27*, 147–180.

Stephenson, J. (2023). Examining disability and work in small island developing states. In S. Nachmias & V. Caven (Eds.), *Inequality and organizational practice* (pp. 235–264). Palgrave Macmillan.

Steven, H., Banks, G., & Scheyvens, R. (2019). Women leadership in business based on customary land: The concept of Wanbel. *Development Bulletin, 81*, 58–61.

Stewart-Withers, R., Sewabu, K., & Richardson, S. (2017). Rugby union driven migration as a means for sustainable livelihoods creation: A case study of iTaukei, Indigenous Fijians. *Journal of Sport for Development, 5*(9), 1–20.

Storey, D., & Connell, J. (2013). Urban challenges. In M. Rapaport (Ed.), *The Pacific Islands: Environment and society* (pp. 310–322). Bess Press.

Storey, D., & Hunter, S. (2010). Kiribati: An environmental perfect storm. *Australian Geographer, 41*, 167–181.

Suaalii-Sauni, T. (2014). Legal pluralism and politics in Samoa: The Faamatai, Monotaga and the Samoa Electoral Act 1963. In P. Butler & C. Morris (Eds.), *Small states in a legal world* (pp. 165–189). Springer.

Sultana, F. (2022). Critical climate justice. *Geographical Journal, 188*(1), 118–124.

Sulu, R., Eriksson, H., Schwarz, A., et al. (2015). Livelihoods and fisheries governance in a contemporary Pacific Island setting. *PLoS One, 10*(11), e143516.

Sutton, P. (2007). Public sector reform in small states: Cases from the Commonwealth Caribbean. In E. Kisanga & S. Danchie (Eds.), *Commonwealth small states. Issues and prospects* (pp. 218–235). Commonwealth Secretariat.

Syddall, V., Thrush, S., & Fisher, K. (2021). Transdisciplinary analysis of Pacific tuna fisheries: A research framework for understanding and governing oceans as social-ecological systems. *Marine Policy, 134*, 104783.

Syddall, V., Fisher, K., & Thrush, S. (2022). Collaboration a solution for small island developing states to address food security and economic development in the face of climate change. *Ocean and Coastal Management, 221*, 106132.

Szadziewski, H. (2020). Converging anticipatory geographies in Oceania: The Belt and Road Initiative and Look North in Fiji. *Political Geography, 77*, 102119.

Szadziewski, H. (2022). Agents of the belt and road initiative or agents of agency? Fijian State and civil society perceptions of Chinese companies. In D. Pavlićević & N. Talmacs (Eds.), *The China question* (pp. 269–289). Palgrave Macmillan.

Szadziewski, H. (2024). *Going elsewhere: Oceania's globalism and the Gulf States.* ANU Department of Pacific Affairs Discussion Paper 2024/5.

Taloiburi, E., & Homasi, L.-A. (2022, August). Financing resilience to mitigate the Pacific's climate vulnerability. *Pacific Economic Monitor*, 36–39.

Tamata, A., & Mohammadnezhad, M. (2022). 'Most nurses in the rural or remote health clinics run away from their job due to no remote allowances and poor working equipment and environment'. Health professional leaders' perception of shortages in the nursing workforce in underserved areas in Vanuatu. *Rural and Remote Health, 22*(3), 7229.

Tandrayen-Ragoobur, V., Banerjee, S., Fauzel, S., & Matadeen, J. (2024). Climate change, equity and sustainable development in small island developing states. In S. Seifi & D. Crowther (Eds.), *Equity and sustainability. Approaches to global sustainability, markets and governance* (pp. 67–90). Springer Nature.

Tarte, S. (2006). Managing Tuna fisheries in the Pacific. In J. Bryant-Tokalau & I. Fraser (Eds.), *Redefining the Pacific? Regionalism, past, present and future* (pp. 89–99). Ashgate.

Tarte, S. (2015). A new Pacific regional voice? The Pacific Islands development forum. In G. Fry & S. Tarte (Eds.), *The new Pacific Diplomacy* (pp. 79–88). ANU Press.

Tate, G. (2025, February 22). It's the biggest small country in the world. *Financial Times*, 15.

Taunisila, L., & Palmieri, S. (2022). *Intractable barriers to women's election in Tonga*. ANU Department of Pacific Affairs InBrief 2022/4.

Tavola, H. (2023). *Tuvalu study on people with disability*. DFAT/Pacific Women Shaping Pacific Development.

Taylor, M. (2019). Foreword. In T. Aqorau (Ed.), *Fishing for success: Lessons in Pacific regionalism* (pp. xiii–xiv). Department of Pacific Affairs, ANU.

Taylor, M. (2025, December 8). The Marshall Islands' experiment with a universal basic income. *Interpreter*.

Taylor, M., & Middleby, S. (2023). Aid is not development: The true character of Pacific aid. *Development Policy Review, 41*(Suppl. 2), e12745.

Taylor, M., Clarke, L., Centella, A., et al. (2018). Future Caribbean climates in a world of rising temperatures: The 1.5 vs 2.0 dilemma. *Journal of Climate, 31*(7).

Taylor, M., Middleby, S., & Vunibola, S. (2025). Aid in the age of Amazon: Imperial logics, Pacific resistance and an alternate paradigm. *Development Policy Review, 43*, e70030.

Techera, E. (2015). Enhancing legal frameworks for biodiversity conservation in the Pacific. *Pacific Conservation Biology, 21*(1), 87–96.

Tekiteki, S. (2024). Reflections on Pacific regionalism. *ANU DevPolicyBlog*, 17 October.

Teiawa, K. (2018). Our rising sea of islands: Pan-Pacific regionalism in the age of climate change. *Pacific Studies, 41*(1), 26–54.

Telesford, J. (2021). Climate change and the case of Grenada's blue growth plan: Using the SDGs to propose a policy planning framework for SIDS's sustainable development. In S. Moncada, L. Briguglio, H. Bambrick, et al. (Eds.), *Small Island developing states: Vulnerability and resilience under climate change* (pp. 307–327). Springer.

Thiault, L., Mora, C., Cinner, J., et al. (2019). Escaping the perfect storm of simultaneous climate change impacts on agriculture and marine fisheries. *Science Advances, 5*(11), eaaw9976.

Thomas, N. (2014). The age of empire in the Pacific. In D. Armitage & A. Bashford (Eds.), *Pacific histories: Ocean, land, people* (pp. 75–96). Palgrave Macmillan.

Thomas, A. (2025). Transformational adaptation to climate change in the Caribbean. In P. Noxolo, K. Rhiney, & R. Cummings (Eds.), *Routledge handbook of Caribbean studies* (pp. 47–60). Routledge.

Thomas, A., & Benjamin, L. (2020). Non-economic loss and damage: Lessons from displacement in the Caribbean. *Climate Policy, 20*(6), 715–728.

Thomas, F., & Tonganibeia, K. (2007). Pacific island rural development: Challenges and prospects in Kiribati. In J. Connell & E. Waddell (Eds.), *Environment, Development and Change in Rural Asia-Pacific: Between local and global* (pp. 38–55). Routledge.

Thomas, A., Baptiste, A., Martyr-Koller, R., et al. (2020a). Climate change and small island developing states. *Annual Review of Environment and Resources, 45*, 1–27.

Thomas, A., Manghubai, S., Fox, M., et al. (2020b). Why they must be counted: Significant contribution of Fijian women fishers to food security and livelihoods. *Ocean and Coastal Management, 205*, 105571.

Thomas-Hope, E. (1993). *Explanations in Caribbean migration*. Macmillan.

Thomas-Hope, E. M. (2002). *Caribbean Migration*. University of the West Indies Press.

Thomas-Hope, E. (2017). Migration, small farming and food security in the Caribbean: Jamaica and St. Vincent and the Grenadines. *International Migration, 55*(4), 35–47.

Thomas-Hope, E. (2025). Migrant remittances, food security and inclusive social development in Jamaica. In J. Crush & S. Ramachandran (Eds.), *New directions in South-South migration* (pp. 63–82). Palgrave Macmillan.

Thompson, C., Cole, S., Moore, S., et al. (2023). Teacher migration and the Jamaican education system: Exploring push and pull factors and policy options. *Journal of Thought, 57*(3/4), 55–81.

Thompson, D. (2015). Disaster logistics in small island developing states: Caribbean perspectives. *Disaster Prevention and Management, 24*(2), 166–184.

Thompson, R. (2025). Implementing SDG 14: Opportunities and challenges in a Caribbean Small Island Developing State. *Marine Policy, 178*, 106713.

Thomson, L. (2020). Looking ahead – Global sandalwood production and markets in 2040, and implications for Pacific Island producers. *Australian Forestry, 83*(4), 245–254.

Thomson, L., Bush, D., & Lesubula, M. (2020). Participatory value chain study for yasi sandalwood (*Santalum yasi*) in Fiji. *Australian Forestry, 83*(4), 227–237.

Thornton, A., Sakai, M., & Hassall, G. (2012). Givers and governance: The potential of faith-based development in the Asia-Pacific. *Development in Practice, 22*(5–6), 779–791.

Thow, A.-M., Swinburn, B., Colagiuri, S., et al. (2010). Trade and food policy: Case studies from three Pacific Island countries. *Food Policy, 35*, 556–564.

Thow, A.-M., Ravuvu, A., Iese, V., et al. (2022). Regional governance for food system transformations: Learning from the Pacific Island region. *Sustainability, 14*, 12700.

Timms, B. (2006). Caribbean agriculture-tourism linkages in a neoliberal world: Problems and prospects for St. Lucia. *International Development Planning Review, 28*(1), 35–56.

Timms, B., & Conway, D. (2012). Slow tourism at the Caribbean's geographical margins. *Tourism Geographies, 14*(3), 396–418.

Tjondronegoro, D., Marshall, A., Hunter, S., & Yuwono, E. (2025). *Bridging digital financial literacy gaps in the Pacific: A data-driven policy roadmap for inclusive growth*. Griffith University.

Tolkach, D. (2023, April 8). Pacific Island Tourism is on the Surge again but its character is changing. *Australian Outlook*.

Tollefson, J. (2020). Why deforestation and extinctions make pandemics more likely. *Nature, 584*, 175–176.

Tomlinson, M., & McDougall, D. (2012). *Christian politics in Oceania*. Berghahn.

Tonga, B., Aguiar, D., Wiegner, T., et al. (2025). Baseline conditions and land-based drivers of nearshore water quality in Pohnpei, Federated States of Micronesia. *Marine Pollution Bulletin, 215*, 117833.

Tourigny, L., Baba, V., & Lituchy, T. (2016). On the retention of younger nurses. *Journal of Nursing and Care, 5*, 3.

Treichel, P., Robertson, M., Wilkinson, E., & Corbett, J. (2024). Scale and access to the green climate fund: Big challenges for small island developing states. *Global Environmental Change, 89, 102943.*

Trotman, A., Mahon, R., Meerbeeck, C., & Riley, J. (2021). *Advancing drought risk management in the Caribbean: A multi-sectoral perspective*. Caribbean Institute for Meteorology and Hydrology.

Trundle, A. (2020). Resilient cities in a sea of islands: Informality and climate change in the South Pacific. *Cities, 97*, 102496.

Trundle, A., Barth, B., & McEvoy, D. (2019). Leveraging endogenous climate resilience: Urban adaptation in Pacific small island developing states. *Environment and Urbanization, 31*(*1*), *53–74*.

Tui, S., & Fakhruddin, B. (2022). Food for thought: Climate change risk and food (in) security in Tuvalu. *Progress in Disaster Science, 16*, 100255.

Turia, N., & Howes, S. (2025). The 2025 Pacific Engagement Visa is open- how did it go in 2014? *ANU DevPolicyBlog*, 18 August.

Turner, R., McConney, P., & Monnereau, I. (2020). Climate change adaptation and extreme weather in the small-scale fisheries of Dominica. *Coastal Management, 48*(5), 436–455.

Tyllianakis, E., Grilli, G., Gibson, D., et al. (2019). Policy options to achieve culturally-aware and environmentally sustainable tourism in Fiji. *Marine Pollution Bulletin, 148*, 107–115.

Uitto, J., Kohlitz, J., & Todd, D. (2017). Evaluating sustainable development in SIDS-lessons from the Pacific and the Caribbean. In R. van den Berg, I. Naidoo, & S. Tamondong (Eds.), *Evaluation for agenda 2030: Providing evidence on progress and sustainability* (pp. 119–133). UNDP.

Underhill-Sem, Y., & Guttenbeil-Likiliki, O. (2025). Challenging colonialism and patriarchy: Oceanic Pacific feminists on development, women and international law. In J. Dawuni, N. Grossman, J. Ramji-Nogale, & H. Fabri (Eds.), *The Oxford handbook of women and international law* (pp. 413–433). Oxford University Press.

UNEP. (2021). *Small island developing states: Waste management outlook*. United Nations Environment Programme.

United Nations. (2024a). *High level panel on the development of a multidimensional vulnerability index*. UN.

United Nations. (2024b). *Surging seas in a warming world*. UN.

United Nations. (2025). *Progress towards the sustainable goals 2025*. UN.

United Nations Office on Drugs and Crime (UNODC). (2023). *Regional report on the existing capacities to measure trafficking in persons in the Pacific Islands*. UNODC.

UNDP, UNODC, and OHCHR. (2024). *Global progress report on sustainable development goal 16 indicators. At the crossroads: Breakdown or breakthrough for peace, justice and strong institutions*. UNDP.

UNESCO & International Task Force on Teachers for Education 2030. (2024). *Global report on teachers: Addressing teacher shortages and transforming the profession*. UNESCO.

UNFPA. (2024). *Violence against women, regional snapshot, 2023*. UNFPA.

UNFPA. (2025). *Pulse on progress: Women in Caribbean Tourism Survey Report*. UNFPA.

UNICEF. (2025). *Multidimensional child poverty in Fiji*. UNICEF.

UN OHRLLS. (2024). *Promising sectors for economic transformation of small island developing states*. United Nations.

UN Women (2018). *Promoting women's economic empowerment: Recognizing and investing in the care economy*. UN.

Urban, M. (2024). Climate change extinction. *Science, 386*, 1123–1128.

Utz, R. (Ed.). (2021). *Archipelagic economies. Spatial economic development in the Pacific*. World Bank.

Uyarra, M., Cote, I., Gill, J., Tinch, R., et al. (2005). Island-specific preferences of tourists for environmental features: Implications of climate change for tourist-dependent states. *Environmental Conservation, 32*(1), 11–19.

Vaha, M. (2025). Responsibility and anxiety in the 'Pacific family': AUKUS as a source of ontological insecurity. *Australian Journal of International Affairs, 79*(3), 406–422.

Vainikolo, L. (2021). Small and isolated Vava'u, Tonga: From weakness to strength during COVID-19. In Y. Campbell & J. Connell (Eds.), *COVID in the Islands: A comparative perspective on the Caribbean and the Pacific* (pp. 207–218). Palgrave Macmillan.

Vakalalabure, E., & Saurara, L. (2022). Fijian education and COVID-19. *Islands Business, 48*(1), 26–27.

Vakaoti, P. (2018). *Street-frequenting young people in Fiji*. Palgrave Macmillan.

Valencia, M. (2007). The East China Sea dispute: Context, claims, issues, and possible solutions. *Asian Perspective, 31*(1), 127–167.

Van Beynen, P., Fenda, A., Akiwumi, A., & Van Beynen, K. (2018). A sustainability index for small island developing states. *International Journal of Sustainable Development and World Ecology, 25*(2), 99–116.

van Tatenhove, J. (2016). The environmental state at sea. *Environmental Politics, 25*(1), 160–179.

van der Grijp, P. (1997). Leaders in squash export: Entrepreneurship and the introduction of a new cash crop in Tonga. *Pacific Studies, 20*(1), 29–62.

van der Plank, S. (2022). Polycentric governance, coordination and capacity: The case of Sargassum Influxes in the Caribbean. *Coastal Management, 50*(4), 285–305.

van der Ploeg, J., Sukulu, M., Govan, H., Minter, T., & Eriksson, H. (2020). Sinking islands, drowned logic: Climate change and community-based adaptation discourses in Solomon Islands. *Sustainability, 12*, 7225.

van der Velde, M., Green, S., Vanclooster, M., & Clothier, B. (2007). Sustainable development in small island developing states: Agricultural intensification, economic development, and freshwater resources management on the coral atoll of Tongatapu. *Ecological Economics, 61*, 456–468.

Vanuatu Chamber of Commerce. (2023). *Vanuatu private sector economic update*. Chamber of Commerce and Industry.

Vanuatu National Disaster Management Office. (2018). *National climate change and disaster-induced displacement policy*. Port Vila.

Varea, R., Varea, R., Kant, R., & Farrelly, T. (2022). *Qi no tu i baba ni qwali* (living down by the river): Impacts of flooding and mining on ecosystems and livelihoods. *Frontiers in Marine Science, 9*, 954062.

Veenendaal, W. (2013). Size and institutional legitimacy: The case of St Kitts and Nevis. *Commonwealth & Comparative Politics, 51*(3), 343–361.

Veenendaal, W., & Corbett, J. (2021). Navigating hyper-fragmentation in small states: A comparison of Solomon Islands and Vanuatu. In L. Briguglio, J. Byron, S. Moncada, & W. Veenendaal (Eds.), *Handbook of governance in small states* (pp. 26–37). Routledge.

Vega, D., Page, T., & Ota, L. (2023). Challenges and opportunities for inclusive value chains of niche forest products in small island developing states: Canarium nuts, sandalwood, and whitewood in Vanuatu. *Journal of Rural Studies, 100*, 103036.

Verlis, K., & Wilson, S. (2020). Paradise trashed: Sources and solutions to marine litter in a small island developing state. *Waste Management, 103*, 128–136.

Vermeij, M., Latijnhouwers, K., Dilrosun, F., et al. (2019). Historical changes (1905-present) in catch size and composition reflect altering fisheries practices on a small Caribbean island. *PLoS One, 14*(6), e0217589.

Vinke, K. (2025). *Redrawing humanity's map*. Springer.

Vogliano, C., Raneri, J., Coad, J., et al. (2021). Dietary agrobiodiversity for improved nutrition and health outcomes within a transitioning indigenous Solomon Island food system. *Food Security, 13*, 819–847.

Volpe, C. (2021). 'High tide by boat, low tide we walk': The everyday digital lives of girls in remote villages of Vanua Levu, Fiji. *Children's Geographies, 19*(6), 766–779.

Voyer, M., Gollan, N., Barclay, K., & Gladstone, W. (2015). 'It's part of me'; understanding the values, images and principles of coastal users and their influence on the social acceptability of MPAs. *Marine Policy, 52*, 93–102.

Vunibola, S. (2023). "Want to help someone? Shut up and listen": Foreign aid, maladaptation, and community development practices in the Pacific. *Development Policy Review, 41*, e12751.

Wade, E. (2025). Fishers' responses to future catch declines in Jamaican small-scale fisheries. *ICES Journal of Marine Science, 82*(5), fsaf070.

Waide, S. (2024, August 13). The Pacific sees a lot of fly-in and fly-out experts-but what would happen if there was a shift to locally led aid? *ABC.*

Wairiu, M. (2017). Land degradation and sustainable land management practices in Pacific Island Countries. *Regional Environmental Change, 17*, 1053–1064.

Waiti, D., & Lorrenij, R. (2018). Sustainable management of deep sea mineral activities: A case study of the development of national regulatory frameworks for the Republic of the Marshall Islands. *Marine Policy, 95*, 388–393.

Walda-Mandel, S. (2016). *'There is no place like home': Migration and cultural identity of the Sonsorolese, Micronesia*. Universitätsverlag Winter.

Wale, C., Nagabhatla, N., & Perera, D. (2021). Institutional and policy analysis: Water security and disaster management in small island developing states. In J. Roberts, S. Nath, S. Paul, & Y. Madhoo (Eds.), *Shaping the future of small Islands: Roadmap for sustainable development* (pp. 205–217). Palgrave Macmillan.

Walker, B., Lee, T., & Li, X. (2021). Sustainable development for small island tourism: Developing slow tourism in the Caribbean. *Journal of Travel & Tourism Marketing, 38*(1), 1–15.

Wallis, J. (2010). 'Friendly islands' in an unfriendly system: Examining the process of Tonga's WTO accession. *Asia Pacific Viewpoint, 51*, 262–277.

Wallis, J. (2023). The enclosure and exclusion of Australia's 'Pacific family'. *Political Geography, 106*, 102935.

Wallis, J., & Corbett, J. (2024, December 9). At first glance, Australia's new treaty with Nauru seems to be a win-win. But questions remain. *The Conversation.*

Wallis, J., & Kemish, I. (2022, August 10). Key to Pacific is education: Ours, not theirs. *Sydney Morning Herald*, 25.

Wallis, J., McNeill, H., Batley, J., & Powles, A. (2022). Security cooperation in the Pacific Islands: Architecture, complex, community, or something else? *International Relations of the Asia-Pacific, 23*(2), 263–296.

Wallis, J., Koro, M., & O'Dwyer, C. (2024). The 'Blue Pacific' strategic narrative: Rhetorical action, acceptance, entrapment, and appropriation? *The Pacific Review, 37*(4), 797–824.

Walshe, R., & Nunn, P. (2012). Integration of indigenous knowledge and disaster risk reduction: A case study from Baie Martelli, Pentecost Island, Vanuatu. *International Journal of Disaster Risk Science, 3*(4), 185–194.

Walters, S. (2025). *Energy poverty among women in Jamaica's low-income communities*. Palgrave Macmillan.

Wang, C., Shu, X., Zhou, S., et al. (2025). Embracing a new era of deep-sea mining: Research progress and prospects. *Marine Policy, 180*, 106778.

Waqairatu-Waqainabete, S., & Kaumaitotoya, L. (2021). Women and the business of aquaculture: A case for women tilapia farmers in Fiji. *SPC Women in Fisheries Information Bulletin, 34*, 4–8.

Waqavakatoga, W., Habru, P., & Koro, M. (2025). How do Pacific Island countries respond to China's statecraft? In J. Wallis, H. McNeill, M. Rose, & A. Tidwell (Eds.), *Power and influence in the Pacific Islands. Understanding statecraftiness* (pp. 201–218). Routledge.

Ward, R. G. (1993). South Pacific Island futures: Paradise, prosperity, or pauperism? *The Contemporary Pacific, 5*(1), 1–21.

Warner, R., & Anatol, M. (2014). Caribbean integration – Lessons for the Pacific? *Asia and the Pacific Policy Studies, 2*(1), 183–196.

Washburn, T., Simon-Lledó, E., Soong, G., & Suzuki, A. (2023). Seamount mining test provides evidence of ecological impacts beyond deposition. *Current Biology, 33*(14), 3065–3071.

Watson, A. (2024, October 29). Pacific Islands need stronger corrections systems to address transnational and organised crime. *Interpreter*.

Watson, D., Howes, L., Dinnen, S., et al. (2023). *Policing in the Pacific Islands*. Palgrave Macmillan.

Watson, D., Pino, N., Amin, S., & Scott, J. (2025). Navigating policing and work-life balance in the South Pacific. *Police Practice and Research, 26*(3), 276–293.

Watson, M., Claar, D., & Baum, J. (2016). Subsistence in isolation: Fishing dependence and perceptions of change on Kiritimati, the world's largest atoll. *Ocean and Coastal Management, 123*, 1–8.

Watt, L. (2024). Relational resource geographies of beche-de-mer under moratorium. *Asia Pacific Viewpoint, 65*(1), 40–54.

Weatherill, C. (2023). Sinking paradise? Climate change vulnerability and Pacific island extinction narratives. *Geoforum, 145*, 103566.

Weatherill, C. (2024). Resisting climate change vulnerability: Feminist and decolonial insights. *International Politics, 61*(4), 661–682.

Weatherill, C. (2025). Colonial fantasies of invulnerability to climate change. *International Feminist Journal of Politics, 27*(1), 34–55.

Webber, S. (2013). Performative vulnerability: Climate change adaptation policies and financing in Kiribati. *Environment and Planning A, 45*(11), 2717–2733.

Weiler, F., Klöck, C., & Dornan, M. (2018). Vulnerability, good governance or donor interests? The allocation of aid for climate change adaptation. *World Development, 104*, 65–77.

Weinberg, H., Bhardwaj, J., Watkins, A., & Kuleshov, Y. (2024). The impact of marine heatwaves on isotherm displacement and tuna distribution in Vanuatu. *Climate, 12*(11), 181.

Weir, T. (2018). Renewable energy in the Pacific Islands: Its role and status. *Renewable and Sustainable Energy Reviews, 94*, 762–771.

Wesley, M. (2023). *Helpem Fren: Australia and the regional assistance mission to Solomon Islands 2003-2017*. Melbourne University Press.

Wesley-Smith, T. (2021). A new Cold War? Implications for the Pacific Islands. In G. Smith & T. Wesley-Smith (Eds.), *The China alternative. Changing regional order in the Pacific Islands* (pp. 71–106). ANU Press.

Wesley-Smith, T., & Finin, G. (2021). US-Pacific engagement and the Biden Presidency: The limits of a China-centred approach. *Journal of Pacific History, 56*(4), 437–458.

Western, J. (1992). *A passage to England. Barbadian Londoners Speak of Home*. UCL Press.

West, P. (2016). *Dispossession and the environment: Rhetoric and inequality in Papua New Guinea*. Columbia University Press.

Westoby, R., Clissold, R., & McNamara, K. (2021). Alternative entry points for adaptation: Examples from Vanuatu. *Weather, Climate and Society, 13*(1), 211–222.

Weston, P., & Greenfield, P. (2022, December 19). 'Crucial' Cop15 deal includes target to protect nature on Earth by 2030. *Guardian*.

Wheatle, S., & Campbell, Y. (2020). Constitutional faith and identity in the Caribbean: Tradition, politics and the creolisation of Caribbean constitutional law. *Commonwealth & Comparative Politics, 58*(3), 344–365.

White, B. (2021, October 29). Barbados: The small but mighty force for climate action. *Open Democracy*.

White, A. (2022, February 18). Island states meet to discuss suing Global North over climate change. *Open Democracy*.

White, I., Falkland, T., & Scott, D. (1999). *Droughts in small coral islands: Case study, South Tarawa, Kiribati*. UNESCO.

Whittingham, N. (2022). *Key motivating factors of worker migration: The migration of experienced Jamaican teachers*. PhD thesis, Northcentral University.

Wilkinson, E. (2023). How a small Caribbean island is trying to become hurricane-proof. *The Conversation,* 29 November.

Wilkinson, E. (2025). Small state innovations to strengthen financial resilience in an era of accelerating climate change. *Small States & Territories, 8*(1), 90–94.

Wilkinson, E., Manley, M., Lindsay, C., et al. (2025). *Fit for Size; rethinking capacity strengthening in small island developing states*. ODI Global.

Willetts, J. (2025). Water at the centre of climate residence: At COP30 and beyond. *ANU DevPolicyBlog,* 25 November.

Williams, A., & Rangel-Buitrago, N. (2022). The past, present, and future of plastic pollution. *Marine Pollution Bulletin, 176,* 113429.

Williams, D., Costa, M., Sutherland, C., et al. (2019). Vulnerability of informal settlements in the context of rapid urbanization and climate change. *Environment and Urbanization, 31*(1), 157–176.

Wilson, K. (2013). Wan Laki Aelan? Diverse development strategies on Aniwa, Vanuatu. *Asia Pacific Viewpoint, 54*(2), 246–263.

Wiseman, V., Lagard, M., Batura, N., et al. (2017). Measuring inequalities in the distribution of the Fiji health workforce. *International Journal for Equity in Health, 16,* 115.

Wolf, F., Filho, W., Singh, P., et al. (2021). Influences of climate change on tourism development in small Pacific Island States. *Sustainability, 13*(8), 4223.

Wolf, F., Moncada, S., Surroop, D., et al. (2024). Small island developing states, tourism and climate change. *Journal of Sustainable Tourism, 32*(9), 1965–1983.

Wong, E., Jiang, M., Klint, L., et al. (2013). Evaluation of policy environment for climate change adaptation in tourism. *Tourism and Hospitality Research, 13*(4), 201–225.

Wood, T. (2021). The cruel ironies of the 2021 Honiara riots. *ANU DevPolicyBlog,* 3 December.

Wood, T., Dornan, M., & Muller, S. (2021). *Change and continuity in Australian aid: What the aid flows show*. ANU Development Policy Centre.

Woodroffe, C. (2008). Reef-island topography and the vulnerability of atolls to sea-level rise. *Global and Planetary Change, 62,* 77–96.

World Bank. (2016). *Pacific possible: Health and non-communicable diseases*. World Bank.

World Bank. (2019). *Using taxation to address noncommunicable diseases: Lessons from Tonga*. World Bank.

World Bank. (2021). *Pacific labor mobility, migration and remittances in times of COVID-19*. World Bank.

World Bank. (2023a). *Improving outcomes of Pacific labor mobility for women, families and communities*. World Bank.

World Bank. (2023b). *The future of Pacific tourism*. World Bank.

World Bank. (2024). *Global rapid post-disaster damage estimation (Grade) report*. World Bank.

World Food Programme. (2025). *Cost of the diet analysis in Samoa*. WFP.

World Health Organisation. (2023). *WHO health workforce support and safeguards list*. WHO.

World Travel and Tourism Council. (2024). *Unlocking opportunities for travel and tourism in Oceania*. WTTC.

Wosnick, N., Curtis, D., & Hauser-Davis, R. (2024). Managing technology-critical elements from electronic waste in small developing Island States: A burden or an opportunity? *Frontiers in Marine Science, 11*, 1459794.

Wright, N. (2024). Assessing the main barriers to student success in the Caribbean, *Focus. ECLAC in the Caribbean*, 8–11.

Wrighton, N., & Overton, J. (2012). Coping with participation in small island states: The case of aid in Tuvalu. *Development in Practice, 22*(2), 244–255.

WWF. (2024). *Living planet report: A system in Peril*. WWF.

Xie, G. (2025). China's strategic narratives in the Pacific. In J. Wallis, H. McNeill, M. Rose, & A. Tidwell (Eds.), *Power and influence in the Pacific Islands. Understanding statecraftiness* (pp. 190–200). Routledge.

Yamamoto, L., & Esteban, M. (2014). *Atoll Island States and international law*. Springer.

Yamano, H., Kayanne, H., Yamaguchi, T., Kuwahara, Y., et al. (2007). Atoll island vulnerability to flooding and inundation revealed by historical reconstruction: Fongafale Islet, Funafuti Atoll, Tuvalu. *Global and Planetary Change, 57*, 407–416.

Yamamoto, T., Sunguya, B., Shiao, L., et al. (2012). Migration of health workers in the Pacific Islands: A bottleneck to health development. *Asia-Pacific Journal of Public Health, 24*(4), 697–709.

Yang, J. (2011). *The Pacific Islands in China's grand strategy*. Palgrave Macmillan.

Yarina, E., & Takemoto, S. (2017). Interrupted atolls: Riskscapes and edge imaginaries in Tuvalu. *The Plan Journal, 2*(2), 461–495.

Yee, M., Piggott-McKellar, A., McMichael, C., & McNamara, K. (2022). Climate change, voluntary immobility, and place-belongingness: Insights from Togoru, Fiji. *Climate, 10*, 46.

Yee, M., Piggott-McKellar, A., McMichael, C., & McNamara, K. (2025). Framing locally led adaptation in a planned relocation in Fiji. *Geoforum, 159*, 104196.

Young, S., Bethancourt, H., Frongillo, E., et al. (2022). Concurrence of water and food insecurities, 25 low-and middle-income countries. *Bulletin of the World Health Organization, 101*(2), 90–101.

Yousefzadeh, M., Lenzen, M., & Tariq, M. (2023). Cooling and power from waste and agriculture residue as a sustainable strategy for Small Islands – A case study of Tonga. *Sustainability, 15*, 537.

Zhang, D. (2018). *China's growing interest in deep sea mining in the Pacific*. ANU Department of Pacific Affairs InBrief 2018/11.

Zhang, D., Leiva, D., & Ruwet, M. (2019). *Similar patterns? Chinese aid to Island countries in the Pacific and the Caribbean*. ANU Department of Pacific Studies InBrief 9.

Zittis, G., Zoumides, C., Zemah-Shamir, et al. (2025). Insular ecosystem services in peril: A systematic review on the impacts of climate change and other drivers. *Climatic Change, 178*, 127.

Index

J. Connell, Y. Campbell, *Island Challenges*,
https://doi.org/10.1007/978-981-95-8177-1

H

I

J

The manufacturer's authorised representative in the EU is Springer Nature Customer Service Centre GmbH, Europaplatz 3, 69115 Heidelberg, Germany. If you have any concerns regarding our products, please contact ProductSafety@springernature.com

Printed and bound by CPI Group (UK) Ltd, Croydon, CR0 4YY

22/09/2026

02220649-0003